Annual Review of Low-Carbon Development in China: 2010

Annual Review of Low-Carbon Development in China: 2010

Edited by

Qi Ye

Tsinghua University, China

Partial funding provided by Climate Policy Initiative

World Scientific

NEW JERSEY • LONDON • SINGAPORE • BEIJING • SHANGHAI • HONG KONG • TAIPEI • CHENNAI

Published by

World Scientific Publishing Co. Pte. Ltd.
5 Toh Tuck Link, Singapore 596224
USA office: 27 Warren Street, Suite 401-402, Hackensack, NJ 07601
UK office: 57 Shelton Street, Covent Garden, London WC2H 9HE

British Library Cataloguing-in-Publication Data
A catalogue record for this book is available from the British Library.

ANNUAL REVIEW OF LOW-CARBON DEVELOPMENT IN CHINA: 2010

ISBN 978-981-4374-18-7
ISBN 978-981-4425-33-9 (pbk)

Typeset by Stallion Press
Email: enquiries@stallionpress.com

Printed in Singapore by B & Jo Enterprise Pte Ltd

Preface

In response to increasingly urgent situation of climate change, the world is experiencing a range of fundamental changes in economic and social development models, where the development of low-carbon economy has become a global trend. The core of low-carbon economy is sustainable development of economy and society under the premise of low energy consumption, low CO_2 emission and moderate environmental pollution, amid higher natural resources productivity and carbon productivity as well as high levels of economic and social development and life standards. Strategies of developing low-carbon economy lies in the development of low-carbon energy technologies, improvement of energy efficiency and energy structure, transformation of economic growth pattern and construction of a low-carbon economic development mode and societal consumption pattern, which should be taken as the essential means to coordinate economic development and climate protection.

The fundamental national conditions and the characteristics in its developing phase impose China severer challenges in addressing climate change than the developed countries, and the task is arduous. Though China is currently in the rapid development period of its industrialization and urbanization, great efforts have been made to attain energy conservation and efficiency. China's scale of energy-saving and emission reduction is the greatest in the world. However, due to the rapid growth of GDP in China, energy consumption and CO_2 emission will continue to be tremendous and growth fast, which is difficult to be changed in a short term. From 1990 to 2009, China's CO_2 emission per unit of GDP fell by 55%, and the rate of decline is rare in the world. But GDP increased 6.6 times over the same period, and CO_2 emission has increased 3.0 times to become the world's largest emitter. CO_2 emission per capita in China was about half of the world average level in 1990 and now it is above the average. From 2005 to 2009, the energy intensity per unit of GDP decreased by 15.6%, but the total energy consumption increased by 30%. It still remains an arduous task to control energy demand and the rise of CO_2 emission.

Climate Conference just concluded in Cancun, reiterated the global temperature control target of no more than 2°C. In this context, all countries in the world will face the challenge of restraint carbon emission and this will become an increasingly serious constraint in China's modernization process. Therefore, a fundamental shift in the mode of development must be achieved. Premier Wen Jiabao pointed in his Government Work Report presented to National People's Congress in 2010 that China shall endeavor to build industrial systems and consumption patterns featured of low carbon emission, which sets out China's basic strategic methodology in addressing climate change and carbon emission mitigation under the framework of sustainable development. Developing a green economy and realizing low-carbon development is a fundamental strategic choice of China in achieving harmony between economic growth and the resources as well as environment, and in addressing climate change within the framework of sustainable development. The key measure of low-carbon development is substantial reduction in CO_2 emission per unit of GDP (carbon intensity). China announced its target at Copenhagen climate conference at the end of 2009 to cut the CO_2 emission per unit of GDP by 40%–45% in 2020 from 2005 level, and this indicated the strategic choice in harmonizing the two subjects of achieving domestic sustainable development and tackling global climate change. Establishing low-carbon industrial systems and consumption patterns is also the fundamental approach in meeting our own emission reduction targets of 2020.

Currently, the major measures of reducing China's carbon intensity and energy intensity per unit of GDP by a significant scale include:

- applying high technology to traditional industries, accelerating the upgrade of industrial technology to improve the additional value of product, the efficiency of energy conversion and utilization, and by promoting technological energy-saving intensity;
- industrial transition by vigorously accelerating the strategic adjustment of industrial structure; structural energy conservation by giving priority to the development of high-tech industries and modern services and by developing low-carbon strategic emerging industries;

- development of new energy and renewable energy with the aims of optimizing energy structure and guaranteeing energy supply while reducing CO_2 emission;
- change of consumer concept and behavior under guidance of the government by developing low-carbon buildings and low-carbon transport and exploring low-carbon city construction models and societal consumption models.

Technology innovation will be an important support for low-carbon development. Urgent need to address climate change and vigorous demand for low-carbon technologies has promoted low-carbon technology innovation and industrial development. Low-carbon technologies mainly includes: energy conservation and energy efficient technologies, such as buildings of ultra low energy consumption, new energy and electric vehicles, waste heat utilization, clean coal utilizing technology; new energy technologies such as wind power and photovoltaic power generation technology, biofuels, advanced nuclear energy, hydrogen energy technologies; CO_2 capture and storage technology (CCS), etc. Large-scale low-carbon infrastructure underlies much new investment in the fields of new energy and energy efficiency, thus becomes a new economic growth point. Low-carbon technologies have also become the forefront and focus of global science and technology competition, as well as the new high-tech fields for major countries to compete in. Master of core low-carbon technologies will be an indicator of a nation's core competitiveness. China's current low-carbon technology research and industrial development has demonstrated a good momentum and there are leap-forward development opportunities for us.

China's current economic development is also constraint by the "bottleneck" of domestic resource supply and capacity of the environment. In this sense, developing a low-carbon green economy is both a fundamental way to address climate change and the requirement for realizing domestic sustainable development. Current efforts to establish a low carbon emission industrial system are in line with the requirements of our basic national policy, that are, conservation of resources, environmental protection and building of a resource saving and environment-friendly society. So, current strategies to address climate change should align with the national sustainable development strategies to achieve win-win of both goals.

A low-carbon development pathway with Chinese characteristics is both an urgent and long-standing arduous task that requires several generations' efforts. The development concept will determine the development pathway, and over the time, differences in the outcomes of different development pathways will be growing. Therefore, we must focus on the long-term and proceed at the present to gradually transmit from a resource-dependent development mode characterized by high emission to a low-carbon technology innovation development mode. It is impossible for China to follow the developed countries' development mode dependent on high resource consumption and carbon emission. We must make leapfrog development to complete the course of industrial civilization in the premise of a low-carbon eco-civilization. A new type of low-carbon development pathway is unprecedented in the history, so it requires our nation's arduous efforts to explore in this field and then create it.

"Annual Review of Low-Carbon Development in China" is a summary and review on China's policies and measures of addressing climate change and achieving a low-carbon development pathway and their implementation. We anticipate, in this report, to explore with the community a green, low-carbon and sustainable development pathway and model as well as a new industrialization and urbanization development framework for China, following the global trend of low-carbon development.

HE Jiankun
Deputy-president of National Expert Committee on Climate Change
Deputy President of China Energy Research Society
December 26, 2010

Foreword

This report is an attempt to summarize the achievements of China's low carbon development, assess low-carbon development related policies and measures and analyze the institutional background and mechanism of the policies. It expects to offer some supports for further policy-making, its optimization and implementation.

Low-carbon development features healthy and sustainable development of the economy and society accompanied by strictly-controlled carbon emission and actively-facilitated carbon absorption.

Carbon emission, in narrow sense, usually refers to carbon dioxide (CO_2) released from fossil energy combustion and industrial process; and in its broad sense, it also includes CO_2 released from decomposition of organic matter in the ecosystem and methane (CH_4) produced by microbial metabolism. When the amount of other greenhouse gases is converted to CO_2 equivalent, they were also included into the wide range discussion of carbon emissions by some people. This mixed usage of the term carbon emission with greenhouse gas emission thus is apt to cause confusions. In this report, the term carbon emission will be used in its narrow, if in the absence of special statement.

Carbon absorption or carbon uptake, sometimes referred to as carbon sink, involves both natural processes and artificial measures. In nature, the ocean is the largest carbon sink with a net absorption of CO_2 in the atmosphere up to 8.4 billion ton per year, followed by terrestrial ecosystems with a net absorption of about 4.8 billion ton per year. Artificial measures most discussed currently include carbon capture and storage (CCS), and methods of increasing marine carbon sinks by microbial fertilization and etc. also often appear in the media and academic literature.

Since the industrial revolution, increasing carbon emission has changed the chemical composition of the atmosphere, thus imposes a huge and far-reaching impact on regional and global climates. Reducing carbon emission is the core consensus of "the United Nations Framework Convention on Climate Change" and "Kyoto Protocol". For a long time, carbon emission and absorption are natural processes without policy intervention. Now, low-carbon development emphasizes active control of carbon emission and carbon absorption, which implies that the nations and the international community will adjust their economic and social processes by intervention of policies, in the hope of mitigating the impact of human activities on the climate system.

China is currently in the large-scale, high-speed development period of industrialization and urbanization. The situation of resource reservation, population, technological level, economic basis and the international division of labor determines that China's fossil energy consumption and carbon emission will inevitably have to undergo a historical stage of rapid growth. However, the choice of energy infrastructure, technology pathway and mode of development will have a profound impact on carbon emission and the quality of economic development, as well as a significant impact on the global carbon emission. Facts show that China can not follow the traditional patterns of industrialization and urbanization, let alone bear the great pressure of the resources and the environment caused thereof. It is Chinese government's solemn commitment to actively develop a low-carbon economy and take a low-carbon development pathway. For China, the low-carbon development lies in the economical use of fossil energy, development and utilization of renewable energy and improvement of energy efficiency. This will bring about reduction in the pressure of the national energy supply, improved efficiency of energy production and usage, promotion of transition of low-carbon model and protection of the future economic competitiveness. Therefore, low-carbon development is an important way to transform economic development path and to keep humankind and nature in harmony, as well as a way towards sustainable development of the economy and the society, and last but not least, an enormous contribution to global environment and economic development.

The core research work of the Climate Policy Initiative, Tsinghua University is assessing the status of low-carbon development in China, summarizing the experience of low-carbon development and analyzing policies related to low-carbon development. Our study focuses on China's economy, the society's energy consumption and carbon emission. We attempt to pin down the changes in energy consumption and carbon emission during China's industrialization and urbanization processes, the factors effect these changes and the mechanism by which they take effects. Our study includes three aspects: performances of China's and its major energy consuming sectors' low-carbon development, policies that influence low-carbon development and systems and mechanisms that play a role in the process when the policies are taking effect. Literally, the term low-carbon development suggests the dialectical unity of the two concepts of low-carbon and development, so neither simply emphasizing low-carbon while neglecting development nor the vise versa are applausible ways of development. Development requires low-carbon, which can facilitate healthy and sustainable development of the economy and the society. We are also concerned about the comparison between China's energy consumption and carbon emission with other countries and regions, and try to answer: at present, at what level are China's energy consumption and carbon emission on a global scale and at what level they are given its specific industrialization and urbanization stage? We study China's energy consumption and carbon emission under the background of its industrialization and urbanization and attempt to find out whether China is repeating the past path of industrialization and urbanization or it is endeavoring to explore a new way out?

"Annual Review of Low-Carbon Development in China", in the form of an annual publication, compiled and refined a year's work carried out by the Climate Policy Initiative, Tsinghua University. The study and drafting of this report adopt two methods, i.e. analyzing and integrating. On the one hand, we collect published official data, and then obtain the data of energy consumption and carbon emission of the major energy consumption sectors of China through calculation and analysis; on the other hand, we come to summaries, analysis and conclusions on the basis of extensive research and literature review. "Annual Review of Low-Carbon Development in China" is a preliminary attempt in this effort. The report focuses on the policies, actions and performance of China's low-carbon development during the "11th FYP" period (2006–2010). Because of limitations of data, analysis and evaluations are largely restricted to the first three years of the "11th FYP" period.

The report is divided into four chapters, namely General Review, the Sectors, Cases and Indexes. Chapter "General Review" gives a brief summary of the overall achievements and challenges of China's low-carbon development during the "11th FYP" period, as well as innovations in the aspects of technology, structure, policy and system. In addition, this section also gives a primary analysis of some key issues, e.g. the response of the enterprises and low-carbon finance. The core content of the report is reflected in the "the Sectors" chapter, where detailed analysis are delivered to the low-carbon development performance, policies and systems of some major sectors, i.e. the electric power, manufacturing, construction, transport, agriculture and forestry and other major sectors. The "Case" chapter embodies the low-carbon development background and frameworks of Baoding city, Hebei province, in the hope of offering a reference for other cities in making their low-carbon development plan. Finally, the Index chapter lists the main indexes for calculation in this report, the results of which are applied for calculating, evaluating and analyzing the performance of different sector's low-carbon development. We do not intend to develop a complete, objective and comprehensive system of indexes for low-carbon development. We include these indexes in the report for, on the one hand, we are obliged to honestly present to the readers and our peers the data and methods used in the calculations and the results obtained thereof; on the other hand, they will be taken as a threshold towards gradual accumulation and improvement of the data system of low-carbon development performance in China. As this report is a preliminary attempt in this field, the four chapters presented here is evitable to be far from complete, or even full of loopholes. We will start from this report and spare no efforts towards improvement of future works in both content and form.

The initiation of this report was inspired by Professor He Jiankun, Professor Thomas C. Heller, Professor Xue Lan and Professor Zhang Xiliang. This report had received wide support and assistance from many experts during its study and drafting processes. Professor Wang Qingyi had given careful guidance to the calculation of energy efficiency and carbon emission as well as preparation of this report during the whole course and he wrote a list of low-carbon technologies special for this report. The Editorial Board and experts in related fields, in particular, Mr. Dai Yande and Mr. Bai Rongchun, carefully reviewed the full report and provided a number of specific and detailed comments and suggestions.

The study and preparation of this report also benefited greatly from support and assistance of many entities and experts. Special gratitudes went to the Energy Foundation, China Sustainable Energy Program, and Ming Xu, Huawen Xiong, Xie Xie, Lei Ren, Minghua Xie, Hongwu Cai, David Nelson, Robert Earley, Dong Ma, Hongsheng Liu, Yihui Zhang and other experts.

Climate Policy Initiative, Tsinghua University, co-founded by Tsinghua University and the international institution, Climate Policy Initiative (CPI), is an academic institution specialized in the research of climate change and low-carbon development policies. We would like to take this opportunity to express our heartfelt thanks to Tsinghua University and CPI for all the assistance they had given to us during the studies.

China has just started the transition to low-carbon economy, and the world is closely watching the development of its green economy. Our long-term follow-up study of China's low-carbon development has just begun. We are in need and also looking forward to the care, support and assistance from our peers in this field.

QI Ye
Director of Climate Policy Initiative at Tsinghua
Professor, School of Public Policy and Management
Tsinghua University
December, 2010

Contents

Executive Summary

In 2006, China began its 11th Five Year Plan (FYP) with the explicit goal of reducing the energy intensity of the Chinese economy by 20%. The goal was set, in part, to address the disturbing reversal in 2002–2005 of the long-term decline in energy intensity, but it also addressed other trends, which increased the energy efficiency imperative: The acceleration of gross domestic product (GDP) growth and the accompanying expansion of energyintensive heavy industry; rapidly increasing energy and commodity prices; rapid expansion of the coal industry, leading to infrastructure strains; and China's emergence as the world's largest greenhouse gas emitter.

Chapter 22 of the 11th FYP set the following strategies to meet the 20% energy intensity reduction goal:

> "*The government shall strengthen policies that induce energy conservation and energy efficiency increase. Energy conservation can be achieved through structural changes (optimizing industrial structure and reducing the share of energy-intensive industries), technology improvement (developing and disseminating energy conservation technologies) and better management practices (institutional development and more effective regulation of energy production, transmission and consumption). Industries with priority for energy conservation are iron and steel, nonferrous metal, coal, electricity, chemistry, building material and other energy-intensive industries. The implementation of vehicle fuel economy policies shall be enhanced and the inefficient old vehicles shall be phased out. Standards for alternative liquid fuels shall be developed to support the alternative fuel industry. The production and consumption of highly energy efficient products shall be encouraged.*"

Under these extraordinary circumstances and with this target in mind, China initiated a series of policies aimed at reducing the energy intensity of the Chinese economy. As China begins the 12th FYP with all of these trends still in play, it is an opportune time to examine the results of the 11th FYP to understand

1. How China performed against its energy intensity targets and how the results break down by sector;
2. What measures and policies China undertook in order to meet these targets;
3. How well these policies performed and why some policies worked better than others; and
4. What lessons could be learned from China's experience and where more investigation into policy effectiveness is warranted.

This study explores these questions, first reviewing China's performance against targets and then describing key policies and institutions that were designed and implemented to achieve the targets. While further work is required on the direct and quantitative connection between policies and their performance, we can offer some initial insights and questions as China gears to meet future goals in the 12th FYP and through 2020.

This study finds that, based on data from the National Development and Reform Commission and from our field survey of local governments and enterprises, energy intensity reduction is in line with the 20% targets. Without closer study of policies and other factors affecting energy intensity, we do not yet attribute all reduction in energy intensity directly to policies, though we do observe a significant level of regulatory and policy activity in support of targets and surmise that at least some of these policies have had an impact on energy intensity. We also make some early observations about the impact and cost of policies as well as remaining opportunities for the plant closure policy, which may have implications for meeting goals set out in the 12th FYP:

- Impact of policies: Throughout 2009, China was on track to meet its energy intensity targets. These targets called for a reversal of the trend of increasing energy intensity experienced between 2002 and 2005. China's carbon emission intensity fell largely as a result of reduced energy intensity, demonstrating the important role of energy efficiency in the transition to a low-carbon economy. With additional carbon-specific policies, China could expect carbon intensity to fall faster than energy intensity in the future.
- Cost of policies: Many of the policies implemented to meet the 11th FYP target are top-down administrative measures. These required significant

financial and human resources and may not be the most cost-effective way to achieve future targets.

- Remaining opportunities: Much of the low-hanging fruit for reducing energy intensity, such as replacing old power plants with new, more energy-efficient plants, has already been picked. Further reductions in energy intensity during the 12th FYP period could impose higher costs on the economy.

Moving forward, key opportunities for China remain in continuing to shift power generation to renewable sources and in reducing the carbon intensity of high energy consuming industries. Doing so in a more cost-effective way may require the development of policy mechanisms that have till date been less used in China, such as effective market systems, regional and local initiatives for low-carbon growth and implementing macroeconomic measures for further sector rebalancing.

China's Performance Against its Energy Intensity Targets

Data for the entire 11th FYP is not yet available; however, according to the data released in July, 2010 by the National Development and Reform Commission, by the end of 2009 the energy intensity of China had decreased by 15.6% compared to 2005, achieving 78% of the 20% target through the first four years of the plan. On a year-to-year basis, the reduction was 2.74% in 2006, 5.04% in 2007, 5.2% in 2008 and 3.61% in 2009. By 2008 the energy intensity had dropped to the 2002 level. In short, the rapidly increasing energy intensity that was evident during the last few years of the 10th FYP period has been curbed.

During the 11th FYP period, China's drive to lower energy intensity affected most sectors of the economy, including the power, industry, building, transport, and agriculture and forestry sectors. Of these, the power, industry, and building sectors are the most important from an energy consumption, Green house gas (GHG) emission, and potential carbon savings standpoint, though agriculture and forestry is also important from a GHG emissions perspective.

For many of the sectors, the energy intensity reduction story is similar, with higher than expected growth providing both challenges and opportunities.

In the **Power sector**, electricity production grew 39% between 2005 and 2008, driven by economic growth and industrialization. This growth was

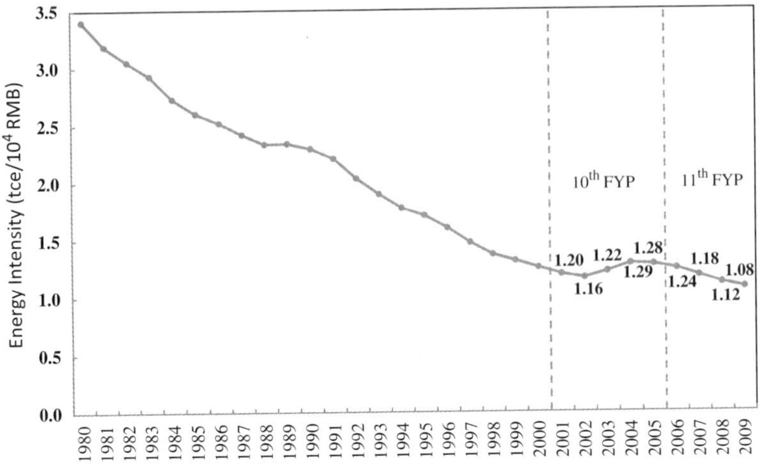

Fig. 1. China's energy intensity (1980–2009).

partly offset by a 5% improvement in the average efficiency of China's coal fired generation capacity and by improvements in the transmission system. The growth in Chinese electricity demand has also helped to improve generation efficiency, as the share of larger and more efficient capacity increased from 47% in 2005 to 69% in 2009. Additionally, China closed a large number of older, smaller, less efficient plants; by 2009, China had already surpassed its FYP target of 50 GW of plant closures. New low carbon generation also added to the effect of higher thermal plant efficiency; by 2009 China had added 90 GW of additional hydro power, almost 25 GW of wind, and 2 GW of nuclear. From a CO_2 perspective, the carbon intensity of electricity production decreased about 6.5% between 2005 and 2008.

In the **Industrial sector**, three factors stand out as drivers of energy intensity: The structure of Chinese industry, the movement toward higher value added products and improvements in the efficiency of production itself. The structure of Chinese industry continued to evolve under the 11th FYP, with heavy industry continuing to grow as under 10th FYP, but at a slower rate relative to other sectors, resulting in the share of energy-intensive industries decreasing in later years of the 11th FYP. As part of this shift, a move towards higher value added products helped improve energy intensity. Through 2008, energy consumption by the industrial sector per unit value added had decreased by 13.3% and carbon intensity by 14.8%. Lastly, many of the large industries improved efficiencies significantly, with energy per ton of production falling from 2005–2009 by close to 5% in the steel industry, 17% in the cement industry, and almost 35% in copper smelting. This improvement was driven by process and technology improvement as well as the construction of new facilities and the closure of older, less efficient ones. The decrease in CO_2 intensity was largely in line with the improvement in energy intensity.

The Building sector accounted for over 18% of China's energy use by 2008 and was growing rapidly, with energy use increasing by 28% from 2005. Construction of new buildings was a major driver of this increase, although energy intensity — that is energy use per square meter — contributed equally, increasing by 16% between 2005 and 2008. Both new construction and the increasing energy intensity of the building stock can be attributed to China's rapid urbanization and rising standard of living. Energy intensity and carbon intensity, as measured by CO_2 per square meter, continued to lag significantly behind the world average, so increasing standards of living, including greater appliance usage and space conditioning, further explain increasing energy and carbon intensity. This trend towards higher energy intensity in the building sector began in 2003, though some slowing of this increasing trend was noted in 2008. One particular area of focus for efficiency has been district heating in Northern China where energy consumed per square meter has continued its notable decline.

Energy consumption in the **Transport sector** grew 25% between 2005 and 2008, with two notable trends. First, the proportion of more energy intensive transport modes such as road and air transport rose. Second, the efficiency of most modes of transport, as measured by energy per ton-kilometer of freight, was either stable or improved slightly.

The Agriculture and Forestry sectors are relatively small consumers of energy, but their impact on CO_2 emissions is significant. Agriculture is the only sector in which direct energy-related emissions declined during the 11th FYP period. While non CO_2 greenhouse gas emissions were stable, CO_2 emissions embedded in fertilizer production grew, leading to a small net increase in overall agriculture-related greenhouse gas emissions. China's forestry development, especially the large scale of afforestation measures, contributed significantly to the building of carbon sinks, adding 420 Mt CO_2 per year on average to the current stock, an amount nearly four times the CO_2 emission from direct fossil fuel combustion in agriculture and forestry.

PART 1

Introduction

CHAPTER 1

China's Move to Low Carbon Economy

1.1 Measures and Policies Undertaken by China

1.1.1 *Policies*

During the 11th Five Year Plan (FYP) period, China developed many laws, regulations, norms, and policies, which constitute a legal and policy framework for its low carbon development. Key policies in the power sector include the "Renewable Energy Law" that came into effect in 2006 and the "Energy Conservation Law" of 2007. The "Renewable Energy Law" was designed to encourage the development of wind and PV generation. In December 2009, the standing committee of the National People's Commission revised the "Renewable Energy Law" to increase national support for grid connection for wind and PV plants. The revised "Energy Conservation Law" identified energy conservation as the country's "Fundamental State Policy" for the first time in China's history.

The "Renewable Energy Law" and "Energy Conservation Law" are central government-driven policies that mandate the phase out of backward technologies and capacity, and their replacement with more energy-efficient ones regulate market access for enterprises in energy-intensive industries, and provide incentives such as tax breaks, financial support and economic instruments, such as tiered pricing for electricity. The central government also endeavored to develop local government capacity and to support the growth of the energy efficient market by providing incentives and assistance for energy service companies.

These policies have two prominent characteristics. First, central government power was fully deployed: It engaged in permitting control as well as the performance evaluation of regional and local government and state-controlled enterprises. Second, the government made significant investments through financial transfers and subsidies, incentives, interest discounts, and favorable loans. The central government motivated local government and state-owned enterprises to expand their inputs, improve their management practices, optimize their structures and innovate their policies, while the economic incentives enabled enterprises to better implement the energy conservation policies. These two policy instruments consumed high levels of both funding and administrative resources.

Several other policies stand out, including the Medium and Long-Term Plan for Energy Conservation, the Ten Key Energy Conservation Projects and a series of fiscal and sectoral policies. The Ten Key Energy Conservation Projects policy focused on ten specific technologies for implementation, such as combined heat and power, efficient lighting, motor systems, and oil conservation; several of the ten projects also included other technological improvement and innovation elements. The sectoral policies included standards for buildings and appliances, intensive enterprise energy audits, energy improvement programmes covering the 1,000 largest energy consuming enterprises (the Top 1,000 Energy-Consuming Enterprises Program), and limits and taxes on inefficient vehicles. Again, these policies were primarily central government-driven or fiscal policies that required significant administrative and financial resources.

1.1.2 *Institutions*

During the 11th FYP period, China established the Energy Conservation Leading Committee, the Climate Change Leading Committee, the National Energy Commission, and the National Climate Change Expert Pool. Local government also established similar organizations.

The central government also developed a system of targets and accountabilities to assess the energy conservation performance of the local governments and major enterprises and to hold the major leaders of the local government accountable for implementing energy conservation policies and achieving the targets. Three schemes for data collection, supervision, and evaluation were developed to support such a system.

There was initial development in the energy conservation market with the support of the government's legal and policy framework, which covered full-purchase of renewable energy generation by law and an electricity wholesale rate-making process. Enterprises were encouraged to participate in CDM projects, invest in renewable energy projects, and implement energy efficiency upgrading projects via public tendering.

1.1.3 *Analysis*

There are three crucial questions regarding the 11th FYP that could provide lessons for developing and improving policies under the 12th FYP and subsequent plans:

1. How much impact did the various policies under the 11th FYP have on energy use and carbon emissions?
2. How much did these policies cost and what resources — material, administrative and financial — did they use?
3. How much opportunity is left for these policies?

1.1.4 *Impact*

The Chinese government and others have estimated the impact of each of the key programs associated with energy conservation and energy intensity improvement under the 11th FYP.

These estimates are calculated against a baseline, where efficiency or energy intensity is held constant from the beginning of the period. For guiding policy, this analysis is a useful start, but it falls short of what is ultimately needed in a number of ways.

First, the baseline comparison against constant efficiency is somewhat arbitrary. Some programs will be implemented in an environment where the existing inefficiency makes a natural improvement inevitable, while other programs might be approaching areas of the economy that are already relatively efficient. Thus, it is difficult to separate out the "savings" that result from policy versus those that result from the benchmark or pre-existing conditions.

A second issue is that these categories reflect multi-policy programs rather than single polices. The Top-1,000 Enterprises Program, for example, includes policies around energy audits, reporting requirements, standards, investments, and incentives. While it may be possible, baseline issues aside, to evaluate the effectiveness of the program as a whole, the overall program package may allow enormously effective, low-cost policy options to mask costly, ineffective ones. To improve policy, the analysis needs to identify which specific policy components, or sets of related policies, were effective.

Where specific policies are directed at specific technologies, we can better analyze impact. For example, with the early closure of inefficient coal plants and their replacement with newer, more efficient coal plants, we can analyze the impact of the policy versus the case where the plants had been left to operate through their normal life. Our initial analysis of carbon and energy savings in this case is consistent with the savings calculated by the Chinese government.

Overall, we find that while energy intensity is indeed in line with targets, carbon intensity is not falling significantly faster than energy intensity, indicating that despite investment in renewable and other low carbon generation sources, China has large potential for progress on greenhouse gas specific actions and policies.

1.1.5 *Cost*

The cost of these policies — an important element in understanding their performance — also remains to be assessed. The cost of implementing a policy includes the incremental investment cost (over what would have otherwise been invested) and the incremental operating/fuel costs or savings, weighed against the carbon or energy saved. In aggregate, many of these policies may effectively be free or create economic value. For example, our initial analysis suggests that in many cases replacing a very old, inefficient coal plant with an efficient new one would save enough energy costs to pay for itself in a few years. Thus, complete evaluation of policy performance will require verified estimates of savings against specific investment and operating costs of the policy change. Furthermore, many of these policies and the investments made may produce savings lasting many years, all of which must be accounted for in the policy evaluation.

At this point, only investment figures for the aggregate of the 11th FYP are available. Estimates vary, but the National Development and Reform Comission (NDRC) has estimated that as much as 2–3 trillion RMB were invested during the 11th FYP for energy conservation and environmental protection, while McKinsey estimated that at least $200 billion, or 1.3 trillion RMB at current exchange rates, was invested for energy efficiency and low carbon development. Tying this spending to specific policies and incorporating the overall economic cost remain issues for ongoing analysis.

Aside from such analysis, we observe that many of the instruments implemented in the 11th Year Plan are top-down regulatory measures that require significant administrative and financial resources, which may not be the most cost-effective way to achieve future targets.

1.1.6 *Remaining opportunities*

While a policy may have been effective and relatively inexpensive, the remaining question for policy makers is whether the opportunities for the policy have been exhausted. In some cases, the answer may be somewhat straightforward. For example, how many inefficient plants remain open? In other cases it may be much more complex. Was the plant really being used? Are there other reasons the lighting cannot be replaced? Does replacing a boiler require a complete (and expensive) system overhaul?[1]

In our initial analysis, we observe that much of the low-hanging fruit for reducing energy intensity has already been picked, for example, replacing old plants with more energy efficient plants. This means that further reduction in energy intensity could mean a higher cost to the economy.

1.2 Further Areas of Study

To arrive at conclusions that can help China improve policy in the 12th FYP and subsequent plans and perhaps help other nations that are implementing or considering policies similar to China's, requires an in-depth look at policy effectiveness. This challenge involves tying specific policies to specific savings, evaluating costs, assessing how much potential remains and then extracting lessons learned.

CPI's approach is to focus on selected policies to isolate some of these effects and learn lessons that can then be spread more broadly. In the near term, CPI will focus on three elements of the Chinese drive towards lower energy intensity:

1. The cost-effectiveness and energy savings from plant closures inside and outside the power sector;
2. The efficacy of China's focus on the top 1000 enterprises under the 11th FYP; and

---

[1] To some extent, this evaluation is beyond the scope of CPI work. However, the evaluation of both the impact and the cost gives insight into this issue of remaining opportunity with a particular policy. The impact evaluation does so if it measures how much of the available savings the policy achieved, the cost because it addresses the fact that costs tend to increase for any policy as the easy opportunities are addressed first.

3. The cost effectiveness of China's focus on specific technologies, particularly as embodied in the 10 Key Projects policy.

1.2.1 *The cost effectiveness and energy savings from plant closures*

Given the important role that plant closures played in the 11th FYP, further study on the cost of this instrument, as well as of the carbon savings derived from it, will be valuable.

Estimating the impact of early plant closures on costs and carbon savings has two components: (1) an estimate of what would have been the output and cost from the closed plant, and (2) similar estimates for the replacement plant (including capital cost estimations). Estimations should account for the life of the plant.

CPI plans to collect data on the efficiency and usage patterns of the older Chinese coal fired power plants as well as cost and operating information on the potential replacement plant. Once we have this data, we can assess whether this instrument has been cost-effective and how much further savings are available from this instrument.

1.2.2 *The efficacy of China's focus on the Top-1,000 enterprises*

The Top 1,000 Energy-Consuming Enterprises Program, which targets the largest industrial energy consumers in China, includes a series of measures including energy audits, mandatory goal setting and reporting requirements, training, incentives, and investment mandates. A closer study of this program will help to determine which components should be continued, which should be strengthened, which could be rolled out to the next set of enterprises, and how much a similar set of policies might contribute to China's future efficiency plans.

CPI plans to disaggregate the Top-1,000 program into a set of specific policies, activities, and initiatives in order to answer the following questions:

- Which elements of the Top-1,000 program have been most successful?
- Which elements cost the most?
- How much of the effect of this policy will be sustainable over the longer term, both and without additional policy support?

- How much more opportunity might be available from a continuation of this program?
- How much opportunity would be available in extending it to the next 9,000 enterprises? Is that practical from a resources perspective?
- Which elements would it make sense to extend given the cost/benefit?

1.2.3 *The cost effectiveness of China's focus on specific technologies*

Another approach to evaluating policy effectiveness is to use technologies rather than policies as the starting point for the analysis. Beyond plant closures, a range of technologies were targets for Chinese energy and carbon efficiency, including coal-fired industrial boilers, district co-generation, motors, lighting, appliances, cement kilns, and no-till farming. CPI will assess the impact of policies for a subset of technologies and specific projects and will use this analysis to draw lessons on policy impact. In particular, the evaluation will look at three issues:

- **Technology lifecycle cost and energy/carbon savings** — Like the plant closure analysis, the cost is a net lifetime cost, which must include capital costs offset by energy and other savings (where applicable). Obviously, for any given technology the costs will vary depending on the application, whether the installation is new or a replacement, and if a replacement, how much life the old unit had remaining.
- **Policy impact** — When a technology is mandated, the policy impact is clear. However, in other cases, the role of the policy may be less straightforward. For example, an energy audit might provide information about lighting options. The objective will be to identify which policy components might have influenced the technology decision and assess the impact of the policy on that decision.
- **Policy cost** — Some policies can, themselves, consume significant planning or management resources. For example, audits will require inspectors as well as resources from the enterprise undergoing the audit. Evaluating these resource requirements as applied to a specific technology decision adds the final piece to the technology review.

The individual evaluations can then be aggregated to provide feedback on policy costs and effectiveness, as well as furthering our understanding of the costs involved in advancing energy efficiency opportunities.

CHAPTER 2

Technology Advancement and Innovations

2.1 Technology Advancement and Innovations During the 11th Five-Year Period

Technology advancement is of great importance in the transition to low carbon economy. Energy saving and carbon reduction actions in China can never be achieved without technological support. Technology advancement during the 11th five-year period is shown in four aspects: promotion of advanced technology, phase-out of obsolete technologies, manufacturing efficient machinery and more inputs to R&D.

2.1.1 *Promotion of advanced technologies*

In power industry, the capacity of power units over 300 MW increased from 46.8% of total thermal power capacity in 2005 to 67.1% in 2009. The supercritical and ultra–supercritical power units spread soon. In 2007, the supercritical and ultra–supercritical power units had outputs of 30 GW, and the average net coal consumption rate of power supply was lower than 300 gce/kWh. By 2008 China had established 120 supercritical power units and one hundred 600 MW power units are under construction; eleven 1 GW-level ultra–supercritical power units was established and in addition to this, over 100 units are under construction or to be constructed.

Meanwhile, the power industry vigorously pushed forward high voltage power transmission. Between 2005 and 2008, the length of over 220 kV power transmission increased by 43.9% from 253,583 km to 364,796 km, of which the length of 500 kV power transmission increased from 62,866 km to 109,642 km and that of 750 kV from 141 km to 536 km. The design, production, construction and operation of China's ultra–high-voltage

(UHV) DC transmission have reached the international advanced level. Yunnan–Guangdong 1,373 km-long ±800 kV UHV DC transmission model project began since 2006, with maximum continuous capacity of 5 GW, and one side has been put into operation in December 2009. Sichuan–Shanghai 2,000 km-long ±800 kV UHV transmission model project began in late 2007 with the maximum continuous capacity of 6.4 GW and it was completed in 2009. In terms of UHV AC transmission, Southeastern Shanxi–Nanyang–Jinmen 654 km-long 1,000 kV UHV AC transmission trial model project was launched since 2006 and put into operation in January 2009.

Advanced technology was as actively promoted in the field of renewables. The over 1 MW wind turbines of the current year increased from 21.5% of the newly-installation capacity in 2005 to 87.9% in 2009. Above 2 MW wind turbines achieved as well, for instance, 3 MW wind turbines successfully rolled off the production line and R&D of individual 5 MW power unit just began. As for nuclear power, 200 MW HTGR (high temperature gas-cooled reactor) model project has initiated since late 2006. China has basically acquired Generation II and improved nuclear power technology and fully acquired Generation II improved PWR (pressurized water reactor) technology and is working actively on import and translation of AP 1,000.

The continuous casting technique spread soon and was widely used in the iron and steel industry. The continuous casting ratio of medium and large-sized enterprises increased from 97.5% in 2005 to 99.2% in 2008. The iron and steel enterprises encouraged recycling and using waste heat and energy as well. Coke dry quenching (CDQ) use rate increased from 35% in 2005 to 70% in 2009. Over 1000 m^3 TRT (blast furnace top gas recovery turbine unit) use rate went up from 81% in 2005 to 100% in 2008. Due to spread of advanced technologies and processes, comparable energy consumption of medium and large iron and steel enterprises decreased from 732 kgce/t in 2005 to 697 kgce/t, narrowing down the gap with international level (610 kgce/t).

In terms of the non-ferrous metal industry, given the China's bauxite resources, the Bayer–mineral processing method, an advanced technology of alumina production, has been successfully developed, and the intensified sintering process for alumina production was further promoted, which helped saving energy of 40%. China successfully developed large-scale 400 kA electrolytic aluminum prebaked cell, leading ahead of the world. New type cathode structure energy saving technology has been successful in trial. The over 160 kA prebaked cell output maximized from

52% of the electrolytic aluminum output in 2000 to 90% in 2009. The electrolytic AC energy consumption fell from 14.7 MWh/t in 2005 to 14.2 MWh/t, lower than the world's 2010 target of 14.6 MWh set by International Aluminum Institute, reaching the international advanced level of 14.1 MWh/t. The promotion of flash smelter technology made comprehensive energy consumption of copper smelting fall from 1,227 kgce/t in 2005 to 509 kgce/t in 2009, reaching the international advanced level of 500 kgce/t.

The new dry process of cement production has been developed and widely used now in the cement industry. Currently, China is able to construct Japan-made 4,000 t, 5,000 t, 8,000 t and 10,000 t production lines with new dry process. The localization dragged down the construction cost of production lines, for instance, the cost for a 4,000 t/d production line fell from 1.5 billion yuan to 400 million yuan with largely shortened construction time. The new dry process cement output was up from 40% of the total cement output in 2005 to 72.5% in 2009. Besides, there were a total of 263 new dry process-based cement production lines equipped with 193 net waste heat generation units with installation capacity of 1.7 GW and annual generation capacity of 11.1 TWh, saving energy of 4.1 Mtce. The bulk rate of cement increased from 37% in 2005 to 46.3% in 2009. The comprehensive energy consumption of cement fell from 167 kgce/t in 2005 to 139 kgce/t in 2009, reducing the gap with the international advanced level (118 kgce/t).

The proportion of float glass output in the flat glass industry jumped to 83.4% in 2009 from 70% in 2005. China's Japan-made 1,100 t extra-large float glass production line has been established and in operation. Its heat consumption is designed as 15%–20% lower than that of Japan-made 600 t production line. The flat glass industry has Japan-made over 500 t float glass (including 700 t and 900 t extra-large production lines) production lines established. The comprehensive energy consumption of flat glass decreased from 22 kgce/weight case to 16.6 kgce/weight case in 2009, reaching the international advanced level (15 kgce/weight case).

The building and sanitary ceramics industry independently researched and developed 7800 t large automatic hydraulic tile presses, 50 t to 100 t ball grinding mills and 2.5–3.0 m-wide and 200 m-long roller kilns, which had been put into application. The scale-up of enterprises and equipment progressed significantly as well. The annual output of over 1,200 building ceramics enterprises exceeded 2.5 million m^2, doubling by the year of 2000. The largest enterprise had an annual output of reaching 100 million m^2. The brick and tile enterprise expanded production lines too. In 2008, 800 more production lines with the annual output of over 30 million were newly established, of which 300 production lines had an annual output of 60 million to 120 million.

Technological equipment, processes, enterprise scale and feedstock line were key factors influencing energy efficiency in the oil and chemistry and chemical industry. Oil refining industry owned 11 enterprises with 10 million tons of capacity in 2008 and two of them had an annual output of over 20 million tons. The comprehensive energy consumption of petroleum processing dropped from 114 kgce/t in 2005 to 106 kgce/t in 2009, though quite a big gap with the international advanced level (73 kgce/t). Between 2005 and 2009, China entered into the prime time of the ethylene industry. In 2007, eight large installations were under construction with the total production capacity of 6.8 Mt and average capacity per installation of 0.85 Mt. In 2010, million tons of ethylene installation worked successfully with the localization rate of reaching 78%. The integration of oil refining and chemical processing was promoted as well. In 2007, the project of integrating oil refining with chemical processing in Quanzhou was officially contracted and put into operation in 2009. As a result, the capacity of oil refining plant expanded to 12 million tons from 4 million tons. It was able to produce 0.8 Mt of ethylene, 0.65 Mt of C ethylene, 0.4 Mt of polypropylene and 1 Mt of aromatics per year. The China's comprehensive energy consumption of ethylene production fell from 1,073 kgce/t in 2005 to 976 kgce/t in 2009. But there was a gap with the international advanced level (629 kgce/t) due to restriction of ethylene feedstock line of China (low light hydrocarbon percentage).

The synthetic ammonia production promoted new processes like exchange, temperature synthesis and optimized synthesis, and fired evaporated slag by FBC (fluidized bed combustion) boilers for generating heat. As a result, the comprehensive energy consumption of synthetic ammonia decreased from 1700 kgce/t in 2005 to 1,591 kgce/t in 2009. The ion-exchange membrane process output was up to 54.7% of the soda output in 2009 from 24.9% in 2000. The comprehensive energy consumption of ion-exchange membrane process fell from 1,067 kgce/t in 2005 to 916 kgce/t in 2009, approaching the international advanced level of 910 kgce/t. Since soda membrane process installation applied anode expansion and modified diaphragm technology, its comprehensive energy consumption fell from 1,448 kgce/t in 2005 to 1,293 kgce/t in 2009, a bit higher than the international advanced level (1,250 kgce/t). Soda ash production relied on two processes: ammonia–soda process and Hou's process. The ammonia–soda process was often used and sometimes a few processing on trona. The comprehensive energy consumption of soda ash fell from 396 kgce/t in 2005 to 323 kgce/t in 2009, a bit higher than the international advanced level (310 kgce/t). The majority of calcium carbide enterprises were small ones. In 2007, there were 404 calcium carbide enterprises, most of which were small plants with the annual output of below 50,000 tons or the average annual output of 34,000 tons. In addition, the closed vessel furnace process has not yet widely used in the calcium carbide industry and the output of closed vessel furnace in 2008 was only 20% of the total. The comprehensive energy consumption of calcium carbide decreased slightly from 3.5 MWh/t in 2005 to 3.4 MWh/t in 2009, 13% higher than the international advanced level (3 MWh/t).

Technology advancement in the building sector was demonstrated as new type building materials and advancement of heat and cold insulation equipment. The new type building materials included new wall materials and low emissivity coated glass. The new material output increased from 28% of the total wall material output in 2000 to 52% in 2009. In 2008, the new wall materials accounted for over 60% of construction areas in the new urban buildings. 2009 witnessed promotion of new energy-saved glasses with production capacity of 28 million m^2, and 2010 will see 50 million m^2. Initiatives of transformation for energy saving and heat insulation mainly included urban heating network and system flow, indoors temperature control, heat meter and building fencing structure improvement.

The agricultural sector started fertilizing on the basis of accurate land measurement. By 2008, a total of 60 million hm^2 farmlands were fertilized through accurate land measurement, reducing nitrogen fertilizer by over 10% and 28,000 tons of N_2O emissions or equivalent to 8.9 Mt of CO_2 emissions.

The transport and transportation sector progressed in energy saving by promoting advanced technologies in public transport, railway and waterway transportation and civil aviation. From the perspective of fuel oil economy, advanced technologies such as heat insulted shell, high pressure combustion, turbo, fraction and weight reduction and high intensity materials helped improve the fuel oil economy of cars and reduced the light vehicle energy consumption per 100 kilometers from 9.11 L in 2002 to 8.06 L in 2006. As alternative fuel-driven cars, natural gas-driven cars grew fast. In 2007, the national output of natural gas cars was 57,000 with the exhaust emissions reaching Europe III standards. Currently, China is researching on the natural gas cars that fit the Europe IV emission standards. Bio liquid fuel for car use also draws attentions. Noncrop bio liquid fuel is the major development direction, but key technology R&D is under way. China's carbon-based synthetic material development has progressed. Carbon-made methyl alcohol and DME production technology came into shape. Pilot works of methyl alcohol for cars were done in some regions and DME-based car model has started in trial. Until 2012, China's new energy cars will reach 1 million, of which hybrids will account for 95%.

Railway transportation reduced electricity consumption of electric locomotives through regenerative braking, frequent control of motor speed in main air blowers, intelligent electric meter and operation assessment and optimization system. Electric consumption of electric locomotives fell from 113.7 kWh/(10,000 t km) in 2004 to 110.6 kWh/(10,000 t km) in 2008. Apart from that, metal friction modifiers application in diesel locomotives, fuel distribution computer management system and diesel thermal insulation and pour-point technology helped reduce oil consumption of diesel locomotives from 25 kg/(10,000 t km) in 2004 to 24.9 kg/(10,000 t km) in 2009.

Waterway transportation enhanced energy efficiency of marine vessels through improving marine power design, optimization technology of machinery and paddles, diesel turbine charged technology, lubricant screw-in injection by electronic timer technology and by using fuel additives and self-polishing marine antifoulant paints. In addition, port and harbor operations applied advanced machinery, equipment and processes to save energy. For instance, container berths applied rail-mounted gantry crane (RMG) to replace rubber-tyred gantry crane (RTG) or ERTG, fuel additives and antifreeze for horizontal transportation equipment, dry-bulk terminals installed the electrical feedback device in gantry cranes to recover potential energy, crude oil reserves applied CHP technology, and cold energy recycling in LNG Terminal.

Civil aviation increased energy utilization efficiency of aircraft through optimization of the whole flight journey and identification of appropriate fuel onboard. Optimized the height, speed, and navigation channel of the whole flight journey; according to analysis of the flight management computer system, identified fuel onboard and optimized extra amount of fuel in need, helping reduce oil consumption per unit of aircraft from 488.5 tce/Mt km in 2005 to 453.7 tce/Mt km in 2008.

2.1.2 *Phase-out of outdated technologies*

While promoting advanced technology, phase-out of obsolete and low production capacity, processing units and technological equipment was the key approach to technology upgrading of China. It involved chiefly 20 high energy-consuming industries (coal, electric power, iron, steel, cement, electrolytic aluminum, glass, papermaking, coke, ferroalloy, alcohol, copper smelting, zinc and lead smelting, calcium carbide, MSG, citric acid, leather, dyeing and finishing and chemical fabrics). Table 2.1 summarizes production capacity utilization rate and phase-out of obsolete production capacities in 11 high energy-consuming industries in 2009.

Table 2.1. Production capacity utilization rate and phase-out of obsolete production capacities of 11 high energy-consuming sectors in 2009.

Category	% production capacity utilization in 2009	Cumulative capacity phased out until 2009	% obsolete production capacities available of the total production capacity in 2009	Target of phase-out in 2010
Coal	82.6	311 Mt	—	200 Mt
Coke	80.1	133 Mt	10.2	21.3 Mt
Thermal power	—	60.1 GW	—	10 GW
Iron	79	81.7 Mt	17.9	30 Mt
Steel	81.7	59.4 Mt	10	8.3 Mt
Cement	81.4	240 Mt	19	91.6 Mt
Flat glass	74.7	25 million weight cases	10.7	6.48 million weight cases
Electrolytic aluminum	64.4	0.8 Mt	5	0.3 Mt
Calcium carbide	65.4	2.9 Mt	27	0.7 Mt
Ferroalloy	63.1	5.2 Mt	10	1.4 Mt
Papermaking	96	6 Mt	—	4.3 Mt

Note: In cumulative obsolete production capacities phased out of coking, primitive coke and modified coking enterprises are 50 Mt and small coking enterprises are 82.5 Mt.
Source: Ministry of Industry and Information Technology, National Bureau of Statistics, China National Coal Association (CNCA), China Coking Industry Association, China Iron and Steel Association, China Building Materials Industry Association, China Carbide Industry Association, China Papermaking Association and Chinese Journal of Environment, 2010–06–09, China Chemical Industry News, 2010–04–08.

2.1.3 *Manufacturing high-efficient equipment*

Technology advancement is demonstrated at localization rate of high efficient equipment. The comprehensive localization rate of 1 GW-level nuclear power units reached 80%. The localization rate of materials used for 1 GW-level ultra–supercritical thermal power units was over 90% and that of 1,000 kV AC and ±800 kV DC UHV was beyond 90%. The localization rate of 10 million tons of oil refining outfit equipment was 90%. The localization rate of 0.3 Mt ethylene outfit equipment was over 80%, and of large coal chemistry outfit equipment was 90%. The localization rate of 4,000 m^3 blast furnace was 95% and of 1,780 mm large-scale cold rolled wide strip production lines was 96%. The localization rate of Japan-made 5,000 t new dry processing cement production lines was over 95%. Some equipment construction was extra large-scale. In 2006 and 2007, there were new over 300 MW thermal power units with the capacity of 153 GW. In 2008, there were 129 new dry process cement production lines established and put into operation with the newly added clinker production capacity of 143 Mt.

But, we have to be aware that China had not yet mastered core contents of some technologies, though the localization rate improved. Take an example of wind turbine, China has owned 1 MW-level production capacity of whole-set and some parts, but the key control system, whole-set and blade design still rely on import. Another example is CDQ technology in iron and steel industry. Though China has been able to design CDQ process and produce some devices, it is not yet able to design and manufacture high pressure CDQ. Analysis showed that China had not yet mastered 43 of 62 key low carbon technologies in six sectors like power, transportation, building, iron and steel, cement, chemistry, and petroleum. It takes to speed up R&D, demonstrate and expand all of these core technologies (UNDP and Renmin University of China, 2010).

2.1.4 *Increased technical inputs*

Between 2005 and 2008, national expenditures on science and technology continuously grew and inputs to scientific and technological researches kept steady growth. The total R&D expenditure increased from 245 billion yuan (accounted for 1.33% in GDP) in 2005 to 461.6 billion yuan (accounted for 1.54% in GDP) in 2008 (See Table 2.2).

In 2008, medium and large-sized energy and manufacturing enterprises' intramural R&D expenditure on energy technologies was 268.13

Table 2.2. Total R&D expenditure.

Items	1995	2000	2005	2006	2007	2008
Total R&D expenditure (billion yuan)	34.9	89.6	245	300.3	366.4	461.6
Percentage in GDP (%)	0.57	0.90	1.33	1.42	1.49	1.54

Note: (1) China's R&D expenditure was 1.7% of the world's in 2005; (2) the world's average scientific research inputs was 1.6% in GDP in 2006 and developed countries were above 2%.
Source: National Bureau of Statistics, 2009.

Table 2.3. Medium and large-sized energy and manufacturing enterprises' intramural R&D expenditure of China in 2008 (unit: billion yuan).

Category	Expenditure
Total of China's medium and large-sized enterprises	268.13
Energy sector	15.36
Coal exploitation and preparation	6.33
Petroleum and natural gas exploitation	3.64
Petroleum processing, coking and nuclear fuel	2.82
Electric power, heat power production and supply	2.57
High energy consuming manufacturing sectors	79.20
Steel and iron	30.13
Nonferrous metal	8.55
Building	4.57
Chemistry	17.66
Chemical fabrics	3.20
Food, beverage and tobacco	5.97
Textile and garment	6.56
Papermaking and paper products	2.56
Low energy consuming manufacturing sectors	113.0
Transport and transportation equipment manufacturing	32.79
Telecom equipment, computers and other electronic equipment manufacturing	48.09
General equipment and specialized equipment manufacturing	32.12
Other sectors	60.57

Source: National Bureau of Statistics, 2009.

billion yuan in total, accounting for 58.1% of the total R&D expenditure of China (Table 2.3). Energy sector (coal exploitation and preparation, petroleum and natural gas exploitation, oil processing, coking and nuclear fuels, electric power and heat power production and supply) spent a total of 15.36 billion yuan, high energy-consuming manufacturing sectors (iron and steel, non-ferrous metal, building, chemistry, chemical fabrics, food, beverage and tobacco, textile and garment, papermaking and paper products) spent 79.2 billion yuan in total. Low energy-consuming manufacturing sectors (transport and transportation equipment manufacturing, telecom equipment, computer and other electronic equipment manufacturing and general and specialized equipment manufacturing) spent a total of 113.0 billion yuan. Other sectors spent all together 60.57 billion yuan.

2.2 Technology Development Issues

Technology development issues during the 11th five-year period were insufficient technological innovation and barriers in technology transfer. Insufficiency of technological innovation was attributable to not enough inputs to science and technology and structural disequilibrium of R&D system. The international technology transfer met difficulties from slow international negotiation and functional failure of international mechanisms. In the context of increasingly urgent climate change, China needed not only independent research and development of high-efficient energy use technology and cleaner energy technology, but also cooperation with and support of the international community.

2.2.1 *Insufficient technological innovations*

Technology advancement in this five-year period met the problem of insufficient innovations. Insufficiency of inputs to science and technology restricted enhancement of innovation capacity. In 2007, China's competitiveness of scientific infrastructure ranked as 15th among the world's 57 major countries and regions; and its competitiveness of technological infrastructure ranked as 27th. China's GDP accounted for 7% of the world's, but its technological inputs was only 2% of the worlds. Seven percent economic development driven by 2% technological inputs would never be sustainable.

Technological R&D inputs by Chinese enterprises accounted for only 0.8% of the national GDP, much lower than the level of developed countries (above 2%). In China, only 25% of medium and large-sized enterprises had R&D units, and only 30% of them had R&D activities. In 2008, the ratio of R&D expenditure to main business revenue in nine enterprises of the manufacturing sector was between 1% and 2%,[1] but it was only 0.56% in enterprises of other industries (it was 0.71% in medium and large-sized enterprises). This ratio in developed countries was often over 5%. In addition, only 3 per 10,000 of Chinese enterprises had decision-making power of intellectual property. From the spending structure point of view, Chinese enterprises' spending on adaptation and localization takes no more than 7% of the total expenditure on imported projects, while South Korea and Japan cost 3–10 times of that. So, Chinese enterprises' capacities for adaptation, localization and innovation remain poor.

Chinese enterprises have not been able to do technological innovations, and scientific research institutes replaced their role. Even if there is corporate participation, enterprise is just playing the supporting role, and so fails to share substantial know-how and technological exchange. When involving multiple cooperation units, problems are even more complicated. Each unit may not let others know the outcomes at different phases and make collaboration even more difficult. In the transformation of technology know-how into practice, participants too often are in a dilemma of "being reluctant" but "having to cooperate" with each other. Therefore, Chinese enterprises have not fostered innovation capacity for a long time.

The structural disequilibrium of the Chinese science and technology research system becomes another barrier for innovation. It is characterized as three aspects: Inappropriate fund management, imbalanced HR distribution and resources decentralization and duplication. Now, R&D fund is managed by different administrations and distributed in an inappropriate way. The amount of single fund for nonscience management departments is sometimes much more than that of science management departments; some lower-rate universities too often could not get R&D funds while top universities sometimes have overabundant funds for R&D, which leads to a waste of money. There are a total of nearly 60 million talents now in China but high-caliber talents are still in short, particularly lack of professionals on energy strategy, auditing and management. HR is a problem geographically. The under-developed regions lack incentives to appeal talents and so their human resources are seriously in short. China's science

[1]Percents of the nine manufacturing industries are respectively: 1.93% specialized equipment, 1.74% medicine, 1.59% general equipment, 1.5% electric-gas machinery and equipment, 1.44% transport and transportation equipment, 1.27% rubber, 1.27% telecom, computer, and electronic equipment, 1.22% instrumentation, culture and office machinery and 1.06% chemical fabrics.

and technology resources are decentralized and duplicated as well. The utilization rate of large R&D equipment is only 25% in China, but 170% to 200% in developed countries. In conclusion, China has not established the mechanism that ensures steady growth of technological development, and the innovation-oriented awareness has not yet been built.

2.2.2 *Barriers of technology transfer*

China's dependence on foreign advanced technology is 50% comparing to 5% of the US and Japan. China's investment on equipment accounted for 40% of the investment on fixed assets, and above 60% of equipment investment relies on import. "Made-in-China" equipment mostly relies on import and 80% of petroleum and chemistry equipment relies on import. 70% of CNC machine and car manufacturing equipment and 85% of integrated circuit chips manufacturing equipment depend on import.

Take wind power as example, though mass production has been made possible for 1 MW-level direct-driven variable speed and constant frequency (VSCF) wind turbines and double-fed VSCF wind turbines, the core technology relies on overseas manufacturers. Though Chinese enterprises have an above 85% share of domestic wind turbine market (Li *et al.*, 2010), applicants for wind power patents in China are mostly international sub-companies in China. Besides, the type of patents applied by Chinese and foreign enterprises is different. Chinese patents are more practical than creative, but nearly 100% of foreign applicants apply for invention patent. In addition, China's initial mass production for 750 kW, 1.5 MW, 2.0 MW, and 3.0 MW occurred seven years later than the international practice. It means that the wind turbine technology from technology transfer is not world's top technology at all now. China has not acquired technological capacity matching its growth, though the overall production has increased beyond anticipation (UNDP and Renmin University of China, 2010).

The difficulty of technology transfer lies in the slow progress of international negotiation. Developed countries have no intention to fulfill their commitment on technology transfer. UN climate change negotiation defined clearly that developed countries should fulfill their obligations on technology transfer to developing countries. *UN Framework Convention on Climate Change*, *Kyoto Protocol* and the *21st Century Agenda* all called upon actions for easier technology transfer. But, these documents have not yet been well translated into action and their contributions to technology advancement of China were nearly zero. Developed countries do not have financial incentives and support to technology hosts. With barriers of IP laws, it is harder for developing countries to get technology transfer in its real sense.

In addition, international transfer mechanisms (such as Joint Implementation, CDM and Emissions Trading) have not functioned well to help technology transfer. The design of the CDM is that developing countries trade emissions reduction for funds from developed countries. Under the CDM, developed countries were concerned more on how to reduce emission costs by CDM rather than technology transfer to developing countries. The CDM does not define what kind of technology can be used for receiving funds (the only assumption is that the technology applied is progressive). The source of technology is determined by project hosts (from overseas or from host country), so the CDM has been criticized for failing to achieve sufficient technology transfer (Forsyth, 2007).

2.3 Outlook of Technology Advancement

In 2007, the Chinese Academy of Science proposed China's strategic map of energy technology development. The recent (until 2010) technology development plan is to prioritize energy saving and clean energy technology for energy efficiency improvement, tackling the key technology of CO_2 capturing, sequestration and utilization, entering into the period of demonstration and large-scale application of coal liquefaction technology, coal-based methanol/ether (DME) and olefin to oil technology, develop safe and clean nuclear power technology and nonhydropower renewables technology, break through the new generation zero emissions, Integrated gasification combined cycle (IGCC) and PFBC-CC (pressurized fluidized bed combustion combined cycle). The intermediate (until around 2030) target is to advance the development of nuclear power and renewables, break through FBR (fast breeder reactor) technology and give commercial demonstration, break through nuclear waste effective utilization and safe disposal technology, break through high-efficient solar energy transformation and integrated solar heating application system, break through the photosynthesis mechanism and select or create high efficient photovoltaic converted species, commercial demonstration and industrialization of agricultural waste and fiber, semi-fiber/biochemistry conversion technology, smart energy network technology, develop hydrogen energy system. The long-term strategy (until around 2050) is to establish the sustainable energy system to ensure the energy storage satisfying the needs of China's economic and social development. The dependence on fossil energy falls below 60% and renewables become one of the mainstream energies. Prioritize scale-up application of renewables technologies and strive to break through nuclear fusion technology.

In 2009, the State Council executive meeting issued the rejuvenation plan for 10 industries such as steel and iron, cars, vessels, petrochemical, textile, equipment manufacturing, light manufacturing, nonferrous metal, electronic information and logistics. It required controlling the total production of the iron and steel industry and phasing out obsolete production capacities, holding back the exclusive expansion project of steel and iron production. The petroleum and chemistry industry shall promote integrated resource utilization and waste-to-energy technology; non-ferrous metal industry shall develop the regeneration system covering the whole society so as to develop cyclic economy and improve practice of integrated use of resources; the textile industry shall phase out obsolete production capacities, environment-unfriendly and high energy-consuming processes and equipment; the equipment manufacturing sector shall go further on restructuring and changing the form of industrial growth.

In October 2010, the CPC issued the suggestions for development of the 12th FYP, stressing that

> "*Stick to technology advancement and innovation as strong support to speed up transformation of economic development, strengthen independent innovation capacity, and accelerate to build an innovation-driven country. The manufacturing sector is required to improve basic processes, basic materials and basic component R&D, and system integration level. Support technological innovations of enterprises, and improve development capacity of new products and brand building capacity. Build up capacity for transforming core technology know-hows into productivity to support national science and technology projects.*"

Currently, China owns the world's largest CDM market. According to statistics of CDM Pipeline, as of December 2010, China has registered 1,065 CDM projects. 159 CDM projects are applying for registration and 1,060 projects are under assessment. In total, there are 2,284 CDM projects, accounting for 40% of the world's CDM projects. The CDM projects in China are bigger than foreign average level. By 2012, GHG emissions reduction in China will reach 1.52 billion Certified Emission reductions (CERs), accounting for 54.3% of the world's CERs. There are 14 types of CDM projects in China. Hydropower and wind energy are main ones, respectively 42% and 25% of China's CDM projects. Specifically, areas of CDM projects are: (1) Energy industries, including wind generation, solar generation, tidal generation, hydropower generation and so forth; (2) other energies, including cement plant waste heat to power generation, CHP of natural gas, straw power generation, and household biogas digesters and so forth; (3) waste treatment, including garbage incineration power generation, large garbage landfill gas to electricity, livestock waste management system improvement and so forth; (4) technological transformation, including industrial boiler transformation, cement industry potential energy, electric industry potential energy, iron and steel industry potential energy, introduction of anaerobic treatment in the industrial waste water treatment facility and so forth; (5) demonstration projects, including green lighting, new burning-free bricks for buildings, hybrid cars, forest plantation and reforestation. Between 2005 and 2010, China has invested a total of 56.8 billion dollars on 1,065 registered projects, accounting for 63.8% of the world's total CDM investment. In the future, CDM will be an important tool for China to import and absorb world's leading technologies. It also takes time to explore new incentives to engage Western regions of China into CDM projects.

In conclusion, technology advance in this period reflects at promotion of first-rate technology, phase-out of obsolete technology, manufacturing high efficient equipment and increasing inputs to R&D. But, China still meets challenges like insufficiency of technology innovation and barriers of technology transfer. Looking into future, technology advancement and innovations will help sharpen competitive edge for future economic development and will be one major topic of China's future development.

CHAPTER 3

Structural Adjustment for Quality of Development

Economic structural change influences the overall energy consumption intensity due to the sectoral difference of energy consumption intensity. If sectors with high energy consumption intensity account for larger proportions in the national economy and grow relatively faster, the total energy consumption intensity will increase. In last 20 years of the 20th century, China made great achievement of quadrupling the national economy while only doubling the energy consumption. Between 1990 and 2005, the energy intensity decreased with an average annual rate of 4.2%, which was mainly attributable to structural factors (here the structural factors mean in a broad sense, including not only industrial structure, but also the energy saving through increasing product's additional value). Quantitative calculation results show that energy saving through structural changes contributed 60%–70% to the total energy saved (Research group on studies of approaches and measures for China to achieve sustainable energy by meeting 20% energy saving target of the 11th FYP, 2008). This means that structural energy saving was the leading factor that drove energy saving in the past. But the situation was contrarily different since 2002. Rather than leading factor towards energy saving, structural factor became the leading factor which caused the increase of energy consumption per unit of GDP. Its influence on energy consumption per unit of GDP was reversed. During this period, though technology improvement helped energy saving to some extent, its positive contribution was diluted by negative contribution of structural factor to energy saving and therefore energy consumption per unit of GDP climbed up. The reason was that the fast growth of energy-intensive industries makes heavy industry come to its prime time in the industrialization process. Since 1999, heavy industry has become the fastest growing sector in the national economy, much faster than light industry and GDP. Its proportion in GDP soared with years. Since energy consumption in heavy industrial sectors are much higher than in other sectors, the reduction of China's energy consumption per unit of GDP relies too much on technological upgrading (Shi, 1999; Shi, 2002; group on studies of approaches and measures for China to achieve sustainable energy by meeting 20% energy saving target of the 11th FYP, 2008).

Over the past five years, with acceleration and intensification of energy saving and emissions reduction, no matter the central government or local governments have all taken industrial structural adjustment as one priority initiative in the 11th five-year energy saving strategy. The *Comprehensive Work Plan for Energy Saving* issued in 2007 by the State Council put industrial structural optimization and adjustment into top priority, which include initiatives such as controlling the excessive growth of high energy-consuming and high-polluting industries, accelerating to phase out obsolete production capacities, improving industrial structural adjustment and so forth. These initiatives helped facilitate the structural adjustment during the 11th five-year period and China's economy is walking towards improving the quality of development by structural adjustment.

3.1 China's Structural Adjustment During the 11th Five-Year Period

3.1.1 *Structural change of primary, secondary, and tertiary sectors*

China's 11th five-year Outline Plan for National Economy and Social Development (hereinafter the 11th FYP) proposes that in this five year, it strives to achieve main socio-economic development objectives such as upgrading and optimizing industrial structure to make industrial, product and corporate structure more appropriate, and increase 3% of the tertiary sector value-added in GDP and 4% of the overall employees in the tertiary sector.

In December 2005, the State Council issued the *Notice on Tentative Regulations for Facilitating Industrial Structural Adjustment* (State issuance No. 40, 2005), with the intent of facilitating industrial structural adjustment. The *Regulations* defined the goals, principles, directions and priorities for industrial structural adjustment now and in the near future, and clarified classification principle and supporting policies for three types as encouragement, restriction, and phase-out in the *Contents of Guidance for Industrial Structural Adjustment*, proposing that the goals of China's industrial structural adjustment are to "advance the optimization and upgrading of industrial structure, facilitate the sound and coordinated development of the primary, secondary, and tertiary sectors, develop an sectoral pattern of agriculture as basis, hi-tech sector as pioneer, basic industries and manufacturing industry as backbones and service industry as full-fledged one. Stick to energy-saved, clean, safe, and sustainable development." The *Regulations* raised directions and priorities: (1) Consolidate and intensify fundamental basis of agriculture, speed up traditional to modernized agricultural transfer; (2) strengthen basic industries and infrastructures building to better guarantee the socio-economic development; (3) develop advanced manufacturing industry, playing its role of supporting economic development; (4) accelerate the hi-tech industry development, as driving force for whole economic growth; (5) stimulate service industry to go into holistic and rapid development, increasing its proportion and level; (6) promote recycling economy in a big way, building a resource-saving and environment-friendly society; (7) adjust regional industrial layout and optimize industrial structure; and (8) pooling resources for domestic development and opening-up to push forward upgrading and optimization of industrial structure.

In 2007, the *Suggestions on Acceleration of Service Industry Development by the State Council* (State issuance No. 7, 2007) pointed out

> *"to accelerate the service industry and increase the proportion of service industry in the structure of tertiary sector and make service industry as the leading industry of national economy as soon as possible, and it is a must-go road for economic restructuring and transforming ways of economic growth."*

Over the past five years, the secondary sector percentage in GDP dropped a bit, which was respectively 47.9%, 47.3%, 47.4%, and 46.3% between 2006 and 2009, falling by 1.6% as a whole. Between 2006 and 2009, the tertiary sector proportion in GDP was respectively 40.9%, 41.9%, 41.8%, and 43.4%, increasing by 2.5% as a whole. Except influences of national policies on industrial structural adjustment, structural changes of three sectors in the first four years of the 11th five years period and global financial crisis attribute to it to some extent.

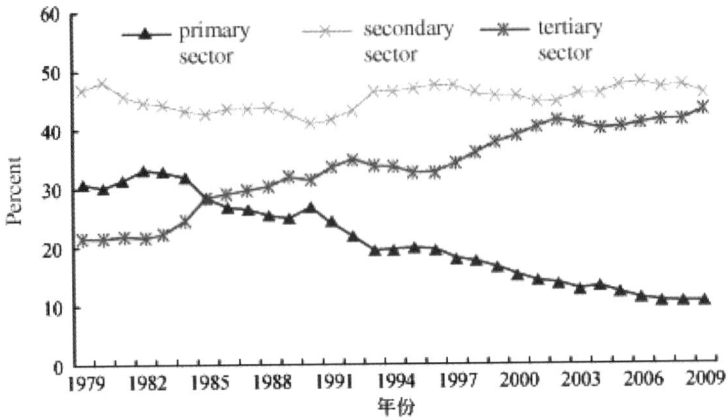

Fig. 3.1. Composition of three sectors in GDP 1979–2009.
Source: National Bureau of Statistics, 2010.

3.1.2 *The proportion of heavy industry in industrial value-added remains high but grows slower*

The 11th FYP notifies to accelerate the transformation of economic growth to achieve new type of industrialization. The *Decision on Implementation of Tentative Regulations for Facilitating Industrial Structural Adjustment* issued by the State Council pointed out to stick to new way of industrialization, "achieve high-technology driven, profitable, low resource-consuming, environment-friendly, safe, and ensured development that maximizes HR advantages and substantial change of economic growth way."

In the first three years of the 11th five-year period, China's heavy industry remained high proportion in the national economy and heavy industry value-added accounted for higher proportion in the total industrial value-added than before. In 2005, China's heavy industry value-added accounted for 68.6% of the total industrial value-added. Between 2006 and 2008, it was respectively 69.5%, 70.5%, and 71.3%, but grew slowly during the 11th five-year period. During the 10th five-year period, the heavy industry value-added was about 6% of total industrial value-added, while in the first three years of the 11th five-year period, it only increased by 1.8%.

3.1.3 *Product structural adjustment and enterprise expansion*

During the 11th five-year period, internal product structure in some manufacturing industry has been optimized to some extent. The proportion of high value-added products output in total outputs increased, like plates, tubes, and strips ratio in the iron and steel industry increased from 49.44% in 2006 to 53.18% in 2008, or by 3.6%; some low energy-consuming product output proportion in total outputs increased, for instance, new wall materials output grew fast and it was 56% of the overall wall materials outputs in 2008, doubled than that of 2000.

In addition, the size of enterprise is the key determinant influencing industrial energy efficiency. In high energy-consuming sectors of China, small and medium enterprises often apply relatively outdated processes and high energy-consuming feedstocks, and their energy consumption per unit of product was 30%–60% higher than that of large enterprises. Over the past five years, some high energy consuming enterprises and manufacturing equipment were expanding, which contributed to structural optimization of high energy consuming industries.

3.1.4 *Structural adjustment's contribution to energy saving*

Policies like controlling excessive growth of high consuming industries, phase-out of obsolete production capacities and improvement of industrial structural adjustment built an enabling environment during the 11th five-year period. China's industrial structure has been adjusted to some extent and contributed to energy saving initially. According to analysis of Mckinsey & Company, technological energy saving of China between 2006 and 2008 was 80% of the total energy saved and will be down to 50% in 2010. In other words, by the end of the 11th five-year period, structural adjustment's contribution to energy saving will reach 50%, equivalent to proportion of technological energy saving.

Energy, Environment and Economy Institute of Tsinghua University calculated the total amount and ratio of structural and technological energy saving of each year during the 11th five-year period (He, 2009). Results showed that by the first half of 2009, structural energy saving was 26.4% of the total energy saved, twisting its negative contribution to energy saving in the years before 11th five-year period (Table 3.1).

3.2 The Factors and Mechanisms Facilitating Structural Adjustment

3.2.1 *Target responsibility system promotes structural adjustment vertically*

Energy target responsibility system as the major institution during the 11th five-year period plays an important role in energy saving of China. In

Table 3.1. The total amount and the changing rate of structural and technological energy saving of each year during the 11th five-year period.

	Structural energy saving		Technology energy saving	
Year	Quantity (Mt)	Percentage (%)	Quantity (Mt)	Percentage (%)
2006	−8.1	−17.8	52.9	117.8
2007	19.7	19.5	81.2	80.5
2008	48.5	35.4	88.5	64.6
First half of 2009	26.8	58.3	19.1	42.7
Total	86.8	26.4	242.6	73.6

Source: He, 2009.

November 2007, the State Council announced *the Notice about Implementation Measures of Statistic, Monitoring and Assessment to Energy Saving and Pollution Reduction* (State issuance No. 36, 2007). In the *Implementation Plan of Assessment System of Energy Consumption Per Unit of GDP*, industrial–structural adjustment and optimization as important item has been listed into the *Scoring Methods of Assessment on Energy-saving Target Responsibility of the Provincial Governments* with 20 scores (full score is 100). The score for increased proportion of tertiary sector value-added in regional gross product, for the proportion of hi-tech industry value-added in regional industrial value-added, and for development and implementation of energy saving assessment and review methods for fixed assets investment program is four, and for meeting current year target on phase-out of obsolete production capacities is eight. The following Chap. 4 will give detailed analysis on implementation of the energy saving target responsibility system.

In January 2007, the State Council issued the *Notice of Suggestions on Accelerating the Shut-down of Small Thermal Power Units* (State issuance No. 2, 2007), requiring "to distribute targets of closing small thermal power units to each province (autonomous region and municipality), sign the target responsibility agreement on shut-down of small thermal power units with provincial people's governments and large state-owned electric power group corporations." In April 2010, the State Council issued the *Notice on Further Phasing Out Obsolete Production Capacities* (State issuance No. 7, 2010), defining that the completion of obsolete production capacity elimination target should be integrated into local government performance review system, and increase proportion of review on the target completion. What is more, the notice also suggested annual targets for phasing out obsolete production capacities be distributed to provinces, autonomous regions, and municipalities. In May 2010, assignments for phase-out of obsolete production capacities in 18 sectors and industries have been appointed to each province.

Given the review system and decomposition of targets, local governments began paying more attentions to industrial structural adjustment. The target responsibilities for phase-out of obsolete production capacities have been distributed to each level of government, facilitating adjustment of industrial structure vertically.

3.2.2 *Industry access system controls high energy-consumption industries at the source*

The *Energy Saving Law* suggests establish energy saving assessment and review system for fixed assets projects in order to control investment projects that do not conform to compulsory energy saving standards and designed specifications, and ultimately contain the blind and excessive growth of high energy-consuming industries. The *Comprehensive Work Plan of Energy Saving and Emissions Reduction* suggests strictly control new high energy-consuming and high-polluting projects, increase the market access threshold of energy saving and environmental protection by controlling lands and project financing, set up combined across-sector mechanism and review and accountability system for newly-established project management, and ensure the conformity to "six necessary terms" in project construction (must comply with industrial policies and market access standards, project approval or filing procedure, pre-view of land use, assessment and approval of environmental impacts, energy saving assessment, review and trust loans, safety, and urban planning).

In December 2005, the *Contents of Guidance for Industrial Structural Adjustment* (*Version 2005*) was approved by the State Council. The contents is the document aligning with *Tentative Regulations for Facilitating Industrial Structural Adjustment*, involving 20 more industries with encouragement type of 539 items, restriction type of 190 items and phase-out type of 399 items. The *Tentative Regulations for Facilitating Industrial Structural Adjustment* specifies review and approval policies for new programs in those three types. For instance, newly-established projects in restriction type are banned from investment and investment administrations will not review the approval, examination or preparation of these projects and the land administrations will neither handle relevant procedures for them.

In September 17, 2010, the National Development and Reform Commission issued the *Tentative Methods of Assessment and Review on Energy Saving of Fixed Assets Investment Projects* (2010 No. 6 Directive, National Development and Reform Commission), which regulated that since November 1, 2010, the new fixed assets investment projects have to be through energy saving assessment and review. Energy saving will be pre-requisite for project assessment, approval, verification and construction. Projects that fail to meet energy saving standards will be vetoed. It is meant to contain inappropriate growth of energy consumption at the source. The Methods marked that the assessment and review system for energy saving of fixed assets investment projects in China would go into official implementation phase and was supposed to exercise binding power on energy utilization of newly launched projects.

The before-mentioned industry access system provides higher threshold for market access, and leads economic structure to develop towards low consuming and polluting production capacities since production, processes, equipment, and products that cause serious pollutions, waste of resources and go against structural upgrading and optimization will not be allowed to enter the market.

3.2.3 *Market pressures constrain high energy-consumption industries from expansion*

During the 11th five-year period, China handled the macro-control of resources-intensive and high polluting and emissions industries by reducing or canceling export tax refund and differential electricity price and so forth.

To address pressures on domestic energy consumption by increasing high energy-consuming products generated from import needs, in September 2006, the Ministry of Finance, National Development Reform Commission, the Ministry of Commerce (MOFCOM), the General Administration of Customs and the State Administration of Taxation jointly issued the *Notice on Adjustment of Some Goods Export Tax Refund Rate and Addition of Contents for Processing and Trade Restriction Type* (Finance No. 139, 2006), which suggested adjust export tax refund rate of some goods. Coals, natural gas, wax, asphalt, silicon, arsenic, stone materials, non-ferrous metal, and wastes were exempted from export tax refunding. The export tax refund rate of steel fell from 11% to 8%, and that of ceramics and some finished leathers, cement and glass fell respectively from 13% to

8% and 11% and export tax refund rate of some nonferrous metal materials fell from 13% to 5%, 8%, and 11%.

To encourage corporate energy saving and punish high energy-consuming enterprises economically, in September 2007, the National Development and Reform Commission, the Ministry of Finance, the State Electricity Regulatory Commission jointly issued the *Notice on Further Addressing Issues around Differential Electricity Price Policy* (National Development and Reform Commission (NDRC) pricing No. 2655, 2007), requiring each province (autonomous region and municipality) to classify local enterprises into encouragement, restriction, and phase-out types, cancel immediately or gradually different preferential electricity prices available. In May 2010, the NDRC, the State Electricity Regulatory Commission and National Bureau of Energy issued the *Notice on Addressing Issues Relating to Preferential Electricity Prices for High Energy-consuming Enterprises* (NDRC pricing No. 978, 2010), demanding the cancellation of preferential electricity prices for high energy-consuming enterprises in some areas after global financial crisis and continuing the differential electricity price policy for eight industries like electrolytic aluminum, ferroalloy, calcium carbide, soda, cement, iron and steel, yellow phosphorus, and zinc smelting. In July 2007, China has cancelled local preferential electricity price initiatives for high energy-consuming enterprises in 22 provinces (including autonomous regions and municipalities), totaling over 1.5 million Yuan.

Export tax refund and differential electricity price policies increase operational costs for high energy-consuming enterprises and market presses some enterprises to save energy spontaneously. Due to more pressures from market, some high energy-consuming enterprises will phase out a few obsolete production capacities and expansion of high energy-consumption industries will be contained by market leverage.

3.2.4 *Governmental energy saving incentives help accelerate structural optimization*

According to *the notice on Publishment of Comprehensive Work Plan of Energy Saving and Pollution Reduction by the State Council* (State issuance No. 15, 2007), the central government set up special funds as incentives for underdeveloped areas that phase out obsolete production capacities by way of special transfer payment during the 11th five-year period. In December 2007, the Ministry of Finance issued the *Tentative Methods of*

Management of Incentives by the Central Government for Phase-out of Obsolete Production Capacities (Finance No. 873, 2007) notified that the central government set up funds as incentives for actions of phasing out obsolete production capacities in underdeveloped areas by special transfer payment. Local governments shall be responsible for eliminating obsolete production capacities and the central government gives incentives according to elimination scale. Incentives are arranged according to local context, applicable for 13 industries like electric power, iron, steel, electrolytic aluminum, ferroalloy, calcium carbide, coke, cement, glass, papermaking, alcohol, MSG, and citric acid. Incentive standards are made according to the average level of investment on phase-out of obsolete production capacities equipment of each industry. Between 2007 and 2009, the central government transferred a total of 15.5 billion Yuan to support the phase-out of obsolete production capacities in underdeveloped areas.

3.3 Factors and Mechanisms Constraining Structural Adjustment

Industrial structural adjustment is complicated and difficult. Constraints are as follows.

3.3.1 *Urbanization promotes heavy and chemical industrialization*

The structure of heavy industry cannot be adjusted within a short time due to market needs generating from urbanization. With rapid economic development and removal of urban-rural barriers, China's urbanization speeds up. Over the past 29 years after reform and opening up (1978–2007), the urbanization rate of China increased by 27%, or an annual increase of 0.93%. Between 2000 and 2007, this rate was up by 1.2% each year (Bai *et al.*, 2009). It is estimated that by 2020, urbanization rate in China will increase by 55% to 60% (Wang, 2005). So, it serves as a great impetus to heavy industry since public infrastructures construction like housing, road, transport and telecommunication, water, electricity, and gas all need large amount of iron and steel, non-ferrous metal, cement, chemicals, and building materials (Wu and Wen, 2006). Urbanization and upgrading of consumer spending demand prosperous development of heavy industry, so heavy and chemical industrialization currently in China is determined by market needs. Study reckons that this industrialization will proceed until around 2020 and the total energy consumption will keep growing fast within a short time (Xu and Wang, 2007; Bai *et al.*, 2009).

3.3.2 *Economic profits stimulate local government's attentions on heavy industry*

Heavy industry creates GDP in a fast way, which is favored by governments of all levels. Usually, high energy-consuming and high-polluting enterprises are also local large taxpayers. Since the tax-sharing system established in 1994, the tax system available now motivates local governments to pay more attentions to these high energy-consuming enterprises for getting more tax incomes.

Local governments are relatively powerful in terms of allocation of local resources. They give clear guidance for industrial patterning and structuring by industrial planning and also invite investment given local favorable policies on land, tax, industry, commerce, and finance (Liu *et al.*, 2010). Driven by the performance review system favoring economic profits and GDP, local governments often take the before-mentioned means to encourage development of high energy-consuming industries.

For example in late 90s of last century, some municipal governments and county governments in Inner Mongolia were rich in coal, electric power and mining resources. They developed favorable policies to attract investment on high energy-consuming industries. They even established industrial parks named as high energy consumption industrial zones (Daily of Inner Mongolia, 2003). This case was often seen in other regions as well.

Another typical case was that in 2008 right after issuance of export tax refund policy and global financial crisis, some high energy-consuming industries like electrolytic aluminum met great shocks. Along with the guideline of promoting local economic development, some provinces (autonomous regions and municipalities) provided different preferential electricity prices for local high energy-consuming enterprises in different ways, stimulating expansion of high energy-consuming industries and new production capacities. As a result, in 2010, NDRC, State Electricity Regulatory Commission and National Energy Administration again jointly issued the *Notice on Addressing the Issues Regarding Preferential Electricity Prices for High Energy-Consuming Enterprises*, with the intent of clearing up the preferential electricity price issues locally.

3.3.3 *Low resource price intensifies investment on high energy-consumption industries*

Distortion of resource price now in China encourages expansion of projects with high resource inputs and low economic efficiency, facilitating the development of heavy industry (Wu, 2008).

In recent years, costs of means of production in China, including domestic labor cost, land cost, raw material cost, energy cost, environmental protection cost, and intellectual property right protection cost were climbing up but resource price still could not reflect the level of scarcity, market need, supply, and environmental damage costs. The per capita capacity of most of natural resources in China is lower than international average level, but prices of natural resources are lower than the general international level, not reflecting the scarcity of resources. For long time, it is generally seen the low tax for mining resources and even tax-free mining, abuse of land resource, and under-standard water resource management and price formation system (Liu *et al.*, 2010). Energy price in China is relatively lower than that of developed countries. In addition, due to economic profits and political achievements, local governments get accustomed to lower the price of means and offer resources to absorb investment, worsening the distortion of resource price.

In this case, investment often flows into heavy industry. Between 2004 and 2007, fixed assets investment of heavy and chemical industries accounted for respectively 84.23%, 83.29%, 82.11%, and 81.95% of the total industrial fixed assets investment. Though the rate fell with years, it remained over 80% (Institute of Industrial Economics, Chinese Academy of Social Science, 2009).

In addition to that, industrial structural adjustment in global terms keeps going. Due to low energy and raw material price in China, it is easy for foreign high energy-consuming industries to move into China. Between 2000 and 2007, no matter the amount of investment or number of enterprises, the foreign investment in manufacturing sector accounted for over half of the total foreign investment, much higher than in other sectors (Su, 2009).

3.3.4 *Short-term effect of employment constrains structural adjustment*

With large population, the supply of labor force is much more than needs for labor force in China and this situation will remain unchanged for long.

Early in the 16th People's Congress of CPC, the employment issue was raised as top priority, same as important as livelihood of Chinese people. The 16th People's Congress Report also wrote that "employment is the fundamental thing for people's livelihood". In April 2004, the *White Paper of China's Employment and Policies* illustrated specifically on employment and its policies in China. It analyzed that at this period, China's population will go on increasing and in the future 20 years, over 16-year old population will grow with an annual increase of 5.5 million. By 2020, the total labor-aged population will reach 940 million. Given the critical time for reform deepening, the employment contradiction will be outstanding for quite a long time.

Global financial crisis undermines the economic growth and ultimately impacts on the employment in China. Between 2005 and 2009, registered urban unemployment population was respectively 8.39 million, 8.47 million, 8.86 million, and 9.21 million or unemployment rate were respectively 4.2%, 4.1%, 4.0%, and 4.3% (National Bureau of Statistics, 2010). Given this situation, local governments paid more attentions on employment. For instance, Heilongjiang province issued in February 2009 the *Further Stabilization of Employment Situation by Implementing More Active Employment Policies by Heilongjiang People's Government* (Heilongjiang government issuance No. 11, 2009), pointing out that global financial crisis impacts on provincial economic development, particularly enterprises are worsening. Employment pressures correspondingly are building up. So we should put employment into top agenda and create employment opportunities by government investment projects, encouraging different enterprises recruiting laid-off workers.

High employment demands rest on the rapid economic growth. Comparing each sector's capacity of creating employment opportunities, heavy and chemical sectors provide 5,000 employment opportunities per 100 million of investment, only one third of light industry. In the prosperous period of light industry, every 1% increase of GDP will address employment of three million people, but if it is the prime time of heavy and chemical industries, every 1% increase of GDP can only offer jobs for 70,000 people (Liu, 2003). From the job offering point of view, the tertiary sector is more advantageous than the secondary one. Since the tertiary sector has become the leading industry that provides employment opportunities, governments of all levels proposed the strategy of industrial structural adjustment to prioritize the tertiary sector for addressing employment issue. The tertiary sector will be the leading direction for employment expansion in the future. But it takes long for industrial structural adjustment to impact on employment and is not easy to see great effects within short term. Phase-out of obsolete production capacities and closure of high energy-consuming enterprises instead lead to cut-down of employees or salaries. So, to avoid employment tension and maintain social stability, some local governments in practice remained encouraging different enterprises to provide jobs, and employment pressures in some areas become one of reasons for not phasing out low production capacities.

3.3.5 *Insufficient technological capacity constrains the development of technology-intensive industries*

Technology-intensive industries are often characterized as high proportion of advanced and complicated technologies and equipment and hi-tech technicians, less labor force, lower raw material consumption, and being environment-friendly. However, weak innovations and lack of core technologies in China constrain the development of technology-intensive industries. The poor technological capacities of China are seen in three aspects: (1) Few invention patents, and fewer the invention patents with core competitiveness; (2) few corporate patent application. Between 1998 and 2003, the Chinese corporate patent application was no more than 30% of all China's patent applications. Throughout China, 99% of enterprises have never applied for any patent; (3) lack of core technologies. China's invention patents in the hi-tech field come mostly from overseas, like wireless transmission, mobile telecommunication, semi-conductor, bio-medicine and computing areas. Foreign and foreign-invested enterprises' patents account for respectively 93%, 91%, 85%, 69%, and 60% in the before-mentioned fields. Due to lack of core technologies, 20%–40% of sales price of China-made cell phones, computers, and CNC were paid to overseas patent holders (Tian and Zhang, 2006).

The lack of core technologies becomes the bottleneck of prosperity of new emerging industries in China.

3.4 Mid and Long-Term Structural Adjustment Outlook of China

With dual action of contributing and limiting factors, though China's strategy of structural adjustment for quality development since the 11th five-year period has laid foundation for structural optimization, the road to structural adjustment is full of challenges. The studies available predicted the industrial structures of China in the mid and long term. Different studies gave different predictions on three sectors and also had different views on the time of inflection of the secondary sector proportion. For example, Bai (2009) believed that in the future 10 years, the secondary sector majoring in industry will remain quite a good growth speed. Its proportion in GDP will not fall significantly like in developed countries. In 2020, the secondary sector will be 50% in GDP and then fall gradually. By 2050, it may fall to below 40%. Liu (2010) thought that China will enter into the later phase of industrial development after 2010 and the inflection of the secondary sector proportion downwards may occur around the year of 2010. Within the secondary sector, Liu (2010) believed that heavy and chemical industries and processing and manufacturing industries would be leading sector in the middle and later phase of industrialization and particularly in later industrialization, the proportion of processing and manufacturing industries will keep going up. After 2018, the growth of manufacturing sector goes mild. It is predicted that by 2025, manufacturing industrial value-added will reach 57.2% of the total value-added of goods.

Though different studies see different futures of China's economic structures, it is unanimously agreeded that the proportion of the tertiary sector keeps going up with that of the primary sector falling and the secondary sector proportion will drop once industrialization is basically achieved. All studies indicate that by 2050, the tertiary sector will exceed 50% and China will proceed on towards optimization of industrial structure.

In October 2010, the State Council issued the *Notice on Speeding up the Cultivation and Development of Strategic Emerging Industries* (State issuance No. 32, 2010), suggesting work further on speed up seven strategic emerging industries like energy saving and environmental protection, new generation information technology, biology, high-precision equipment manufacturing, new energy, new materials, and new energy-based cars. It also proposed that by 2020, strive to increase the proportion of strategic emerging industrial value-added to 15% in GDP. This will provide another tide of opportunities for development of industrial structure in China.

3.5 Conclusion

During the 11th five-year period, China witnessed changes of economic structures including industrial, sectoral and product structures etc. According to the study by Energy, Environment and Economy Institute of

Tsinghua University (He, 2009), by the first half of 2009, structural energy saving accounted for 26.4% of all energy saving, twisting its negative contributions to energy saving over the previous five years.

Contributing factors and work mechanisms for structural energy saving in China can be simply put as: Target responsibility system promotes structural adjustment vertically, industry access system controls high energy-consumption industries at the source, market pressures constrain high energy-consumption industries from expansion, governmental energy saving incentives help accelerate structural optimization. All of these actions are government-led and administrative directives function well. Therefore, the country, society, and enterprises pay high costs for it.

The structural adjustment during the 11th five-year period lays important foundation for the 12th five-year period, mid and long-term low carbon development of China. But, with one more look, we will see that the foundation is not solid. Main factors that hinder economic structural optimization in China include urbanization, local governments attentions on economic profits, low resource prices, employment pressures, and insufficient technological capacity. As it is hard to change the situation overnight, decarbonization of China's economic structures meet huge challenges.

With dual action of contributing and limiting factors, China will walk steadily and slowly towards industrial structural optimization. It takes efforts for China to transform ways of economic growth and push forward industrial structural adjustment, optimization, and upgrading.

CHAPTER 4

Energy Saving Target Accountability System

The "Energy Saving Target Accountability System" is a management system under which "energy saving agreements" is signed between upper-level and lower-level governments, between governments and enterprises and between management teams and implementation teams within enterprises. Such agreements define accountability for achieving the energy saving targets assigned, and specify the time framework for achieving such targets, as well as data collection and monitoring for performance evaluation. The system has four components, decomposition and allocation of energy saving targets, agreement signing, energy saving data collection and monitoring, and performance evaluation/accountability review. The target decomposition identifies energy saving target for each entity; agreement signing holds people in charge accountable for achieving the targets, data collection and monitoring enables transparent evaluation, and performance evaluation reviews accountability and ensures motivation for completing the targets.

4.1 Implementation Roadmap of the Accountability System

4.1.1 *Decomposition of energy saving targets*

On September 17, 2006, the State Council issued the *Reply on the Plan of Energy Consumption Reduction Per Unit GDP for Local Entities during the 11th Five Year Period*, which approved in principle the Plan of Energy Consumption Reduction *Per Unit GDP* for Local Units during the

11th Five Year Period, and required each provincial governments to strictly implement and ensure achieving the targets stated in the plan. See Table 4.1 for targets voluntarily proposed by each province and the reply from the State Council.

Some provinces have already decomposed their energy saving targets before the State Council issued the *Reply on the Plan of Energy Consumption Reduction Per Unit GDP for Local Entities during the 11th Five Year Period.* Shanghai municipality, for example, has allocated the 20% target evenly among its districts and counties as early as in May 2006. From November 2006–2008, other provinces such as Shanxi, Fujian, Hainan, Henan, Beijing, and Inner Mongolia completed their target decomposition. Targets sometimes were allocated to individual administrative region as well as to different industries, as in the case of Shanxi, Shanghai, Beijing, and Inner Mongolia. Targets allocated to the secondary industry are higher than those to other industries. In addition, some provincial governments, such as Hubei, Hebei, and Yunnan, allocated the targets to different administrative departments.

4.1.2 *Agreement for the accountability system*

Though the performance evaluation system wasn't established by end of 2007, agreements for the Accountability System were signed between central and provincial governments, between provincial and municipal governments, and between municipal and county governments before that. On July 26, 2006, on behalf of the State Council, National Development and Reform Commission signed the agreements with 30 provinces and 14 central government-owned enterprises. A few months later, these provincial governments signed agreements with municipalities in their jurisdiction, and the latter followed the suit and signed agreements with local districts, county governments, and major energy-consuming enterprises. After 2007, signature of similar agreements became a prevailing practice between governments and industries and was institutionalized.

These agreements usually contain two components — the allocated target to be achieved and performance evaluation. The first component specifies target for reduction in energy consumption per unit (10,000 yuan) GDP at given area, in energy consumption per unit of value-added of industries and enterprises, absolute reduction among key energy-consuming enterprises, etc., which the second component specifies people accountable for achieving the target, evaluation index, and reward and sanction

Table 4.1. Targets voluntarily proposed by each province and the reply from the state council.

Commitment lower than national targets			Commitment equivalent to national targets			Commitment higher than national targets		
	% reduction in energy consumption per unit GDP			% reduction in energy consumption per unit GDP			% reduction in energy consumption per unit GDP	
Region	% Proposed	% Approved	Region	% Proposed	% Approved	Region	% Proposed	% Approved
Hainan	−7.5*	12	Hebei	20	20	Shandong	22	22
Guangxi	10	15	Liaoning	20	20	Shanxi	25	25
Qinghai	10	17	Heilongjiang	20	20	Inner Mongolia	25	25
Yunnan	12	17	Shanghai	20	20	Jilin	30	30
Guangdong	13	16	Jiangsu	20	20			
Tibet	13	12	Jiangxi	20	20			
Beijing	15	20	Henan	20	20			
Tianjin	15	20	Hubei	20	20			
Zhejiang	15	20	Hunan	20	20			
Ningxia	15	20	Chongqing	20	20			
Shaanxi	19	20	Sichuan	20	20			

(Continued)

Table 4.1. (*Continued*)

Commitment lower than national targets			Commitment equivalent to national targets			Commitment higher than national targets		
	% reduction in energy consumption per unit GDP			% reduction in energy consumption per unit GDP			% reduction in energy consumption per unit GDP	
Region	% Proposed	% Approved	Region	% Proposed	% Approved	Region	% Proposed	% Approved
Fujian	–**	16	Gansu	20	20			
			Xinjiang	20	20			
			Anhui	20	20			
			Guizhou	20	20			

*According to Hainan's 11th Five Year Plan, it is at the initial stage of industrialization, and with the increasing percentage of heavy industries, overall energy consumption will continue to grow at a stable rate.

**Fujian's 11th Five Year plan did not specify a number for reduction percentage, instead it stated to achieve "notable improvement in resource utilization efficiency and energy consumption per unit GDP remains lower than the national average level". As a result it is categorized as "lower than national targets".

Source: Department of Development Planning, National Development and Reform Commission, *National and Regional 11th Five Year Plans for Economic and Social Development*, China Market Press, Beijing, 2006.

mechanisms. These agreements are usually signed between the head of upper-level government and the head of lower-level government.

4.1.3 *Data collection and monitoring*

On November 17, 2007, the State Council issued the Circular on Approving and Transmitting the Implementation Program and the Measures for Statistics, Monitoring and evaluation of Energy Saving and Emission Reduction, specifying the institutional arrangements for statistics, monitoring and evaluation of energy saving. There are three plans relating to energy saving, the *Implementation Plan of the Statistical Indicator System of Energy Consumption Per unit GDP*, the *Implementation Plan of the Monitoring System of Energy Consumption Per unit GDP*, and the *Implementation Plan of the Review and Evaluation System of Energy Consumption Per unit GDP*.

The *first one* specifies methodologies for data collection and statistics on energy consumption. Overall, it aims at setting up energy statistics and investigation system for both the supply and the consumption of energy on the basis of calculating total energy consumption of each level of jurisdiction. The *second one* specifies methodologies for energy consumption monitoring. Energy saving indicators and data quality shall be approved and monitored by statistics administrations in upper-level government, which monitors energy consumption reduction realized, local energy consumption per unit GDP, and the reduction rate, etc.

4.1.4 *Review of energy saving targets*

The *Implementation Plan of the Review and Evaluation System of Energy Consumption Per unit GDP* specifies objects, contents, methodologies, procedure, and award and sanction measures, hence provides specific approaches for implementing the target accountability system. The review indicators fall into two categories: Target achieving indicators and policy implementation indicators. The former is the critical one — as long as the government fails to meet the target for energy saving, it will fall into the category of "target not achieved".

The Implementation Plan also clarifies specific award and sanction measures. For provincial governments, the evaluation results will be filed at human resource department and will serve as an important reference to evaluate for the performance of responsible officials. Accountability System and

Veto System will apply in line with policies including "*Tentative Methods of Comprehensive Review on Local CPC Leaders and Government Official*". *Veto* System means that once a provincial government falls into the category of "target not achieved" in energy saving evaluation, its officials will be denied with any opportunity to receive any awards and honorary titles, moreover, the national government will suspend evaluation and approval of any new energy-intensive projects proposed to be built in that region.

Till date, energy saving target review and evaluation system is basically established at both national and local (provincial, municipal, prefectural, and county) level. Local government developed their review systems mostly using the national review plan as the blueprint, as a result, their evaluation system and award and sanction measures are in general similar to the national system defined by *Implementation Plan of the Review and Evaluation System of Energy Consumption Per unit GDP*.

4.2 Performance Review on the Target Accountability System

4.2.1 *Completion of energy saving targets*

(1) *Overall performance*

2006 was the first year of 11th Five Year period, when China's energy consumption per unit GDP reduced by 2.74% compared to the previous year. In 2007, 2008, and 2009, the reduction relative to the previous year was respectively 5.04%, 5.2%, and 3.61%. In the first four years of 11th Five Year period, reduction of energy consumption reached 15.5% relative to 2005 level.

(2) *Provincial governments*

At the provincial level, only Beijing met its target of energy intensity reduction rate in 2006. Except Qinghai, 29 out of 30 provinces and autonomous regions (excluding Tibet Autonomous Region) achieved some reduction in energy consumption per unit GDP. In 2007, 6 provinces (Beijing, Tianjin, Liaoning, Shanghai, Jiangsu, and Shandong) were rated as "exceeded the reduction target"; Seven provinces (Hebei, Shanxi, Inner Mongolia, Hainan, Guizhou, Ningxia, and Xinjiang) were rated as "Target unmet". In 2008, seven provinces (Beijing, Tianjin, Hebei, Liaoning, Jiangsu, Shandong, and

Yunnan) were rated as "exceeded the reduction target". Due to natural disaster like earthquakes, Xinjiang and Sichuan failed to meet their targets of 2008 and were rated as "Target unmet". In 2009, 18 provinces (Beijing, Tianjin, Hebei, etc) were rated as "exceeded the reduction target"; 10 provinces (Jilin, Heilongjiang, Zhejiang, etc) were rated as "Meeting the reduction target"; Tibet Autonomous Region basically met its target; Guizhou and Xinjiang were rated as "Target unmet" (Table 4.2).

As energy saving policies and instruments being developed and implemented, the number of provinces failing to meet their annual targets reduces year by year. Most provinces failed to meet their targets in 2006 because many energy saving policies were issued in mid and late 2006 or even in 2007. Meanwhile, energy saving investments were not in place yet at that time and there is usually a time lag for investments to yield results.

Noticeably, since 2006 annual target for each province was proposed by such province itself and submitted to National Development and Reform Commission (NDRC) for approval. When NDRC conducted evaluation for target achieving in 2006, the annual target was the average value of the Five Year target, that is to say, the reduction target was 20% for the five years and annual rate was 4.4%. Under this system, all provinces except Beijing failed to achieve the annual target. In 2007, all provinces except four provinces (Beijing, Shaanxi, Yunnan, and Guizhou), set lower annual targets than the average ones. Consequently all provinces except seven (Hebei, Shanxi, Inner Mongolia, Hainan, Guizhou, Ningxia, and Xinjiang) succeeded in meeting their annual targets in 2007. Most provinces continued this strategy in 2008 and set annual targets lower than the average level. Some provinces that failed to meet 2007 target lowered their annual target, for example Guizhou from 4.95% to 4.1%, and Hainan from 2.5% to 1%, and successfully met their lower 2008 target by 2009. In addition, Shanxi, Inner Mongolia, and Jilin, set higher target than the required 20% during 11th Five Year period (Shanxi and Inner Mongolia 25%, Jilin 30%), and later, when they were facing challenges to meet the previous high targets, reduced the target to 22% with the approval by the central government in 2008. These three provinces met their lower targets in that year. Considering the prevailing practice of setting lower annual targets among provinces and to ensure to meet cumulative energy saving targets during 11th Five Year period throughout China, the NDRC added the progress of cumulative energy saving as part of review in 2008.

Table 4.2. Completion of energy saving targets by provincial governments during the 11th five year period.

Region	Target (1) of 11th five year period (%)	2006 (2)	2007 (3)		2008 (4)		2009 (5)		Cumulative 2006–2009	
		Reduction rate %	Annual reduction target %	Actual reduction rate %	Annual reduction target %	Actual reduction rate %	Annual reduction target %	Actual reduction rate %	Cumulative reduction rate %	Target completion during 11th five year period %
Beijing	20	5.25	5.00	6.04	5.0	7.36	4	5.57	23.34	119.14
Tianjin	20	3.98	4.40	4.90	4.5	6.85	4	6.03	20.07	100.39
Hebei	20	3.09	4.40	4.02	4.5	6.29	5	5.02	17.21	84.62
Shanxi	25	1.97	5.60	4.52	4.4	7.39	5.6	5.72	18.28	81.23
Inner Mongolia	25	2.50	5.26	4.50	5.0	6.34	5	6.91	18.82	83.9
Liaoning	20	3.20	4.00	4.01	4.0	5.11	4	5.08	16.64	81.55
Jilin	30	3.32	4.00	4.41	4.2	5.02	5	5.98	17.47	77.29
Heilongjiang	20	3.04	4.00	4.09	4.5	4.75	5	5.61	16.39	80.23
Shanghai	20	3.71	4.00	4.66	3.6	3.78	3.6	6.17	17.12	84.14
Jiangsu	20	3.52	4.20	4.28	4.4	5.85	4.6	5.17	17.51	86.28
Zhejiang	20	3.52	4.00	4.18	4.0	5.49	4	5.41	17.36	85.42
Anhui	20	3.44	4.00	4.11	4.0	4.52	4	5.13	16.13	78.82
Fujian	16	3.20	3.50	3.51	3.5	3.70	3.2	3.52	13.22	81.32
Jiangxi	20	3.18	4.00	4.01	3.6	5.53	4	4.54	16.68	81.79
Shandong	22	3.46	4.50	4.54	4.5	6.47	5.45	5.46	18.51	82.39
Henan	20	2.98	4.0–4.5	4.11	5.1	5.10	4.8	6.02	17.03	83.65
Hubei	20	3.21	4.00	4.06	4.4	6.29	4.4	5.84	18.46	91.43
Hunan	20	3.39	4.00	4.43	4.0	6.72	4.5	5.02	18.2	90.02
Guangdong	16	2.93	3.00	3.15	3.5	4.32	3.7	4.14	13.77	84.99

(Continued)

Table 4.2. (*Continued*)

Region	Target (1) of 11th five year period (%)	2006 (2) Reduction rate %	2007 (3) Annual reduction target %	2007 (3) Actual reduction rate %	2008 (4) Annual reduction target %	2008 (4) Actual reduction rate %	2009 (5) Annual reduction target %	2009 (5) Actual reduction rate %	Cumulative 2006–2009 Cumulative reduction rate %	Cumulative 2006–2009 Target completion during 11th five year period %
Guangxi	15	2.50	3.20	3.31	3.5	3.97	3.4	4.43	13.48	89.1
Hainan	12	1.17	2.50	0.80	1.0	2.55	1	2.78	7.12	57.75
Chongqing	20	3.41	4.00	4.46	4.2	3.55	3.8	5.83	16.36	80.05
Sichuan	20	2.10	4.00	4.44	4.2	3.55	3.8	5.83	16.36	80.05
Guizhou	20	1.85	4.95	3.97	4.1	6.11	4.3	3.95	15	72.84
Yunnan	17	1.52	3.80	3.98	4.4	4.79	4.3	4.6	14.11	81.63
Tibet	12	—	—	—	2.5	2.50	2.5	2.67	9.6	78.99
Shaanxi	20	3.39	4.50	4.54	4.0	5.92	4.5	4.56	17.24	84.78
Gansu	20	2.61	4.00	4.09	4.5	4.53	5.2	6.82	17.32	85.21
Qinghai	17	−1.51	2.00	2.20	4.0	4.18	5	6.46	12.53	71.84
Ningxia	20	1.01	4.00	3.52	4.0	6.79	4	6.04	16.36	80.04
Xinjiang	20	1.06	4.00	3.08	4.0	3.15	2	1.53	8.55	40.05

Source: (1) *Reply on the Indicator Plan of Energy Consumption Reduction Per unit GDP for Local Units during the 11th Five Year Period by the State Council*; (2) *Announcement on Indicators like Energy Consumption Per unit GDP of Provinces, Autonomous Regions and Municipalities* in 2006; (3) *Announcement by the National Development and Reform Commission of the People's Republic of China*, No. 55 in 2008; (4) *Announcement by the National Development and Reform Commission of the People's Republic of China*, No. 13 in 2009; (5) *Announcement by the National Development and Reform Commission of the People's Republic of China*, No. 8 in 2010.

4.2.2 *The role accountability system played in energy saving during the 11th five year period*

(1) *The development of energy saving management system at central and local level in the government*

Before the formation of the 11th Five Year Plan Outline in 2006, institutions involved for energy saving were limited to development and reform commissions or economic and trade commissions at provincial level. In June 2007, the State Council issued the *Notice on Printing of the Comprehensive Work Plan for Energy Saving and Emissions Reduction by the State Council*, further stressing

> *"the top priority at present is to establish accountability system for energy saving and emission reduction at each level of government to ensure relevant policies in place are thoroughly implemented. Local governments of all levels shall be in general charge of energy saving and emission reduction at their administrative areas and principal heads of governments are the primary responsible persons."*

Right at the same time, the central government formed the "National Leading Group on Climate Change, Energy Saving and Emission Reduction" as the coordination institution for national climate change, energy saving and emission reduction programs, on the basis of the National Coordination Group on Climate Change. Provincial government followed the suit and established their leading institution shortly after the national government (Qi *et al.*, 2008). Some municipal developed similar institutions as well. For example, Putian city of Fujian province established its "leading group on energy saving and emission reduction" (task force for addressing climate change) in November 2007, and Huzhou city of Zhejiang province formed its "leading group on climate change, energy saving, and emission reduction" in December 2007.

Most leading groups have set up their secretariats for coordination on regional energy saving and emission reduction programs. The secretariats are usually based in development and reform commissions or economic and trade commissions. Initiated by such leading groups, local governments developed their implementation plan for statistics, monitoring, and review system on energy saving and emission reduction. Such institutional development enabled China to develop a relatively coordinated system for monitoring, reporting, and verification. During the review of energy consumption reduction target achieving by local governments, most provinces scored high in organizational building. Even provinces failing to meet their targets have succeeded in the organizational development. The four level (national, provincial, municipal, and county level) institutional arrangement for energy saving management system was developed. Some provinces also set up monitoring and inspection center to supervise key energy-consumers in their region.

(2) *The target accountability system effectively incentivized local governments to pursue energy saving and emission reduction*

The key to incentivize local government is the promotion of government officials. The National *Implementation Plan of the Review System of Energy Consumption Per unit GDP* specifies that once the review and evaluation results on energy saving of each province is verified, they will be sent to HR department to serve as a key indicator for evaluating the overall performance of government officials according to the *Tentative Methods of Comprehensive Review on Local CPC Leaders and Government Officials*. *Veto* system is applied in this case. Local government developed similar or even stricter arrangements to incentivize their lower-level counterparts. In May 2008, Shijiazhuang province issued the *Notice on Printing of the Implementation Plan of Review of Key Counties (Cities) and Enterprise for Energy Saving and Emissions Reduction*, stating "With failures to achieve his target in three consecutive years, principal heads of key counties (cities) shall resign from their position, corporate representatives of state-owned enterprises shall be fired right away, and private-owned enterprises shall be penalized with production suspension." Target accountability system requires failure to achieve targets as *veto* for any promotion offers has tremendously changed the original incentive system for government officials. In addition, some national and provincial leader's commitments also motivated their stuff for achieving energy saving. Shousheng Xu, governor of Gansu province, made commitment that if failing to achieve the energy saving and emission reduction that he is accountable for, he would resign.[1] Shandong's governor Daming Jiang was even more outspoken, "What is compulsory indicator? It is the project of 'kicking people off their positions'. If Shandong fails to meet our energy saving and emission reduction target,

[1]China website, Shousheng Xu, governor of Gansu province: I would resign if failed to achieve the energy saving and emissions cutting target (2007), http://www.china.com.cn/city/txt/2007-05/24/content_8298233.htm (accessed on 28 July 2010).

I would resign from my office."[2] Zushan Lv, governor of Zhejiang province said "It does not matter GDP grows by 8% or 10%, this is just an indicative number. But the target for 20% reduction is a compulsory one and a must-do." Zushan Lv wrote to 11 mayors in Zhejiang provinces to ensure that they will "achieve it by all means". An official in charge of energy saving and emission reduction in a city of Zhejiang province said that his mayor warned him "If I am kicked off from my position for you failing to meet our target, I would kick you off your position before I step down."[3] In May 2010, Premier Jiabao Wen required to "use iron hand" to phase out backward capacities in a video conference with local officials. Some local governments translated it into taking iron fists to achieve their energy saving and emission reduction targets. Such determination is unprecedented for energy management in history.

The principal responsible person of the target accountability system is the head of local government, who controls usually significant power in local development and have strong influence on local development patterns. The intrinsic design of this accountability system compels local governments to prioritize energy saving as part of local development. Given China's context, energy saving goals often conflicts with economic development interests. The target accountability system served as an effective driving force for energy saving and energy efficiency improvement during the 11th Five Year period.

(3) *The target accountability system induced energy efficiency improvement in enterprises*

Pressured by the target accountability system, local governments paid attentions to major energy-consuming enterprises and forced them to conduct energy saving practices. Statistics shows that during the 11th Five Year period over 1,500 provincial-level key enterprises signed agreement on energy saving with local government.

Such major energy-consuming enterprises taken into the review system are subject to direct monitoring and supervision of local governments.

[2]News China website, Governors of Shandong and Gansu provinces expressed: if failing to achieve the target, they would resign (2008), http://news.sina.com.cn/c/2008-03-11/185715125432.shtml (accessed on 28 July 2010).
[3]Huaxi City News, One mayor of given city in Zhejiang warned an official on energy saving to fire him first if he was fired for failing achieving energy saving target (2010), http://news.sohu.com/20100926/n275260025.shtml (accessed on 28 September 2010).

On the one hand, governments provide funding for enterprises seeking technology innovation and upgrading in order to achieve energy saving. For enterprises that met or exceeded their targets, the government provides awards in some cases too. Such awards are very attractive to managers as they will enable the manager to get more chances for promotion (in the case of state-owned enterprises) or enable the company to get more social recognition and political support (in the case of private companies). On the other hand, governments also punish enterprises for excessive energy use. Governments will formally criticize enterprises that failed to meet their target and deny them with any opportunities for winning annual award or honorary titles. Such punishment means losing promotion opportunities for leaders in State-owned enterprises (the *veto* system for energy saving), and provides effective incentives for them to meeting their energy-saving targets. Moreover, governments do not provide favorable policies or political support for enterprises failing to meet their target, sometimes suspending the approval of newly proposed energy-intensive programs or of new industrial land use, or even applying tougher policies such as differential electricity price, limiting or cutting off power supply, or suspending production. These strict measures motivated enterprises to do whatever they can to achieve the targets.

With the direct monitoring and supervision of local governments, completion of energy saving targets among key energy-consuming enterprises was rather satisfying. At the end of 2008, nationwide the "one thousand enterprises" have together achieved total energy saving of 106.2 Mtce, meeting 106.2% of the 11th Five Year target and two years ahead of scheduled time. A total of 443, or 48.05% of them have met or exceeded their targets.[4] At provincial level, key energy-consuming enterprises also met their targets. For example, in the first three years of the 11th Five Year period, 200 key energy-consuming enterprises of Shaanxi province have achieved cumulative energy saving of 6.6 Mtce, meeting the target of 6.4 Mtce ahead of schedule.[5] At municipal and prefectural level, it was optimistic as well.

[4] *Source*: National Development and Reform Commission, Announcement of review results of energy saving target of thousands of enterprises (2009), http://www.sdpc.gov.cn/zcfh/zcfbgg/2009gg/t20091124.tm (accessed on 6 August 2010).

[5] *Source*: National Development and Reform Commission, 200 key energy-consuming enterprises of Shaanxi province over fulfilled the 11th five-year target for energy saving (2009), http://www.sdpc.gov.cn/dffgwdt/t20090813_296234.tm (accessed on 6 August 2010).

For example, Wenzhou city of Zhejiang province reviewed 506 enterprises with annual energy consumption of/above 1,000 tce in 2008, and 94% of them met their target while 46% exceeded their target.[6]

4.2.3 *Limitations of the target accountability system*

(1) *Capacity of local governments does not match with their responsibilities*

The energy saving target accountability system features its hierarchical decomposition of energy saving targets from central, provincial, municipal to county governments. Under this mechanism, the role of provincial and municipal governments is to urge lower-level government and local key energy-consuming enterprises to meet their targets. County and lower-level governments are burdened with most of the pressure for implementation and meeting their target.

Currently, China's organizational setup for energy saving is basically established. Since the 11th Five Year period, leading groups on energy saving and emission reduction were all formed from central to local level along with specialized organizations for implementation of the target accountability system. However, at local level the capacity is weaker, and the resource is less available. Unlike central and provincial governments, county governments usually find it difficult to get sufficient funding to support enterprises with technology transformation. Within their administrative areas, enterprises are mostly small and medium-sized and have less funding for implementing energy saving measures, unlike those large corporations. Meanwhile, it is difficult for them to acquire government funds for technology transformation as the saving potential is usually too small to qualify for subsidies. Local governments also have no way to control and supervise state-owned enterprises in their region since the latter are of higher political hierarchy than local governments. As for large private companies, foreign-invested or joint venture enterprises, due to their backbone role in local economy, local governments are often caught in a dilemma between economic development goals and energy saving tasks. As a result, local

[6] *Source*: Wenzhou municipal government, Announcement on energy saving review of 2008 throughout the city by the office of Wenzhou People's Government (2009), http://www.wenzhou.gov.cn/art/2009/7/22/art_4752_161.html (accessed on 6 August 2010).

governments sometimes lose their supervision on energy saving practices of these enterprises. In the meanwhile, county and lower-level governments are weak at delivering energy saving technology services, making it hard for enterprises within their administrative areas to meet the assigned targets. Therefore, though the target accountability system regulates local energy saving, lower-level governments undertake much more pressures than their upper-level counterparts due to limited capacities.

(2) *Local governments lacked efficient means for achieving structural energy saving*

Both central and local governments take industrial restructuring as one of the top priorities in energy saving strategy for the 11th FYP period. But it is very difficult to achieve tremendous reduction through changing industrial structures within such a short time as five years. As of first half of 2009, contribution rate of structural and technological energy saving took up respectively 26.4% and 73.6% of total energy savings. In the limited structural energy saving, product mix restructuring, particularly upgrading product value chain and increasing value-added play a determinant role in structural energy saving (He, 2009).

Local governments have very limited means to achieve structural energy saving beyond phasing out backward production capacities, applying differential electricity price and controlling rapid growth of energy-intensive industries. Such means have a great impact on local GDP and ultimately undermine the completion of "growth retention" task. Local governments often prefer technological energy saving by reducing energy consumption per unit of product of major energy-consuming enterprises. Some local governments, rather than restricting, even invite investments by offering preferential electricity price for energy-intensive industries. In meeting energy saving targets, governments could assert technological energy saving easier than structural energy saving. In the first three years of the 11th Five Year period, most major energy-consuming enterprises generally met their targets and so local governments had to shift their energy saving priority to structural changes, which contributed more to energy saving later. In 2006, the contribution rate of national structural energy saving was 17.8%, and it increased to 19.5% in 2007, 35.4% in 2008 and 58.3% in 2009 (He, 2009).

Towards 2010, energy saving measures were getting tougher. In May 2010, Premier Jiabao Wen, required, in a TV conference of national energy saving and emission reduction work to use "iron hands" in phasing out backward production capacities and shutting down all listed enterprises with obsolete production capacities in the third quarter. Meanwhile, the *Notice on Further Ensure to Achieve 11th Five Year Energy Saving and Emission Reduction Targets by the State Council* issued in May 2010 stated that

> "*Provincial energy saving administrations organize inspection teams at each level to review and audit energy consumption of main products of major energy-consuming institutions during the previous 18 month before end of June 2010 and propose a list of enterprises and products exceeding energy (electricity) quota. For production with unit energy consumption exceeding twice of the quota, they will be charged with the 'sanction electricity price', the same price applied for backward capacities to be phase-out.*"

This is the first time that Chinese government proposes "punishment electricity price", and is tougher than differential price before. Afterwards, most of provinces issued similar policies. To achieve 11th Five Year targets in the last year, electricity and production limitation were applied in many places, some even put limitation on household electricity consumption. Since the end of July, Zhejiang, Jiangsu, and Shanxi limited and even cut off electricity supply for energy-intensive and highly polluting industries.[7] In order to achieve energy saving and emission reduction targets, on September 3, Anping county in Hengshui city of Hebei province categorized the whole country into three groups and put limiting quota on electricity supply for the three groups with 22-hour for each group. Not only households were affected, even hospitals and traffic lights were power off.[8] Though officials of the NDRC disapproved such policy of Anping county, the case indicates that local governments lacked effective means to achieve energy saving, particularly structural energy saving. Power-off and production suspension could yield energy saving, yet the negative impact on the society was costly. Given the context of pursuing economic growth, these measures are just makeshifts that local governments take, suggesting that they find

---

[7]Finance News Website, Crazy for meeting energy saving targets, enterprises pay for government's delinquency 2010, http://news.sohu.com/20100920/n275118374.shtml (accessed on 21 September 2010).

[8]China Broadcasting Website, Hebei Anping limited electricity for energy saving and civilians were asked to be powered off for 22 hours every three days (2010), http://news.qq.com/a/20100905/000352.htm (accessed on 8 September 2010).

no other effective means to achieve energy saving and meet the assigned targets.

(3) *Political incentives could hardly translate into intrinsic motivation for energy saving*

The *veto* system did not change the GDP-based performance review system for government officials. Once energy saving target is achieved, performance review still relies largely on the GDP. Contrarily, if GDP target is not achieved, the review result might look bad even if energy saving target is met to its best standards. Given such system design, most local governments set lower targets for energy saving. As Table 4.1. shows, since targets were proposed by each province of their own will, only four provinces proposed targets higher than the national ones. A total of 15 provinces proposed targets equivalent to the national ones, and 12 provinces proposed lower targets than the national ones. This case is in sharp contrast to GDP planning of each province. In the 11th FYP, most provinces proposed much higher GDP growth rates than the national one so that NDRC had to issue documents in order to urge them to cool down the fever for GDP growth. While decomposing energy saving targets, NDRC had to coordinate with provinces that set lower targets and ensure them to raise the targets so the country could achieve its national targets. Local governments meet the same issue while decomposing the targets and assigning them to lower level governments. Some provinces raised an important principles for target decomposition — "municipal and prefectural targets in general shall not be lower than the provincial one." Due to lack of internal driving force for, local governments take energy saving targets as big burden. They question the target decomposition and complain that they are assigned with more than what they should be accountable for. In terms of implementation of energy saving measures, local government's inaction went beyond the central government's expectation and the latter found the former's performance quite disappointing. According to the *Monitoring and Supervision Report on Implementation and Settlement of Electricity Price Policies* by the State Electricity Regulatory Commission, some provinces and municipalities provided preferential electricity price to support the development of energy-intensive enterprises in their region; some slowed down the process of identifying the list of enterprises that shall pay differential/sanction electricity price, some energy-intensive enterprises that should have been restricted or phased-out have not yet been required to pay differential/sanction electricity price. All those cases reflect that local governments lack intrinsic motivation for achieving energy saving and therefore tolerated or even encouraged the expansion of energy-intensive enterprises.

The *veto* in energy saving target accountability system did not translate into effective intrinsic motivation for energy saving among local governments. In fact, they often choose measures with less economic costs such as working with key enterprises, phasing out obsolete production capacities and promoting energy-efficient products. For quite a long time, local governments hesitated to take tough policies on energy saving in order to maintain economic growth. As a result, in the last year or specially the last six months of the 11th Five Year period, in order to achieve energy saving targets, local governments had to take extreme measures such as suspending production or even powering off the city, which incurred tremendous cost for the society.

CHAPTER 5

Enterprises' Response

5.1 Chinese Enterprise's Response to Low-Carbon Development

To ensure the transition to low carbon economy and society, Chinese government issued a suite of policies and measures during the 11th five-year period. The core of low-carbon policies is to improve energy utilization efficiency; compulsory energy-saving targets pushed the high energy-consuming enterprises to improve their technologies or withdraw from the market; Benefiting People Project (also known as "Hue-Min" Project) stimulated the increase of output and adoption of energy-efficient home appliance by providing subsidies for the purchase; policies promoting renewables built up enterprise's investment confidence and encouraged them to expand their investments to compete for market shares. Measures taken by enterprises are always for their own benefits. Except for meeting the energy-saving targets, low-carbon policies in the 11th FYP help enterprises to concern on energy saving and emissions reduction and they started the first step towards low-carbon development. From the perspective of enterprises, this chapter will analyze the external environmental changes brought by low-carbon policies and the corresponding measures taken by enterprises.

5.1.1 *High energy-consuming enterprise's technological transformation reduced energy intensity*

Energy saving is the core of low-carbon policies during the 11th five-year period. As main energy-consuming bodies of China, high energy-consuming enterprises are key subjects in energy saving review. Central and local governments set energy saving targets for "Top 1,000 enterprises" — and local energy-consuming enterprises respectively, and pushed and monitored their execution through energy conservation supervision and annual review. The major means for energy saving of high energy-consuming enterprises is to improve energy utilization by technological transformation. Investments on processes and equipment upgrading shall rely on analysis of technological stability and feasibility of the investment. Ten energy saving projects and energy saving technology policies provide rich introductions and guidance on energy saving technologies for enterprises, and they can make benefit analysis based on specific equipment investment according to their current conditions. Besides financing capacity and return rate on investment, their investment decision-making is influenced by payback period. Shorter payback period means less risk. Enterprises with good profitability and great market prospects often prefer increasing corporate benefits by enlarged investments. The input of Energy saving investment for enterprises is implementation of energy-saved technology transformation projects and the return is the amount of cost reduction by saving energy. In recent years, the proportion of energy-saved equipment could be manufactured locally improved helps reduce the amount of investments greatly. Meanwhile, rise of energy price helps increased returns of enterprises. So, the return on investment improves largely than previous years, and the payback period shortened significantly. This drove the internal motivation of enterprises to invest. But when the return on energy-saved equipment investment is lower than benefits coming from expansion, it is hard to put technology advancement in place at enterprises. In response to that, governments take specific measures, for example, if they fail to pass the review for energy saving, except being asked to improve by deadline, government administrations stop reviewing and approving new high energy-consuming investments and land use within one year. Restricted by the government's control by means of production, scale of production, and space of development, even reluctant enterprises have to follow government's instructions to invest on energy saving so as to exchange more development space. Expansion of financing ways like green projects financing enlarges sources of funds for enterprises. Central government's special funds and some local government's subsidy budgets also helped reduce investment costs of enterprises according to energy saved, stimulating their motivations of investment on technology advancement. "Top 1,000 enterprises" are most outstanding corporate groups with energy saving achievements during the 11th five-year period. By the end of 2009, 1,000 enterprises have saved energy 132 Mtce in total. Within four years, they surpassed the energy saving target of 11th five-year period by

32%, reducing 798 Mt of CO_2 emissions comparing to the base number of 2005.

5.1.2 *Enterprises with obsolete production capacities exit the market*

The broad market in China asks for large amount of low-end products and leaves space for enterprises with obsolete production capacities. This kind of enterprises is influenced greatly by market demand and energy price. They have low capacity utilization and generally poor economic profits. During the 11th five-year period, the Ministry of Industry and Information Technology set targets on the basis of the enterprise registration data urged local government to close down enterprises and phase out obsolete production capacities by administrative directives. Since this kind of enterprises helps increase local financial revenues by tax and afford local employment, governments often hesitate to close them down. With pressures of performance indicators from top and quota restriction of total amount for given industries, local governments had to make innovation in relative institutions. They often compensate the exit enterprises by subsidies and quota trading, to encourage the enterprises to exit the market positively. Take Henan province as example, enterprises that phased out obsolete power units can be compensated in other ways except the incentive fund from the central government. Enterprises closing down internal power plants get extra preferential electricity. According to calculation of installation capacity of power units shut down, the enterprise could gain 5,500 kilowatt hour preference every kW each year,[1] and the price of electricity with preferential price is 0.36 Yuan/kWh, 28% – 40% lower than the normal, which is 0.5–0.6 Yuan/kWh. The earlier the plants were closed down, the more preferential price they will gain. Plants closed down before 2006 will be offered preferential price for three years and after 2007 will be offered for only two years.[2] The public power plants get compensation by way of trading the generation plan quota to other large power plants. With incentives of central and local governments, many enterprises chose to phase out obsolete production capacities and exit the market. In the first four years of 11th five-year period, enterprises in the power, iron and steel, and cement sectors

[1] 1 kilowatt hour = 1 kWh.

[2] Data from the record of the field visit for this study.

saved energy of about 110 Mtce by way of phasing out obsolete production capacities (Zhou, 2010).

5.1.3 *Active investments in renewables market*

The implementation of the *Renewable Energy Law of the People's Republic of China* in 2006 provided unprecedented development opportunities for enterprises to invest on renewables market and helped facilitate the investment and prosperity of some enterprises in this field. During the 11th five-year period, enterprises invested in renewables both in direction of equipment manufacturing industry and renewables utilization. Power enterprises are main representatives investing in renewables utilization. Installation capacity of renewables was increasing quickly and hydropower installation capacity reached 168 GW in 2008; wind generation grew with annual rate of 100% with three straight years since 2006 and at the end of 2008 the total installation capacity reached 12.27 GW. In 2008, excluding traditional way of biomass utilization, the annual renewables utilization is 246 Mtce, about 8.6% of the total primary energy consumption, increasing by 2% than 2005. Hydropower provided energy of 200 Mtce and wind, solar, and modern biomass energy utilization provided 45 Mtce of energy (Beijing Oriental Environmental Institute and Center for Renewable Energy Development, Energy Research Institute of the NDRC, 2009).

The Benefiting People Project stimulated the demand for solar water heater and enterprise's motivations of investment and production by government subsidies. Expansion and adoption of solar water heater scaled up quickly. In 2008, the solar water heater in China exceeded 140 million m^2, accounting for 66% of the world's total and covering 50 million households and 200 million population. The China's successful model of solar power commercialization is a good practice in global terms and was taken as one of the solutions for future world energy issue by International Energy Agency (IEA) (World Wide Fund (WWF) China, 2009).

With joint engagement of low-carbon policies and market, the renewables manufacturing industries grow fast. By March 2009, there were over 100 enterprises that manufacture wind power equipment or related components in China, among them 80 are set equipment manufacturers. The localized wind power equipment takes increasing shares of domestic market. In 2008, the localized equipment accounted for 75% of all manufactured wind power equipment. At the end of 2008, capacity of producing solar photovoltaic cells was over 2 GW, 12.3 times of 150 MW in 2005, making China the top photovoltaic cells producer. In 2008, over 10 solar photovoltaic cells manufacturers had capacity of over 100 MW, and four of them ranked top 16 in the world. There are over 3,000 solar water heater manufacturers, of which over 10 have sales incomes of one billion RMB (Beijing Oriental Environmental Institute and Center for Renewable Energy Development, Energy Research Institute of the NDRC, 2009).

5.1.4 *Accelerating upgrading of energy-saved home appliance*

Exclusive mid and long-term energy saving plan issued in the year 2004, set out the direction for promoting high efficient energy-saved home appliances and lighting, encouraging investing in R&D of high efficient home appliances. Due to high production costs, the price of energy-saved products is not competitive in the market and the return rate on investment is relatively lower, undermining the motivation for R&D on core technology of energy-saved products. During the 11th five-year period, further regulation of energy efficiency standards and logos of some home and office appliances helped raise the competitiveness of high energy-efficient products in market, stimulating motivation of enterprises to develop and produce high energy-efficient appliances and lamps etc. The "green lighting project" in 10 energy saving projects and 2009 "Energy-saved Products Benefiting People Project" prompted production and sales of energy-saved products by financial subsidies. According to statistics of Ministry of Finance, at the end of April 2010, the Energy-saved Products Benefiting People Project had distributed over seven billion Yuan of governmental subsidies, which enlarged market space for high energy-efficient home appliances and lighting enterprises and helped improvement and development of relevant industries (WWF China, 2009). Take high energy-efficient air-conditioner as example, with every one level of energy efficiency rate lifted, enterprises increase cost of 100–150 Yuan. If a rate-5 air-conditioner upgrades into rate-2, it costs RMB 300–500 or more. The high cost causes high price and therefore lost the competitiveness of energy-saved products. By end of 2008, one report from China's home appliance market study indicated that, sales of rate-4 and rate-5 products accounted for 95% of all sales of air-conditioners. The Energy-saved Benefiting People Project started since June 1, 2009. The *Implementation Rules for Promotion of High-efficient Energy-saved Room Air-conditioners by the Energy-saved Benefiting People Project* sets up the limit of promotional price, and gives subsidies of 300 – 650 Yuan for each rate-2 air-conditioner and 500 – 850 Yuan for each rate-1 air-conditioner. The subsidies are directly allocated to enterprises, which sell products with the price after deducting subsidy from original price. High efficient energy-saved air-conditioner's price fell by 50% on average, which increased their market shares quickly. Statistics from the Ministry of Finance indicated that market share of high-efficient air-conditioners increased from 5%–50% after promotion. Optimistic market prospect helped build up enterprise's confidence on investment. Changhong planned to put 30 million Yuan in 2010 to upgrade high-efficient air-conditioner production line and manufacturing equipment in a comprehensive way. Chunlan, within one year, puts into market 468 new products with nine series like environment-friendly fluorine-free air-conditioner, direct-current and variable-frequency (DCVF) air-conditioner, noise-free air-conditioner and so forth. Chunlan has also owned the intellectual property right of direct-current and variable-frequency control technology. Qianhu Wang, assistant to secretary of China's Customers Association, remarked that "Energy-saved Products Benefiting People Project" greatly changed environment for consuming energy-saved products, facilitating upgrading of consuming structure and also helped adjustment of industrial structure (Sang and Dong, 2010).

5.1.5 *Stepping into the market of energy services*

The low-carbon development policies help incubate the energy service market. Relevant preferential policies facilitate the fostering and development of this kind of enterprises. The number of enterprises, investment size, output value, and energy savings grew fast. The energy service companies (ESCo) operating energy saving projects by energy performance contracting (EPC) increased from 515 in 2008 to 683 in 2009, with a year-on-year rate of 32.6%; the gross output of energy service sector increased from 41.73 billion Yuan in 2008 to 58.768 billion Yuan in 2009, with a year-on-year rate of 40.83%; the comprehensive energy-saving investment amounted from 1.148 billion Yuan in 2003 to 36.037 billion Yuan in 2009. Investments on EPC programs increased from 11.67 billion Yuan in 2008 to 19.532 billion Yuan in 2009, with a year-on-year rate of 67.37%. Calculating on the basis of comprehensive energy-saving investments, the energy-saving capacity in 2009 was 17.6 Mtce, 22.33 times of 2003 (0.75 Mtce); 2009 saw a 11.3 Mt of CO_2 emissions reduction, 22.27 times of 2003 (0.49 Mt) (EMCA, 2010; Yao, 2010). Operational modes of EPC companies became diversified as well. First emerged in China's market was share-based energy management service contract, and later a hybrid with two more modes, commitment-based energy management contract and energy costs and management services contract. Except equipment manufacturing companies and the third party service companies who are active investors in energy service market, many large high energy-consuming groups and enterprises are tapping into this market, such as Shenhua Group and State Grid, who have invested on the energy service companies.

5.2 Factors and Mechanisms Affecting Energy Savings of Enterprises

China is in the middle phase of industrialization and industrie's percentage in GDP remains high. The growth of comprehensive energy consumption between 2002 and 2005 is faster than national economic growth. Rapid economic development consumes large amount of energy. According to statistics of *China Statistical Yearbook* in 2004, industrial value-added accounted for 45.9% of national economic output (National Bureau of Statistics, 2005), but industrial energy consumption accounted for 70.5% of all national energy consumption, and high energy-consuming enterprises consumed large proportion of energy, which accounted for 59.6% of total energy consumption and 84.6% of industrial energy consumption (National Bureau of Statistics, 2006). The comprehensive energy consumption of 1,008 enterprises with over 0.18 Mtce of comprehensive energy consumption was 670 Mtce in the whole year, accounting for 33% of all national energy consumption and 47% of industrial energy consumption (NDRC etc, 2006). The economical structure of China determines that the core of low-carbon development at this period is to improve energy utilization of high energy-consuming enterprises. So the priority of energy saving policies in the field of industry during the 11th five-year period energy is governance on energy saving of high energy-consuming enterprises.

5.2.1 *Constraints and incentives faced by enterprises*

(1) *Constraints of energy-saving target accountability*

The governance on energy consumption of existing high energy-consuming enterprises is the focus of energy policy during the 11th five-year period. The target accountability system is the major measure the government has taken to control high energy-consuming enterprises. Except decomposition of energy consumption per unit of GDP, some provincial governments also decomposed the target of energy saving for high energy-consuming enterprises to local governments, urging them to take effective monitoring and supervision on high energy-consuming enterprises. "1,000 enterprises" and provincial key energy-consumption enterprises have to go through energy saving target review every year by government sectors. Failing enterprises would face suspension of production and exclusion from governmental approval of their new projects. The restriction on energy efficiency indicators does constrain the survival and development of enterprises, urging them to take effective measures to do much as they can to attain the requirements of the government.

(2) *Pressures on phase-out of obsolete production capacities*

Though high energy consumption increased the cost of enterprises, enterprises using high energy-consuming and obsolete production capacities and electrical-mechanical equipment had no incentive to phase out them initiatively for strong market demand, made the survival of the enterprises impossible. Government sectors urged them to shut down and update equipments by administrative commands. The central government sets out key sectors to phase out obsolete production capacities and targets for each province according to the registration of different sectors and enterprises. Local governments make specific and locally suitable measures according to the local current conditions. Enterprises phase out obsolete production capacities and exit the market, face the problem of settling down the laid-offs, so local governments usually need make institution innovations to combine mandatory and incentive measures to push the enterprises to exit the market. Taken power industry as example, government sectors allow enterprises to trade electricity plan quotas as economic compensation within three years after phasing out obsolete production capacities according to cap incentives policies to help them exit the market steadily and meanwhile solve employment problems.

(3) *Affection of developmental space*

Regional developmental space control is the main restraint that enterprises met on the process of expansion. Control of developmental space includes restraints like regional environmental capacity, industrial structural adjustment's control of newly launched programs, installation capacity of power units, and so forth. The affection on high energy-consuming and high polluting enterprises are especially significant. Control of environmental capacity is the important indicator of expansion for enterprises in the cement, iron and steel, and power industry. Local governments pursue regional economic output by simply phasing out obsolete production capacities and give resources to low energy-consuming enterprises that have wider space of development. Local governments could sell the quota of developmental space produced by exiting in an auction to attract enterprises with stronger competitive strength. The financial income helped to subsidy the settlement of the exiting enterprises.

(4) *Guidance of energy price*

During the 11th five-year period, the main market tools used by central government to intervene on enterprises are adjustment of electricity price and on-grid price of renewables generation. From the perspective of energy production, to increase the proportion of renewables in energy consumption and support the development of renewables industry, the country gives favorable on-grid price for electricity from wind power, biomass, and photovoltaic solar energy. These preferential policies help encourage investments of renewables generation enterprises and therefore perfect technology in practice. From the perspective of energy utilization, the central government implemented punishing electricity price for enterprises in the list of restriction and phase-out type and excessive energy-consuming enterprises in order to push them to take effective measures for energy saving. Since local governments are responsible for making specific electricity price plans, some did not follow the suit of above-level governments, since they would like to encourage local economic development. So, market tools are quite useful for motivation of renewables enterprises during the 11th five-year period but not so effective for promoting energy saving of other enterprises.

(5) *Financial incentives for energy-saving technology transformation*

If enterprises start the technology transformation projects meeting some requirements of 10 energy-saving projects, as long as the amount of energy saved is over 10,000 tce, they can apply for incentives funds supplied by central government. Enterprises could submit the feasibility report beforehand to local development and reform commissions for incentives, meanwhile they need submitting the information of energy consumption before transformation, measures for energy saving, estimated amount of energy savings, measurement and testing methods. Provincial-level energy saving administrations work with financial sectors to review, identify and summarize proposals and then submit to NDRC and the Ministry of Finance, who were responsible for organizing advisory group to review proposals. The NDRC sets the amount of incentives based on standards and Ministry of Finance distributes 60% of incentives identified in the initial review to proposed enterprises through the financial system according to the implementation plans for technology transformation projects issued by the NDRC. During the project implementation, local energy saving administrations work with financial sectors to take necessary measures to urge the implementation of technology transformation projects and ensure timely completion of energy saving targets. Upon completion of the projects, the NDRC and the Ministry of Finance take second review of projects and distribute the rest of money to enterprises according to actual amount of energy saved.

(6) *Preferential tax for comprehensive resource utilization*

Along with low-carbon development, tax policies issued by central government during the 11th five-year period include preferential tax for comprehensive resource utilization projects and lowering export tax rebate rate of some high energy-consuming products. Crops, straws and bran-based (including grain straws, agricultural economic crop straws, grain bran, and corns) products, electricity, thermal power, and gas enjoy preferential income tax; for electrical or thermal power produced from waste fuels, valued-added taxes will be returned once collected; for electricity produced from wind power, value-added taxes will be 50% returned once collected. The way of tax preference encourages enterprises to access to renewables market and save energy by comprehensive utilization of resources.

5.2.2 *Factors affecting enterprise's response*

Enterprise's response to energy saving policies is affected by many factors. The following will give analysis from three perspectives: Ownership, size, and category of sector.

(1) *The impact of ownership*

The ownership determines the way of intervening. The property owner of state-owned enterprises (SOE) is the nation, The State-owned Assets

Supervision and Administration Commission of the State Council is authorized to motivate and restrict the SOEs' behaviors by reviewing on performances of the leaders. While pursuing profits, SOEs, particularly the central SOEs, need to undertake social responsibility for energy saving and emissions reduction. Since the performance review on SOE leaders rely on indicators include not only economic achievement but also target for energy saving. Motivated by the direct pressure, SOEs become the leading bodies of energy saving and emissions reduction during the 11th five-year period. The State-owned Assets Supervision and Administration Commission of the State Council managed SOEs by classification. According to actual energy consumption, pollutants emission level and category of industry, central SOEs are classified into 3 clusters, including key, focus, and general. The SOEs were supervised dynamically, and were required to establish corresponding organizational management systems for energy saving and emissions reduction (Supervision and Administration Commission of the State Council, 2010). The *veto* from State-owned Assets Supervision and Administration Commission of the State Council and review pressures from local governments prompt SOEs to be actively responding to demands for energy saving. A large amount of SOEs are included in the list of "Top 1,000 enterprises" and key energy-consuming enterprises at each province. Their excellent performances of achieving energy saving targets lay foundation for the "Top 1,000 Enterprises Energy Saving Program" to achieve the target ahead of schedule. Other high energy-consuming enterprises have to maintain cooperative connection with local governments because they are supervised and controlled by local governments in terms of production factors and development space. If their energy saving performances cause the failure of local governments to achieve targets, it is no good for them to continue getting resources from local governments. Non-SOEs take corresponding solutions according to their own status due to less intervenes from governments. If they make good profits and need expansion to win space for development, generally they would actively transform technology to meet energy saving target; but for enterprises in difficulties caused by low energy efficiency, they could choose to exit the market based on government-led compensation. The government sector's compensation measures for solving the actual problems played a determinant role in guiding the exit of the enterprises.

Except difference of responding mechanisms, the internal systems of SOEs still have the trait of planning economy, which is convenient for the government to monitor and review. It takes efforts of enterprises to do well on organizational management, energy metering and statistics system to be important evidence of review by government sectors. SOEs have advantages in this sense. In contrast, similar organizations in private enterprises cannot be equipped with qualified staffs since they are not profitable units, targets cannot be well decomposed and achieved.

(2) *The impact of size*

The targets of energy saving policies during the 11th five-year period are chiefly made for high energy-consuming enterprises. The bigger the size is, the more constraints there are. The *Energy Saving Law of the People's Republic of China* defines enterprises with annual comprehensive energy consumption over 5,000 tce as key energy-using units. The target of energy saving accountability during the 11th five-year period was mainly aimed at this kind of enterprises. Likewise, incentives from national financial sector only were open to the projects of over 10,000 tce energy savings. If the amount of energy saved is between 5,000 and 10,000 tce, EPC companies could assist applying for central and local financial subsidies. This means key subjects of national energy saving governance are still medium and large-sized energy-using enterprises. The government motivate and restrict enterprises by promoting capacity building of energy saving and review on energy saving performance so as to ensure size effect of enterprises and improve energy utilization; for smaller enterprises, since they have relatively less energy consumption and were weakly controlled by the government, energy saving depends on market demands. Due to large investment and high fixed costs of high energy-consuming enterprises, production expansion helps improve energy utilization and reduce costs per unit of product. So for large-sized enterprises, since their energy consumption level is closely related to their profits, they have stronger motivations for energy saving. Comparatively, some smaller enterprises mostly use second-hand equipment phased out by other enterprises. They have smaller investments and low fixed costs and usually locate their factories in areas with lower energy price. Relatively higher energy consumption per unit does not increase costs for enterprises, so these kind of enterprises have now good market space. Since they have relatively obsolete processes and equipment in general, if putting too much investments on technology transformation, profits would be greatly reduced, so they do not have drives or capacities for energy saving.

(3) *The impact of sector*

Restriction on energy consumption level during 11th five-year period is mainly on the production process of high energy-consumption industries. This governance model presses high energy-consuming enterprises to take measures on technology transformation to improve processes and reduce energy consumption per unit of product so as to win space for further development. During the 11th five-year period, increased localization of large amount of energy-saving equipment greatly helps reduce enterprise's investment on energy saving and shorten the payback period of investment; meanwhile, pressures of increasing energy price stimulate most of enterprises to take actions to reduce costs by saving energy so as to improve competitiveness. The main stream of high energy-consuming enterprise's response to low-carbon policies in this period is to grasp opportunities brought by market and favorable policies and transform pressures into driving forces. But for enterprises in other sectors and service industry, low-carbon policies are less motivated and restrictive, due to the cost-effectiveness of governance on energy saving. Compared with high energy-consuming enterprises, the total consumption is high, they could save enormous energy within a short time by governance and see visibly effective outcomes of energy saving and GHG emissions reduction. But enterprises in other sectors (not high energy-consuming) and service enterprises have generally less energy consumption levels and therefore smaller potentials for energy saving. Governance on energy saving for these enterprises meets great difficulties and also brings about expensive administrative costs. In this situation, enterprises in non-high energy-consuming sectors still take profits as objectives. Though low-carbon policies may cause impact on enterprise's strategy of development, unlike high energy-consuming enterprises externally motivated and restricted, these enterprises are slower to take actions. They prefer concern on related policies and take actions cautiously.

5.2.3 *Problems and shortcomings*

Though government-led monitoring and review on energy saving of enterprises during the 11th five-year period achieved good performance, there still exist some problems and shortcomings. The whole process of government-led review takes huge administrative costs without relation to energy saving, which undermines utilization of government's funds. Government sectors set a uniform indicator of 20% as reference target without considering the original energy efficiency of each enterprise. This creates much more pressures on high energy-efficient enterprises, which may not achieve the set target. In addition, since monitoring and review prioritize medium and large enterprises, it is too busy to keep eyes on small enterprises or illegal workshops. Once a large amount of obsolete equipment phased out flow into these workshops and are put into production, the actual energy consumption may climb up outside of official statistics. It takes to build the energy consumption management system covering wider scope and further improve energy price development mechanism in order to push forward comprehensive energy saving of enterprises.

CHAPTER 6

Low Carbon Financing

To achieve the target of reducing the energy consumption per unit of GDP by 20% during the 11th five-year period, it takes up to 800 billion RMB of annual investment as a total (Yang and Gong, 2009). The central budgetary investment of 33.3 billion yuan and the central government expenditure of 50 billion yuan could hardly satisfy the needs of low carbon investment. Therefore, most of energy saving and emissions reduction funds come from other channels: investments from international organizations, bank loans, venture capitals, and private equity. Compared to traditional energies, low carbon projects demand much more funds. However, due to uncertainty of new technologies and low expectations on market shares, low carbon projects are often hard to attract enough funds. Even though some projects have higher return on investment than traditional projects, banks and other investors remain cautious to carry on large-scale and long-term investment to avoid possible risks.

China's corporate financing is characterized as: (1) Currently, low carbon sector has not yet established a holistic R&D and innovation system. The key manufacturing equipment nearly all relies on import. The risks with technology and operation are both high; (2) manager's skills and qualifications vary from each other, and their awareness of creditability is weak; (3) the financial system in small and medium enterprises is not well established. The information transparency of enterprise is poor, leading to poor prestige and credibility; (4) enterprises lack property as guaranteed mortgage. Based on the above characteristics, large state-owned enterprises (SOEs) are in a much better situation of low-carbon financing than small and medium private enterprises. This situation is mainly due to the fact that

SOEs can easily get financial support from the government. The majority of their low-carbon investment funds come from central and local governments. In addition to that, large SOEs have easier access to bank loans than small and medium private ones. On the contrary, most of small and medium private enterprises have been meeting difficulties of low carbon financing for the following reasons: These enterprises do not have endogenous financing capability; commercial banks afraid of risks are very cautious on low carbon project loans; venture capitals lack interests in low carbon projects as well. Therefore, many small under-commercialized low-carbon projects with high cost stay in difficulty of financing. Though low carbon projects are favored by policies, they are coldly welcomed by real world investors.

A good low-carbon financing environment and enterprise's financing capacity are necessities of low carbon investment. Although low carbon investment has problems such as high risks and low economic return, government's incentive initiatives and relevant mechanisms may help change the environment for low carbon financing. This chapter discusses channels, mechanisms, and financial products for low carbon financing, as well as policies that promote low carbon financing in China.

6.1 Low-Carbon Financing Channels in China

6.1.1 *Government expenditure*

In the first four years of 11th five-year period, the central government spent 128.5 billion yuan on energy saving and environmental protection projects. In 2010, the expenditure was 83.3 billion yuan. According to Zhenhua Xie's speech, central government spent over 200 billion yuan, which accounted for 10%–15% of the total investment on energy saving and environmental protection (Xinhua website, 2010). This implies that the total investment on energy saving and environmental protection during this period was over 2 trillion yuan. The central government expenditure on energy saving and environmental protection are specific as follows:

- In the first three years of 11th five-year period, the central government arranged 33.6 billion yuan of central budgetary investment and 50.5 billion yuan of central budget on construction of key projects on energy saving and emissions reduction.
- In 2008, the central government funded 27 billion yuan on energy saving and emissions reduction, of which 12.5 billion yuan was for energy saving.

Specifically, 6 billion yuan was incentive for energy-saving technological innovation, 4 billion yuan was incentive for phase-out of obsolete production capacities, 1.5 billion yuan was subsidy for energy saving in buildings and 1 billion yuan was to promote products like high efficient lamps and support capacity building of energy saving. In addition, the central government took 14.8 billion yuan out of the central construction budget for energy saving. Therefore, the total government expenditure for promoting energy saving and emissions reduction in 2008 was 41.8 billion yuan (Xinhua.net, 2008).
- In 2009, the central government funded 49.5 billion yuan on energy saving and emissions reduction.

6.1.2 *Green financing of commercial banks*

The nature of commercial banks determines that risk control is crucial for banks. Compared to traditional sectors, low carbon sector involves large amount of new emerging technologies, and its business model is very original and unique. Thus banks face higher risks in low carbon financing. In addition, enterprise's upgrading may lead to higher operational costs and lower profitability. Therefore, in the initial phase of low carbon development, low carbon projects often have social benefits but lack economic profits, leading directly to increase of bank loan risks. Commercial banks may moderate risks by commercial means, but that will increase operational costs of banks. Therefore, banks have a lot of concerns about the under-commercialized low carbon projects with relatively new technologies.

At the national level, commercial banks do not have any uniform or written loan standards for loans on low carbon projects. Most of the banks provide services to energy saving and emissions reduction projects in traditional way of loans. Due to risks of fund and technology, it is very difficult for most of energy saving and emissions reduction projects to pass through traditional project risk assessment process of banks. Thus, most banks choose not to finance energy saving and emissions reduction projects. Some even treat them as poverty alleviation projects. Therefore, the total amount of loans from commercial banks on low carbon projects is small and hard to meet financial needs of low carbon projects.

Though green financing of commercial banks met mountainous difficulties, some banks have started studying and delivering loan products for energy saving sector and achieved a bit in supporting financing of low carbon projects. According to data from the China Banking Regulatory

Commission, five Chinese banks have loaned 106.3 billion yuan to industrial energy efficiency projects. According to the report by China Securities Journal, the total loan amount on energy saving and emissions reduction projects in 2009 reached 856.046 billion yuan, accounted for 8.93% of total bank loan amount of the year. In which:

- Shanghai Pudong Development Bank provided loans of 100 billion in recent three years to support low carbon financing, including FGD in power plants, waste-to-energy plants and wind power projects (Baiying Xing, 2010).
- Industrial and Commercial Bank of China provided loans of 102.9 billion yuan to new energy development projects.
- Agricultural Bank provided loans of 209 billion yuan, increasing by 53.4 billion yuan from early the year. Of which, 134.5 billion yuan was invested on hydropower, wind and nuclear power projects.
- Bank of China provided loans of 166.1 billion yuan, increasing by 48.3 billion yuan from early the year, of which 115.6 billion yuan was invested for clean energy projects.
- Bank of Construction provided loans of 227 billion yuan, increasing by 36 billion yuan from early the year. The loans supported nearly 1,000 clean energy projects with installation capacity of 120 GW.
- Beijing branch of Industrial Bank Co. Ltd set up the first sustainable specialized financing organization, specializing in energy efficiency financing, environmental financing, and low carbon financing products development and promotion.

Up-to-date news indicates that as of end of the first quarter of 2010, the balance of credit extension loans of Bank of China in energy saving and emissions reduction has cut through 180 billion yuan. As of end of June 2010, the balance of loans of Bank of Beijing in energy saving and emissions reduction was 5.226 billion yuan, increasing by 810 million yuan from early this year (State-owned Enterprises Energy Conservation and Emission Reduction Website, 2010a).

6.1.3 *Low carbon financing of non-bank financial institutions*

(1) *Venture capitals and private equity investment*

Commercial banks often select to invest on projects with clear and stable expected return on investment, so some projects with new and under-commercialized technologies yet rely on venture capitals, which too often have no interests in low carbon projects. For example, the article of Baiying Xing in China Securities Journal mentioned the suspension of policy on on-grid price of photovoltaic power made quite a few venture capital investors on photovoltaic projects worry for long. One venture capital manager complained the period of investment projects might be postponed indefinitely due to uncertainty of policies. Many other sectors are in similar situation as photovoltaic sector. Because of lack of with supporting policies, venture capital investors rack brains a lot (Baiying Xing, 2010).

However, financing channels are opening wide step by step. Since China loosened restrictions for international private equity companies, they are launching RMB funds with unprecedented size in an effort to get favorable conditions to reduce risks of foreign exchange control and avoid risks from government approval requirements. Meanwhile, private equity companies in China are expanding and local funds are increasing as well. Expansion and depth of investment activities are greatly furthered. With regulatory and market powers available now in China, clean technology has become most popular among venture capital and private equity projects and its investments are continuously rising. From 2006 to the first quarter of 2010, there were a total of 114 entrepreneurship investments and private equity trades in the field of clean technology in China, with total disclosed amount of 2.57 billion dollars. In the primary industrial sector, investments focus on energy power generation, energy efficiency, transport, agriculture, and energy storage. Popular investments in the secondary sector are solar power projects (31 trades), natural pesticide projects (10 trades), water treatment and waste water treatment projects (10 trades), high-energy battery projects (9 trades), lighting projects (9 trades), and wind power projects (8 trades) (Qin De, 2010).

(2) *Energy saving financial leasing*

Financial leasing is a modern investment and financing business that lessors provide appointed energy saving equipment for lessees, who only pay installments and can flexibly deal with the balance after contract is finished. This is a kind of cooperation with dual function of monetary and goods financing. Rental of financing leasing on energy saving comes from benefits generated from energy saving projects, that is to say, the amount of rental depends on benefits of energy saving. As of end of 2006, there were over 360 professional financial leasing companies, banks and other financial institutions operating financial leasing business. Joint-ventured leasing companies imported foreign capital of 5 billion dollars over past 15 years, supporting over 4,000 enterprises and 6,000 technological transformation projects and construction projects nationally.

6.1.4 *Endogenous financing of enterprises*

Endogenous financing refers to finance for energy saving and emissions reduction projects by internal financial resources of enterprises. Endogenous financing is characterized with endogeneity and autonomy. It has relatively lower costs and does not have any payback risks or pressures of debt service due or dividend payout. But it is limited by the amount of funds of enterprises. Since there are other needs competing for internal funds like product R&D, marketing, and equipment upgrading, energy saving and emissions reduction projects do not have priority right and enterprises have very limited budgets for energy saving and emissions reduction, which too often needs higher trading and development costs. That is why small and medium private enterprises fail to address low carbon financing by endogenous financing. But some large SOEs are able to rely on their own money to invest on low carbon projects. A latest typical case is that China Power Investment Corporation won seven projects in the bidding initiated by the Phase II authorized photovoltaic projects in 2010. Estimated by 20 yuan/W of building and operational cost, China Power Investment Corporation in the future three years shall put 3 billion yuan to construct 150 MW photovoltaic power plant. To maximize the project profitability, it is planned to use its own money without any bank loans (Xinjiang Energy Website, 2010).

6.2 Low Carbon Financing Mechanisms in China

6.2.1 *EMC financing*

EMC

in China has three modes: Share-based energy management, commitment-based energy management, and energy costs management. The first one is mostly utilized in practice. This mode is useful for energy saving and emissions reduction from the perspective of financing. Because by using this mode, clients do not need to put financial investment in the project. In addition, this mode helps to reduce risks caused by uncertainties of technology and project workflow. Nevertheless, EMC financing was still difficult in reality until the end of 2009. A report from the Ministry of Science and Technology showed that 92% of China's energy service companies met financing difficulties (Sun, 2009). This situation is mainly due to the following reasons: (1) The capacity of assessment on energy efficiency projects in investment institutions is insufficient; the financing procedure is complicated and inefficient; the risk management and control on energy saving projects are difficult; (2) financing costs of energy efficiency projects are relatively high, especially for small and medium enterprises. Most energy service companies in China are small and medium ones; (3) it is hard for energy service companies to get loan guarantees. The payback period of investment on energy saving is longer. Usually it takes about 2 5 years. The advance is the primary condition of EMC mechanism. With more projects, the amount of advance is larger. Energy service companies could not bear the break-off of cash flow and they crave for financing. But energy service companies available now in China are often small and weak without good credit history in the financial system. Therefore, they resort to guarantee companies for successful bank loans. Now, the solution to get loan guarantee is to apply for EMC project guarantee fund through China National Investment & Guaranty Co. LTD. In fact, even the guarantee is not easily acquired. By the end of 2008 of nearly 400 energy service companies operating EMC, only 40 companies got loan guarantees; (4) venture capitals and private equity have no interests in energy service companies operating EMC either. Considering IPO feasibility and return on investment, these companies are really less attractive due to long development period, poor profitability within a short time, high debt ratio, and lack of core technology etc. Though venture capitals do have their eyes on many energy service companies, no more than 10-plus companies have been invested.

Due to the lack of reputation and strength, financing was very difficult for small and medium energy service companies over the past decade. In 2008, the total investment by energy service companies was less than 20 billion yuan. This difficult situation hindered greatly the promotion of technology and expansion of energy service companies, as well as further development of EMC mechanism. Luckily, new policy on EMC issued in April 2010 was expected to change the status quo. On May 18, 2010, Beijing Environment Exchange together with Energy Service Industry Committee of China Association of Energy Saving launched the press release on China EMC investment and financing platform, officially announcing that EMC investment and financing platform would be officially launched and operational on June 5. This is an important initiative to push EMC by market mechanism, and also a powerful tool to push energy saving and emissions reduction by creative financial instruments (China-ESI.com, 2010).

6.2.2 *Guarantee mechanism of international financial organizations*

International Financial Company (IFC has started cooperation with several banks (Industrial Bank Co. Ltd, Bank of Beijing, Shanghai Pudong Development Bank etc.), jointly providing commercial loans to energy saving and emissions reduction projects of Chinese enterprises. In 2006, IFC China Utility-Based Energy Efficiency Finance Program (CHUEE), which was jointly funded by IFC, Global Environment Fund, Ministry of Employment and the Economy of Finland and Norway Agency for Development and Cooperation, was officially established. It is expected to achieve the target of reducing 200 Mt of CO_2 and other GHG emissions per year before 2010 within six-year implementation period. It is also expected to provide 5–10 billion yuan of financing on energy efficiency programs to improve external financing environment for private enterprise's energy efficiency programs (Low carbon economy website, 2009). As of December 31, 2008, IFC CHUEE has approved 97 loans with a total of 3.3 billion yuan. The program delivery supported financing in the field of energy efficiency of small and medium enterprises in China and broadened financing channels of low carbon projects. For example, between 2006 and 2008, IFC and Industrial Bank Co. Ltd signed Phase I and II collaborative agreements on energy saving and emissions reduction financing projects. In Phase I agreement, Industrial Bank Co. Ltd provided a loan of 900 million yuan to 46 projects. In Phase II agreement, IFC provided a loan of 100 million dollars as risk-sharing fund to support Industrial Bank Co. Ltd distribute loans of 1.5 billion yuan to energy saving and emissions reduction projects. In June 2007, IFC provided risk guarantee of 130 million yuan for Bank of Beijing to support its loan of 300 million yuan to energy efficiency projects. In June 2008, IFC provided 500 million yuan of risk-sharing fund for Shanghai Pudong Development Bank to the loan of 1 billion yuan to China's energy saving and emissions reduction projects. According to Cunfeng Dai, program officer of China's energy saving and emissions reduction financing in IFC, IFC plans in the next step to use 100 million dollars of loans to support project financing of small and medium enterprises and would like to engage more urban commercial banks into this program. Meanwhile, it plans to use about 200 million dollars of mezzanine finance to make up with the fund shortage of small and medium renewables programs so as to help them get bank loans (State-owned Enterprises Energy Conservation and Emissions Reduction Website, 2010b).

The energy efficiency financing program of AFD (French Development Agency) is a creative financial product for financing of energy saving and emissions reduction projects. As a state-owned agency and specialized bank set up according to French bank regulations, AFD is the backbone power of French government to give official development aids. The French government authorizes AFD to sign the agreement with the Ministry of Finance of China to launch the phase I intermediate financing facility. Shanghai Pudong Development Bank and other banks provide funds with favorable interest rates and technical aids to China's renewables and energy efficiency projects. With aids of AFD, Shanghai Pudong Development Bank finished distribution of sub-loans for three energy saving and emissions reduction projects in Guangdong, Shandong and An'hui in 2009, with all together 9.9 million Euros and also provided additional 44 million yuan as co-financing to support these projects. The projects are respectively in chemistry, electric power, and cement sector. It is predicted that three projects will save an annual 0.14 Mtce of coal and reduce about 0.36 Mt of CO_2 emissions each year (Stockstar.com, 2009).

Asian Development Bank announced in May 2008 to provide a total of 800 million yuan (about 115 million dollars) for China as part of credit guarantee to support energy efficiency in buildings projects in Southern and Eastern China. The Asian Development Bank prepared to increase its business inputs on energy efficiency improvement and clean energy promotion to 1 billion dollars each year (State-owned Enterprises Energy Conservation and Emissions Reduction Website, 2010c). According to report of Japan-Sino Economic Correspondence in January 2009, Asian Development Bank determined to provide 800 million yuan again for the credit guarantee scheme of China to support energy saving financing of small and medium private enterprises. Standard Chartered is one of cooperative institutions in this financing scheme. The financing will be first and foremost used for projects focusing on energy saving in buildings (Chinafpn.com, 2009).

6.2.3 *Clean development mechanism (CDM)*

In 1999, World Bank set up Prototype Carbon Fund, which was the earliest fund for CDM operation. It was composed of six national governments and 15 private companies with total fund of 1.8 billion dollars. This mutual fund was controlled by World Bank to purchase carbon emissions reductions. Later, World Bank and other organizations established more funds that purchase GHG emissions reductions. The establishment and operation of these funds greatly boosted trading and financing in carbon market. Therefore, carbon market along with low carbon technology and product markets riveted attentions of capital market and experienced investors. Since the Kyoto Protocol taking effect in 2005, CDM and other carbon trading markets went into track of fast development. In 2009, the size of whole international carbon market including primary CDM market was over 144 billion dollars.

CDM market grows fast in China in recent years. Large internationally well-known carbon financing companies are tapping into China's market. Two large buyers of China's CDM emissions reductions are Europe and Japan. China is now expected to overpass other developing countries such as India and Brazil to become the largest seller. Now, CDM in China has financing models chiefly as follows:

(1) Advance payment contract financing: Upon signature of emissions reductions sales contract, international buyers advance part of or all agreed funds and pay back advances after emissions reductions are met.
(2) Foreign investment contract: Emissions reductions are directly sold to international buyers, who directly buy shares.
(3) Offshore loans: Apply for offshore loans and reduce interests of loans by using currency exchange rate and sales of emissions reductions.

Nevertheless, CDM financing for domestic small and medium enterprises has been difficult. When small and medium enterprises develop CDM projects, most of their funds rely on bank loans. However, since banks set up relatively higher threshold and require an 8% of project profitability, it is almost impossible for most of CDM projects that never have over 8% of profitability. It is widely advocated now that financial institutions shall increase their knowledge of and give more supports to CDM projects. It is also urgent to race up launching green projects financing for carbon trading, green insurance business and exclusive carbon trading exchanges. Banks shall be aware that CDM project risks are controllable and loan assessment system shall be connected with carbon trading system. Since domestic banks are not well aware of the point, domestic carbon financing is still hard. Though some banks would like to give it a go, only one or two banks go further and deeper up to now and Industrial Bank Co. Ltd is one of them. Bin Cao, manager of products development, Business

Department of Industrial Bank Co. Ltd, said that in the next three years, the Bank would at least input 10 billion to participate in energy saving and emissions reduction, and carbon market to build the Bank as the best and largest carbon trading financing platform in China. Industrial Bank Co. Ltd started a comprehensive research on CDM market since 2008 and prepared to launch the first carbon financing product kit to market in 2010. The kit was designed to provide diversified banking services for principal stakeholders in CDM catering for different needs (China Clean Development Mechanism Website, 2010).

The price of emissions reductions in carbon trading fell from 26 Euros in 2007 to around 10 Euros now and global carbon trading shrunk greatly. The price shrinking of carbon emissions reductions put profits down as well. Given this context, businesses of domestic small and medium enterprises dropped. Difficulties of approval and certification of CDM projects mounted up and financing issues for CDM project development are even outstanding. It takes mutual efforts of financial institutions and governments to address the issues in order to push CDM implementation.

6.3 Low Carbon Financing Models in China and Relevant Financial Products

Innovation of low carbon financial products in China now is just starting and low carbon financial products are not diversified. As pointed out by Bingzhu Yan, director general of Bank of Beijing, in his proposal to Chinese People's Political Consultative Conference, the, advancement of green projects financing in China now is meeting barriers, including over-generalized standards for green projects financing, lack of specific guidance contents for credit loans and environment risk rating standards, and banks lacking of specialized organizations to implement green projects financing etc. He suggested active R&D of creative products for interaction between environment and finance, creating innovative environmental financial products step by step to address the issue of lack of diversity of environmental financial products. (Hualong Website, 2010).

Though at the initial stage, overseas financial institutions and some domestic banks have already tried new financing models and relevant financial products in low carbon field, for example:

(1) IFC CHUEE was the first new sustainable financing mechanism cooperating with the Ministry of Finance of People's Republic of China to

support domestic industrial and commercial enterprises and institutions to improve energy efficiency, utilize clean energy, and develop renewables projects. The CHUEE had four specific components: EMC financing for energy service companies, financing for end-user enterprise energy-saved technology transformation projects, equipment suppliers co-financing, and public utility service providers co-financing. From the implementation phase since 2006 to now, the CHUEE has supported a total of three partnership banks to provide loans of over 4 billion RMB for 128 energy saving and renewable energy projects throughout over 20 provinces (autonomous regions and municipalities). The implementation of these projects will help reduce about 15.8 Mt of CO_2 and other GHG emissions each year (China Energy Saving Service Website, 2010).

(2) Industrial Bank Co. Ltd provides a "7 + 1" financing model targeting on different clients and projects. They are respectively: Energy saving technological transformation project financing model, CDM project financing model, energy saving service provider financing model, energy-saving equipment supplier buyer credit model, energy-saving equipment manufacturer capacity enhancement model, public utility service provider financing model, financial leasing model, and non-credit financing model (China Energy Saving Service Website, 2010).

6.4 China's Policies and Measures on Promoting Low Carbon Financing

To boost development of low carbon economy, central and local governments allocated a large amount of funds to support investments on energy saving and emissions reduction projects by ways of central government expenditure and project subsidy, energy saving funds established by provincial governments, project financing incentives supported by central and local governments (including loan interest subsidy, loan guarantee, and green projects financing) and consumer subsidy (see detailed data in the appendix of this chapter).

Since low carbon financing is still a relative new concept, though there are no laws or regulations directly on low carbon financing, the *Suggestions on Accelerating to Promote Energy Management Contract to Boost the Development of Energy Service Industry* and the *Promotion of Cyclic*

Economy Law of People's Republic of China issued in April 2010 set out a few operational supporting policies, including:

(1) Give direct investment, subsidy or incentives.
(2) Provide supporting tax policies.
(3) Encourage banks and other financial institutions to simplify application and approval procedures for energy service companies.
(4) Provide more supports by way of loan interest subsidy etc.
(5) Give diversified project financing supports including credit loans, and deliver corresponding financial services like investment counseling, fund settlement and cash management etc.
(6) Develop innovative project financing products. Expand scopes of collateral/guarantee, innovate guarantee ways, work on businesses such as accounts receivable, charging right pledge, and intangible assets pledges including know-how, licensed patents and copyrights.
(7) Support direct financing tools like issuance of enterprises (companies) bonds, convertible bonds and short-term financing bills, medium term notes and so forth.
(8) Encourage different guarantee agencies to provide guarantee services for debt financing products.
(9) Play the role of equity investment funds and angel capitals to support resources recycling enterprises to seek financing by listing in the stock market.
(10) Support application of eligible cyclic economy projects for loans from international financial institutions and foreign governments.

To intensify energy saving and emissions reduction technology transformation of carbon-based industries and low carbon economy development, some provinces (autonomous regions and municipalities) recently issued relevant documents. For example:

(1) Fujian: Local government spent billions yuan each year to encourage establishment of energy saving and emissions reduction projects and regulated that all energy saving and emissions reduction projects in the encouragement list could enjoy interests discount and be offered project financing support.
(2) Guangxi: Set up and improved sustainable green projects financing mechanism, clients environmental protection classification and identity

system, create and foster new growth points of green projects financing. Developed multiple low carbon financing products like carbon emission rights pledge financing etc. Explored to develop financing products relating to carbon emission rights and businesses like commissioned carbon management funds.

(3) Shanghai: Required financial institutions to support energy saving and environmental protection, low carbon economy and cyclic economy development, and expanded 'green financing'. Banks and financial institutions were prohibited from issuing blind loans to enterprises and projects in the sectors with excess production capacities identified by the country, while actively supporting enterprise's technological transformation and phasing out obsolete production capacities. Gave further and more financial inputs to energy saving, emissions reduction and environmental protection projects, and supported low carbon economy. Encouraged banks and financial institutions to develop diversified financial products on low carbon and support enterprises and projects in line with national energy efficiency and environmental protection requirements in the principle of green projects financing. Explored to set up and improve client's environmental protection classification and identity system, support development of cyclic economy, control financial supports to high energy and resource-consuming and high-polluting enterprises and projects in a tougher way.

(4) Jiangxi: Not allowed to give any form of new support credit to any projects (even under way) that breach national energy saving and emissions reduction regulations; did not give any new floating loans to launched projects, and took measures to ensure security of bank loans issued.

6.5 Conclusion

The move to low carbon economy takes larger investments than direct/indirect investments of governments and international organizations. Therefore, long-term development of low carbon economy will need to rely on market capitals and equity capitals. However, current investments still count on traditional energies to pursue high profits. At this stage, the development and production costs of low carbon technologies are high and profits are relatively slim and grow slowly. Given the small size of projects, low recognition, barriers of feasibility assessment on technology, difficulty of testing and measurement of energy saved, low carbon economy is hard to appeal spontaneous market and private investments. Before new low carbon technologies become commercialized and appeal market and private capitals, low carbon economy development relies mainly on policy support and financial subsidy of governments. With years of strict implementation of low carbon policies, low carbon financing will be definitely strengthened. The planned investments in low carbon market now is much more than expected before, but regional distribution of low carbon financing is not even and current low carbon financing has not got some important emissions reduction opportunities. Therefore, there is a need for further studies on how low carbon financing works and how to promote development and utilization of new technologies in a better way.

References

Q. Bai *et al.* Prospect of China's 2050 economic and social development// China sustainable energy topic group, *China Sustainable Energy: China 2050 Road to Low Carbon Development* (Science Press, Beijing, 2009).

Beijing Oriental Environment Institute, Center for Renewable Energy Development, Energy Research Institute of the National Development and Reform Commission, Effective Implementation and Industrial Development Study of Renewable Energy Law Review on Implementation of Renewable Energy Law (Second), China Sustainable Energy Program Report (Beijing, 2009).

Qin De. Deqin China Research and Observation Center 7th (2010).

Low carbon economy website, China energy saving and emissions reduction financing (2009), http://zselzl.chinaw3.com/jienenjianpai/jienenjianpairongzi/200912071328.html (accessed on 25 November 2010).

Energy Research Institute of the National Development and Reform Commission, *Studies on Significant Issues of Energy Saving during the 11th 5-year Period* (The National Development and Reform Commission, Beijing, 2009).

National Bureau of Statistics, *China Statistical Yearbook 2006* (China Statistics Press, Beijing, 2006).

National Bureau of Statistics, *China Statistical Yearbook 2005* (China Statistics Press, Beijing, 2005).

National Bureau of Statistics, *China Statistical Yearbook 2010* (China Statistics Press, Beijing, 2010).

J. He. The assessment on energy saving in 11th 5 year period and suggestions on energy saving effort in the 12th 5-year period, (2009), http://efchina.org/csepupfiles/workshop/20091130423446 96.9134111265911.pdf/1.He%20Jiankun.pdf (accessed on 28 October 2010).

Hualong website, Bingzhu Yan: Creative environmental financing products to boost development of low carbon economy (2010), http://www.cqnews.net/lc/lc6/201003/t20100312_4183980.tm (accessed on 25 November 2010).

UNDP, Sustainable Future: Towards a Low Carbon Economy and Sustainable Society: China Human Development Report, Renmin University of China (China Translation and Publishing Corporation, Beijing, 2010).

S. Liu, X. Wang and D. Yuan *et al.* Risks and solutions for industrial structural upgrading in China, *Economic Research*, (13) (2010) 2–43.

S. Liu. China's entering into phase of heavy and chemical industries and its implications on macro-economy (2003).

W. Jinglian. *China's Options of Growth Model*, added edn. (Yuandong Publishing House, Shanghai, 2008).

X. Liu. Good Effects of Investment in Energy Saving and Emissions Reduction and Energy Consumption Per Unit of GDP Reducing at First Half of the Year (2009), http://www.cenews.com.cn/xwzx/zhxw/qt/200908/t20090811_620482.html (accessed on 25 November 2010).

Inner Mongolia Daily, Inner Mongolia: Joys and Worries of the Prosperity of High Energy-consuming Industrial Parks (2003), http://www.nmgcz.gov.cn/nqn-sii.ASP?id=1346 (accessed on 28 October 2010).

Y. Qi *et al.* *Study on China Environmental Monitoring System* (Joint Publishing, Shanghai, 2008).

Q. Jing website, Sinopec Group invested 14.5 billion Yuan within 5 years to achieve 11th 5-year energy saving target ahead of schedule (2010), http://search.10j1ka.com.cn/snapshot/20100312/M_L3yDrd2xIoHs8k-GHKV_CIoYH4m2AW18KZzfgaNdNs∼.html (accessed on 1 December 2010).

X. Sang and F. Dong. Benefiting people project opening the energy saving market, *China Consumer Journal*, **B**(3), (2010).

D. Shi. Structural change is the major factor that influences on energy consumption in China, *China Industrial Economy*, (11) (1999) 38–43.

D. Shi. Improvement of energy use efficiency in the process of economic growth of China, *Economic Study*, (9) (2002) 49–56.

J. Su. International transfer mechanism of high energy-consuming industries and practical studies, Master thesis of Hangzhou University, Hanghzhou (2009).

T. Sun. Energy management contract: Looks beautiful (2009), http://finance.sina.com.cn/leadership/mroll/20091208/19137074324.shtml (accessed on 5 November 2010).

Y. Tian and Z. T. Zhang. China Invention Patent: Short "life span" and low "gold content" (2006), http://news.xinhuanet.com/mrdx/2006-06/29/content_4765896.htm (accessed on 28 October 2010).

Y. Wang, Features of social and economic development during the 11th 5-year period, *Dynamics of Economics*, (11) (2005) 44–49.

G. Wu. *Strategic Study on China Energy Saving and Emissions Reduction* (Economic Science Publishing House, Beijing, 2009).

J. Wu. *China's Options of Growth Models*, added edn. (Shanghai Far East Publisher, Shanghai, 2008).

L. Wu and R. Wen. "Light and heavy" discussion of industries since 1949 of China, *Economic Study*, (9) (2006) 39–49.

Xinhua.net. Ministry of Finance arranged 27 billion Yuan to push energy saving and emission reduction in 2008 (2008), http://news.xinhuanet. com/newscenter/2008-03/24/content_7849963.htm (accessed on 1 December 2010).

Xinhua.net., Five-year action plan of energy saving and emissions reduction put on a sprint (2010), http://finance.people.com.cn/GB/12877667.html (accessed on 1 December 2010).

Xinjiang energy website, China power investment corporation became winner of photovoltaic power generation program (2010), http://www.tianshannet.com.cn/energy/content/2010-09/19/content_5253924.htm (accessed on 1 December 2010).

B. Xin. Commercial banks all made commitment to support low carbon financing (2010), http://www.cs.com.cn/xwzx/jsbd/201004/t20100411_2390468.htm (accessed on 3 January 2011).

H. Xu. Boundless potentials of cross-national low carbon investment — UNCTAD issued the World Investment Report (2010), http://paper.ce.cn/jjrb/html/2010-07/28/content_112906.htm (accessed on 3 January 2011).

X. Xu and Y. Wang. Discussions on Peak Value of Iron and Steel Consumption of China, *Metallurgy Information Journal*, (1) (2007) 4–10.

L. Yang and P. Gong. The way of funding for energy saving (2009), www.chinaequip.gov.cn.

W. Yao, Energy management contract generates energy service industry, *Energy Saving and Environmental Protection*, 3 (2010) 12–15.

Stockstar, Shanghai Pudong Development Bank cooperated with AFD to successfully financing for energy saving and emissions reduction projects. (2009) http://stock.stockstar.com/JL2009072300002181.shtml (accessed on 5 November 2010). China energy saving industry website, China first EMC financing and investment trading platform initiated in Beijing (2010), http://www.chinaesi.com/industrial/HTML/9341.html (accessed on 1 December 2010).

China energy saving industry website, Developing sustainable financing for pushing energy saving and emissions reduction financing projects (2010), http://www.emca.cn/bg/rzfw/qtrzfw/20100610042317.html (accessed on 25 November 2010).

Group on studies of approaches and measures for China to achieve sustainable energy by meeting 20% energy saving target of the 11th 5-year (Science Press, Beijing, 2008).
Chinafpn.com, ADB provided energy saving financing to small and medium-sized enterprises of our country (2009), http://www.chinafpn.com/2009/0120/15949.html (accessed on 5 November 2010).
China CDM website, Carbon trading financing myth waits to be explained (2010), http://cdm.ccchina.gov.cn/web/NewsInfo.asp?Newsld=4478 (accessed on 25 November 2010).
Institute of Industrial Economics, Chinese Academy of Social Science, China Industrial Development Report 2009 (Economic Management Press, Beijing, 2009).
National Bureau of Statistics, China Statistics Yearbook 2006 (China Statistics Press, Beijing, 2006).
State-owned Enterprises Energy Conservation and Emissions Reduction Website, Bank of Beijing actively promoting green projects financing for emissions reduction (2010a), http://114.255.43.243/news_view5.asp?lm2=15&id=2274 (accessed on 5 November 2010).
State-owned Enterprises Energy Conservation and Emissions Reduction Website, Asian Bank financing for energy saving (2010c), http://114.255.43.243/news_view6.asp?lm2=15&id=407 (accessed on 5 November 2010).
State-owned Enterprises Energy Conservation and Emissions Reduction Website, World Bank gave 100 million dollars to support small and medium-sized enterprises financing for energy saving and emissions reduction (2010b), http://114.255.43.243/news_view5.asp?lm2=15&id=2293 (accessed on 5 November 2010).
R. Zhou and Zhenhua Xie. Explained efforts of emissions reduction (2010), http://www.chinanews.com.cn/ny/2010/09-29/2564422.shtml (accessed on 26 October 2010).
Denguang.com, Shanxi Taigang Stainless Steel Co. LTD spent 5.5 billion Yuan of energy saving investment during the 11th 5-year period (2010), http://www.denguang.com/html/21/n-629921.html (accessed on 1 December 2010).
EMCA, China Energy Service Industry Development, Report 2009 (2010), http://www.emca.cn/bg/emca/emca/20100310021037.html (accessed on 5 December 2010).
T. Forsyth. Promoting the development dividend of climate technology transfer: can cross-sector partnership help? *World Development* **35** (10) (2007) 1648–1698.

Y. Qi, L. Ma and H. B. Zhang *et al.* Translating a global issue to local priority: China's local government response to climate change, *The Journal of Environment & Development*, **17**(4) (2008) 379–400.

WWF China, *Low Carbon Business Leadership Focusing on New Opportunities for Future Economic Development*, (WWF China, Beijing, 2009).

PART 2

Sectoral Report

CHAPTER 7

Energy

7.1 Low Carbon Development in the Energy Sector

Power industry is the major one accountable for energy consumption and emissions in the energy sector and hence the top priority field for achieving energy saving and emission reduction in China. In 2008, power industry was accountable for an energy consumption of 930 million tons of coal equivalent (Mtce) and 2.5 Gt of CO_2 emissions, respectively 32% of all energy consumption and 42% of energy-related CO_2 emissions. This chapter analyzes low carbon development in the power industry.

In 2002, China initiated its power sector reform by separating generation and transmission and restructuring the state-owned enterprises (SOEs). The previous State Power Company was restructured into five independent power generation companies (Huaneng, Dateng, Guodian, Huadian, and China Power Investment Corporation, known as "The Big Five"), and two grid companies (State Grid Corporation of China and China Southern Power Grid which provides transmission for five provinces in the south — Guangdong, Guangxi, Yunan, Guizhou, and Hainan).

In 2008, installed generation capacity reached 791 GW and electricity production reached 3466.9 TWh (Table 7.1), an increase by 53.1% and 38.7% respectively compared to that of 2005 or by an annual growth rate of 17.7% and 12.9% respectively. In terms of energy mix, from 2005 to 2008, the installed capacity of hydro, nuclear, and wind accounted for 23% of the total installation capacity, a decrease 1%, while the generation of hydro, nuclear, and wind power from 18.1% to 19.2% at the same period.

Net coal consumption of thermal power generation witnessed a significant decrease from 370 gce/kWh in 2005 to 340 gce/kWh in 2009, achieving the 11th Five Year Plan (FYP) target of 355 gce/kWh ahead of schedule.

Table 7.1. Installation capacity and generation (2005–2008).

	2005	2006	2007	2008
Total capacity (GW)	517	621	713	791
Thermal (GW)	391	484	554	601
Hydro (GW)	117	129	145	172
Wind (GW)	1.1	1.9	4.0	8.9
Nuclear (GW)	6.9	6.9	8.9	9.1
Total generation (TWh)	2500	2866	3282	3467
Thermal generation	2047	2370	2723	2790
Hydro generation (TWh)	397	436	485	585
Wind generation (TWh)	1	3	6	13
Nuclear generation (TWh)	53	55	62	68

Note: Data on installed capacity is calculated from statistics on the website of China Electricity Council.
Source: (1) China Electricity Council 2005–2008 National Power industry Statistics Brief; (2) Energy Statistics Department of National Bureau of Statistics, *China Energy Yearbook 2009* (China Statistics Press, Beijing, 2010).

Station service electricity consumption rate of power plants was 5.76%, a decrease by 0.14% than 2008; line loss rate was 6.72%, a decrease by 0.07% than 2008 (State Electricity Regulatory Commission, 2010).

China's overall technological level in power industry has become among the top international ones. The total installed capacity was the world's second largest, only secondary to the United States. China is among the leading countries in terms of technology and equipment of large ultra super-critical (USC) units, large circulating fluidized bed (CFB) boilers, air-cooling units, large hydropower units, ultra high voltage transmission, etc., (Wang, 2009).

7.1.1 *Power industry achieved significant reduction in terms of energy intensity and carbon emission intensity*

Energy intensity is generally defined as energy consumption per unit of output produced, and in this case the coal equivalent[1] of all energies consumed

[1]Energy intensity of thermal power industry = energy consumption for thermal generation/actual thermal generation. This chapter introduces the concept of energy intensity of thermal power generation and calculates energy saved from thermal power generation based on its production so as to match the carbon emissions calculated by the same production data.

Table 7.2. Energy intensity and carbon intensity of the power industry 2005–2008.

Thermal generation energy consumption (Mtce)	720	830	935	928
Power industry CO_2 emission ($GtCO_2$)	2	2.2	2.5	2.5
Thermal generation energy intensity (gce/kWh)	352	350	344	333
Power industry CO_2 emission intensity (gCO_2/kWh)	766.8	772.4	760.3	716.6
Thermal generation carbon emission intensity (gCO_2/kWh)	935.8	933.2	914.7	887.8
Thermal generation CO_2 reduction ($MtCO_2$)	0	6.4	57.4	133.9
Thermal generation energy saving (Mtce)	0	3.6	22.1	52.7

Note:

A. Assume nonfossil generation carbon emission factor is zero, sectoral emission = $\sum$ energy consumption for thermal generation × corresponding emission factor of different kinds of fossil energy consumed.

B. Thermal generation energy intensity = Thermal generation energy consumption/actual thermal generation.

C. Power industry CO_2 emission intensity = Thermal power CO_2 emission/total generation of power industry.

D. Thermal power industry carbon emission intensity = Thermal power carbon emission/Thermal power generation.

E. Take 2005 as the base year, Thermal power industry CO_2 reduction in a year = thermal generation of that year × (Thermal power industry CO_2 emission intensity in that year — thermal power industry CO_2 emission intensity in 2005).

F. Take 2005 as the base year, Thermal power industry energy saving in a year = Thermal generation in that year × (Thermal power industry energy intensity in that year − Thermal power industry energy intensity in 2005).

per kWh generated. Between 2005 and 2008, the energy intensity of thermal generation remained decreasing. It fell to 333 gce/kWh in 2008, a decrease by 19 gce/kWh than 2005 (Table 7.2). Take energy intensity of 2005 as the baseline, energy saving of 2008 in thermal power industry reached 52.7 Mtce and cumulative energy saving between 2006 and 2008 reached 78.3 Mtce. With chain method calculation, i.e. energy intensity of the year minus that of the previous year, the amount of energy saved in thermal power industry in 2008 was 30.1 Mtce and cumulative energy saving between 2006 and 2008 was 51.6 Mtce.

Carbon intensity is generally defined as carbon emission per unit of output produced, in this case the CO_2 emissions per kWh generated. To decarbonize China's electricity system it is critical to improve thermal generation efficiency and to increase the share of non-fossil energies in energy mix. In fact China has committed tremendous efforts in these two fields and has managed to decrease carbon intensity in the power industry quickly from 766.8 gCO_2/kWh in 2005 to 716.6 gCO_2/kWh in 2008 (Table 7.2), a decrease by 6.5%. Carbon intensity of thermal generation decreased by 48 gCO_2/kWh from 2005–2008, a decrease by 5.1% within three years. Take carbon intensity in 2005 as the baseline number, CO_2 emissions reduction in thermal generation in 2008 was 134 Mt and cumulative CO_2 emissions reduction between 2006 and 2008 was 198 Mt. With chain method calculation, i.e. carbon emission intensity of the year minus that of the previous year, CO_2 emission reduction in 2008 was 75.1 Mt and cumulative reduction between 2006 and 2008 was 132 Mt.

7.1.2 *Changes of major energy consumption indicators of the power industry*

7.1.2.1 *Gross coal consumption and net coal consumption of thermal generation*

In 2009, net coal consumption of thermal generation decreased to 340 gce/kWh, 30 gce/kWh less than that of 2005, achieving the 11th FYP target of 355 gce/kWh by end of 2010 ahead of schedule. As Fig. 7.1 shows, within 10 years between 2000 and 2009, both net and gross coal consumption were decreasing year to year. 2007 and 2008 witnessed the largest decrease of 11 gce/kWh in net coal consumption and 10 gce/kWh in gross coal consumption compared to the previous years.

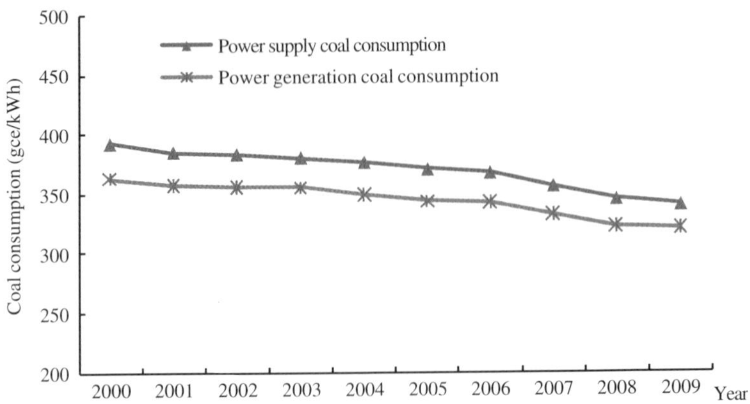

Fig. 7.1. Changes of net and gross coal consumption in thermal generation.

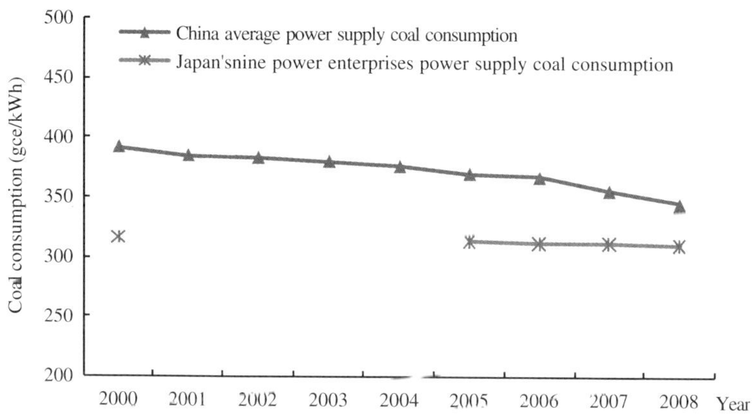

Fig. 7.2. Average net coal consumption of thermal generation in China and among Japan's nine major power companies.
Source:

(1) China Electricity Council, *Abstract of Power Industry Statistics* (China Electricity Council, Beijing, 2010).

(2) Energy Statistics Department, National Bureau of Statistics, *China Energy Statistics Yearbook 2009* (China Statistics Press, Beijing, 2010).

During this period, China's net and gross coal consumption of thermal generation is catching up with those of the developed countries. In 2000, China's net coal consumption was 392 gce/kWh, 76 gce/kWh more than that of Japan's nine major power companies (316 gce/kWh). Such difference reduced to 56.3 gce/kWh in 2005 and to 35.1 gce/kWh in 2008. Figure 7.2 shows the gap narrowing down.

7.1.2.2 *Station service electricity consumption rate of power plants*

From 2000–2005, the station service electricity consumption rate of power plants decreased from 6.28%–5.87% and then increased again to 5.93% in 2006. After a mild kink, it went down to 5.8% in 2009. The rate went up mainly because after 2006, a large amount of flue-gas desulphurization (FGD) and air cooling equipments came into operation, driving up electricity consumption by power plants. In addition, the increase in new capacity led to a sharp reduction of load factor of power plants. All these factors contributed to the fluctuation of station service electricity consumption rate. If this continues, in order to achieve the target of 4.5% by the end of 11th

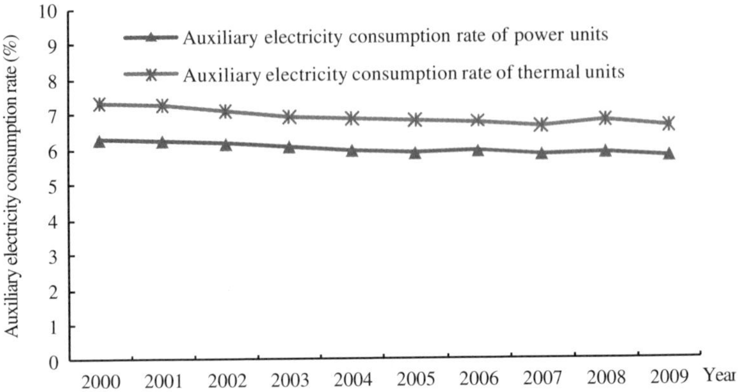

Fig. 7.3. Changes in station service electricity consumption rate of power plants (2000–2009).
Source: China Electricity Council, *Abstract of Power Industry Statistics* (China Electricity Council, Beijing, 2010).

FYP period, the station service electricity consumption rate of power plants shall drop by 1.26% in 2010 compared to 2009 (Fig. 7.3).

7.1.2.3 *Average generation hours (capacity factor)*

Figure 7.4 shows the generation hours of power plants between 2000 and 2009 in China. It increased from 4,517 hours in 2000 to 5455 hours in 2004. The number reduced gradually after 2004, to the 4,546 hours in 2009. This is largely because that influenced by the global financial crisis China's economic growth speed slowed down, and electricity consumption experienced a notable decrease. In addition, the increase of newly added capacity of thermal power units led to a sharp drop of units utilization rate, hence average generation hours of thermal power plants reduced significantly as well. However, the average generation hours of hydro power increased by large between 2006 and 2008, suggesting the success of China's effort in dispatching renewable generation with priority.

7.1.2.4 *Line loss rate*

Nationwide line loss rate decreased since 2004 (Fig. 7.5). According to the 11th FYP for Energy, line loss rate of the national grids should reduce to 7% by end of 2010. As a matter of fact, such target of 7% was met as early as in 2007, and the number further reduced to 6.79% and 6.72% respectively in

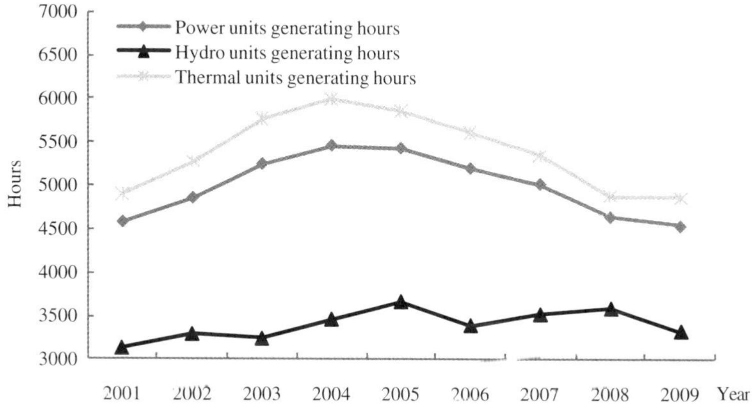

Fig. 7.4. Changes of generation hours 2000–2009.
Source: China Electricity Council, *Abstract of Power Industry Statistics* (China Electricity Council, Beijing, 2010).

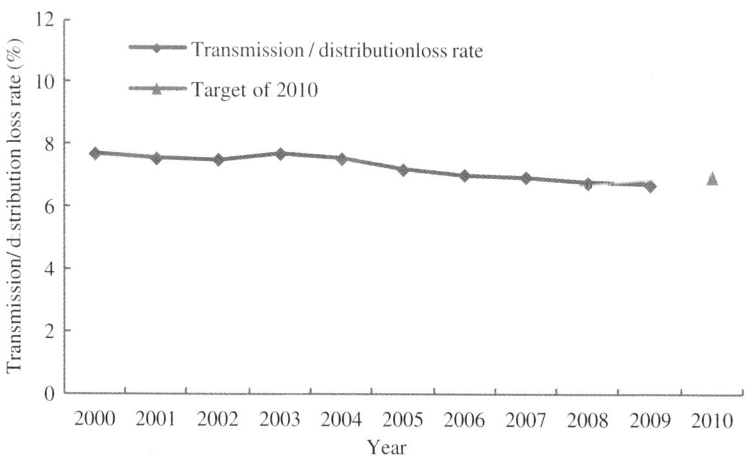

Fig. 7.5. Line loss rate of the grid (2000–2009)
Source: China Electricity Council, *Abstract of Power Industry Statistics* (China Electricity Council, Beijing, 2010).

2008 and 2009. However, some experts pointed out that China's statistics of line loss rate did not include losses of low voltage power distribution in rural area, which were often significant. If that is included, nationwide line loss rate would be higher than the current number.

Box 7.1. Low voltage grid loss in rural area.

Line loss is defined as all energy consumption and losses occurred from power plant transformers to end-users. Line loss rate in China did not include the significant low voltage power distribution losses in rural area. Distribution grids in rural areas are mostly of 380/220 V, 10 kV and 35 kV; and are characterized as numerous and wide-spread distribution stations, long diameter, and numerous branch lines. In 1998, the rural low voltage line loss rate was about 20%–30%; after the nationwide rural grids upgrading project, it went down to about 12% in 2005 (Wang, 2005). The number further decreases in recent years, with the implementation of new rural electrification projects. For example, the average overall line loss rate of counties where State Grid promoted electrification went down from 5.9%–5.2% and the low voltage line loss rate dropped from 8.4%–7.5%; and this rate at village level dropped to 8.0% from the previous 9.8%. In general, rural low voltage line loss rate remains higher than the national average line loss rate.

Take the 2005 line loss rate as baseline, energy saving through the decrease of line losses in 2008 was 1.8 Mtce, equivalent to 4.9 Mt of CO_2 emission reduction; and the cumulative energy saving between 2006 and 2008 was 3.4 Mtce, equivalent to 9 Mt of CO_2 emission reduction. Calculating with the chain method, energy saving from the decrease of line losses in 2008 was 0.8 Mtce, equivalent to 2.1 Mt of CO_2 emission reduction; and the cumulative energy saved between 2006 and 2008 was 1.7 Mtce, equivalent to 4.5 Mt of CO_2 emission reduction.

7.1.3 *Energy mix and grid structure optimized*

7.1.3.1 *Diversified energy mix*

During the 11th five-year period China witnessed fast development of renewable energy and nuclear power in the country, changing the previous energy mix. In 2009, China's newly installed renewable energy capacity was 37 GW, or 46.3% of the world's new capacity. The total capacity of renewable plants reached 226 GW, ranking number one in the world (REN21, 2010). Hydro capacity increased by the largest amount (20 GW new capacity each year, 2.6 times of the annual growth during the 10th five year period) and total hydro capacity exceeded 200 GW in 2010. Wind capacity grew fastest with the annual average growth of 1.3 times between 2006 and 2009. Currently, total wind installed capacity reached 25.8 GW, accounting for nearly one third of the world's total. To put this into the context, China's total installed capacity of wind in 2004 was 8 GW, only 2% of the world's total. Nuclear will experience the most eye-catching development in the next FYP period. By end of 2009, a total of 12 nuclear units are in commercial operation with the total capacity of 10.2 GW; and 22 new nuclear power units are under construction, accounting for 40% of the world's new nuclear units under construction.

In terms of generation, the share of renewables and nuclear power in 2008 accounted for 19.2% of the total. The generation from hydro, wind, and nuclear units increased from 451.4 TWh–666.7 TWh, or by 47.7% compared to that in 2005, exceeding the growth of coal-fired thermal generation (36.3%). Noticeably, the share of wind generation increased from 0.05% in 2005 to 0.75% in 2009 (Table 7.3).

Infrastructure investment shows a structural change at the same time. For example, in 2009 completion rate of thermal infrastructure investment drop by 11.11% compared to the previous year, while the rate of nuclear and wind power went up by 74.91% and 43.9% respectively during the same period. At the same year, the share of new wind capacity in all new capacity increased by 4.58%, while the share of new thermal capacity decreased by 3.39%.

The rapidly increasing generation from renewables and nuclear power resulted in the reduction in fossil energy consumption and consequently in

Table 7.3. Electricity generation of different sources (2005–2009) Unit: %.

	2005	2006	2007	2008	2009
Hydro	15.88	15.21	14.79	16.88	15.88
Thermal	81.9	82.9	83.0	80.5	81.8
Nuclear	2.12	1.91	1.89	1.97	1.90
Wind	0.05	0.09	0.17	0.38	0.75

Source: (1) 2005–2008 data: Energy Statistics, National Bureau of Statistics, *China Energy Statistics Yearbook 2009* (China Statistics Press, Beijing, 2010); (2) China Electricity Council, *Abstract of Power Industry Statistics* (China Electricity Council, Beijing, 2010).

Table 7.4. Emission reduction and alternative energy saving from hydro, wind and nuclear power 2005–2009.

	2005	2006	2007	2008
Avoided emission reduction ($MtCO_2$)	422	460	506	592
Reduction relative to 2005 (Mt CO_2)	0	37.9	83.4	16.9
Reduction with chain calculation (Mt CO_2)	0	37.9	45.5	86.1
Avoided energy consumption (Mtce)	154.8	168.7	183.6	214.7
Alternative energy saving relative to 2005 (Mtce)	0	13.9	28.8	59.8
Alternative energy saving with chain calculation (Mtce)	0	13.9	14.9	31.1

CO_2 emission. Renewable and nuclear generation resulted in a reduction[2] of 420 million tons of CO_2 emission in 2005 and 590 million tons in 2008 (Table 7.4), increasing by 40% over three years. Take emission reduction in 2005 as the baseline, the CO_2 emission reduction in 2008 was 170 million tons, and the cumulative emission reduction between 2006 and 2008 was 290 Mt. Calculating with the chain method, which is deducting the emission reduction in the previous year from that of the current year, then the emission reduction in 2008 was 90 million tons of CO_2 and the cumulative emission reduction between 2006 and 2009 was 170 million tons of CO_2.

The rapid development of renewable and nuclear generation resulted in significant reduction in fossil fuel consumption. Energy saving[3] in 2005 was 150 Mtce and about 210 Mtce in 2008 (Table 7.4), an increase by 38.7% over three years. Take energy saving in 2005 as the baseline, additional energy saving in 2008 was 60 Mtce and cumulative saving between 2006 and 2008 was 100 Mtce. Calculating with chain method, additional energy saving in 2008 was 30 Mtce and cumulative saving between 2006 and 2008 was 60 Mtce.

[2]Calculation of emissions reduction: Energy consumption of different energies (see the China Energy Statistics Yearbook 2009) × corresponding emission factor = total carbon emissions of thermal power, the value divided by thermal power generation gives the carbon emission intensity of thermal power. Electricity production generation of other energies × the carbon emission intensity of thermal power = emissions reduction generated from alternative energies.

[3]Calculation of energy saving: Energy saved from nuclear power, hydropower, wind and other powers = corresponding electricity production generation × coal consumption of thermal power generation in the current year.

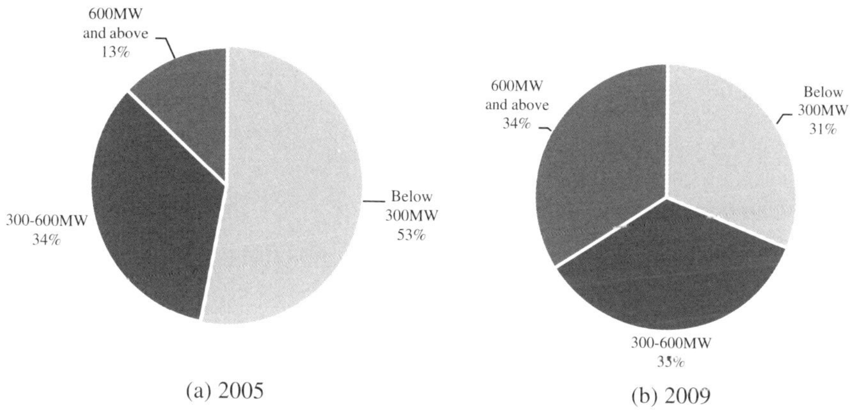

Fig. 7.6. Structure of China's thermal power units.
Source: (1) H. Xu, China's power industry status and development trend (2007), http://www.docin.com/p-22655048.html [2010-80 = −20]; (2) 2009 brief of energy saving and emission reduction of power enterprises issued by State Electricity Regulatory Committee and etc.

7.1.3.2 *Structural optimization of thermal units*

Fast growth of highly efficient large units and the widespread shutting down of small and low-efficient units throughout China are the two major contributing factors that drove the restructuring of thermal units. By end of 2009, the average capacity of thermal units was 103.1 MW, while units of 300 MW and above accounted for 69% of all thermal units. Units of/above 600 MW accounted for 34% of all thermal units (Fig. 7.6b). To put such progress in the context, at the end of 2005, the average unit capacity of all China's thermal generation was only 60.9 MW (China Energy Conservation Association CPGC, 2009), while units of 300 MW and above accounted for only 47% of all thermal units (Fig. 7.6a) (Xu, 2007).

(1) *Developing large and highly-efficient units*

At the end of 2009, China has built 158 supercritical units, 21 GW-level ultra supercritical units have been put into operation with 24 more under construction. China ranks the top one in the world in terms of the number of GW-level units under construction and in operation (State Electricity Regulatory Commission etc, 2010). In the 2009 review by State Electricity Regulatory Commission, units of 100 MW has an average net coal consumption of 371 g/(kWh), while the average net coal consumption of 1 GW units is as low as 287 g/(kWh). In 2009, China had GW-level ultra-supercritical units with total capacity of 21 GW. According to alternative thermal power generation with gross coal consumption of 340 g/(kWh) and the annual operation hours of 4865 of that year, the avoided energy consumption was about 5.4 Mtce. By developing highly-efficient large units, the structure of thermal units experienced dramatic improvement.

(2) *Closing small thermal power plants*

Starting from the 10th five year period, China was restructuring its thermal generation structure by closing highly inefficient and highly polluting small coal fire plants. In 2009, China closed 26.2 GW small thermal power plants small thermal power plants, exceeding its target of 15 GW for the year. As of July 2010, a cumulative 70.8 GW thermal power plants were closed, exceeding the 11th five year target of 50 GW ahead of schedule. Phasing out small thermal power plants has also generated significant energy saving and environmental benefits. Assuming to replacing the small units of 60.4 GW closed down between 2006 and 2009 with large units of 600 MW and above (excluding GW-level units), energy saving would be about 47.6 Mtce each year, CO_2 emissions reduction would be 129 million tons, and avoided SO_2 emission would be about 1.33 million[4] tons.

7.1.3.3 *Grid structure adjustment*

Starting from the 11th five year period, investment for grid construction expanded, and there was a national push for developing high voltage power transmission. Between 2006 and 2009, the country invested totally 1.135046 trillion Yuan on grids, and 2009 budgets increased by 34.7% over 2008. Between 2005 and 2009, lines of 220 kW and above increased from 253,700 km to 399,400 km, or by 57.6%, among which 500 kV lines increased from 62,900 km–121,900 km and 750 kV lines increased from 141 km–2,747 km (Table 7.5). Meanwhile China has become the leading country on engineering, researches, key equipment manufacturing,

[4]Equation: Energy saved = capacity shut down × operation hours × (gross coal consumption of large units − gross coal consumption of small units); CO_2 emission reduction = energy saved × 2.71; SO_2 emission reduction = energy saved / 0.7143×0.01×2; operation hours in 2009 was 4,865 hours (China Electricity Council, 2010); net coal consumption of 0.6 ∼ 1 GW large units is 321 gce/kWh (State Electricity Regulatory Commission, 2010); net coal consumption of small units closed down is 483 gce/kWh (NDRC and National Energy Administration, 2007); average sulfur content in China's commercial coal is 1% (Zhou and Wang, 2002). The CO_2 emissions factor of standard coal is 2.71 and 1 kg raw coal is equivalent to 0.7143 kgce in terms of heat value.

Table 7.5. Line of 220 kV and above (Unit: km).

Voltage	2000	2005	2006	2007	2008	2009
1000 kV	—	—	—	—	—	640
±800 kV	—	—	—	—	—	1,375
750 kV	—	141	141	511	630	2,747
500 kV	25,910	62,866	77,092	92,567	107,993	121,939
330 kV	8,524	13,059	13,762	15,795	16,717	19,156
220 kV	122,597	177,617	195,392	219,337	233,558	253,573
Total	157,031	253,583	286,387	328,210	358,896	399,430

Source: State Grid Energy Research Institute, *China Energy Saving and Electricity Saving Analysis Report 2010* (China Electric Power Press, Beijing, 2010).

technological standards development, and operation monitoring for ultra high-voltage projects (State Grid Energy Research Institute, 2010).

7.1.4 *Development of renewables*

7.1.4.1 *China ranked number one on renewable energy development in 2009*

In 2009, China's total utilization of renewables was 259.1 Mtce, ranking number one in the world, of which the new installed capacity of renewable energy units was 37 GW, accounting for 46% of the global market. Total installed capacity reached 226 GW in 2009. Comparing to 2005, renewable energy has increased in each technology (Table 7.6). The total renewable energy consumption in 2009 increased by 56% than 2005, with installed hydro capacity increased by 78%, grid-connected wind capacity increased 13 times, installed photovoltaic capacity increased 3.3 times, biogas production increased by 93%, coverage of solar water heaters went up by 81%, ethanol production increased by 67% and bio-diesel production increased five times. The 11th FYP targets for hydro, wind, photovoltaic generation, and bio-diesel production have all been met ahead of schedule.

China has become the big country for renewable energy industries. In 2009, China has produced 40% of the world's solar photovoltaic products, 30% of world's wind generators (increase by 10% than 2007) and 77% of world's solar water heaters. At the same year, China's investment in renewables went up to 33.7 billion dollars, an increase by 53% than 2008 and ranking number one in the world. About 81% or 27.2 billion dollars of the

Table 7.6. Renewable energy utilization (2005 and 2009).

	2005	2009	Coal equivalent (Mtce)
Power generation			
(1) Medium and large hydro (GW)	60	114.2	145.8
(2) Small hydro (GW)	50	55.1	51.2
(3) Grid-connected wind (GW)	1.3	17.6	8.9
(4) Small off-grid wind (10,000)	25	11.3 (2008 data)	3
(5) PV generation (MW)	70	298	0.2
(6) Biomass and waste-to-energy (GW)	2	2.4	20.3
Gas supply			
Biogas (billion m^3)	8	15.4	11.1
Rural household biogas digesters (million households)	18	35.07	0
Heat supply			
(1) Solar water heaters (million m^3)	80	145	17.4
(2) Geothermal etc. (Mtce/a)	2	2.2	2.2
Fuel			
(1) Ethanol (Mt)	1	1.7	1.6
(2) Bio-diesel (Mt)	0.05	0.3	0.5
Total (Mtce/a)	166		259

Note: (1) Small hydro refers to hydro plants with capacity below 50 MW; (2) energy provided by solar water heaters and solar cookers was respectively 120 kgce/(m^2 a) and 500 kgce/(m^2 a); (3) wind refers to the cumulative grid-connected in-operation wind capacity, and the cumulative installed capacity was 25.8 GW; (4) coal equivalent is calculated using the gross coal consumption of thermal generation of the year.
Source: Zheng (2009); Wang (2009).

investment was in the wind industry (UNEP and Bloomberg New Energy Finance, 2010).

7.1.4.2 *The development of small hydro plants increased rural electrification rate*

Small hydro refers to hydro plants with capacity below 50 MW. China has wide spread resources for small hydro plants. Resource technologically available in the country was totally 128 GW, mostly distributed in under-developed mountain areas where state grid finds it hard to reach and residents cook with fired woods.

China relies on small hydro for its rural electrification. In 2009, the installed capacity of small hydro reached 55.1 GW, and its generation was over 160 TWh, accounting for 26% of the total hydro generation. By end of 2007, China had developed 653 counties with primary rural electrification (defined as counties with per capita annual electricity consumption and per household annual daily electricity consumption above 200 kWh) and 400 counties with rural electrification (defined as per capita annual daily electricity consumption above 150 kWh). Development of small hydro provided electricity access to a total of over 300 million rural population.

In January 2009, China officially approved the *Plan for National Small Hydro as Fuel Alternative Project 2009–2015*, suggesting the small hydro development is no longer at its piloting phase but will be generally implemented. According to this plan, there will be about 250,000 new rural households starting to use hydro as alternative fuels annually, and by the end of 2015, about 1.7 million rural households in 543 counties across 24 provinces will use hydro as their daily fuel source. In places covered by the plan, at least 80% of all rural households could use hydro as a long term steady alternative fuel supply. The new installed capacity will reach 1.7 GW and the whole project investment will reach 14.1 billion yuan, of which central government will invest 5.7 billion yuan (Yu, 2010).

7.1.4.3 *Wind generation increased the fastest*

Of all renewable energies, wind generation developed the fastest. Between 2004 and 2009, except 85.85% in 2008, all the other years witnessed per annum 120% growth rate of new capacity (Fig. 7.7). New capacity in 2009 accounted for one third of world's wind power market. At the same year, China's total installed capacity of wind reached 25.8 GW, accounting for 16.2% of world's total. This number was only 760 MW in 2004, which was only 1.6% of the world's total at that time.

According to the latest data on wind resource released by Central Meteorological Bureau at the end of 2009, capacity available for development of onshore wind resource above level 3 at 50 m height was about 2,380 GW, and about 200 GW of near-shore wind resource at 5–25 m underwater depth and 50 m height above sea level. Assuming the total wind resource to be 2,400 GW, as of 2009, China's total installed capacity of wind was 25.8 GW,

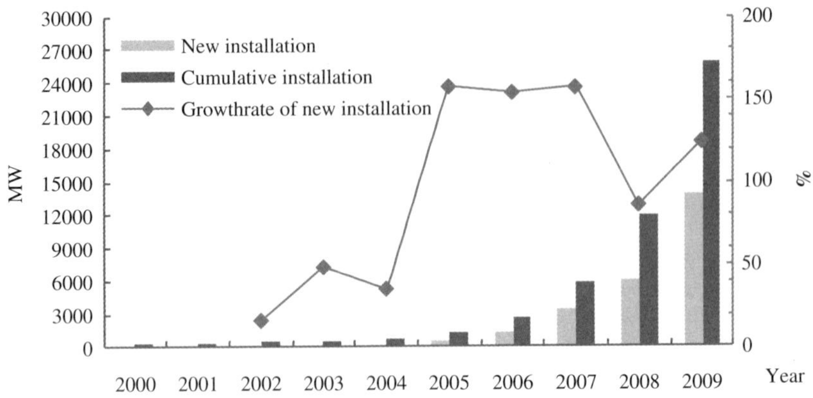

Fig. 7.7. China's wind power development status 2000–2009.
Source: CWEA, 2009 installed wind power capacity statistics. Wind power, 1 (2010) 29.

only 1% of the total resource potential. The remaining resource has large potential for development.

7.1.4.4 *Solar power industry has grown strong*

Capacity of China's solar photovoltaic industry ranked world's top one. In 2009, the world's total solar cells production was 9.34 GW, while factories of China mainland and Taiwan have taken up 49% of the market shares (Soalrbuzz, 2010). Within a few years, China's photovoltaic industry has developed its complete production chain from poly-silicon material production to photovoltaic cell packaging. In 2009, China had over 40 poly-silicon manufacturers, over 60 ingot/silicon chip manufacturers and over 60 solar cell manufacturers, 38 membrane cell manufacturers, and 330 photovoltaic components packaging (including non-crystal silicon) enterprises (Beijing Oriental Environment Institute *et al.*, 2010).

China's domestic photovoltaic market has been focusing on rural electrification. Since the implementation of Renewable Energy Law in 2006, grid-connected photovoltaic generation was offered opportunities to grow up and solar photovoltaic installed capacity in 2009 reached 298 MW. Due to high production costs, though China's solar power resources are rich, it is very difficult to disseminate solar generation in China within short time. At present, China's photovoltaic products are mostly exported for overseas markets. However, when costs for photovoltaic generation fall to certain level, the giant manufacturing capacity of photovoltaic industry will provide strong support for China's photovoltaic generation in the future.

7.1.5 *Summary*

Energy consumption of power industry increased by 29% in 2008 compared to 2005. However, at the same time power generation, transmission, and distribution efficiency experienced significant improvement, as a result energy and carbon intensity kept decreasing. Decarbonization trend of the power industry was evident. Energy intensity of thermal power industry reduced from 352 gce/kWh in 2005 to 333 gce/kWh in 2008, or by 5.4% over three years. Take energy intensity of 2005 as the baseline, energy saving in thermal power industry in 2008 was 52.7 Mtce, and cumulative energy saving between 2006 and 2008 was 78.3 Mtce; calculating with chain method, energy saving was 30.1 Mtce in 2008 and 51.6 Mtce cumulatively between 2006 and 2008.

Carbon intensity for thermal generation reduced from 935.8 gCO_2/kWh to 887.8 gCO_2/kWh, or by 5.1% over three years. Improvement in efficiency and development of nonfossil energies were the two key factors that contributed to the decarbonization in China's power sector. Two factors made the carbon emission intensity of the power industry fall quickly from 766.8 gCO_2/(kWh) in 2005 to 716.6g CO_2/(kWh) in 2008, or by 6.5%. Take carbon emission intensity of 2005 as the baseline, emission reduction of thermal power sector was 134 million tons of CO_2 in 2008 and 198 million tons of CO_2 between 2006 and 2008. Calculating with chain method, the emission reduction of thermal power industry was 75.1 million tons of CO_2 in 2008, and that 132 million tons of CO_2 between 2006 and 2008.

Fast development of renewable energy generation and nuclear power improved energy mix and helped to reduce fossil energy consumption, which translated into energy saving and emission reduction. Take 2005 as the base year, emission reduction was 170 million tons of CO_2 in 2008 and 290 million tons of CO_2 cumulatively. Calculating with chain method, it was 90 million tons of CO_2 in 2008 and 170 billion tons of CO_2 cumulatively between 2006 and 2009.

Fast growth of large and highly-efficient thermal units throughout China and shutting-down low-efficient and small units were the two main factors that contributed to the improvement of thermal generation fleet. By end of 2009, the power units with the average unit capacity of thermal reached 103.1 MW, and units of 300 MW and above have become dominating ones in China. In 2009, China had GW-level ultra supercritical units with total capacity of 21 GW, and at the same year it closed 26.2 GW of small thermal power units.

In 2009, the total utilization of renewables reached 259.1 Mtce, ranking number one in the world. The newly installed renewable capacity was 37 GW and total installed capacity was 226 GW. Comparing to 2005, renewable energy utilization witnessed significant increase.

7.2 Policy and Institutional Analysis on Low Carbon Development of Power Sector

7.2.1 *Overview of low carbon development policies*

The 11th FYP (Chapter 22, Session 1) requires to "achieve structural energy saving by optimizing industrial structure, particularly reducing the share of energy intensive industries; achieve technological energy saving by promoting energy saving technologies; achieve management energy saving by institutional improvement and monitoring in the production, transportation and consumption of energy".

These policies greatly contributed to reduction of energy consumption and carbon intensity of power industry during the 11th FYP. Avoided emission by renewable and nuclear generation was 592 million tons of CO_2 in 2008, and avoided energy use was 215 Mtce; carbon emission of power industry kept decreasing and reduced by 50.1 gCO_2/kWh in 2008 relative to 2005; By July 2010, 70.8 GW of small thermal power plants were closed, reducing 164 million tonnes of CO_2 emissions annually; gross coal consumption dropped by 21 gce/kWh in 2008 relative to 2005; carbon intensity of thermal generation decreased by 48 gCO_2/kWh in 2008 relative to 2005. Line loss rate kept decreasing too, reaching 7.21% in 2005 and 6.79% in 2008.

As Fig. 7.9 shows, policies that contribute to low carbon development in the power industry included five categories: Closure of small thermal power plants, development and application of low carbon energy technologies, renewable generation development and policy on, nuclear development, and energy efficiency dispatch. Among these five categories, closure of small thermal power plants and renewable and nuclear development aims at achieving structural energy saving, development and application of low carbon energy technologies aims at achieving technological energy saving, and energy efficiency dispatch aims at achieving management energy saving.

<table>
<tr><td colspan="5">Policies that influence on low carbon development in the power industry</td></tr>
<tr><td>Performance</td><td>energy mix upgrading; increasing share of renewable and nuclear generation</td><td>net coal consumption improvement; Carbon intensity reduction for generation</td><td>Station service electricity consumption Reduction</td><td>line loss rate Reduction</td></tr>
<tr><td rowspan="3">Structural energy saving</td><td colspan="2">expand renewable generation</td><td></td><td></td></tr>
<tr><td colspan="2">Develop nuclear power</td><td></td><td></td></tr>
<tr><td></td><td colspan="2">Close small thermal power plants</td><td></td></tr>
<tr><td rowspan="6">Management energy saving</td><td colspan="2">Financial subsidy</td><td></td><td></td></tr>
<tr><td colspan="2">Tax incentives</td><td></td><td></td></tr>
<tr><td colspan="2">Development plans</td><td></td><td></td></tr>
<tr><td colspan="2">Standards and regulations</td><td></td><td></td></tr>
<tr><td colspan="4">Dispatch of energy efficient plants</td></tr>
<tr><td colspan="3">Generation price leverage through electricity market</td><td></td></tr>
<tr><td rowspan="4">Technical energy saving</td><td>renewable generation technologies</td><td></td><td colspan="2"></td></tr>
<tr><td>nuclear technologies</td><td></td><td colspan="2"></td></tr>
<tr><td></td><td colspan="2">technology for large and highly efficient thermal units</td><td></td></tr>
<tr><td colspan="3"></td><td>high voltage transmission technology</td></tr>
</table>

Fig. 7.8. Policies influencing low carbon development of the power industry.

7.2.1.1 *Policies on the closure of small thermal power plants*

The power industry is a technology-intensive and capital-intensive one. Energy consumption rate of thermal generation depends on technologies of the units. Generally speaking, large capacity, high-parameter units have lower gross coal consumption and higher efficiency. Small thermal power plants, however, could not meet energy saving standards technically. During electricity shortage China once encouraged the development of small coal plants, resulting in much obsolete generation capacities. With the rapid development in power industry during recent years, electricity supply is no longer in shortage, while resource and environment concerns are growing, therefore, closing small thermal power plants becomes imperative.

Spectrum of low carbon development policies for the power sector			
Category	Structural energy saving policies	Management energy saving policies	Technical energy saving policies
Policies on closure of small thermal power plants	Notice on State Economic and Trade Commission's Suggestions on Issues Relating to Closing Small Thermal Power Plants transferred by the General Office of the State Council (General Office of State Council, No. 44, enacted on 26 May 1999)		
	Suggestions on Accelerating to Close Small Thermal Power Plants by Energy Department of National Development and Reform Commission (NDRC) approved and transferred by the State Council (State Council, 2007, No. 2)		
	Notice on Promoting to Close Small Thermal Power Plants by Reducing On-grid Electricity Price by NDRC (Pricing of NDRC, 2007, No. 703)		
Policies on development and application of low carbon energy technologies	Notice on Printing and Issuing the Energy Saving Mid and Long-term Plan by NDRC (NDRC, 2004, No. 2505)		
	The 11th Five-Year National Plan Outline for National Economy and Social Development of People's Republic of China (approved in the 4th session of the 10th National People's Congress 14 March 2006)		
	Notice on Printing and Issuing the National Plan for China's Response to Climate Change (State Council, 2007, No.17)		
	The 11th Five-Year Plan for Energy Development (NDRC, enacted in April 2007)		
			Notice on Revision of the Policy Outlines for China's Energy Saving Technologies by the Office of NDRC (NDRC, enacted 2 July 2004)
			Notice on Printing and Issuing the Plan on R&D of Significant Technical Equipment and Industrial Technology During the 11th five year Period by the Office of NDRC (NDRC, 2008, No. 162)

Fig. 7.9. Spectrum of low carbon development policies for the power sector.

Policies on development of renewable generation	Notice on Printing and Issuing the Mid and Long-term Development Plan for Renewable Energy by NDRC (NDRC, 2007, No. 2174)		
	Notice on Printing and Issuing the 11th FYP for Renewable Energy Development by NDRC (NDRC, 2008, No. 610)		
	The Renewable Energy Law of People's Republic of China (2009 amended version) (Directive of Chairman No. 23, enacted on 1 January 2006)		
Nuclear development policies	National Thematic Plan for Nuclear Development (2005–2020), enacted in 2007		
Policy on energy efficiency dispatch		Methods of Energy efficiency dispatch (tentative) by NDRC and Other Sectors transferred by the General Office of the State Council (State Council, 2007, No. 53)	

Fig. 7.9. (*Continued*)

7.2.1.2 *Policies on development and application of low carbon energy technologies*

Low carbon development in the power industry cannot be achieved without low carbon energy technologies. Currently, the focus is mainly on clean energy (wind power, solar power, and nuclear power etc.) technologies, clean coal technologies (supercritical, ultra-supercritical units, Integrated Gasification Combined Cycle (IGCC) technology, and carbon capture and storage technology) and advanced grid technologies (ultra high-voltage transmission technology and smart grid technology) and so forth. These technologies have a significant impact on main performance indicators of the power industry, such as net coal consumption, station service electricity consumption, and line loss, etc. Policies on development and application of low carbon energy technologies directly or indirectly promote the development and application of all those technologies in China's power industry.

7.2.1.3 *Renewables generation and nuclear development policies*

China's coal-based energy mix determines that its carbon intensity will be relatively higher, and its fast urbanization and industrialization process is demanding higher amount of energy supply as well.

China has been stressing importance on renewable energy development. In 2009, the total installed capacity of renewables (excluding large hydropower) ranked number one in the world. China was officially promulgated its Renewable Energy Law in 2006, and developed a number of supporting regulations including policies supporting wind generation, photovoltaic generation, and biomass utilization, which together effectively promoted the renewable energy development.

China has been actively exploring the development of new energy and renewable energy generation. It started developing nuclear power since 1970s, and built its first nuclear power plant in 1985. China has developed its *Mid and Long-term Development Plan of Nuclear Power 2005–2020* to guide the development of nuclear power.

7.2.1.4 *Policy on energy efficiency dispatch*

Energy efficiency dispatch is defined as under the condition of ensuring reliable power supply, apply principles of economic and energy efficiency and prioritize the dispatch of units with less energy consumption and less pollutant emission. Such dispatch system aims at achieving energy saving by better management. It indirectly influences the generation energy mix and the application of low carbon energy technologies. Energy efficiency dispatch reforms the traditional way of dispatch. Instead of distributing generation quota under administrative planning process, it relates generation quota to energy efficiency. This dispatch system tries to impact generation fleet by dispatching, which is a typical administrative approach.

7.2.2 *Policy analysis on closing small thermal power plants*

7.2.2.1 *Policy changes for small thermal power plants*

Policies on closing small thermal power plants have gone through three phases: "Launching small, medium, and large thermal plants all together", "closing small thermal power plants", and "launching large units and closing small ones". These three phases are closely related to two "electricity shortage" in the history of China's power supply. Electricity shortage

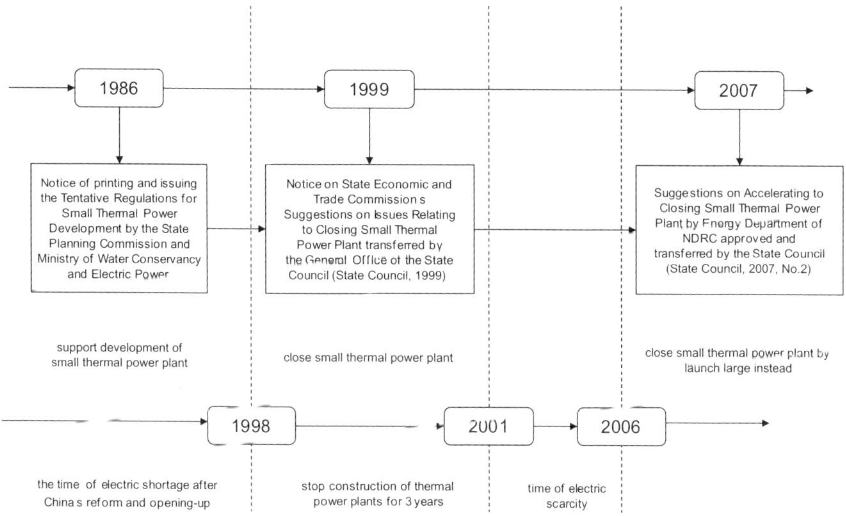

Fig. 7.10. Policy changes on closing small power units.

necessitated the development of small thermal power plants small thermal power plants, while the transition to low carbon economy required to phase out obsolete capacities like small thermal power plants (Fig. 7.10).

Power supply in China has been in the shortage until 1998. In 1986, the National Planning Commission issued the *Tentative Regulations on Development of Small Thermal Power Plants*, encouraging local government to raise fund for building economically and administratively independent small thermal power plants with capacity below 25 MW. Small thermal power plants prospered at that time.

After 1998, power supply was able to meet the demand of economic and social development. Influenced by international financial crisis, electricity demand kept falling nationwide. Policies at that time were made to address the "surplus electricity". The general office of State Council approved and issued the *Suggestions on Issues regarding to Closing Small Thermal Power Plants*, which required to shut down condensed steam generators below 25 MW that were obsolete, to be replaced, or coming to the end of its life cycle by 1999, and shut down ordinary medium and low pressure coal (oil)-fired units with unit capacity below 50 MW by 2000. It can be said that policies of that time reflected awareness of the importance of structural energy saving. However, with overestimation of power supply capacity in 1998, China discouraged construction of new thermal units for three years and as a result, in 2002, it was hit by greater electric shortage again. During that period, China had to put the thermal power plants yet to be closed back into operation again, which brought negative impacts on the future closing of small thermal power plants.

In 2007, to support structural energy saving policy in the 11th FYP, the *Suggestions on Accelerating to Shut Down Small Thermal Power Plants* was approved by State Council and issued by NDRC and State Energy Office, again re-arranging to shut down small thermal power plants. It required to "launch large ones and limit small ones", meaning to launch large thermal units and shut down small thermal power plants, while approval of newly proposed units is conditional to the closing of small thermal power plants. While building large-capacity, high-parameter, low-consumption, and low-emission units, provinces are required to shut down small thermal power plants at the same time. "Launch large ones" creates market environment for "close small ones", while "close small ones" leaves space for "launch large ones". To motivate local government and enterprises to "launch large ones and limit small ones", relevant coefficient was developed. When building plants with unit capacity of 300 MW, they are required to shut down small units of 80% of the new capacity, For units of 600 MW the coefficient is down to 70% and for units of 1 GW it is further down to 60%. There was also another methodology of calculating based on the coal consumption.

It can be concluded that the premise of low carbon development in the power industry is that power generation can meet growing demands of national economy. China has the advantage of strong national control in investment and financing system for power plants, which is very beneficial for fast development and decarbonization of the power industry. As long as China takes advantage of such system, making good judgment on the trend of power development and maintaining the steady growth of electricity supply, policies on closing small thermal power plants can be smoothly implemented.

7.2.2.2 *Implementation of the policy on closing small thermal power plants*

See Fig.7.11 for implementation factors of the *Suggestions on Accelerating to Shut Down Small thermal power plants* issued in 2007.

Policies on Closing small thermal power plants are "a series of efforts". NDRC took the lead to be in charge. It assigned decomposed targets to each province and signed target accountability agreements with provincial

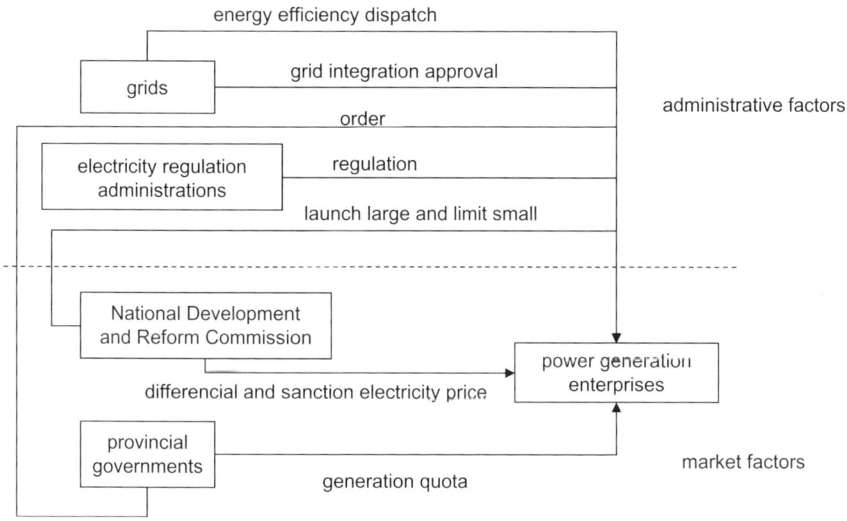

Fig. 7.11. Implementation factors of policies on closing small power units.

governments and large state-owned power companies. Regulators and grid companies worked together closely to develop corresponding measures for implementation.

In terms of administrative control, the central government required grid enterprises to strengthen supervision and management of prioritizing the dispatch of renewables and high-efficiency and clean units and restricting energy intensive and highly-polluting units. On the other hand, it required the regulators to strengthen supervision and management over small thermal power plants, establish supervision and management information system and suspend the license to those who fail to meet engineering standards and other relevant regulations. Meanwhile, for small thermal power plants that are supposed to be but refuse to shut down, provincial and central government shall order them to shut down and suspend approval for new unit development proposed by the same company until they shut down the small units that are required to be shut down. The government encourages restructuring, merging or acquisition of small plants, and allows constructing large units after closing the small ones. NDRC will approve construction of additional capacity for each province according to its obsolete capacity shut down. For cross-provincial closing, capacity shut down could be accounted for in the province where the shut down small units were and would be considered in calculating new capacity to be approved for this province.

In terms of market leverage, provincial government strengthened to-grid electricity price management of small thermal power plants, lowered the to-grid price of all coal (oil)-fired small thermal power plants to the same as or below local benchmark price, no additional subsidies were allowed to be provided for such plants; for the units with price already lower than the benchmark level, the original price should be kept at the low level. On the other hand, for units listed in the 11th FYP Closing list of each province, they are granted with certain generation quota before the deadline for closing (maximum three years) and could get economic compensation by selling generation quota to large units. Such generation quota shall decrease annually during the grace period closing.

7.2.2.3 *Implementation and performance*

Starting from the 11th five year period, installed capacity of small thermal power plants shut down each year between 2006 and 2009 was respectively 3.1 GW, 14.4 GW, 16.7 GW, and 26.2 GW, and the cumulative capacity of small thermal power plants shut down was 60.4 GW, exceeding the target of 50 GW for the 11th FYP set by the State Council (Fig. 7.12). In 2010, the power industry further implemented the policy of phasing out obsolete production capacities. Between January 1 and July 15, a total of 468 small thermal power plants were shut down with total capacity of 10.7 GW,

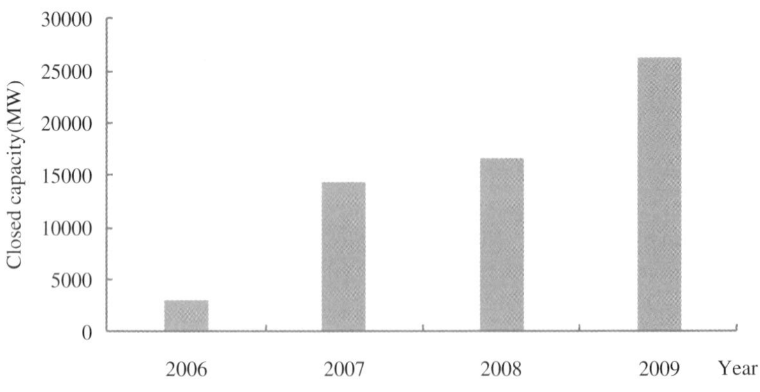

Fig. 7.12. Number of small thermal power plants closure during the 11th FYP. *Source*: 2008–2010 brief of energy saving and emission reduction of electric power plants issued by State Electricity Regulatory Committee and etc.

achieving the target of 10 GW for 2010 ahead of schedule. Between January 2006 and July 2010, China has shut down cumulatively 70.8 GW of small thermal power plants (National Electricity Regulatory Commission, 2010).

During the 11th FYP, "the Big Five" generation companies played an important role in closing small thermal power plants. They have commitment of closing a total of 22.3 GW small thermal power plants. Huaneng Group, besides fulfilling the commitment to shut down 63 small thermal power plants of 2.5 GW, offered to further shut down 1.1 GW more units, resulting in the shutdown of all together 80 units with 3.6 GW. Datang Group also reset the target of 74 units with 4 GW to 92 units with 4.7 GW, and achieved the target for 2008–2010 as early as in 2007. In total Datang Group shut down 2.6 GW units. Table 7.7 shows the small thermal power plants shut down by "the Big Five".

Effective implementation of closing small thermal power plants helped upgrading the generation fleet. The share of large units increased, while the share of small ones, particularly the energy intensive and highly-polluting ones decreased. By end of 2009, thermal power units with average unit capacity of 103.1 MW, 300 MW and above have become major generators of thermal generation fleet in China, taking up 69% of all installed thermal capacity. In the new fleet 344 units with totally 219 GW, or 34% were units of 600 MW or above, and 706 units with totally 226 GW, or 35% were units of 300 MW level. China had 21 ultra-supercritical units with GW-level capacity and with 14 more under construction. The number of GW-level ultra-supercritical units in operation or under construction together ranked as top one in the world. Among the new capacity of 65.9 GW added in 2009, 126 units each of. 300 MW and above accounted for 59.7 GW or 91% of the

Table 7.7. Comparison of capacities shut down by "the big five" (unit: GW).

Year	Huaneng	Datang	Guodian	Huadian	China power investment
2006–2008	2.5	4.6	2.4	3.4	3.4
2009	1.4	1.3	2.6	1.7	1.5
Total	5.2	7.3	5.7	5.6	7

Source: Z. Long, Outstanding achievements of five power enterprises on Closing of small thermal power plants and significant change of structure in thermal power (2010), http://www.chinanews.com.cn/ny/nes/2010/02-10/2118816.shtml (accessed on 22 October 2010).

total, while 50 units each of 600 MW and above accounted for 35.5 GW or 54% of the total. (State Electricity Regulatory Committee *et al.*, 2010).

Thermal generation efficiency improved by large through the "replacing small units with large ones". In 2009, net coal consumption of all China's large units with capacity of 6 MW and above decreased to 340 g/kWh, 5 g/kWh less than 2008. In the first four years of the 11th five year period, results of energy saving and emission reduction were evident with the cumulative net coal consumption dropping 30 g/kWh (State Electricity Regulatory Committee *et al.*, 2010).

Closing small thermal units also led to reduction in energy intensity and carbon intensity of the power industry. Assuming the generation quota of small plants shut down was transferred to large units of 600 MW–1 GW (excluding 1 GW), taking average operation hours of thermal power units in 2009, closing small thermal power between 2006 and 2009 would result in energy saving of 47.6 Mtce every year and emission reduction of 129 million tons of CO_2 and 1.3 million tons of SO_2 (Table 7.8).

Table 7.8. Energy saving and emission reduction associated with closing small thermal plants during the 11th FYP.

Year	Actual shut down (GW)	Per annum energy saving (Mtce)	Per annum CO_2 emission reduction (Mt)	Per annum SO_2 emission reduction (Mt)
2006	3.1	2.5	6.7	0.07
2007	14.4	11.3	30.7	0.32
2008	16.7	13.2	35.6	0.37
2009	26.2	20.6	55.9	0.58
Total	60.4	47.6	128.9	1.3

Note: Calculation formula: Energy saved = capacity shut down × hours of power generation × (gross coal consumption of large-sized units − gross coal consumption of small units); CO_2 emission reduction = energy saved × 2.71; SO_2 emission reduction = energy saved/0.7143 × 0.01 × 2; (State Electricity Regulatory Commission, 2010); net coal consumption of 600 MW–1 GW large units is 321 gce/(kWh); 1 kg raw coal is equivalent to 0.7143 kgce in terms of heat value; net coal consumption of small thermal power plants is 483 gce/(kWh) (NDRC).

Source: (1) 2007–2009 brief of energy saving and emission reduction of electric power plants issued by State Electricity Regulatory Committee and etc; (2) China Electricity Council 2005–2008 National Power industry Statistics Brief; (3) the brief on closing small thermal power plants in 2007 issued by the State Electricity Regulatory Committee and NDRC in 2008.

All of these successes are of great importance to upgrading China's electricity generation fleet, the development of the power industry, and to achieving the target of 20% reduction of energy consumption per unit GDP during the 11th FYP.

7.2.2.4 *Obstacles during the implementation of closing policies*

Policies on closing small thermal power plants achieved great success, however, there was various obstacles during the implementation.

(1) It is hard to avoid "clear-cutting" while selecting small thermal units to closing

Currently, there are many generators with unit capacity of 125 MW or below 200 MW in China. Although they have negative impact energy intensity and carbon intensity, it is difficult to shut them all at once. The selection of plants to be shut down will become more and more difficult as policy being implemented. Units that cannot be shut down at once are divided into five categories: (1) Units recently constructed. China had not limited on the development of small thermal power before 2006 and even supported it before 1999, so there are quite a few small units built not too long ago. These small units are in rather good condition and have embedded energy in the plants. If they were dismantled right away, energy savings might be negative from life cycle perspective. (2) Regional grids need certain small units for load following or for balancing the fluctuation of nonhydropower renewable energy generation in the region. Closing small units for variable loads, large units would have to take over such tasks and bear the efficiency lost for providing such ancillary services. (3) Some small combined heat and power plants in northern China are actually very energy efficient when they generate power and supply heat in winter and play an important role in reducing energy consumption for heat supply in Northern China. Closing them while new large power plants are not CHP ones, coal-fired boilers would be needed for heat supply, which may lead to great increase of energy consumption and overall reduction in energy efficiency. (4) Complementary power plants of enterprises should not be shut down. Some large enterprises (iron and steel, cement, glass, coal chemical, petroleum, and chemistry etc.) developed their own complementary power plants that generate power with waste heat and pressure. Such plants are energy inefficient judging from technical parameters such as coal consumption, however they save energy in the whole process. Currently, grids pay such power plants a discriminatory price in order to force them to close down. This may undermine motivations of enterprises to utilize waste heat and pressure. (5) Units located at the end of grids or in independent grids and are of backbone importance for the local grids shall not be shut down, in order to ensure secure and reliable power supply or enable emergency power supply in the region.

(2) *It's hard to keep equality among different interests during implementation*

Policy of closing small thermal power plants requires closing small ones while building large ones, and re-distribution of interests becomes inevitable. Inequality is seen in three areas as follows: (1) Inequality of distribution of interests between generation enterprises of different size. Closing small units and replacing them with large ones could be a good opportunity for the big five generation companies and other large state-owned generation enterprises to expand their territory. They are able to launch large units at the same time when small generators are shut down by purchasing small power enterprises. But many small generation enterprises were often unable to finance large generators once their small ones are shut down, and some of them were even unable to qualify for new generation permits. This encouraged centralization of the generation market and reinforcing of monopolizing. (2) Interest distribution is not equal among different regions. Since small generators shut down were not in the same area where new large generators will be located, some place may lose their generation quota and need to arrange cross-region transmission. This would inevitably have an impact on local economy and employment. (3) Inequity to local people and private enterprise. Some of the shut down small units are owned by small local private enterprises, which are usually overstaffed and have slim profits. They usually go bankrupt once their plants are closing. The number of staff who can be re-hired in new large units is always limited. It is often very difficult and costly to re-allocate all the involved staff.

Box 7.2. Unemployment associated with closing small plants.

By calculating the number of employment positions needed for unit capacity, we could estimate how closing small thermal power plants impacted on employment. According to official data of NDRC, Shanxi province

(*Continued*)

Box 7.2. (*Continued*)

thermal power statistics, data from survey on thermal power plants by China Power Investment Corporation and the *Information of Closing Small Thermal Power Generators (2009 No. 4 announcement)* (March 2009) issued by NDRC, National Energy Administration, Ministry of Environmental Protection, State Electricity Regulatory Committee, and the *Plan of Closing the First Group of Thermal Power Generators before 2010* (2005 No. 50 announcement, August 2005) issued by NDRC, for every 10 MW closing, on average 62 staffs need to be reallocated. Since national policies restrict the reconstruction of general coal-fired generators with unit capacity below 300 MW, power enterprises will not build new generators below 300 MW. Then if taking 300 MW as standard capacity for large unit, according to survey on installed capacity and staff number by the end of 2008 in Huaneng group, it can be estimated that 6.03 positions are created per installed new capacity of 10 MW (Pan, 2010). Assuming there is no time lag between when large units are built and when small one are closed, for every 10 MW of small thermal power unit closing, it would cause 56 cases of unemployment, and the total 70.8 GW closing would result in 396,000 people becoming unemployed.

7.2.3 *Policy analysis for development and application of low carbon energy technologies*

Technology is the key of economic reform and low carbon energy technologies are the key to decarbonize the power industry. No matter clean and efficient generation or highly-efficient grid transmission, neither of them could be achieved without application of new technologies. From generation perspective, nuclear power technologies, renewable energy technologies and clean coal generation technologies all contributed to generation efficiency improvement and emission reduction; from transmission and distribution perspective, smart grid development will enable security, stability, and high efficiency for grid operation and help reduce line loss.

7.2.3.1 *Policy's direct influence on the development of low carbon technology*

Guided by the principal policies of the 11th *five year Plan Outline* and *Mid and Long-term Energy Saving Plan*, policies supporting the low carbon energy technologies were developed and implemented continuously, bringing vitality to the development of low carbon energy technologies in China. Some policies directly addressed application of low carbon energy technology, and others influenced the development of low carbon energy technology in indirect ways. Table 7.9 shows policies that have direct impact on low carbon technologies.

It can be concluded that most of these policies are at the macro level and guide low carbon technological development in the power industry. Among structural, technological and management energy saving, technological energy saving is the most fundamental and direct one. Emphasis on low carbon energy technologies among the policies is the factor that directly pushes forward low carbon development of power industry.

7.2.3.2 *Policy's indirect influence on development of low carbon energy technology*

Figure 7.13 shows the interaction between key policies and key technologies in the low carbon development of power industry. The application of technologies helps achieving policy goals, while policy implementation helps promote the development and application of technologies.

Take the policy on closing small thermal units as an example. In its implementation, "launch large ones and limit small ones" necessitates to apply clean coal generation technologies in highly-efficient large thermal units to replace small ones that are phased out. Apart from that, geographically China's small thermal power plants are sparsely distributed and newly launched large ones will to some extent converge, therefore the regional balance of power supply and demand may be influenced. As a result, newly built grids using cross-regional inter-connection grid technology will play a key role. At present, China has high-voltage and ultra-high-voltage power transmissions of 500 kW, ±660 kV and in some place even up to 750 kV and ±800 kV. Meanwhile, long-distance and large-capacity grid transmission technology is maturing. Consequently, power supply balance can be addressed by adjusting grids of different voltage, while application of these mature technologies helps facilitate the effective implementation of policies.

Another example is that policies supporting the development of wind and nuclear power will encourage investments in these two areas and investment in return helps technologies to mature. On the other hand, as technologies mature, their costs decreases and electricity production become

Table 7.9. Policies directly impacting development and application of low carbon energy technology.

Policies	Policies directly relating to development and application of low carbon energy technology: Clean energy technologies	Advanced grid power transmission technologies
Notice on printing and issuance of the mid and long-term energy saving plan by the NDRC (No. 2505, 2004)	• Develop ultra-supercritical and supercritical generators and large combined circulating units of 600 MW and above. • Adopt highly-efficient and clean power generation technology, renovate existing thermal power generators and improve power generation efficiency of generators. • Develop CHP, CCHP and combined heat, power and gas.	• Advance cross-regional grid connection and implement grid economic operation technologies. • Adopt advanced power transmission, transformation and distribution technologies and equipment, phase out energy intensive and obsolete equipment and reduce losses in power transmission, transformation and distribution.
The 11th FYP Outline of National Economic and Social Development of People's Republic of China (approved by the 4th session of the 10th CPC Congress on March 14, 2006)	• Prioritize to develop large highly-efficient and environment-friendly thermal units and build large ultra-supercritical power station and large air-cooling power station. • Push forward clean coal power generation, build circulating liquidizing beds power station with unit capacity of 600 MW, and initiate the IGCCC power station project.	—

(*Continued*)

Table 7.9. (*Continued*)

Policies	Policies directly relating to development and application of low carbon energy technology	
	Clean energy technologies	Advanced grid power transmission technologies
	• Promote development of nuclear units. Prioritize to build GW nuclear units, build capacity for independent design, construction and operation of pressurized water reactor nuclear power station.	
Notice on printing and issuance of the national plan of China's response to climate change by the State Council (State issuance, 2007, No. 17)	• Nuclear power technology: Research and acquire fast neutron reactor design and its core technologies, relevant nuclear fuels and structural materials technologies, break through key technologies like sodium cycle and participate in construction and researches on international thermonuclear experimental reactor. • Renewable energy technologies: Focus on researches on low-cost and sizable renewable energy use and application technologies, develop large wind power generation equipment,	• Power transmission and distribution and grid security technologies: Focus on research and development of large-capacity and long-distance DC power transmission technology and ultra high-voltage AC power transmission technology and equipment, intermittent grid integration of electric sources and transmission and distribution technology, electric power quality monitoring and control technology, large-scale interconnected grids security guarantee technology, key technologies in the project of borrowing electricity from Western to Eastern China, grid

(*Continued*)

Table 7.9. (*Continued*)

Policies	Policies directly relating to development and application of low carbon energy technology	
	Clean energy technologies	Advanced grid power transmission technologies
	high cost-effective solar photovoltaic cells and application technology, solar thermal power generation technology, BIPV technology, biomass and geothermal energy development and application technologies. • Clean and highly-efficient coal use technologies: Key researches on development of highly-efficient power generation technologies and equipment like medium gas turbine, IGCC, high parameter ultra-supercritical and supercritical units, supercritical large circulating liquidizing beds etc. Develop and use liquidized and multiple combined production technologies, develop transformation technologies of CO_2 capture, use and storage technologies etc.	dispatch automation technology, highly-efficient power distribution, and power supply management information technology and system.

(*Continued*)

Table 7.9. (*Continued*)

Policies	Policies directly relating to development and application of low carbon energy technology	
	Clean energy technologies	Advanced grid power transmission technologies
The 11th five year energy development plan (NDRC, taking effect in April 2007)	• Focus on the development of GW large advanced pressurized water reactor nuclear power technology. • Focus on low-cost and large renewable development and application: Large wind turbines, biomass generation, biogas generation, ethanol, bio-diesel and solid biomass briquette, and key solar power development and application technologies. • Focus on development of clean and highly-efficient coal power generation technology.	• Focus on development of ultra large-sized power transmission and distribution and secondary system of grids: Flexible power transmission systems, high voltage power transmission, intermittent grid integration of electric sources and transmission and distribution technology, power quality monitoring and control technology, large-scale interconnected grids security guarantee technology, grid dispatch automation technology.
Notice on amendment of the outline of China's energy saving technology policies by the office of NDRC (NDRC, July 2, 2004)	• Develop and utilization of wind power. Build large, medium, and small wind generators at the same time. Build grid-connected wind farms where wind resources are rich and close to grid; build island power supply system combining wind	—

(*Continued*)

Table 7.9. (*Continued*)

Policies	Policies directly relating to development and application of low carbon energy technology	
	Clean energy technologies	Advanced grid power transmission technologies
	and diesel generation or solar photovoltaic generation. Develop wind turbine industry step by step. • Develop solar power utilization. Solar thermal utilization includes expansion of applying highly-efficient solar water heaters, solar power agricultural greenhouse, passive solar house, solar power dryer etc. Research and develop medium and solar thermal power utilization technologies; promote highly-efficient and low-cost small and medium photovoltaic generation system and solar and wind supplementary generation system • Improve and optimize energy mix and upgrade coal combustion technology to achieve clean coal combustion step by step. Strengthen scientific research and develop fluidized bed combustion generation, new coal-fired combined	

(*Continued*)

Table 7.9. (*Continued*)

Policies	Policies directly relating to development and application of low carbon energy technology	
	Clean energy technologies	Advanced grid power transmission technologies
	cycle power generation technology and new technique of gas, heat, and electricity co-generation.	—
Notice on printing and issuing the significant technological equipment R&D and exclusive industrial technological plan during the 11th five year period by NDRC (Hi-tech, NDRC, 2008, No. 162)	• R&D and production of key technologies and equipment: 1 GW nuclear power key equipment. • Development of key industrial technologies: Key technologies on wind power, large offshore wind power key technologies and equipment; development of key solar technologies.	—

more affordable while application becomes safer and more reliable, which lays solid foundation for effective implementation of relevant policies.

As a result, some policies indirectly facilitate the development of low carbon energy technologies.

7.2.3.3 *The impact of low carbon policies on low carbon development*

During the 11th five year period, technology development and application policies played an important role in promoting low carbon technologies in the power industry. Nearly all the low carbon development achievements in the power industry were made possible by these policies.

(1) Net coal consumption of coal units edged into the advanced level in the world. On the basis of having achieved the 11th five year target for net

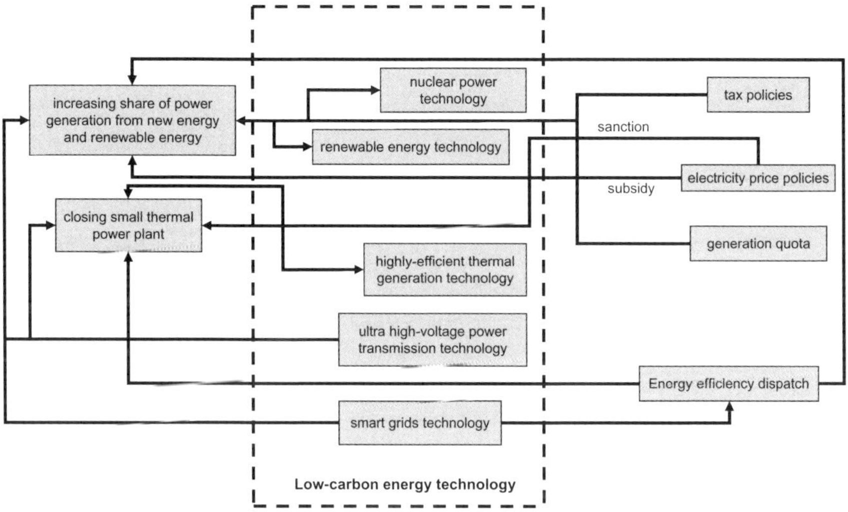

Fig. 7.13. Interactions between policies and low carbon energy technologies.

heat rate (in 2008) two years ahead of schedule, 2009 witnessed further reduction to 340 g/kWh for units of 6 MW and above, 5 g/kWh lower than 2008 level. In the first four years of 11th five year period, net coal consumption cumulatively fell by 30 g/kWh. Such achievements were significant for energy saving and emission reduction.

(2) Station service rate and power line loss rate decreased year to year. Power enterprises attached importance to energy saving and emission reduction. They renovated technologies and strengthened operational management, achieving reduction in station service power rate drop year by year. In 2009, station service rate of units of 6 MW and above was 5.67%, 0.14% lower than 2008 level. Station service rate of thermal plants was 6.62%, 0.17% lower than 2008 level. Investment for grid development, for the first time, exceeded the investment for generation. Cross-regional and cross-provincial power transmission has been upgraded, and provincial grid framework and urban/rural grid distribution network has been improved. In 2009, national grid line loss rate decreased to 6.72%. In the first four years of 11th five year period, line loss rate cumulatively dropped by 0.32%. Achievements of energy saving and emission reduction were significant (State Electricity Regulatory Commission, 2010).

Box 7.3. Shanghai Waigaoqiao ower Generation Co. Ltd applied the low carbon energy technology to save energy.

In July 2005, Shanghai Waigaoqiao Power Generation Co. began constructing two largest ultra supercritical coal units (1 GW) in China with highest parameters, which were put into commercial operation on March 26, 2008 and June 7, 2008 respectively.

In 2009, the actual operational coal consumption of the two generators fell to 282 g/kWh, breaking the previous world's record created in 2008, which was 287 g/kWh. Comparing to the data from China Electricity Council, the coal consumption of most advanced GW-capacity ultra supercritical units in the developed countries in 2009 was around 300 g/kWh. It suggests that these two units have exceeded the level of its peers in developed countries.

Compared to most other advanced domestic generators, these two can further save over 0.2 million tonnes of steam coal and reduce over 0.4 million tons of CO_2 emissions annually. Apart from that, Shanghai Waigaoqiao Power Generation Co. is now working on the innovative project of new turbine generators, which can further reduce energy consumption by about 5% without changing the unit production cost of ultra supercritical units on the basis of currently available materials and technologies.

These two GW-capacity ultra-supercritical coal units are the typical representatives of China's practice of using low carbon technology to promote energy saving and emission reduction. It reflects that technologies are of importance and the necessity for achievements and policies are of same importance to technologies in low carbon development in power sector.

7.2.4 *Institutional analysis of low carbon development in the power sector*

7.2.4.1 *Stakeholders in low carbon development of the power sector*

In the power industry, there are three major players in low carbon development — the electricity market, enterprises, and the government.

Electricity market before the reform was wholly monopolized. The power industry was operated by government or government-owned enterprises. Power generation, transmission, distribution and sales were vertically integrated and there was only one seller without any competition. By bringing in competition and breaking monopoly, the market-oriented reform brought some changes in the power sector. Separation of generation and transmission enabled the initial formation of power generation market. Increasing independent power generation enterprises stimulated intense competition in the power generation market and increased productivity. The establishment of demand side management system and the implementation of direct access policy helped users get access to information on electricity price in a timely manner and enabled them to join in competition which therefore developed the initial power market, improving efficiency, and profit margin for the power industry. However, though regional grids are most important carrier of this electricity market, they were not independent yet. Regional grid companies are not independent corporate persons and lacked corporate rights. Given the spread of ultra high-voltage power transmission, the roles of regional grid companies were weakening both in terms of technology and economy. Regional electricity markets in China have not yet formed and businesses transmission and distribution are still monopolized.

Power enterprises are cells of electricity market. Power-related enterprises can be divided as generation enterprises, grid enterprises, and consumer enterprises. Their role in the market is determined by its capacity of resource distribution. Low carbon development in the power industry needs tremendous amount of inputs. The resource distribution in China's electricity sector is evolving from the planning manner to market-driven manner. The size of an enterprise determines its capacity of resource distribution and also its profits acquired in the transition to low carbon development.

Governments are policy-makers and institutional designers. Governments design the roles of stakeholders in the power industry and also fulfill the obligation of institutional improvement.

Table 7.10 shows the role of different stakeholders in low carbon development. On the left it shows the list of stakeholders, on the top it shows low carbon oriented policies, and the remaining part of the table shows different roles of each stakeholder under different policy and institutional arrangement. It can be concluded from the table that low carbon development relies on the joint participation and collaboration of markets, enterprises, and governments.

Table 7.10. Low carbon development in the power industry.

System stakeholders		New energy and renewable generation policies	Policies on closing small thermal generators	Low carbon energy technology policies	Policies on energy efficiency dispatch	Policies on electricity price
Markets	Generation market	Subsidized electricity price	Sanction electricity price and trading of generation quota	—	—	Bidding for access to grids, energy-saved electricity price leverage
	Power sales market	Electricity price surcharge	Consumer enterprises enjoy preferential price if they shut down their own generators	—	—	Differential electricity price and electricity price of direct purchase
Enterprises	Big generation companies	Vigorous development	Launch large ones and limit small ones	Investment and development	—	Bidding for access to grids*
	Small generation companies	—	Acquired and closed down	—	—	

(*Continued*)

Table 7.10. (*Continued*)

System stakeholders	New energy and renewable generation policies	Policies on closing small thermal generators	Low carbon energy technology policies	Policies on energy efficiency dispatch	Policies on electricity price
Grid companies	Prioritize the dispatch of new energy and renewables, build new lines to enable grid-connection	Restrict dispatch of small units or forbid its connection to grids, build new lines to enable balance	Investment on development	Implementation	Power demand side management
Large consumers	—	Enjoy preferential rate if closing down own generator	—	—	Have the right to purchase electricity directly
Other social capitals	Entry	—	Entry	—	—

(*Continued*)

Table 7.10. (*Continued*)

System stakeholders	New energy and renewable generation policies	Policies on closing small thermal generators	Low carbon energy technology policies	Policies on energy efficiency dispatch	Policies on electricity price
Governments	Generation quota for renewables, and tax relief	Order to close down and approve trade of generation quota	Policy incentives	Develop provincial generation schedule	Develop benchmark electricity price, appointed price and reference price

*The mechanism of bidding for access to grids is an important demonstration of market mechanism. However, due to the current electricity shortage, fossil energies are greatly influenced by international market, while China asserts strict control on electricity price, as a result this mechanism has not achieved substantial progress in reality.

7.2.4.2 *Institutional analysis of policies on closing small thermal power*

Policies on closing small thermal power generators are key in low carbon development of the power industry. Its smooth implementation relies on cooperation and labor distribution among different stakeholders.

(1) Electricity reform provides the appropriate opportunity for policy implementation:

Through the market-oriented reform started in 2002, generation and transmission was separated and "the Big Five" power generation companies were established. This has resulted in much improved capacity for distribution of social resources in the generation side. China's newly installed capacity for generation maintained fast growth for five straight years between 2004 and 2008, providing reliable power supply for the fast growing GDP. In 2007, *Notice on Suggestions on Acceleration of Closing Small Thermal Power Generators* issued by the State Council was a strategically smart move by Chinese government — it was right at the time for its implementation when financial crisis at that time slowed down the electricity shortage. Such policy achieved the double goal of both restructuring and maintaining growth.

(2) China's economic growth provided the costs associated with closing small thermal plants:

With China's fast growing GDP and stronger capacities of resource distribution by both governments and enterprises in recent years, solid support were available for closing small thermal power generators. Capacities of resource distribution are reflected in three aspects:

(1) Generation companies provided compensation for closed plants, especially by those big companies: In recent years, annually an average of 70 GW capacity was newly added, which often involved clean energies and large units and required intensive financial and technology input.
(2) Grid enterprises achieved balance among generation resources: During the 11th five year period, the overall grid investment and construction accelerated, particularly construction of high voltage and ultra-high-voltage power transmission, addressing the needs of power supply. Closing small thermal power generators is related to launching large thermal power generators. Small generators were often sparsely distributed, while newly built large ones would unavoidably converge in location, and this influences the geographical balance of power supply.

Acceleration of grid development enabled cross-regional grid connection and helped achieve balance among generation resource.

(b) Central government committed financial resources: The central government directly funded RMB 2 billion yuan to address a series of issues including reallocation of staff who became unemployed from closing small thermal generators. All of these provided enabling environment for policies on closing small thermal generators to be successfully implemented.

(3) Closing small thermal generators relied on stakeholder's' close cooperation:

(1) The government has attached importance and attention to policy implementation. In 2007, government work report clearly specified the target of closing small thermal generators during the 11th five year period. In the following three years, leaders of the State Council attached high importance to phasing out obsolete production capacities in the power industry, provided guidance and instructions on many related issues.

(2) The success was achieved jointly through market instruments and administrative power. The State Council issued *Notice on Suggestions on Acceleration of Closing Small Thermal Generators* and supporting policies, specifying the major principles, roles and responsibilities, incentives, and guarantee system for phasing-out small thermal generators. In terms of market instruments, the to-grid electricity price of all coal (oil)-fired small thermal plants was reduced and no subsidies was allowed; meanwhile, generators that were listed in the provincial shut down plan during the 11th five year period and was shut down as scheduled could get electricity production quota within certain period and get economic compensation by selling such quota to large generators. In terms of administrative power, the government encouraged to shut down small generators by restructuring, merger or acquisition, and launch large projects at the same time; on the other hand, the government required grid companies to strengthen dispatch monitoring and management, prioritizing the dispatch of renewable generation, highly-efficient and clean units and limiting the generation of energy intensive and highly-polluting units. Meanwhile, for small units supposed to be but refused to shut down, provincial government shall order them to shut down immediately and suspend approving any new project proposed by the same enterprise. These policies reflect both market principles and administrative power, combining both economic and social benefits, integrating current and long-term interests, and achieving the double goal if phasing

out obsolete capacity and upgrading the generation fleet. This pragmatic approach fit into the reality of the power industry, laying good foundation for smoothly closing of small generators.

(3) There was clear responsibility division for governments and enterprises. As required by the State Council, NDRC signed agreements with 30 provincial governments and major power enterprises, specifying targets of each province during the 11th five year period. Meanwhile, they clarified roles of concerned departments of the State Council, local governments and enterprises and work procedures to ensure different issues or contradictions occurred during the implementation can be addressed correspondingly.

7.2.5 *Conclusion*

(1) Policies that contribute to low carbon development in the power industry included five categories: Policies on closing of small thermal power plants, policies on development and applications of low carbon energy technologies, policies on renewable generation development, policies on nuclear power development, and policies on energy efficiency dispatch. The first three aims at achieving "structuring energy saving"; policy on development and applications of low carbon energy technologies aims at achieving "technological energy saving", and energy efficiency dispatch policy aims at achieving "management energy saving" All of these policies contributed to significant reduction in energy consumption and carbon emission intensity of the power industry during the 11th five year period.

(2) Closing small thermal units was achieved by a series of administrative and market approaches like "launch large ones and limit small ones", electricity price regulation, trading of generation quota, grid-connection permit, and energy efficient dispatch. Small inefficient generators were shut down. The structure of generation fleet was effectively improved. The proportion of highly-efficient large generators increased, ultimately leading to lower energy intensity and carbon emission intensity of the power industry. It is of importance for optimizing the structure of the power sector, promoting sectoral development, and achieving the target of 20% reduction of energy consumption per unit of GDP in the 11th five year period.

(3) There were obstacles during the implementation of closing small power units. It was difficult to avoid closing some useful small generators with clear-cutting, such as newly-built small generators, variable load following

units, CCHP generators, generators using recovered waste heat or pressure, and other important small generators. Meanwhile, policy implementation may bring inequality of interest distribution to large and small enterprises, different regions, and staff affected by the closing down.

(4) The development of low carbon energy technologies greatly contributed to the achievements of low carbon development in the power industry. It directly impacted the major energy saving indicators such as average gross coal consumption, station service electricity consumption rate, line loss rate, etc. Policies played an important role in facilitating the development and applications of low carbon energy technologies.

(5) Low carbon development in the power industry is the result of joint effort by different stakeholders of the power industry. Taking policy on closing small thermal power generators as example, the smooth and effective implementation of this policy relied on the combination of market instruments and administrative measures in the power industry. Each stakeholder worked closely to implement the policy and ensured its efficient and effective implementation under appropriate time and conditions.

7.3 Policy and Institution Analysis on Low Carbon Development of Renewable Energy

Renewable energy development in China started by 1980s initially as part of the solution for rural electrification and providing power for the vast rural areas in western China. In January 1, 2006, the Renewable Energy Law officially took effect, followed by a series of supporting policies, regulations, and standards. Since 2009, local governments in China issued a series of policies with the intent of promoting the development of new energy sector in their region. Currently, China's legal framework for renewable energy covers different aspects including rules and regulations, finance and subsidies, tax incentives, pricing, planning, R&D, technical standards, industrial support, etc.

The Renewable Energy Law set out five institutions: the target of total capacity, compulsory grid connection, preferential electricity price, cost-sharing, and special fund. The revised version of Renewable Energy Law of the People's Republic of China taking effect in 2010, changing the previous compulsory grid connection system to the system of guaranteeing complete purchase of renewable generation, and updating the "renewable energy

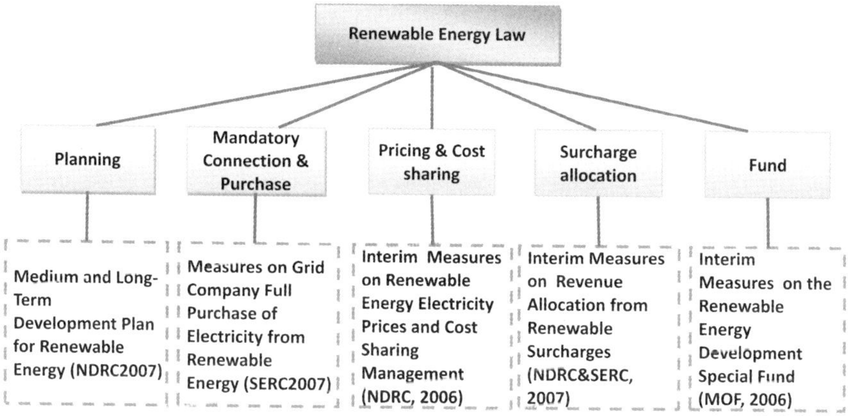

Fig. 7.14. Renewable energy law and supporting policies.

development special fund" to "renewable energy development funds". All these improvement provides good institutional environment for renewable energy development in China.

7.3.1 *Main supporting policies for the renewable energy law*

Figure 7.14. reflects five components and their supporting policies. This section conducts analysis and assessment on implementation of supporting policies for four components. The component on preferential electricity price for renewable generation will be analyzed in Sec. 7.3.3.

7.3.1.1 *Implementation of the Medium and Long-term Development Plan for Renewable Energy*

The *Medium and Long-term Development Plan for Renewable Energy* issued in 2007 specifies guiding principles, major tasks, development objectives, key areas, and guarantee measures for China's renewable energy development until 2020. Current statistics shows that most of 2010 targets in the *Medium and Long-term Development Plan for Renewable Energy* have been achieved or mostly achieved ahead of schedule (Table 7.11).

It will be challenging for China to meet the target of "10% renewable of total energy consumption by 2010" in the plan. In spite of the fast renewable energy development, China's total consumption also grew as fast as 8.8% annually since 2005 along with increasing urbanization and

Table 7.11. Comparison of targets in the plan and actual development.

Program	Planned objectives	Actual development
Development targets	By 2010 renewable energy consumption reaches 10% of total energy consumption	Renewable energy consumption in 2008 accounted for 8.6% of total energy consumption
New technology industrial development	By 2010 basically build local capacity of manufacturing equipment	Already acquired capacity of building wind, nuclear, photovoltaic power equipment locally
Hydro power	By 2010 installed hydropower capacity reaches 190 GW, of which large hydropower is 140 GW and 50 GW is small hydropower	Installed hydro capacity reached at 196 GW by end of 2009, installed small hydro capacity was 55.1 GW
Biomass energy	By 2010 the total installed biomass generation capacity reaches 5.5 GW, annual utilization of biogas reaches 19 billion m^3, ethanol reaches 2 Mt and bio-diesel reaches 0.2 Mt	In 2009 installed biomass capacity reached 2.4 GW, biogas utilization was 15.4 billion m^3, ethanol utilization 1.7 Mt and bio-diesel 0.3 Mt
Wind power	By 2010 the total installed wind capacity reaches at 5 GW	In 2009 the total installed wind power capacity was 25.8 GW
Solar power generation	By 2010 the total installed photovoltaic capacity reaches 300 MW	In 2009 total installed photovoltaic capacity was 300 MW
Solar thermal power utilization	By 2010 total area with solar water heaters reaches 150 million m^3	In 2009 total area with solar water heaters reached 145 million m^3

Source: Q. Wang, *References of China's sustainable energy programs-2009 energy data* (Energy Foundation, Beijing, 2009).

industrialization, as a result the share of renewables in the total energy consumption did not experience the increase as planned. With the lack of fuels, biomass generation came across barriers. There is limited data on overall resource of biomass energy, or on biomass generation, biogas, application of stalk particles, stalk return to soil, etc. The target of installed biomass power generation capacity in the plan was not visionary.

7.3.1.2 *Implementation of the measures on grid company full purchase of electricity from renewable energy*

The Measures on Grid Company Full Purchase of Electricity from Renewable Energy issued in 2007 aims at promoting grid-connection of renewable generation and regulating grid companie's purchase of renewable generation. With implementation of this policy, grid companie's purchase of renewable generation was growing yearly (Table 7.12). Hydro connection increased stably, while wind and solar connection increased rapidly. The amount of generation purchased from hydro and wind was equivalent to 80.0 Mtce in 2000, and it reached 136.7 Mtce in 2005; The amount of generation purchased from hydro, wind, and solar was equivalent to 192.7 Mtce in 2008 (biomass generation is not included because some biomass generation needs to co-fire with coals, and there is no data on this).

The government does not subsidize grid connection projects built by grid companies. But it does subsidize grid connection projects built by renewable generation companies proportionally to its generation amount with the fund from renewable surcharge in electricity rate: 0.01 yuan/kWh for lines shorter than 50 km, and 0.02 yuan/kWh for 50–100 km lines and 0.03 yuan/kWh for lines longer than 100 km. There were only five subsidized

Table 7.12. 2000–2009 Renewable generation purchased by grid companies (Unit: TWh).

Program	2000	2005	2006	2007	2008	2009
Hydro	222	397	436	485	585	—
Wind	0.3	2	3	6	13	28
Biomass	0	21	22	27	31	22
Solar	0	0	0	0	0.2	3
Total	223	419	461	518	630	—

Note: (1) 2000–2008 hydropower data from China Energy Statistical Yearbook 2009; (2) 2000 wind power generation data from Q. Wang, *References on China Sustainable Energy Program-2009 energy data* (Energy Foundation, Beijing, 2009); 2005–2009 wind power generation data from China Electricity Council; (3) 2005–2008 biomass power generation data from Beijing Oriental Environment Institute *et al.*, *Effective Implementation of the Renewable Energy Law and Industrial Development Study* (Energy Foundation, Beijing, 2009); 2009 data from Beijing Oriental Environment Institute *et al.*, *Follow-up Assessment of Practices of the Renewable Energy Law and Relevant Environment Policy Study* (Energy Foundation, Beijing, 2010); (4) photovoltaic power generation 2008 data from Q. Wang, *References on China Sustainable Energy Program-2009 Energy Data* (Energy Foundation, Beijing, 2009); 2009 data from Beijing Oriental Environment Institute *et al.*, *Follow-up Assessment of Practices of the Renewable Energy Law and Relevant Environment Policy Study* (Energy Foundation, Beijing, 2010).

Table 7.13. 2006–2009 Renewable generation grid-connection projects and subsidies.

Time	Number of projects subsidized	Additional project number	Amount of subsidies (Million yuan)
2006	5	—	1.2
2007.01–2007.09	35	30	15.1
2007.10–2008.06	82	47	48.5
2008.07–2008.12	134	52	57.5
2009.01–2008.06	189	55	115.6
2009.07–2009.12	281	92	153.9
Total	—	—	391.8

Note: Data sorted out by 2006–2009 renewals-generating electricity price subsidy and cap and trade plan announced by the NDRC.

project like this in 2006 with total subsidy amount of RMB1.156 million yuan, but the number increased to 281 projects with 270 million Yuan subsidies in 2009 (Table 7.13).

By end of 2009, State Grid invested nearly 40 billion Yuan for wind integration projects and brought a total of over 16 GW wind capacity onto grids (Zhang, 2010), yet there were still some wind farms un-connected yet. According to Chinese Wind Energy Association, by end of 2009 China's total installed wind capacity was 25.8 GW, of which 17.6 GW was connected to the grid according to China Electricity Council. This suggests that except wind generators emanate to be off-grid, there are 20%–30% of total wind projects yet to be connected to the grid.

7.3.1.3 *Implementation of interim measures on revenue allocation from renewable surcharges*

Implementation of Interim Measures on Revenue Allocation from Renewable Surcharges specified source of funding, collection, settlement, and distribution of subsidies for renewable generation subsidy. According to the methods, the incremental cost (renewable generation cost minus regular generation cost-local desulphurization coal generation benchmark price) shall be recovered through a surcharge in electricity rate collected for electricity sales by provincial and national grid. Such surcharge was set to be 0.001 yuan/kWh in 2006, 0.002 yuan/kWh in 2008, and 0.004 yuan/kWh in 2009. In 2008, the national renewable energy surcharge revenue was 3.088 billion yuan, of which 3.077 billion yuan were paid as subsidies of renewable generation (State Electricity Regulatory Commission, 2008). In 2009, the revenue was 4.236 billion Yuan and subsidy paid was 4.558 billion yuan (State Electricity Regulatory Commission, 2009).

Currently, there are a number of issues in implementing the above-mentioned two methods, for example, public fund levied by government for developing renewables is treated as corporate revenues, surcharge collection and distribution does not match with current fiscal and tax systems; there have been repetitively levying value-added tax and income tax on renewable energy surcharge which shrinks the subsidy revenue of generation companies; surcharge distribution is too complicated and the cycle of distribution could be as long as over six months, in addition, surcharge management is not in line with the current electricity price approval system (Beijing Oriental Environment Institute *et al.*, 2010)

7.3.1.4 *Implementation of interim measures on the renewable energy development special fund*

The *Interim Measures on the Renewable Energy Development Special Fund* promulgated in May 2006 set out the priorities and rules for application and approval, financial management, review, and monitoring of special fund. It required special fund to prioritize the support for development and utilization of renewables for oil alternatives, building heat supply and cool ing, power generation etc. But the methods did not specify the amount of fund or standards for project application procedures or fund using. Later on, the government issued "management measures on special fund" for specific technologies which initially formed a special fund management system for renewable energy development. These management methods include the *Interim Methods of Special Fund for Renewable Energy Application in Buildings*, the *Interim Methods of Bio-energy and Non-food Bio-chemicals Prize Fund Management*, the *Management Methods of Rural Biogas Project Construction Fund*, the *Interim Management Methods of Subsidized Fund of Bio-energy and Bio-chemical Raw Material Base*, and the *Interim Management Methods of Special Fund of Industrialization of Wind Power Generation Equipment* etc.

Now, the main issue on special fund system is unclear fund source and amount, uncertain scope of fund using, lack of transparency on budgets and fund allocation, and lack of coordination between special fund and other fund. For example, in 2008 the central government had special fund budget of 8.5 billion Yuan for renewable energy development, of which actual allocation was 3.9 billion yuan (part of fund were not in pace due to some new policies issued) (Beijing Oriental Environment Institute, 2010).

The revised *Renewable Energy Law* clarified to build Renewable Energy Development Fund through merging "electricity surcharge", "special fund", and other fund. Renewable Energy Development Fund will be used to pay incremental cost of renewable generation, grid-connection cost, scientific researches on development and application of renewables, standards development and demonstration projects etc. It aims at adjusting the way of fund management under the condition of unchanged fund channel and developing a government-led renewable energy development fund to improve fund using efficiency.

7.3.2 *Policy on financial subsidy for renewables*

Financial subsidy for renewables in China is divided as central government subsidy and local government subsidy. Subsidized projects cover rural small hydropower construction, rural grid renovation, rural biogas project construction, bio-energy and bio-chemical base construction, biomass energy utilization, industrialization of wind power equipment and solar power generation projects etc. At national level, policies relating to financial subsidy will be introduced in the following section.

7.3.2.1 *Policy on industrialization and R&D of renewable energy equipment*

The *Interim Management Measures of Special Fund of Industrialization of Wind Power Generation Equipment* issued in 2008 was the first central government special fund intended to support industrialization of renewable energy equipment with the "awarding instead of subsidizing" approach. According to this policy, central government provides award of 600 yuan/kW for the first 50 wind turbine equipment produced to qualified wind turbine manufacturing enterprises, of which half are whole turbine manufacturing enterprises and half are key components manufacturing enterprises. The fund for industrialization must be spent exclusively for R&D of new products of wind power equipment. Qualification for such award include — possessing independent IP rights and brands, unit capacity of wind power generators of 1.5 MW or above, turbines are produced, installed and commissioned domestically and are in operation over 240 hours, wind power generators have to be certified, etc. According to this subsidy standard, every whole turbine or component manufacturing enterprises with unit production capacity of 1.5 MW and above, their top 50 products will get a total of 45 million Yuan subsidy from the government finance, which each whole turbine manufacturing enterprises gets 22.50 million Yuan (Beijing Oriental Environment Institute *et al.*, 2009). The subsidy covers about 10% of production cost of wind power equipment (UCI, 2010).

During the 11th five year period, the Advanced Technology R&D Plan (National 863 Plan) set four topics-"hydrogen and fuel cell technology", "highly-efficient energy saving and distributed energy supply technology", "clean coal technology", and "renewable energy technology". The focus of renewable energy technology is to research and develop wind, solar, ocean, and geothermal energy technology and increase the percentage of renewables in the energy mix. In addition, the project application guideline 2008 also included topics like new material technology and advanced technology production. Between 2006 and 2008, the 863 Plan cumulatively supported 138 research projects relating to renewable energy with a total of 200 million yuan.

7.3.2.2 *Subsidy for off-grid renewable generation project*

China has become the country with largest coverage of independent renewable energy system in the world which provides remote areas access to power supply. By the end of 2009, installed capacity of off-grid photovoltaic generation was about 95.6 MW, taking up 32% of all installed capacity (China Semiconductor Industry Association, 2010). China's subsidy for off-grid renewable energy pays both for investment and for electricity price, the latter allocating renewable energy surcharge revenue to subsidize the operation and maintenance of the power supply system.

Starting from 2000, China began to provide electricity to households in remote areas through off-grid renewables. In this period, financial subsidy is mainly for the investment. For example, Inner Mongolia Lighting Program Pioneering Project started to subsidize each household 3,000 Yuan with provincial public revenue (Ma, 2005). The "Electricity to Rural Areas" project began in 2001 was totally funded by central and local governments and the total investment reached 4.7 billion Yuan. International cooperative programs often provide grants that are matched with supporting fund by Chinese government (Table 7.14). These projects did not only provide power

Table 7.14. Off-grid renewable energy power supply projects.

Project name	Implementation time	Installed capacity	Funder	Project investment
Pilot Program of "Brightness" Projects(1)	2000	210.5 kW solar PV and 155.4 kW wind	NDRC, provincial government	40 million yuan
Ali Lighting and Electricity Project, Tibet(1)	2002	220 kW solar PV, 165 kW solar PV generation station and 30 kW water pump, 16 kW solar satellite ground receiving stations, 431 kW totally	Tibet provincial government	60 million yuan
Township Electrification(2)	2001–2003	Built 268 small hydropower stations, 721 PV and wind-PV supplementary power station, all together 17 MW	NDRC provincial government	4.7 billion yuan
China-Germany "Western Solar Power" project(3)	2001–2010	Installed 1300.5 kW solar power stations in villages	German government and Chinese government	Total investment 364 million yuan, German grant 26 million Euros and China's matching fund 104 million yuan

(Continued)

Table 7.14. (*Continued*)

Project name	Implementation time	Installed capacity	Funder	Project investment
Silk Road Brightness Project[4]	2002–2006	Installed 1,970 kW of user system totally	Dutch government, Xinjiang provincial government	Dutch grant 13.79 million Euros, China's matching fund 10 million Euros.

Source: (1) China renewable energy development project office, China photovoltaic industrial development research report 2004, (2004); (2) J. Li, S. Wang and M. Zhang, China photovoltaic development report 2007, (China Environmental Science Press, Beijing, 2007); (3) S. Ma, Make contributions in the frontline of renewables generation consultation services (2008), http://www.fchot.com.cn/Website/index.php?NewsID=6121 (accessed 12 September 2010); (4) H. Wu, "Silk road lighting project" lit up houses of 30,000 rural and nomadic people in Xinjiang (2005), http//news.sohu.com/20050617/n225982724.shtml (accessed 12 September 2010).

supply for residents in remote areas, but also contributed to local economic development and social security.

China also subsidized electricity price to encourage the development of off-grid renewable energy supply system. The government pays the incremental operation and maintenance cost from the surcharge revenue. From 2006 to date, NDRC has totally issued six renewable energy electricity price subsidy and quota trade plans. The plans issued in October 2007 unified subsidy standard for the off-grid renewable energy power supply system as 4,000 yuan/kW/year for projects built after January 2006 and projects built in Qinghai, Xinjiang, Sichuan, and Inner Mongolia under "Electricity Delivery to Rural Areas" program. By 2009, the government subsidized off-grid renewable energy system with installed capacity of 15 MW, the subsidy totaling 54.105 million yuan.

7.3.2.3 *Golden sun program*

On 22 July 2009, the Ministry of Finance, Ministry of Science and Technology, and State Energy Administration jointly promulgated the *Notice on Implementation of Pilot Golden Sun Programs*, planning to support a minimum of 500 MW photovoltaic generation demonstration projects with subsidy within two or three years, and in principle each province gets a quota of 20 MW for this subsidy.

(1) Scope of support

The Golden Sun program subsidizes investment for photovoltaic generation projects, including to-grid photovoltaic generation for user sides like large mining enterprises, pubic institutions, demonstration projects in remote areas with shortage of electricity, and large to-grid demonstration projects in areas with rich solar resource. Implementation of demonstration projects of solar photovoltaic application in buildings is subject to the *Tentative Management Methods of Subsidy for Solar Power Photovoltaic Application in Buildings*; projects already get subsidies from solar photovoltaic buildings and national renewable energy price-sharing shall not be included in order to avoid cross-subsidizing.

(2) Conditions

The contractor has a minimum of 100 million yuan assets. Project equity fund equals 30% of total investment in minimum. Unit project installed capacity is 300 kW or above. Key equipment (photovoltaic components, controllers, inverters, and batteries) must be certified by state-approved certification authorities, and to-grid projects have to meet relevant technical standards and requirement for grid connection.

(3) Subsidy standards

Subsidy for to-grid photovoltaic projects will cover 50% of total investment on photovoltaic generation system and supporting transmission and distribution systems, or 70% of total investment for independent photovoltaic generation projects in remote areas with shortage of electricity. For photovoltaic power generation key technological industrialization and infrastructure projects, subsidies or discounted interests will be provided.

(4) To-grid electricity price

The principle for photovoltaic power generation projects is to produce electricity for its own use. In case there are surplus electricity produced and sent to public grids, the grid company shall purchase such electricity with full amount at the benchmark electricity price for local desulphurization coal-fired generators approved by the central government.

The *Notice on Implementation of Pilot Golden Sun Program* issued on November 9, 2009 enlisted 275 projects with a total of 640 MW in the Golden Sun demonstration program. Initial estimate of total program investment was nearly 20 billion Yuan and it was planned to be paid within 2 to 3 years. Around 222 projects were for large industrial and commercial users with installed capacity of 298 MW in total, and in principle electricity produced would be used for its own production; 18 projects were independent photovoltaic generation in remote areas with shortage of electricity with installed capacity of 46 MW, providing electricity supply for over 300,000 households; 35 projects were large to-grid power generation projects with installed capacity of 306 MW, and all electricity produced would be sent to grid. The Notice also set out higher technical standards for demonstration projects and a price cap for key equipment, and prohibited local government from requiring project developers to only purchase products produced locally.

7.3.3 *Subsidized electricity price for renewable generation*

Policy on feed-in tariff (FIT) for renewable generation is a key factor that influences investment motivation of developers and shapes the market development. To promote renewable energy development, China provides subsidies for electricity from renewable generation (excluding hydropower). The "Management Guidelines for Renewable Generation To-grid Price and Cost Sharing" enacted in 2006 set out reference feed-in tariff of wind generation after public tendering (higher than the benchmark to-grid electricity price of local desulphurization coal-fired generation). The policy specified feed-in tariff for biomass as benchmark price of local desulphurization coal-fired generation plus a subsidy of 0.25 Yuan/kWh, and such subsidy is paid for the first 15 years of its operation. The government also set feed-in tariff for solar generation, ocean power, and geothermal power generation, and the FIT is set by the pricing administration of the State Council under the principle of reasonable costs plus reasonable profits.

7.3.3.1 *Feed-in tariff for wind generation*

Since 2003, China has conducted five public tendering for onshore wind power projects and gradually identified the FIT of wind generation. At end of July 2009, NDRC issued the *Notice on the Improvement of Policy on Feed-in tariff of Wind Power*, identifying the regional fixed feed-in tariff based on their different resources endowment. The policy divides the whole country into four regions according to wind power resources and project construction conditions, and set the benchmark FIT accordingly as 0.51 yuan/kWh, 0.54 yuan/kWh, 0.58 yuan/kWh, and 0.61 yuan/kWh. Wind power projects approved before August 1, 2009 will continue to apply its original FIT.

The FIT policy set uniform standard for wind power development so investors could better estimate their return on investment. Such policy regulated and shaped wind development market and set China on track of mature and large scale development of wind power.

China's first offshore wind power demonstration project — Shanghai Dongdaqiao wind farm project, was provided after tax FIT of 0.987 Yuan/kWh. On January 22, 2010, State Energy Administration and State Ocean Administration jointly issued "Tentative Management Methods of Offshore Wind Power Development", regulating offshore wind power planning, market access, project approval, sea occupancy for construction use, environmental protection, project construction, completion check and approval, operation information management etc. On May 18, 2010, State Energy Administration initiated the first round of official public tendering for franchised offshore wind power projects, where Jiangsu province tendered four projects with a total of 100 MW capacity.

Comparing to EU countries, China's onshore wind power FIT [0.055–0.065 Euro/kWh] was lower. It is slightly lower than that of Germany [0.05–0.09 Euro/kWh] and Denmark [0.078 Euro/kWh], the two countries with large amount of wind power. The FIT of Shanghai Dongdaqiao wind power project [0.105 Euro/kWh] was a bit higher than that of Denmark [0.078 Euro/kWh] and Spain [0.073 Euro/kWh] and a bit lower than that of Germany [0.13 ~ 0.15 Euro/kWh].[5]

7.3.3.2 *Feed-in tariff for solar generation*

China is tendering for grid-connected solar power generation project and is not ready for setting a reference FIT yet for solar generation. Due to the current high costs of solar generation, the government tendered phase I and phase II projects respectively in 2009 and 2010 and intended to help reduce costs of solar power generation by tendering.

In March 2009, China's first solar generation project- Dunhuang 10 MW grid-connected solar generation project in Gansu province, was finally approved with the FIT of 1.09 yuan/kWh (tax included). In April 2010, the FIT of four solar generation projects was temporarily set as 1.15 Yuan/kWh. On June 23, 2010, NDRC and State Energy Administration issued the announcement on the second round of solar generation project tendering. 13 projects totaling 280 MW in Shanxi, Qinghai, Gansu, Inner Mongolia, Ningxia, and Xinjiang provinces bidded. All FIT of tendered projects was lower than 1 Yuan/kWh (Table 7.15).

China's bidding FIT for solar power generation projects are 1.09 yuan/kWh and 1.15 yuan/kWh (0.117 Euro/kWh and 0.123 Euro/kWh), much lower than that of European countries, for instance the 0.3194 Euro/kWh in Germany (Europe's Energy Portal, 2010) or the 0.36–0.43 Euro/kWh in Italy. Though China has large photovoltaic production base and production cost would be lower, it can be concluded that tendering would result in prices lower than average level.

Some local governments also issued FIT for solar generation. Jiangsu province is the strongest supporter for solar generation and set FIT for ground, roof, and BIPV. The feed-in tariffs are also declining year by year to encourage reduction in production costs (Table 7.15). Such local policies greatly promoted the development of local photovoltaic industry and the solar market.

[5]The exchange rate was 1 Euro = 9.3177 CNY, on October 11, 2010.

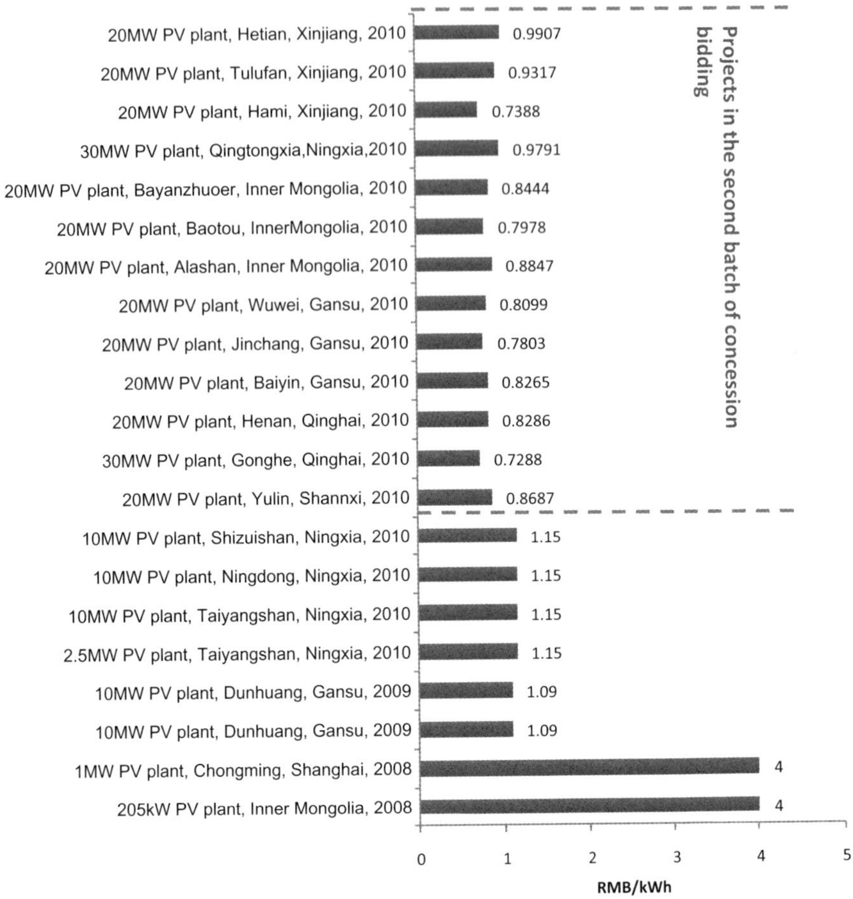

Fig. 7.15. Electricity prices of solar generation projects approved by NDRC 2008–2010. *Source*: Solarzoom,(2010). http://www.spvchina.com/news.asp?news_id=3247 (accessed on 15 August 2010).

As fossil energy stock is limited, renewables will be important alternative energies in future. However in order to achieve large-scale application of renewables, costs reduction and technology maturity will be necessary. The subsidized FIT for renewable generation improves investor's expectations for cost and profit and attracts large amount of investment, while increased investment resulted in tremendous improvement of equipment manufacturing and the expansion of installed capacity.

Table 7.15. Feed-in tariff of solar generation issued by local governments [Unit: yuan/kWh].

Province	Year	Ground	Roof	BIPV
Jiangsu	2009	2.15	3.7	4.3
	2010	1.7	3	3.5
	2011	1.4	2.4	2.9
Shandong	2010	1.7	—	—
	2011	1.4	—	—
	2012	1.2	—	—

Note: Shandong's policy for solar power roof and BIPV stations funded by central or provincial government is encourage using their generation to meet their own loads, and for excess electricity to be sent to grid, it gets the same rate as national coal-fired benchmark electricity price; for full independently funded solar power roof and BIPV stations, they are encouraged to apply for subsidies from different sources.
Source: *Suggestions on Pushing Forward the Solar Generation in Jiangsu Province*; *Suggestions on Acceleration of Development of Photovoltaic Industry in Shandong Province.*

7.3.4 *Institutional analysis of renewable energy development*

Renewable energy development has a different institutional environment than the conventional energy. Due to high costs, the development of renewable generation was very slow until the Renewable Energy Law set up a system beneficial for renewable energy development. The government committed tremendous efforts in promoting renewable energy development to accelerate their commercialization and scaling-up. The government's role is to develop policies and legislation that could help form institutional environment, create and regulate market for renewables. The Renewable Energy Law institutionalized five important components — the renewable portfolio standard, compulsory grid connection, categorized FIT, cost-sharing, and special fund. In the revised Renewable Energy Law, "compulsory grid connection system" was changed to "guaranteed full amount purchase of renewable generation", and the renewable energy development special fund was integrated as "renewable energy development fund".

7.3.4.1 *Renewable portfolio standard*

Renewable portfolio standard (RPS) refers to compulsory regulations a country or a region developed that require certain share of its energy mix to be renewables (Ren *et al.*, 2009). RPS regulations usually specify overall amount, time schedule, appropriate technologies, cost calculation, and regional distribution etc.

Article 4 of the Renewable Energy Law stated:

> "*Development of renewables is the top priority of energy development. The government encourages development of renewable energy through passing standards and regulations that specifies renewable portfolio and action plans. All kinds of economic entities are encouraged to participate development and application of renewables and the government shall protect legal rights of developers and users of renewable energy*".

To set up the renewable portfolio standard, China developed its *Mid and Long-term Renewable Energy Development Plan* and *the* 11th *five year Renewable Energy Development Plan*. The former specified overall goals and specific development targets of renewable energy development by 2020; and the latter set out the overall goals and main development indicators of renewable energy development during the 11th five year period.

From the perspective of market development, an overall target is actually the government's plan on current or mid and long-term market capacity, while renewable portfolio standard legalizes the plan and ensures achieving the targeted market capacity. Setting up renewable portfolio standard is helpful for investors to evaluate the development speed, size, market potentials, profitability of renewable energy, and the overall social, economic, and ecological environment for developing renewable energy, which guarantees the long-term development and sustainability of renewable energy market.

7.3.4.2 *Compulsory grid connection*

Compulsory grid connection refers to that it is mandatory for grid companies to purchase generation from qualified renewable energy plants at the FIT approved by the government. The amount of purchase is determined by market and the system needs to be supported with pricing policies and cost-sharing mechanism. Compulsory grid connection policies aim at encouraging investment for renewable energy development and promote commercialization and industrialization of renewable energy development.

Renewable generation is generally intermittent. Considering security, technological and economic factors, grid companies are usually reluctant to integrate renewable energy into their grids. With current technologies and economic conditions, except for small hydropower, very few renewable energy products could compete with conventional energy products, and grid integration is usually a challenge. As a result, with transmission and distribution network being monopolized, it is necessary to implement compulsory grid connection.

Article 13 of the Renewable Energy Law stated "the government encourages and supports to-grid renewable generation"; and Article 14 set out "grid enterprises shall sign integration agreements with renewable power generation enterprises that have legally obtained administrative license or for which filing has been made, and purchase all of the to-grid generation in the coverage of their grid network, and provide grid-connection services for renewable power plants". To regulate grid companie's purchase of renewable generation, the *Supervision Methods of Grid companies Purchasing Renewable generation* specified roles and responsibilities of generation and gird enterprises and of regulators.

In the revised Renewable Energy Law enacted in 2010, "compulsory grid connection system" was changed to "guaranteed full amount purchase of renewable generation" and strengthened restrictions for grid and generation enterprises. The construction of renewables generation will be put into the national plan of renewable energy development to ensure grid integration and funding, synchronizing the development of generation farms and of the grids to ensure the purchase could happen. On the other hand, it also requires renewable energy development to comply with planning, and for plants built outside the plan, grid companies could refuse to purchase their generation. In addition, to-grid renewable generation shall meet technical requirements for grid integration without hindering the grid security.

With implementation of the compulsory grid connection policy, the amount of renewable generation purchased by grid companies increased rapidly. The total amount of purchase was 417 TWh in 2005 and 630 TWh in 2008, an increase by 51% over three years.

7.3.4.3 *Categorized feed-in tariff*

Categorized FIT policy refers to set FIT different renewable energy technologies (small hydropower, wind power, solar power, and biomass etc.) according to their average costs and announce the tariff to the public. This policy aims at simplifying project approval procedures, increasing transparency for return on investment, reducing costs and restricting vicious competition for renewable energy development projects, to ensure different projects in different areas, times and resource conditions getting similar level of return on investment.

Article 19 of the Renewable Energy Law set out "FIT of renewable generation shall be set by the pricing administration of the State Council considering different technologies and different regional conditions with the principle of encouraging renewable energy development and enabling reasonable economic returns, and the tariff shall be adjusted in response to development of renewable energy technologies". The *Tentative Management Methods of Price of Renewable Generation and Cost-sharing* issued in 2006 was the specific translation of the categorized FIT system. The method set out the pricing way and level of on-grid electricity according to different features and economy of renewable energy technologies.

The key of commercialized development of renewables is power generation technology, which is mainly impacted by FIT. At present, renewable generation costs are higher than conventional generation and cannot economically compete with the latter. Considering generation costs vary for different types of renewables, it makes sense to develop categorized FIT system that can facilitate the development of different renewable energy technologies and expand renewable generation market quickly.

Categorized FIT system is the result of long time exploration and practice of wind power development in China. This system replaced the previous project-based FIT system, and provided similar return on investment for different kinds of renewable energy development, ensuring their balanced development. This created market coordination for renewable energy development, ensuring the joint implementation of government control and market competition.

7.3.4.4 *Cost-sharing*

The core of cost-sharing system is to require consumers of different regions to share in a relatively reasonable way the incremental cost for developing renewables. The system can effectively address equality of cost sharing between regions and between enterprises so as to boost the scaling-up of renewable energy development (Ren *et al.*, 2009).

Article 20 of the Renewable Energy Law sets out "the incremental cost incurred for grid companies when they are purchasing renewable generation as mentioned in Article 19 compared to their purchase of conventional generation shall be recovered through the electricity sales. Price authorities of the State Council shall prepare specific methods for such cost-sharing." The *Tentative Management Methods of Feed-in Tariff of Renewable Generation and Cost-sharing* set clear regulations on the cost-sharing and payment transfer methodologies and specified "the incremental cost between renewable and conventional shall be shared by provincial and national electricity sales." The *Tentative Method for Distribution of Renewable Energy Surcharge* set out the distribution of the balance between the renewable energy surcharge levied by each provincial grid companies and expenses paid to generation companies for their renewable generation.

The cost-sharing system successfully addressed incremental cost for renewable generation. Consumers of different regions paid for such cost in a balanced way, reflecting the equality principle. Subsidy paid for renewable generation from renewable energy surcharge was 260 million yuan in 2006, 3,077 billion yuan in 2008 and 4.558 billion yuan in 2009. This is the evidence that cost-sharing system boosted the scaling-up of renewable energy development.

7.3.4.5 *Special fund system*

Though the cost sharing system addresses incremental cost for renewable generation, there are still other economic bottleneck obstacles to be address. The Renewable Energy Law required the central and local governments to set up special fund for renewables, which are used for subsidy and other forms of financial support for renewable energy projects. The special fund for renewables comes from renewable energy surcharge levied by grid companies and government grants.

The Renewable Energy Law clearly set out

> *"the central government shall set up the special fund for renewable energy development, which could support the following activities: scientific and technological researches, standards development, and demonstration projects; renewable energy projects for household power supply for rural and nomadic people; independent renewable generation system in remote areas and islands; source mapping and exploration, assessment and relevant information system building; localization of renewable energy development equipment manufacturing."*

The Revised Renewable Energy Law enacted in 2010 elaborated on the special fund. It adjusted fund management without changing fund channels, and developed a uniform government-led "renewable energy development fund" to improve efficiency of fund using.

The five components of Renewable Energy Law have successfully created a good policy and institutional environment for wind power development in China by improving the market and the government leverages.

7.3.5 *Case Study: Analysis on institutions for wind power*

New wind capacity grew by 136% per year between 2005 and 2009 and reached 25.8 GW in 2009, ranking second in the world. The success of China's wind power industry couldn't be achieved without policy and institutional support.

China's total energy consumption has been growing by 9.5% per year since 2000. Data released by IEA in 2010 indicated that China had surpassed the U.S. and become the top energy consuming country in the world. In the 2009 Copenhagen Conference, China made the commitment to reduce its CO_2 emissions per unit of GDP by 40% to 45% by 2020. The huge energy demand and the pressure for emission reduction urged the government to take bold actions to develop new energies and renewables.

From the perspective of domestic environment, the central government faced double pressures of increasing energy supply and reducing carbon intensity as well as developing its economy and improving people's life standards. On the other hand, China is strong with macro-control and resource distribution through the central government. This explains the necessity and possibility for China to pursue bold policies for renewable energy development. Local government faced double pressure of economic development and emission reduction, particularly the vast majority areas of Western China, which are not only rich with wind power resources but also are underdeveloped areas with strong urge to pursue economic development. As a result local government was even more motivated to encourage investment to develop brand-new industries. From the perspective of enterprises investing in renewables generation, wind power generation has relatively mature technology and production costs is decreasing while the FIT provides predictable return on investment. Therefore there is a good prospect for developing wind power.

From the perspective of institutional environment, there were six factors that stimulated the fast development of wind power in China.

7.3.5.1 *Renewable Energy Law created good policy environment for wind power development*

The Renewable Energy Law enacted in 2006 shaped five fundamental systems — renewable portfolio standard, compulsory grid connection, categorized FIT, cost-sharing, and special funding, all together creating good institutional environment for renewable energy development in China. The renewable portfolio standard identified the overall goal of wind power development and its market shares; the compulsory grid-connection system guaranteed grid integration of wind power; categorized FIT clarified the FIT of wind power and the return on investment of wind power development enterprises; the cost-sharing system required incremental cost of wind power to be covered by end users of all grids; the special fund supports wind power development and its sustainability issue. These five systems created good policy and institutional environment for wind power development and helped forming the wind power market in China, laying foundation for fast development of wind power.

7.3.5.2 *Feed-in tariff reduced the risks for wind power development*

Before 2008, price for wind generation was set in three ways: bidding price, fixed price, and approved price. The uncertainty of electricity price increased risks for investors. For example, there were risks in the process of approval, and the risks of price getting too low due to severe competition in tendering. With the rapid development of wind power, the bidding and approval way of pricing could no longer meet the demands for wind development market and government control. When wind power was scaling-up throughout China, the shift to the benchmark electricity price mechanism became imperative.

At the end of July 2009, NDRC issued the *Notice on Improving Policy of the Feed-in tariff of Wind Power*, specifying the resource-specific benchmark FITs. That was a sign that China has developed its market leverage mechanism in this area. The policy also set the first standardized pricing mechanism for renewable generation. The benchmark FIT enabled investors to better assess their returns and encouraged the prioritization of developing good quality resources. At the same time it motivated enterprises to pursue management and technological innovations to reduce their cost and increase their profit. The benchmark price is also helpful for regulating price of wind generation and improving the efficiency of government approval.

7.3.5.3 *Compulsory market share targets motivated large energy enterprises to vigorously pursue wind power*

The *Mid and Long-term Plan of Renewable Energy Development* set out compulsory market share target for power generation from non-hydropower renewables: By 2010 and 2020, power generation from nonhydropower renewables in large grid-covered areas should account for 1% and 3%

respectively in total electricity production; for energy companies with total equity installed capacity 5 GW and above, the share of equity installed capacity of nonhydropower renewables should account for 3% and 8% respectively of its total equity installed capacity. According to Energy Research Institute of NDRC, there were 25 energy companies fitting into this policy, all of them being state-owned energy enterprises and only half of them being able to achieve the target of 3% by 2010. Of "the big five" power generation groups, except China Power Investment Corporation, all the rest of them have achieved the target of 3% one year ahead of schedule (Li *et al.*, 2010).

The targets specified in the *Mid and Long-term Plan of Renewable Energy Development* were not decomposed and distributed for each province and the Plan did not evolve into a binding policy. However, the compulsory market share targets embedded some indication for the quota system. Though China has not yet officially issued the policy on quota system, the government has been working on it for a long time. It once considered to incorporate the renewable energy quota system in the Renewable Energy Law; in 2009 the Proposal on Quota Standards for Power Generation from Renewables drafted by Energy Research Institute of NDRC suggested that the big five power generation groups and the two grid companies should be assigned with proportional renewable quota (Guotai Jun'an Securities Research Institute, 2009). In early response to the anticipated quota system policy, some large energy groups started working on incorporating renewable generation from nonhydropower sources. China has rich wind power resources, while wind turbine was the most mature renewables generation technology and ready to be commercialized in large scale, therefore it becomes the first choice of large energy enterprises among renewables.

"China Wind Power Development Report 2010" pointed out that in 2009 wind power field became the hot spot for energy investors. Of all installed capacity in 2009, energy investment enterprises accounted for up to 90%, while central-government owned ones invested over 80%, the "Big Five" generation groups over 50%, and the total of all other state-owned investors, foreign and private enterprises accounted for less than 10%. In 2009, the overall wind power development of the "Big Five" generation groups accelerated — their newly installed capacity and cumulative installed capacity accounted for 54.8% and 54.4% respectively of national new and total installed capacity.

Wind power projects for the first few years require large amount of capital and has long pay-back period and little return on investment.

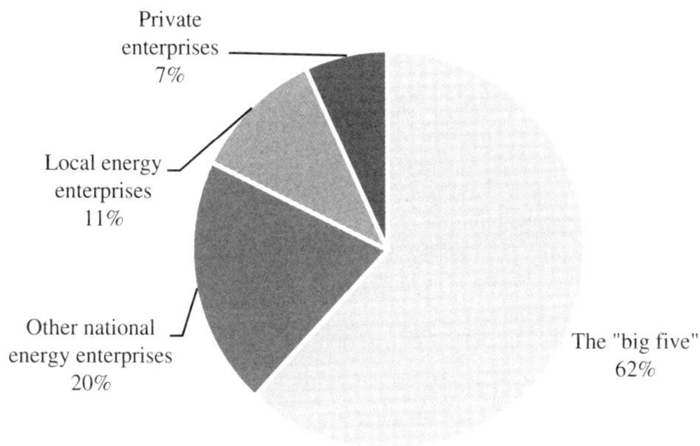

Fig. 7.16. Shares of newly added installed wind power capacity over 100 MW in 2009 (total: 12,157.32 MW).
Source: J. Li, P. Shi and H. Gao, China Wind Power Development Report 2010 (Hainan Press, Haikou, 2010).

Difficulties and risks became barriers for small and medium private enterprises. But for large state-owned energy enterprises, which have adequate capital and better access to bank loans, have incomparable advantages in terms of developing wind power programs. Compulsory market share targets resulted in the convergence of wind projects in large energy enterprises.

7.3.5.4 *The program approval system lacks disciplinary mechanism and resulted in the over-heated development of wind power*

"The Regulation on Management of Renewable Generation" clearly stated "wind power programs above 50 MW shall be approved by NDRC, while the rest are subjected to the approval of provincial governments and approval list shall be submitted to NDRC for record". This approval system stimulated an abnormal "50 MW phenomenon", when investors dismantled large projects into several small ones or invest only small ones less than 50 MW in order to qualify for provincial approval rather than national approval of NDRC, since the latter is more risky and much slower.

The installed capacity of the 72 to-grid wind power projects of China in 2007 was all under 50 MW each (China New Energy and Renewable Energy Website, 2008), which is to say, these projects only need approval by local governments. Between July and December 2009, China subsidized

268 onshore wind power projects with total installed capacity of 14.2 GW, of which 39 projects were of 50 MW or above with total installed capacity of about 4.5 GW and 229 projects below 50 MW with total installed capacity of about 9.7 GW. Both the number and the installed capacity of wind power projects approved by local governments were much more than that approved by central government.

The two-level approval system is good for shortening approval time and improving administrative efficiency. However in practice, it also resulted in lack of coordination with national targets and even violation of approval rules. The undisciplined approval practice resulted in the gaps between central planning and local development. The approved installed capacity of wind power projects exceeded by far the overall national target. Wind power development went out of control. The fast turbine installation and the slow progress of grid development and other supporting projects resulted in various problems such as obstacles in grid connection, challenges in load following, uncertainty of power sales, and difficulty in operational management of wind plants, etc.

According to the report on wind power development by the investigation team of State Electricity Regulatory Commission, the impact of large-scale wind integration on power quality and system security was becoming evident. As wind generation has variability and intermittency, its integration would impact the voltage, frequency, and stability of the system both theoretically and in practice. By end of 2008, wind accounted for 8.26% in Inner Mongolia and 7.8% in Jilin province. With rapid growth of wind integration and generation, there would be increasing impact on the safe operation of the grid.

7.3.5.5 *Local government's urge for economic growth also stimulated wind power development*

Wind power development is an important channel for local government to increase investment, develop industry and increase tax revenue. Wind power investment is capital-intensive and could effectively stimulate local economy. Take Jiuquan 10 GW wind power base in Gansu province as an example, 8.8 GW wind turbines will be installed in two phases. Assuming a capacity cost of 7,000–9,000 Yuan per kW, total investment on wind turbines would reach 60–80 billion yuan. Jiuquan Municipal Government built the industrial park exclusively for manufacturing wind equipment, which accommodated many well-known manufacturers, like Sinovel Wind Co. and

Gold Wind etc. As of 2009, sales revenue of Jiuquan wind power equipment manufacturing industry reached 6.4 billion yuan, accounting for 80% of all sales revenue of equipment manufacturing industry of Gansu province (Zhao, 2010). Tax incomes from wind farms and wind power industry were also one important source for local government revenue. Western and central China is relatively underdeveloped with rich wind resources. Local government there endeavors to pursue economic growth through developing new industries, and such urge has boosted wind power development in those regions. Of the seven large 10 GW wind power bases, five are located at central and western China.

When the central government called upon wind power development local government seized the chance and issued local new energy plans and favorable polices. For example, the *Inner Mongolia Green Energy Development Plan 2009–2015* set out the target until end of 12th five year period (2015) — regional total installed wind power capacity reaching 30 GW. Shandong's *Policies on Facilitating Rapid Development of New Energy sector* set out to accelerate the development of offshore wind power industry and advanced wind power equipment manufacturing industry, and specified FIT for wind power between 2010 and 2012 as in principle 0.7 yuan/kWh. *Jiangsu New Energy Industrial Adjustment and Rejuvenation Plan* set out to make use of its current industrial advantages to boost industrialization of wind power equipment by scaling-up wind farms construction, building wind generation and wind equipment manufacturing bases, and aims at achieving sales revenue of wind equipment of 80 billion yuan by 2011, and building 4 GW complete machine manufacturing capacity. All of these plans reflected local government's urge and determination to pursuing economic growth and developing new industries.

7.3.5.6 *Expected increasing profits and a series of domestic and international favorable policies also drove the wind power development*

In the wake of rapid development of China's wind manufacturing industry, wind power capacity cost kept decreasing and wind power projects became profitable. In addition, installation of renewables help improve the images of energy-intensive and highly-polluting enterprises and better image may translate into economic benefits when the company is listed into market or applying for bank loans. Moreover, wind power development is inevitable for China's new industry development strategies. From the long-term point of view, wind capacity cost will definitely keep falling. If entering into the market until wind power technology fully matures, it would be too late and too hard to seize large market share and access better wind resources. Therefore, as profit margin is expected to increase, energy enterprises vied to jump into competition for seizing wind power resource and market.

CDM is one of the three international cooperation mechanisms in Kyoto protocol. It allows developed countries listed in annex one to invest GHG emission reduction programs, in the nonannex one developing countries to meet their compulsory targets set out in the Protocol. At the same time, CDM projects will bring profits for hosting countries (developing countries). By June 2010, 481 wind power projects with total capacity of 27.5 GW in China applied for CDM, accounting for 23% of all types of Chinese projects applied for CDM (Junfeng Li *et al.*, 2010). CDM provides a subsidy of about 0.1 yuan/kWh for wind projects, which increases the economic viability of wind projects. In addition, wind power projects also enjoy preferential value-added tax and income tax, preferential loan terms from foreign governments and international financial institutions. Such international and domestic favorable conditions effectively reduced risks for wind power development and improved the economy of wind power.

The above-mentioned factors together inevitably stimulated the rapid development of wind power and of wind power industry and supporting industries, setting an example and reference for other renewable energy development; on the other hand, the fast development of wind power brought about challenging issues such as grids integration and power sales etc., which demands facilities construction, development of rules, regulations and policies, giving the impetus for development, and improvement of renewable energy policies and institutions in China.

7.3.6 *Conclusion*

China started developing renewables since 1980s of 20th century to promote electrification in remote rural areas. The *Renewable Energy Law* enacted in January 1, 2006 and the series of supporting policies together stimulated the renewable energy development. Since 2009, local governments were also busy with composing and implementing new energy development plans and supporting policies. At present, China has basically established its renewable energy legal framework which is being improved.

The *Renewable Energy Law* and its supporting policies set up the legal framework for renewable development in China. The *Mid and Long-term Development Plan of Renewables* set out the overall goal and specific targets for renewable energy development. Till date, the targets for 2010 have been completed or nearly completed. The *Supervision Methods of Grid companies Purchasing Renewable Generation* promoted to-grid renewable generation and the amount purchased by grid companies was mounting up with years. The *Tentative Methods of Distribution of Renewables Surcharge* helped achieve distribution and balance of fund by transferring payment among provincial-level grids. However during implementation, there have been issues such as levied fund and its distribution not matching with the current fiscal and tax system etc. The *Tentative Methods of Special Fund for Renewable Energy Development* came across challenges during implementation as instable sources of fund, uncertain scopes of fund using, lack of transparency of budgets and allocation etc. It was replaced already by the "renewable energy development fund".

China's renewable energy subsidy policies have been in place for long time and cover a large variety and amount of projects. At present, the policy shifted its original investment focus and started to provide subsidy to more areas such as electricity pricing, tax, and finance. The subsidy also started to support renewable energy equipment industrialization, technological R&D, off-grid power supply systems, BIPV, to-grid photovoltaic systems, and household systems.

China has been improving its policies on FIT for renewable generation. FIT for wind is categorized according to four regions with different wind resources. Currently, FIT policies for wind effectively standardized the previously unorganized wind generation price systems. FIT for solar generation projects is set by tendering currently. Due to high cost, the industry is not ready for a unified FIT yet. There are some problems on wind power benchmark FIT, for example, it did not specify duration for the tariff to apply, in addition, as domestically manufactured wind equipment quality and life span still have room for improvement, and the price system for recovering future turbine replacing investment is still to be developed.

The five components of Renewable Energy Law have successfully created a good policy and institutional environment for wind power development in China. The renewable portfolio standard ensured market shares of renewables; the compulsory grid-connection (guaranteed complete purchase) guaranteed the survival and development of renewables generation enterprises; categorized FIT system ensured the return on investment of renewable energy investment enterprises; the cost-sharing system provided solutions to recover incremental cost for wind generation relative to conventional fossil generation; the special fund system provided financial support for wind power development and its sustainability.

7.4 Policies on Low Carbon Development of Nuclear Energy

7.4.1 *Nuclear development accelerated and significantly contributed to energy saving and emission reduction*

7.4.1.1 *Increasing nuclear capacity and generation*

China's nuclear power sector experienced accelerating development. In 2008, China's total installed capacity and generation of nuclear power was respectively 9.1 GW and 68.3 TWh, accounting for 1.1% of national total installed capacity and 1.97% of total electricity generation (Table 7.16). Comparing to international peers there is large space for China's nuclear power development In recent years, nuclear power experienced a large increase in absolute terms, effectively easing the tension of power supply

Table 7.16. China's nuclear power installed capacity and generation.

	Installed capacity		Power generation	
Year	Installed capacity (GW)	Year-on-year increase (%)	Generation (TWh)	Percent in total generation (%)
2000	2.1	0	17	1.26
2001	2.1	0	17.5	1.25
2002	4.5	112.86	26.5	1.62
2003	6.2	38.48	43.9	2.31
2004	6.8	10.5	50.1	2.28
2005	6.9	0.15	53.1	2.13
2006	6.9	0	54.3	1.90
2007	8.9	29.2	62.6	1.91
2008	9.1	2.82	68.3	1.97

Source: State Statistics Bureau, *China Statistics Yearbook* (Beijing Statistics Press, Beijing, 2009); M. Cui, *China Energy Development Report 2008* (Social Science Academic Press, Beijing, 2008).

shortage. In the near future, several new nuclear power stations will come into operation. Judging from the current trend, the target of 40 GW by 2020 in the *Mid- and Long-term Nuclear Power Development Plan 2005–2020* will be achieved ahead of schedule. In fact, the Government was also adjusting the original target of 40 GW and increased it to "70 GW completed and 30 GW under construction". That is to say, in the coming 10 years, China will construct GW-level nuclear power stations with the total installed capacity equivalent to 100 of the existing Daya Bay nuclear power station, with total investment exceeding 600 billion Yuan.

7.4.1.2 *Achievements on energy saving and emission reduction were significant*

Nuclear power is generally considered as zero carbon emission, as some argues that energy from nuclear power comes from nuclear fission, i.e., the operation is zero carbon emissions. But using life cycle analysis to consider the complete cycle of energy generation from nuclear fuel production, construction of nuclear power plants and supporting facilities, carbon emissions incurred. The emission factor is about 350 gCO_2/kWh (ECOFYS, 2009). In 2008, carbon emission intensity of thermal power was 887.8 gCO_2/kWh, as a result, comparatively speaking, no matter zero emission or emission of 350 gCO_2/kWh, nuclear power is one of the most effective ways to reduce CO_2 emissions

Table 7.17. Energy saving and emission reduction of nuclear power.

Year	Energy saving and emission reduction with zero emission		Energy saving and emission reduction with emission of 350 gCO2/(kWh) (MtCO2)
	Emission reduction ($MtCO_2$)	Coal saved (Mtce)	
2005	51.00	1.80	32.68
2006	51.14	1.87	31.96
2007	56.89	2.09	34.88
2008	62.23	2.23	38.1

Note: The above numbers are calculated with two assumptions: Zero emissions or emission of 350 $gCO_2/(kWh)$ according to data from the *China Energy Statistics Yearbook* and *China Electric Power Yearbook*.

Box 7.4. Energy saving and emission reduction of Qinshan nuclear power station.

Qinshan nuclear power station has a 300 MW pressurized water reactor independently designed and built by China. It has been in secure operation for 19 years, known as "Honor of the State".

Qinshan nuclear power station has five nuclear power units with total installed capacity of 3 GW. Once Phase II expansion is completed, its total installed capacity will reach 6.3 GW, and it will become the largest nuclear power base of China. In 2009, total generation of Qinshan was around 23.7 TWh, avoiding over 7 million tons of coal equivalent coal burning and over 14.2 Mt of CO_2 emissions. Environmental monitoring indicated that radiation level in the surrounding area remained the same as it was before the station was built. In the near future, the total installed capacity of Qinshan nuclear base will be 6.3 GW, avoiding 19.5 Mt of coal firing and 39 Mt of GHG emissions each year.

Source: Xinhua.net, China's nuclear power meets the time of strategic opportunities (2010), http://news.china.com/zh_cn/news100/11038989/20100315/15854117_1.html (accessed on 15 March 2010).

7.4.2 *Nuclear power development policy: Mid and long-term plan for nuclear power development 2005–2020*

In October 2007, NDRC promulgated the *Mid- and Long-term Plan for Nuclear Power Development 2005–2020* (hereinafter referred as the Plan), which was another guiding document for nuclear power development in China after the *Key Points of Nuclear Power Technology Policies*.[6] It played an important role of leading nuclear power development in the long run at the time when China's nuclear power moved into scaling-up phase and marked that China's nuclear power development entered into new stage.

[6]To implement the directive by the State Council to early develop policies for nuclear power technology development in the future 20 years, the National Planning Commission (previous NDRC) and State Science and Technology Commission held the review meeting of nuclear power development technology policies with over 70 participating units and 108 experts (later it was called Huilongguan Meeting) from January 12 to 18, 1983. Later, two Commissions jointly drafted the *Key Points of Nuclear Power Technology Policies* and submitted to the State Council.

7.4.2.1 *Promulgation of the Plan is an important milestone in the development of China's nuclear power*

The Plan set out to achieve the installed nuclear power capacity of 40 GW by 2020 and increase the percentage of nuclear power in all installed capacity from 2% to 4%. Later, the Plan was readjusted. In April 2009, Guobao Zhang, director of State Energy Administration wrote in the journal *Seeking for Truth* that the State was adjusting the mid and long-term nuclear power development plan, which was intended to strengthen the nuclear power development and planned nuclear power construction in a more scientific way in inlands, and strived to increase the percentage of nuclear power in all installed capacity to over 5%. The identification of development goal directed all the follow-up work around nuclear power development. It is informative and beneficial for fund raising and appropriate deployment of nuclear power programs, standardization of nuclear power design and distribution of design capacity, technology import and innovation, deployment of generation in each link of nuclear fuel circle, and systematic cultivation of professional teams for nuclear power industry.

The Plan arranged initially for nuclear power construction programs by 2020 and raised a series of requirements for project construction, such as economy, standardization, bidding system, and operation indicators etc. It also directed for operation of nuclear power construction projects and nuclear power stations to ensure security of nuclear power and increase economic profits.

The Plan raised new requirements for localization capacity and technological level of nuclear equipment manufacturing industry system. This is good for revitalizing the nuclear power equipment manufacturing industry and helps the Chinese nuclear power walk into the world stage and join in the competition of international nuclear power market.

In line with the *Key Points of Nuclear Power Technology Policies*, the Plan stuck to the technological path of closed nuclear fuel cycle for the nuclear fuel supply insurance system and raised new important principles of utilization of foreign resources and appropriately excessive development of the nuclear fuel industry. This is of importance for the development of nuclear fuel cycle industry and reliable supply of coal fuel and also an important innovation and development guideline for nuclear fuel industry.

The Plan set out again to accelerate the legislation of laws, rules and regulations and to complete as early as possible the *Atomic Energy Law* and its supporting rules. That was what the *Key Points of Nuclear Power*

Technology Policies did not cover. The *Atomic Energy Law* is the parent law of China's nuclear energy development and the establishment of the *Atomic Energy Law* is meaningful for China's nuclear power development.

7.4.2.2 *Supporting policies: The program for standard PWR nuclear power plant system and rules of standard construction of nuclear power*

To implement the national nuclear power development plan, catch up with the trend of rapid development of nuclear power in China and facilitate and support independent development of nuclear power, State Energy Administration officially issued the *Program for Standard PWR Nuclear Power Plant System* and *Rules of Standard Construction of Nuclear Power.* These two documents helped guide the construction of standard nuclear system suitable for China's industrial basis. The *Program for PWR Nuclear Power Plant Standard System* (the Program hereinafter) is the guiding document for standard nuclear power construction. The Program set out overall framework of China's standard nuclear power system, and specifically listed the standard nuclear power programs that China needs to establish lately, including standards in eight areas involved in the whole process of nuclear power construction, which are 680 standards about general use and basis, preliminary preparation of nuclear power, project design, equipment, construction, commissioning, operation, and out of service etc. In the wake of new technological development of nuclear power, particularly independent construction of third generation nuclear power, State Energy Administration will go on supplementing and improving the Program. The *Rules of Standard Construction of Nuclear Power* (the Rules hereinafter) is the fundamental principle on management of standard nuclear power construction. To ensure the smooth standard nuclear power construction, the Rules clarified organizations and their roles and responsibilities in the process. The 'leadership group of standard nuclear power construction' is responsible for organization and leading of construction work and its subordinate "standard nuclear power construction office" is in charge of daily work, directing and coordinating with standard revision work. The office is located in the energy saving and science and technology equipment department of State Energy Administration. There is an "advisory consultation group on standard nuclear power construction" responsible for providing technical consultation services for the leadership group and standard construction. The Rules set out specific regulations for program set-up, drafting, review, submission for approval and publication procedures of nuclear industrial standards.

7.4.2.3 *Implementation*

(1) Nuclear power station construction: Gradually developed into a pattern of multiple bases

From the establishment of Haiyan of Zhejiang province and Daya Bay of Guangdong province since the 80s of 20th century to construction of Tianwan of Jiangsu province, Yanjiang of Guangdong province, including nuclear power stations in Haiyang of Shandong province, Sanmen of Zhejiang province and Hongyan River of Liaoning province under construction and preliminary preparation of inland nuclear power stations in Fujian, Hubei, Hunan, and Jiangxi province etc, all of these helped develop a pattern of multiple bases in the China's nuclear power industry. Meanwhile, number of nuclear power units increased and nuclear power's shares in the electric source structure in China kept increasing. At the end of 2008, there were a total of 11 nuclear power units put into commercial operation and 11 units are under construction (Table 7.18). In 2009, there were eight new nuclear power generators put into operation and the generation of nuclear power stations with above GW-level installed capacity was up to 6.8 GWh and their installed nuclear power capacity reached 8.8 GW (Tables 7.19 and 7.20). On September 20, 2010, Guobao Zhang, director of State Energy Administration expressed in the high-level forum on nuclear power development that up to now, the State Council has approved 34 nuclear power units with installed capacity of 37 GW, of which 25 units are working with installed capacity of 28 GW. China is now the country with largest-scale nuclear power in the world (Xinhua.net, 2010). In particular, No. 1 unit of Ling'ao nuclear power station Phase II project has been smoothly put into use and No. 3 unit in Qinshan Phase II expansion project has successfully connected to grids, ending years of standstill of nuclear power installation.

(2) Nuclear power technology: China has acquired 2nd generation PWR technology and is importing and localizing the 3rd generation technology

Currently, China is actively building the national-level R&D platform and has set up 10 national-level nuclear power R&D centers. It has developed initially the capacity to independently build large PWR units and

Table 7.18. Nuclear power stations completed and under construction in China (as of end of 2008).

Status	Name		Type of reactor	Normal rated power (MWe)	Construction commencement date	Date of commercial operation
In Operation	Qinshan nuclear power plant		PWR	310	February 21, 1985	April 1, 1994
	Daya Bay nuclear power plant in Guangdong province	No. 1 unit	PWR	2 × 984	August 7, 1987	February 1,1994
		No. 2 unit			April 7, 1988	May 6, 1994
	Qinshan No. 2 nuclear power plant	No. 1 unit	PWR	2 × 650	June 2, 1996	April 15, 2002
		No. 2 unit			April 1, 1997	May 3, 2004
	Ling'ao nuclear power plant of Guangdong province	No. 1 unit	PWR	2 × 990	May 15, 1997	May 28, 2002
		No. 2 unit			November 28, 1997	January 8, 2003
	Qinshan No. 3 nuclear power plant	No. 1 unit	HWR	2 × 700	June 8, 1998	December 31, 2002
		No. 2 unit			September 25, 1998	July 24, 2003
	Tianwan nuclear power plant	No. 1 unit	PWR	2 × 1060	October 20, 1999	May 17, 2007
		No. 2 unit	VVER		September 20, 2000	August 16, 2007

(Continued)

Table 7.18. (*Continued*)

Status	Name		Type of reactor	Normal rated power (MWe)	Construction commencement date	Date of commercial operation
Under construction	Guangdong Ling'ao nuclear power plant	No. 3 unit	PWR	2 × 1,080	December 15, 2005	October 2010
		No. 4 unit	2nd generation +		June 15, 2006	2011
	Qinshan No. 2 nuclear power plant	No. 3 unit	PWR	2 × 650	April 28, 2006	December 2010
		No. 4 unit			January 28, 2007	2011
	Hongyan River nuclear power plant	No. 1 unit	PWR	2 × 1,080	August 18, 2007	2011
		No. 2 unit	2nd generation +		March 38, 2008	2013
	Ningde nuclear power plant	No. 1 unit	PWR	2 × 1,080	February 18, 2008	December 2012
		No. 2 unit	2nd generation +		November 3, 2008	2013
		No. 1 unit	2nd generation +	1,080	November 21, 2008	2013

(*Continued*)

Table 7.18. *(Continued)*

Status	Name		Type of reactor	Normal rated power (MWe)	Construction commencement date	Date of commercial operation
		No. 1 unit	2nd generation +	1,080	December 16, 2008	April 2013
		No. 1 unit	2nd generation +	1,080	December 26, 2008	2013
Total	In operation	11	—	9,078	—	—
	Under construction	11	—	11,020	—	—

Note: "2nd generation +" means the second generation improved PWR: CPR-1000.
Source: X. Xiao, S. Gao and W. Han, Advantages, challenges and suggestions for nuclear power development in China, China Energy (3) (2009) 7–9.

Table 7.19. Nuclear power stations with GW capacity in China.

Name of power plant	Region	Composition (number of units × GW)	Equipment capacity at the end of time (GW)	Electricity generation (GWh)	Using hours (hour)
Total	—	—	8.8	6.8	—
Jiangsu nuclear power company ltd	Jiangsu	2 × 1	2.1	1.4	1,092
Ling'ao nuclear power station	Guangdong	2 × 1	2	1.6	7,812
Daya bay nuclear power station	Guangdong	2 × 1	2	1.6	8,320
Qinshan No. 3 nuclear power company ltd	Zhejiang	2 × 0.7	1.4	1.2	8,052
Nuclear power Qinshan joint venture company ltd	Zhejiang	2 × 0.7	1.3	1	7,647

Source: China Electricity Council, *Annual Development Report on China Electrical Power Industry* (China Electric Power Press, Beijing, 2010) p. 250.

other reactors. Through construction of Daya Bay, Ling'ao Phase I and Phase II nuclear power projects, China has developed capacity on the 2nd generation and improved 2nd generation PWR nuclear power technology. China is currently importing and localizing the 3rd generation PWR technology represented by AP1,000 and ERP and working steadily towards its construction of demonstration projects. China's first AP1,000-based nuclear power projects, such as Sanmen Phase I No. 1 and No. 2 units project in Zhejiang province, and Haiyang nuclear power station No. 1 unit in Shandong province, have been commenced in 2009. The Phase I project of the Taishan nuclear power station — the nuclear power station with the world's largest unit capacity — has commenced its construction. The project adopts the European advanced EPR technology in its two nuclear power units with unit capacity of 1.75 GW each. The advanced nuclear power technology R&D, like the high temperature gas-cooled reactor CPR 1,400 etc., is also achieving important outcomes at this phase.

(3) Nuclear fuel cycle: China has established relatively complete supply security system

Table 7.20. Key nuclear power programs newly commenced in 2009.

Name of program	Construction site	Size of construction: Number of units	Size of construction: Capacity (GW)
Liaoning Hongyan River nuclear power Phase I	Dalian, Liaoning province	2	2.2
Sanmen nuclear power Phase I project	Taizhou, Zhejiang province	1	1.3
Qinshan Phase I expansion project	Jiaxing, Zhejiang province	1	1.1
Shandong Haiyang nuclear power	Haiyang, Shandong	2	2.5
Guangdong Taishan nuclear power project	Jiangmen, Shandong province	1	1.8
Yangjiang nuclear power station project	Yanjiang, Guangdong provinec	1	1.1

Source: China Electricity Council, *Annual Development Report on China Electrical Power Industry 2010* (China Electric Power Press, Beijing, 2010) p. 216.

The projects of nuclear fuel cycle industry initiated quickly and production capacity went up fast, particularly China succeeded in R&D, independent production and industrialization of key uranium enrichment equipment, becoming one of a few countries of the world that acquired advanced uranium enrichment technology and put it into production. The advanced nuclear fuel cycle could improve utilization of uranium resources by several times, ensure large-scale and sustainable development of nuclear power, minimize nuclear wastes volume and reduce final nuclear waste to its minimum. Therefore, the central government committed tremendous efforts for researches on advanced nuclear fuel cycle and accelerated research progress on uranium enrichment, advanced fuel components, thermal reactor spent fuel reprocessing, and preparation of fast reactor fuel, fast reactor spent fuel reprocessing. Currently, the fast neutron reactor has reached criticality in the China Institute of Atomic Energy, which is important for nuclear fuel supply security for China.

(4) Nuclear waste processing: Research on ADS has made important progress

After construction of renovation project of Beijing electron-positron collider (BEPC) and several large accelerators, China's accelerator science and engineering technology has progressed by large and reached the world's advanced level in various areas. Chinese scientists are currently researching on accelerator driven sub-critical system, expecting to transfer highly radioactive wastes into materials with much shorter half-life by combusting such wastes.

(5) Capacity building: developed high-level nuclear power technological service teams

After the 11th five year plan identified the nuclear power development goal, China has started building nuclear power talents since 2007. Each year, 1,200 graduates get nuclear degrees and started to work on relevant subjects such as mechanics, dynamics, electronics, civil science, and environment etc. Currently, China has 30 universities that have nuclear power departments, capable to provide sufficient professionals for nuclear power development. It established the comprehensive R&D system, with innovation capacity and capacity of localizing foreign advanced technology. In addition, many enterprises attached high importance to capacity building and started early the education of key talents like controllers and expand channels to get life-long training.

7.4.2.4 *Limitations*

The Plan only set out the overall goal by 2020 but did not specify number of stations to be built each year, where, by whom, with what qualifications etc; the Plan set out "uniform technological path of nuclear power development", but did not provide any definition for the word "uniform"; the Plan set out "stick to the technological path of developing GW-level advanced PWR nuclear power" but did not define "advanced". Nuclear power construction shall comply with the comprehensive principle of safety, economy and sustainable development rather than "being advanced for its own sake"; the Plan required to reduce operational costs on the basis of ensuring safety and reliability and make nuclear power FIT competitive with electricity price of desulphurization coal-fire power plants in the same region, but as a matter of fact, the most important and effective measure to improve the economy of nuclear power is to reduce construction costs, yet the Plan only stressed "reduce operational costs" without mention anything about "reduce construction costs".

In conclusion, the Plan laid out overall deployment, program planning, technology, and favorable policy for China's nuclear power development. It made a series of important decisions and played an important role in implementing the independent development strategy, scientifically schedule and pace the construction, selecting and reserving sites, establishing safe operation technical service system, and ensuring the supporting fuel cycle and technology R&D for China's nuclear power sector.

7.4.3 *Institutions for nuclear development*

China has a 20-year history of nuclear power development since the construction of its first Qinshan nuclear power station in 1985 and made many achievements. But in general, nuclear power development is rather slow, largely because of the national government's uncertain and unclear strategy for nuclear power development. To some extent, institutional hurdles were the key factor that hindered the rapid development of nuclear power.

7.4.3.1 *Institutional factors that influenced China's nuclear power development*

(1) Lack of long-term and stable national-level strategic plan

In the history of nuclear power development, there were different opinions on the role of nuclear power in the national economic development, the safety of nuclear power, nuclear power's influence on ecological system, relationship between military use and civil use of nuclear power and most importantly different opinions on technological path of nuclear power development. The disagreement influenced the decision-making on important issues such as strategy, plan and technological path of nuclear power development. For a long time, China did not formulate the mid and long-term development plan for nuclear power and did not make any specific guidelines for scale deployment, patterning, reactor type, design, construction, management system, staff training, and financing etc.

(2) The management and decision-making system of nuclear power development did not meet the requirements of mass and large-scale nuclear power development.

China's nuclear power management administrations changed several times and organizations involved in management of nuclear power industry includes the follows: Nuclear Power Work Leadership Group of the State Council that is responsible for coordinating nuclear power industrial development; State Atomic Energy Organization that is in charge of peaceful use of atomic energy development plan and making industrial standards etc; NDRC that is responsible for approval of nuclear power project construction, State Energy Administration that is in charge of energy strategic planning and development; the National Nuclear Security Administration of Ministry of Environmental Protection that is responsible for nuclear security, radiation environment, nuclear materials control, nuclear equipment monitoring etc; State Electricity Regulatory Commission that is in charge of guidance and monitoring of each step after power generation; China Electricity Council that is responsible for independent operation of power enterprises and industrial disciplinary regulation etc. Such decentralized, multi-sectoral, and crossed management mechanism obviously did not meet the needs of rapid development of nuclear power industry and undermined the management efficiency. The overlapping, multi-sectoral and low efficient management is a prominent challenge for China's nuclear power development. The nuclear power construction administrations, management institutions, and roles and responsibilities are basically set up according to the State Council's plan and by leaders' directives, but specific regulations are not in place yet.

Particularly, the long-term absence of the Atomic Energy Law, the parent law in the nuclear energy area, resulted in the missing of legal insurance for decision-making procedure in the nuclear power development and failure of guaranteeing sustainability and authority of decisions, ultimately impacted on follow-ups and sound development of nuclear power.

(3) Partition in nuclear power market and incomplete market mechanism made impossible the rapid, large scale, and highly-efficient development of nuclear power

First of all, there was partition between nuclear system and power system. In 1980s, the power industry has been basically formed in China, but nuclear power stations construction was intermittent. During the 10th five year plan period, there was no voice of nuclear power and not any new nuclear power projects built within the whole five years. The construction was limited in its scale. Second, the management and operational mechanism of nuclear fuel industry did not fit for demands of large-scale nuclear power development. China managed its nuclear power industry in the way of military use and the Government gave direct investment and conducted monopolized business. Due to historical reasons, ideas of the planned economy were not totally eliminated yet, and the stagnation of management and operational system reform in the nuclear fuel industry made it hard to fit for the demands of market economy and scale-up of nuclear power.

7.4.3.2 *Nuclear power faces good opportunity in the* 11th *five year period*

(1) Nuclear power was integrated in national energy development plan

During the 11th five year period, NDRC assigned China Power Engineering Consulting Group Corporation (CPECC) to develop the 2020 plan for national nuclear power. By 2020, China's nuclear power installed capacity is expected to reach 40 GW, accounting for 4% of all installed capacity. This topic was not in any five-year plans in the past, and hence was a critical point to ponder on the strategic role of nuclear power in China's energy mix. Nuclear power was facing an unprecedented opportunity.

(2) Shift from multi-reactor and multi-capacity construction to the path of using advanced and uniform technology

China shall import advanced nuclear power technology and build independently-designed GW-level nuclear power stations after localization and technology renovation, including localization and construction of level 1 and 2 nuclear equipment. The nuclear power technology is expected to be secure, economic, suitable, reliable, and relatively advanced to push forward the scaling-up of nuclear power industry of China. The State also offered international tender for 3rd generation nuclear power technology with the intent of unifying the technological path of China's nuclear power development and starting at a higher point. This important decision laid foundation for the professional service system of nuclear power development of China. In nature, the scaled-up development is the key step of reducing construction costs of nuclear power, which would help reduce electricity price and enable nuclear power enter development in a positive cycle. In doing so, public perception on nuclear power would get clear that nuclear is a form of energy, and a safe, clean, and reliable energy form.

(3) Shift from a decentralized and segregated management system for project construction to one leadership system

The leadership group is responsible for developing and approving nuclear power development plan and technological direction and path and implementing institutional reform and composing policies that standardize, monitor, and review nuclear power development. Under this leadership, nuclear power, covered by the uniform industrial framework and scientific management system, will enter a sound development phase supported by holistic policy (such as tax and electricity price), and develop into an independent and complete (from design, construction, manufacturing to operation) system. Mass production and localization will be made possible and construction costs and electricity price of nuclear power station will be further reduced. The nuclear power will become the best alternative energy for fossil and hydro energies. To ensure sustainable development of nuclear power, the sector must better understand it as a form of "electricity generation", and integrate it into national plan for electricity. The key contradiction of developing nuclear power has shifted from people's concerns on safety to reduction in construction cost and electricity price.

7.4.4 *Conclusion*

The development of nuclear power industry is no short term solutions, but long term strategy for China and requires visionary consideration. Given the global shortage of energy supply and pressure for environmental protection, nuclear power, as a clean, safe, and highly-efficient energy, has promising future. Accelerating the development of China's nuclear power industry will not only help address domestic electricity demands, optimize China's energy mix, ensure energy safety, and ease pressure on environment, but will also boost the development of nuclear power globally. Nuclear power is a brand-new industry in China. Given the history and reality of the current nuclear management system, the urgent priority is to build a nuclear power industry that caters for socialist market economy, effectively coordinates between relevant sectors and the industries in a low-cost way, combines nuclear administration practice and market practice of nuclear power, and further strengthens the role of market-led distribution of resources. Therefore, China shall ensure consistency between nuclear power development policy and inputs; attach importance on nuclear security to fully guarantee uranium resource and nuclear fuel supply and improve capacity of nuclear fuel supply; tackle key problems in large-scale advanced PWR technology to achieve breakthrough of core technology; optimize and restructure nuclear power R&D resources to combine resource advantages and HR and put more efforts on breakthrough of the 4th generation nuclear power technology and nuclear fusion technology. The key to success lies in better institutional arrangement.

Appendix

Table 7.21. List of low carbon policies in the power industry.

Name of policy	Type	Announcement date	Announcement authorities	Subject	Main contents
Mid and Long Term Renewable Energy Development Plan (Energy, NDRC, 2007, No. 2174)	Plan	September 4, 2007	NDRC	Renewables	The renewable portfolio standard, renewables account for 10% and 15% respectively in 2010 and 2020.
The 11th five year plan for renewable energy development	Plan	March 18, 2008	NDRC	Renewables	Until 2010, renewables percentage in energy consumption reaches at 10%, increasing by 2.5% than 2005
The Renewable Energy Law of the People's Republic of China (directive of Chairman of the People's Republic of China, No. 33)	Law	February 28, 2005	Chairman of the State	Renewables	Clarified the renewable portfolio standard system, compulsory grid connection system, categorized FIT system, cost-sharing and special fund system
Revised Renewable Energy Law of the People's Republic of China	Law	April 1, 2010	Chairman of the State	Renewables	Government-guaranteed complete purchase system and renewable energy development funding system

(Continued)

Table 7.21. (*Continued*)

Name of policy	Type	Announcement date	Announcement authorities	Subject	Main contents
Guidance Contents of the Development of Renewable Energy sector (Energy, NDRC, 2005, No. 2517)	Notice	November 29, 2005	NDRC	Renewables	Contents of industry
Ministry of Finance printing and issuing the notice on the Tentative Methods of Special Fund for Renewable Energy Development (Building, MoF, 2006, No. 237)	Notice	May 30, 2006	Ministry of Finance	Renewables	Management on special fund for renewable energy development
Management regulations on power generation of renewables (Energy, NDRC, 2006, No. 13)	Notice	January 5, 2006	NDRC	Renewables	Regulated the power generation project management methods, responsibilities of grid companies and power generation enterprises

(*Continued*)

Table 7.21. (*Continued*)

Name of policy	Type	Announcement date	Announcement authorities	Subject	Main contents
Tentative Methods on Renewable Energy-generating Electricity Price and Cost-sharing Management (Pricing, NDRC, 2006, No. 7)	Notice	January 4, 2006	NDRC	Renewables	Pricing electricity, cost payment and sharing
Tentative Methods of Distribution of Renewables Surcharge (Pricing, NDRC, 2007, No. 44)	Notice	January 1, 2007	NDRC	Renewables	Clarified the responsibilities of central grid companies, power generation enterprises and relevant sectors in the distribution process and flows of distribution
Regulatory Methods of Grid companies Complete Purchase of Renewable Generation (State Electricity Council directive, No. 25)	Meeting solution	July 25, 2007	State Electricity Commission	Renewables	Clarified the responsibilities of grid companies complete purchase of Renewable Generation, and responsibilities of grids regulatory enterprises and regulatory measures, and responsibilities of grid companies and power generation enterprises

(*Continued*)

Table 7.21. (*Continued*)

Name of policy	Type	Announcement date	Announcement authorities	Subject	Main contents
Notice on the Adjustment of Electricity Price of Shaanxi Grid by NDRC (Pricing, NDRC, 2006, No. 1263)	Notice	June 30, 2006	NDRC	FIT	Set out that except Tibet and electricity for agricultural use, levy renewable energy surcharge of 0.001 yuan/kWh in electricity sales of users
Notice on Increasing Electricity Price of Northern China Grid by NDRC (Pricing, NDRC, 2008, No. 1677)	Notice	June 29, 2008	NDRC	FIT	Set out that except residential use of electricity and electricity use for fertilizer production, increase surcharge on electricity from renewables to 0.002 yuan/kWh
Notice on the Adjustment of Electricity Price of Southern Power Grid by NDRC (Pricing, NDRC, 2009, No. 2926)	Notice	November 18, 2009	NDRC	FIT	Raise the renewables-generating electricity surcharge standard to 0.004 yuan/kWh

(*Continued*)

Table 7.21. (*Continued*)

Name of policy	Type	Announcement date	Announcement authorities	Subject	Main contents
Implementation Opinion on Accelerating the Use of Building Integrated Renewable Energy by the Ministry of Housing and Urban-Rural Development and Ministry of Finance (MoHURD, 2006, No. 213)	Suggestion	August 25, 2006	The Ministry of Housing and Urban–Rural Development and Ministry of Finance	Building integrated renewable energy	Promoted the application of renewables in buildings
Notice on printing and issuing the Management of Special Fund for Renewables Application in Buildings by the Ministry of Finance and Ministry of Housing and Urban-Rural Development (MoF, MHURD, 2006, No. 460)	Notice	September 4, 2006	The Ministry of Finance and Ministry of Housing and Urban–Rural Development	Special fund for renewables application in buildings	Set out the principle of fund use, priority supporting areas, scope of fund use, application, approval, assessment and review etc, and issues like fund advance and management

(*Continued*)

Table 7.21. (*Continued*)

Name of policy	Type	Announcement date	Announcement authorities	Subject	Main contents
Appraisal Methods of Demonstration Projects of Renewables Application in Buildings (MoF, MHURD, 2006 No. 459)	Notice	September 4, 2006	The Ministry of Finance and Ministry of Housing and Urban–Rural Development	Building integrated renewable energy	Set out the project application, review procedure and methods, specified the scoring standards and requirements for advisory review group
Notice on Strengthening Management of Demonstration of Renewables Application in Buildings (MoF, MHURD, 2007, No. 38)	Notice	February 23, 2007	The Ministry of Finance and Ministry of Housing and Urban–Rural Development	Building integrated renewable energy	Strengthened supervision in the process of demonstration projects, and management on fund subsidized for demonstration projects of renewables application in buildings

(*Continued*)

Table 7.21. (*Continued*)

Name of policy	Type	Announcement date	Announcement authorities	Subject	Main contents
Notice on printing and issuing the Contents of Renewables Technologies Applied in Buildings during the 11th five year Period by the Ministry of Housing and Urban-Rural Development (MHURD, 2007, No. 216)	Notice	August 30, 2007	The Ministry of Housing and Urban–Rural Development	Building integrated renewable energy	Promote the application of renewables in buildings during the 11th five year period, lead the technological development of renewables application in buildings, including ground source heat pump technology, groundwater heat pump technology, solar hot water preparation technology, solar power heating/cooling technology, solar power green lighting technology, integrated building solar power generation technology, integration technology of solar power and other energies, biomass gasification heat and gas supply technology, and waste incineration and waste-to-energy technology

(*Continued*)

Table 7.21. (*Continued*)

Name of policy	Type	Announcement date	Announcement authorities	Subject	Main contents
Implementation Plan for Accelerating the Renewables Application in Buildings of Rural Areas	Notice	July 6, 2009	The Ministry of Finance and Ministry of Housing and Urban–Rural Development	Building integrated renewable energy	Promoted demonstrations of renewables application in buildings of rural areas (county as unit); each year, each province's (autonomous region, municipality) application for demonstration county shall be over four. In 2009, subsidy standard for renewables application in buildings of rural areas was: 60 yuan/m^2 for ground source heat pump technology application, 15 yuan/m^2 for integrated solar heat application, subsidize newly built solar shower and house as 60% of inputs. Maximum subsidy for each demonstration county shall not be over 18 million Yuan.

(*Continued*)

Table 7.21. (*Continued*)

Name of policy	Type	Announcement date	Announcement authorities	Subject	Main contents
Implementation Suggestions on Accelerating Solar PV for Building Application (MoF, MHURD, 2009, No. 128)	Suggestion	March 23, 2009	The Ministry of Finance and Ministry of Housing and Urban–Rural Development	Building integrated renewable energy	Implement the solar roof plan to push the PV building application initiation in domestic market and amplifying demonstration effects for large-scale promotion
Tentative Methods of Management on Financial Subsidized Fund for Solar PV Building Application (MoF, MHURD, 2009, No. 129)	Notice	March 23, 2009	The Ministry of Finance and Ministry of Housing and Urban–Rural Development	Building integrated renewable energy	Set out to provide quota subsidies for the urban IBPV and PV building application in rural and remote areas. In 2009, the standard cap was 2009 yuan/Wp. Clarified scope of fund to be subsidized, technical requirements, priority conditions and application procedure etc.

(*Continued*)

Table 7.21. (*Continued*)

Name of policy	Type	Announcement date	Announcement authorities	Subject	Main contents
Guidelines for Application of Solar PV Building Application Demonstration Projects (2009, No. 34)	Notice	April 16, 2009	The Ministry of Finance	Building integrated renewable energy	Set out types of applying projects, subsidy standards, applicants and application standards etc.
Notice on Application of Solar PV Building Application Demonstration Project in 2010 (2010, No. 29)	Notice	April 12, 2010	The Ministry of Finance and Ministry of Housing and Urban–Rural Development	Building integrated renewable energy	Subsidy standard in principle is set as 17 yuan/W for building materials and components IBPV projects; for roof, walls IBPV projects, it is 13 yuan/W; specific subsidy standards will adjust according to progress of the PV industry in 2010

(*Continued*)

Table 7.21. (*Continued*)

Name of policy	Type	Announcement date	Announcement authorities	Subject	Main contents
Notice on Construction of Large-scale Grids-connected Demonstration PV Stations (Energy, NDRC, 2007, No. 2898)	Notice	November 22, 2007	NDRC	Solar power	Decided to construct the large-scale grids-connected demonstration PV station, requiring: construction size no less than 5 MW; occupy mainly untilled lands like deserts, Gobi deserts and wastelands; easy access to grids and easy to connect onto grids, and consider constructing with large wind farms; through open bidding select PV station investors by FIT and the excess between its price and local average FIT could be shared by renewable energy surcharge levied from national power sales

(*Continued*)

Table 7.21. (*Continued*)

Name of policy	Type	Announcement date	Announcement authorities	Subject	Main contents
Notice on Implementation of Golden Sun Program (Finance, 2009, No. 397)	Notice	July 16, 2009	Ministry of Finance, Ministry of Science and Technology, and State Energy Administration	Solar power	Submitted Golden Sun Program implementation plan of each province before August 31, 2009, and in principle, the total construction size of demonstration project of each province (autonomous region, municipality) shall not exceed 20 MW
Notice on Ensuring Implementation Quality of Golden Sun Program (Finance, 2009, No. 718)	Notice	November 9, 2009	Ministry of Finance, Ministry of Science and Technology, and State Energy Administration	Solar power	Set out specific implementation requirements for demonstration projects and announce the content of the Phase I demonstration projects

(*Continued*)

Table 7.21. (*Continued*)

Name of policy	Type	Announcement date	Announcement authorities	Subject	Main contents
Tentative Methods of Management on Land Use for Wind Farm Project Construction and Environmental Protection	Notice	August 9, 2005	NDRC, Ministry and Land and Resources, and State Administration of Environmental Protection	Wind power	Set out calculation of construction occupancy for wind farm, responsible sectors for approval of construction occupancy and related environmental protection issues
Implementation Suggestions on Facilitating Development of Wind Power Industry (Energy, NDRC, 2006, No. 2535)	Suggestion	November 1, 2006	NDRC and Ministry of Finance	Wind power	Set out six key work and arranged corresponding units responsible for submitting plans, of which wind power resource census has started already

(*Continued*)

Table 7.21. (*Continued*)

Name of policy	Type	Announcement date	Announcement authorities	Subject	Main contents
Notice on printing and issuing the Tentative Methods of Management on Special Fund for Industrialization of Wind Power Equipment by the Ministry of Finance (Energy, NDRC, 2006, No. 2535)	Notice	August 11, 2008	Ministry of Finance	Wind power	Set out supporting subjects, ways, conditions, and application scope of subsidy standards, fund application ,and notification etc.
Notice on Improvement of Wind Power FIT Policy by NDRC (Pricing, NDRC, 2009, No. 1906)	Notice	July 20, 2009	NDRC	Wind power	Classified the country into four kinds of wind resource areas, and benchmark electricity prices are respectively 0.51–0.61 yuan/(kWh)

(*Continued*)

Table 7.21. (*Continued*)

Name of policy	Type	Announcement date	Announcement authorities	Subject	Main contents
State Energy Administration announced the Notice on Tentative Methods of Offshore Wind Power Generation Management (2010, No. 29)	Notice	January 22, 2010	State Energy Administration and State Oceanic Administration	Wind power	Set out procedures and requirements for the offshore wind power development plan, offshore wind power project grant, project approval, sea occupancy for construction use, environmental protection, commencement, completion and reception of construction, and operational information management
Notice on Strengthening Bio-fuel Ethanol Project Construction Management and Promoting Sound Industrial Development by NDRC and Ministry of Finance	Notice	December 14, 2006	NDRC, Ministry and Finance	Bio-liquid fuel	Stressed to standardize the management on bio-fuel ethanol project construction, strict market entry standards and policies, and strict project construction management and approval etc.

(*Continued*)

Table 7.21. (*Continued*)

Name of policy	Type	Announcement date	Announcement authorities	Subject	Main contents
Implementation Suggestions on Development of Bio-energy and Bio-chemistry Financial Supporting Policy by the Ministry of Finance, NDRC, Ministry of Agriculture, State Administration of Taxation, State Forestry Administration (2006, No. 702)	Suggestion	September 30, 2006	Ministry of Finance, NDRC, Ministry of Agriculture, State Administration of Taxation, and State Forestry Administration	Bio-liquid fuel	Set out the principle of supporting policies, contents, and organizations etc. including flexible subsidy for loss, subsidy for raw material base, demonstration subsidy, and taxation preference etc.
Ministry of Finance notice on the printing and issuing the Tentative Methods of Prize Fund Management for Shift to Non-food-based Bio-energy and Bio-chemistry Development (2007, No. 282)	Notice	July 12, 2007	Ministry of Finance	Bio-liquid fuel	Set out the scope that prize fund support, application conditions, way of subsidy, and fund allocation etc.

(*Continued*)

Table 7.21. (*Continued*)

Name of policy	Type	Announcement date	Announcement authorities	Subject	Main contents
Ministry of Finance and Ministry of Agriculture Notice on printing and issuing the Management Methods of Fund for Rural Biogas Project Construction	Notice	September 8, 2007	Ministry of Finance and Ministry of Agriculture	Biogas	Set out the budget management and government procurement, fund management, financial management, and supervision and review etc.
Ministry of Finance notice on printing and issuing the Tentative Methods of Management on Subsidy Fund for Raw Material Base for Biomass and Bio-chemistry (2007, No. 435)	Notice	September 20, 2007	Ministry of Finance	Bio-liquid fuel	Set out compulsory requirements for raw material bases, construction and operational model, and subsidy standards etc.
Notice on printing and issuing the Tentative Methods of Management on Subsidy Fund for Straw-to-Energy Utilization	Notice	October 30, 2008	Ministry of Finance	Biomass Straw-to-energy utilization	Set out to provide subsidies for utilization of, and clarified supporting subjects and requirements

(*Continued*)

Table 7.21. (*Continued*)

Name of policy	Type	Announcement date	Announcement authorities	Subject	Main contents
Notice on the Contents of Preferential Income Tax of Enterprises for Comprehensive Resource Utilization by the Ministry of Finance, State Administration of Taxation and NDRC (2008, No. 117)	Notice	January 1, 2008	Ministry of Finance, State Administration of Taxation, and NDRC	Tax preference	Wood alternative products, electricity, and heat and gas from crops straws and residue (including grain crops straws, agricultural crops straws, crop residue, and corn cob) enjoy tax preference
Notice on Issues Around the Implementation of Contents of Preferential Enterprise Income Tax Public Infrastructures Building Project by the Ministry of Finance, State Administration of Taxation (2008, No. 46)	Notice	September 23, 2008	Ministry of Finance and State Administration of Taxation	Tax preference	Hydropower, nuclear power, new grids construction projects, wind power, solar power, marine power, and geothermal power all enjoy preferential policies of "three tax-free and three half-tax".

(*Continued*)

Table 7.21. (*Continued*)

Name of policy	Type	Announcement date	Announcement authorities	Subject	Main contents
Notice on the Policy of Value-added Tax for Comprehensive Resource Utilization and Other Products by the Ministry of Finance, State Administration of Taxation (2008, No. 156)	Notice	December 9, 2008	Ministry of Finance and State Administration of Taxation	Tax preference	Value-added tax is free for electricity or thermal generated by wastes; for electricity generated from wind power value-added tax will be returned 50%

CHAPTER 8

Manufacturing

8.1 Performance Evaluation of Manufacturing

The development of China's manufacturing has been defined by the trend of its industrialization, urbanization and globalization. At present, China, in general terms, is in the second half of the mid-industrialization stage (CASS, 2010). The characteristics of mid-industrialization stage determine that China's development mainly relies on heavy industry and that the proportion of industrial value added to China's GDP is too high. Meanwhile, rapid urbanization leads to increased demand for energy-intensive materials, such as steel, cement and building materials. Therefore, China has to invest heavily on energy-intensive infrastructure and real estate industries to cope with the demand of massive urbanization. Besides, several developed countries have moved their energy-intensive industries to China along with the global industrial division. China's energy consumption and greenhouse gas emission are determined by all the above mentioned factors.

China's total energy consumption and CO_2 emission has increased rapidly since 2000, but its cumulative CO_2 emission per capita is much lower than the world average level. From 1850 to 2005, China's cumulative emission per capita was 45 tons, about 1/4 of the world average. As a developing country, China's primary task is to realize rapid economic growth and reduce its poverty-stricken population. According to analysis, China's heavy industrialization will continue until 2020 thereby increasing its total energy consumption and greenhouse gas emission rapidly in the short term (Xu and Wang, 2007).

China's industry is divided into light and heavy industry according to their productive character and products usage. Light industry produces consumer goods and hand tools, or can be divided into farm products and industrial products by their raw materials. Heavy industry provides means of production based on materials and technology to various sectors of national economy, including the extractive (logging) industry, raw materials industry and processing industry. From the perspective of energy consumption, some light industries are energy intensive, such as papermaking, synthetic fiber and domestic ceramics production, while some heavy industries are not, such as communications equipment, computers and other electronic equipment manufacturing. The energy consumption of six energy-intensive industries (steel, building materials, oil, chemistry and chemical engineering, nonferrous metal, papermaking and textiles) accounts for 70% of the total industrial consumption, which is the focus of our book.

Energy consumption of the manufacturing was 1.21 billion tce[1] in 2008, accounting for 66% of the final energy consumption and 44% of China's total energy consumption; CO_2 emission of the manufacturing was 3.94 billion ton, about 62% of China's total energy-related CO_2 emission. Coal is China's primary energy source, accounting for 64.9% of the total consumption, while the proportions of oil, natural gas, heat and electric power are 7.6%, 3.4%, 4.1%, and 20% respectively. Total energy consumption and CO_2 emission had increased from 2006 to 2008, but with apparent improvement in energy efficiency. Performance evaluation of the manufacturing sector consists of four parts: Energy performance and energy savings, technology development and structural adjustment, clean production and recycling economy, and investment in energy-saving and emission reduction.

8.1.1 *Energy performance and energy savings*

(1) Energy consumption and energy savings per unit of industrial value added

Energy efficiency of the manufacturing sector could be measured by two indicators: Energy consumption per unit of industrial value added reflecting energy efficiency of the manufacturing, and the energy consumption per unit produced reflecting technology advancement. With respect to energy consumption per unit of industrial value added, energy consumption per RMB 10,000 of industrial value added of the manufacturing reduced from 2.10 tce

[1]Energy consumption of the manufacturing is calculated according to *China Energy Statistics Yearbook 2009*. Deducting energy consumption in production, processing, transfer, transmission and transportation from the total industrial energy consumption results to the energy consumption of the manufacturing.

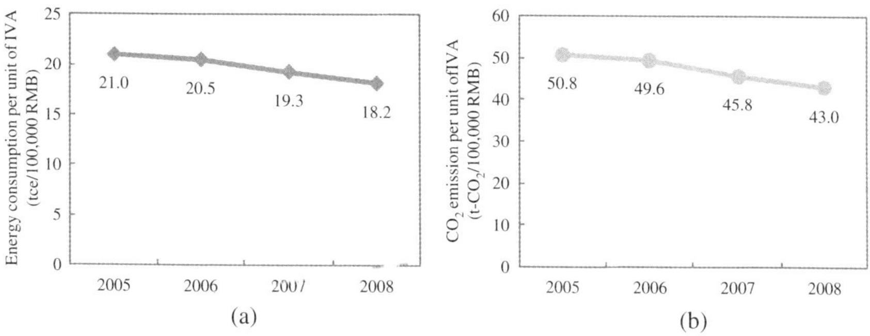

Fig. 8.1. Energy consumption and CO_2 emissions per unit of industrial value added of the manufacturing from 2005 to 2008.
Source: *China Energy Statistics Yearbook 2009* and *China Energy Statistics Yearbook 2010*.

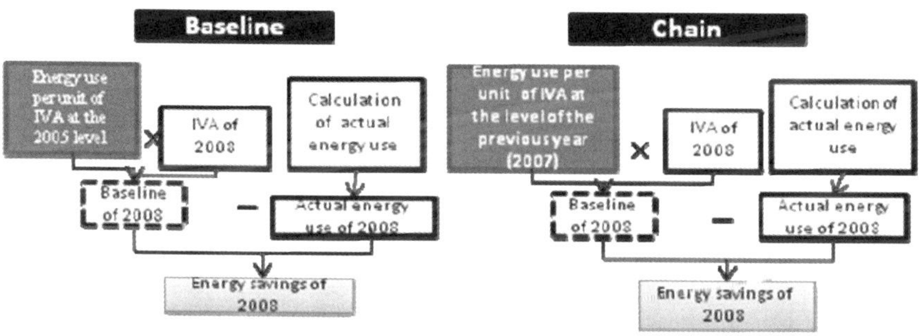

Fig. 8.2. Calculation process of energy savings.

in 2005 to 1.82 tce in 2008, decreasing by 13.3%, with an annual reduction rate of 4.6% [Fig. 8.1(a)]. During the same period, annual reduction level of energy consumption per RMB 10,000 of GDP was 4% annually, lower than that of the manufacturing sector. CO_2 emissions per RMB 10,000 of industrial value added reduced from 5.08t in 2005 to 4.30t in 2008, decreasing by 14.8%, with an annual rate of 5.4% [Fig. 8.1(b)].

Based upon the changing energy consumption of per unit industrial value added, the report shall calculate energy savings of the manufacturing between 2006 and 2008 through constant proportion method (Fig. 8.2). By this method, energy consumption of per unit industrial value added in 2005 is the base point, which is then multiplied by the industrial value added of the current year (2006, 2007, and 2008), and the result is the quantity of energy consumption of the selected year. The result by subtracting the real energy consumption of each selected year is the energy savings of the year. Cumulative energy savings between 2006 and 2008 is the addition of the energy savings of 2006, 2007, and 2008. The constant proportion method reflects the energy savings of the *11th Five Year Plan* with and without the compulsory target of decreasing 20% energy consumption per unit GDP, which is suitable to the report. Meanwhile, the chain method reflects the improvement of energy efficiency year on year instead of the energy savings without the compulsory target. Figure 8.2 is the calculation process of constant proportion method and chain method.

According to the constant proportion method,[2] energy-savings of the manufacturing sector in 2006, 2007 and 2008 were 34 Mtce, 140 Mtce and 248 Mtce respectively, and the cumulative energy savings was 421 Mtce [Fig. 8.3(a)]. Energy savings of the manufacturing sector from 2006 to 2008 amounted to 58.7%[3] of the total. From 2006 to 2008, CO_2 emission reduction of the manufacturing sector was 80 Mt, 398 Mt and 695 Mt respectively, and the cumulative amount was 1.17 billion ton [Fig. 8.3(b)].

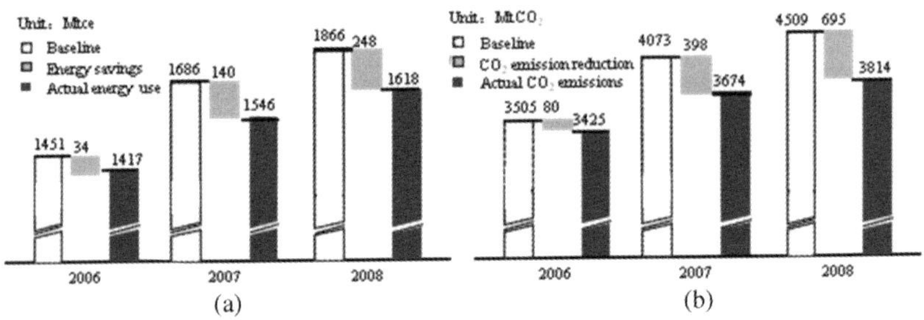

Fig. 8.3. Energy savings and CO_2 emissions reduction of the manufacturing from 2006 to 2008.
Source: *China Energy Statistics Yearbook 2009* and *China Energy Statistics Yearbook 2010*.

[2]According to the link relative method, energy savings of the manufacturing in 2006, 2007, and 2008 was 34 Mtce, 100 Mtce, and 93 Mtce, and the cumulative amount was 227 Mtce. Reduction of CO_2 emissions was 80 Mt, 305 Mt, and 254 Mt in 2006, 2007, and 2008, and the cumulative amount was 639 Mt.

[3]According to the constant proportion method, energy savings of the entire country in 2006, 2007 and 2008 were 72 Mtce, 231 Mtce, and 414 Mtce, and the cumulative amount was 717 Mtce; reduction of CO_2 emission of the entire country in 2006–2008 was 147 Mt, 544 Mt, and 1.06 billion ton, and the cumulative amount was 1.76 billion ton. According to the link relative method, energy savings of the entire country in 2006–2008 was 72 Mtce, 148 Mtce, and 161 Mtce, and the cumulative amount was 381 Mtce; reduction of CO_2 emission of the entire country in 2006–2008 was 140 Mt, 416 Mt, and 427 Mt, and the cumulative amount was 984 Mt.

CO_2 emission reduction of the manufacturing sector from 2006 to 2008 made up 65.7% of the total amount.

(2) Technological and structural energy savings

Energy savings of the manufacturing could be divided into technological energy savings and structural energy savings. The former consists of the technological energy savings by improving energy efficiency within the industry, and the power contribution due to improved energy efficiency of the energy sector. According to the analysis, structural transformation of the manufacturing sector contributed in a moderately positive way to energy savings in 2006, while the contribution was negative in 2007 and 2008 (Table 8.1).

(3) Energy consumption per unit of energy-intensive products has decreased. From 2006 to 2008, cumulative energy savings of the manufacturing sector was 218 Mtce, and the cumulative CO_2 emissions reduction was 523 Mt. In 2009, energy savings of the products was 169 Mtce, and cumulative CO_2 emissions reduction was 411 Mt.

Reduction of energy consumption per unit produced reflects technological improvement. From 2006 to 2009, energy consumption per unit of 13 energy-intensive products (crude steel, electrolytic aluminum, copper, cement, plate glass, crude oil, ethylene, sodium hydroxide, sodium carbonate, calcium carbide, synthetic ammonia, paper and cardboard, chemical fiber) of six energy-intensive industries had decreased, narrowing the gap with the world advanced level (Table 8.2). According to the constant proportion method,[4] by improving the energy efficiency of 13

Table 8.1. Technological and structural energy savings of the manufacturing sector (unit: 10,000 tce).

Item	2006	2007	2008
Technological energy savings	3,286	16,663	25,009
Technological energy savings in the manufacturing sector	3,128	14,648	20,988
Power contribution	158	2,015	4,021
Structural energy savings	92	−2,691	−225
Energy savings of the manufacturing sector	3,379	13,972	24,783

Source: *China Energy Statistics Yearbook 2009* and *China Energy Statistics Yearbook 2010*.

[4]Energy savings of products: Taking energy consumption per unit produced in 2005 as the datum quantity, and energy savings of current year = (energy consumption per unit produced of current year-energy consumption per unit produced in 2005) × production in current year.

Table 8.2. Production of energy-intensive products and energy consumption of per unit produced.

Category	2005	2006[5]	2007	2008	2009	World advanced level[6]
Steel						
Crude steel (Mt)	353.2	419.2	489.3	503.1	572.2	—
Comprehensive energy consumption of the industry (kgce/t)	1,019	998	976	975	973	—
Comprehensive energy consumption of large and medium-sized enterprises[7] (kgce/t)	760	745	733	729	717	—
Power consumption of large and medium-sized enterprises (kWh/t)	452		433	445	434	
Comparable energy consumption of large and medium-sized enterprises (kgce/t)	732	729	718	709	697	10
Electrolytic Aluminum						
Production (Mt)	7.79	9.27	12.34	13.17	12.89	—
AC power consumption[8] (kWh/t)	14,680	14,671	14,488	14,323	14,171	14,100

(*Continued*)

[5]In 2006, China's coal consumption for power generation method changed into electro-thermal equivalent method. In order to compare with previous energy consumption per unit produced and the world advanced levels (coal consumption for power generation method is adopted by both Japan and EU), power consumption of the total energy consumption in 2007 was calculated by the coal consumption for power generation method [about 350 gce/(kWh)] in *China's Energy Balance Sheet*; and the energy consumption per unit produced for some products in 2008 and 2009 were calculated according to the reduction rate of previous year.

[6]World advanced level refers to the average rate of the world-leading countries and areas.

[7]China's comparable energy consumption for steel is calculated according to the data of large and medium-sized enterprises. China had 70 large and medium-sized steel enterprises in 2008, whose production amounted to 83% of the total production. In 2008, the comprehensive energy consumption of the large and medium-sized steel enterprises calculated according to the coal consumption for power generation method and electro-thermal equivalent method were 730 kgce/t and 630 kgce/t respectively, and the power consumption per ton was 445 kWh/t. The world advanced level takes Japan's value, which is calculated according to its crude steel production and energy consumption of the steel industry in 2007.

[8]AC power consumption of electrolytic aluminum of Canada's Alam Company is 12,900 kWh/t.

Table 8.2. (*Continued*)

Category	2005	2006	2007	2008	2009	World advanced level
Copper						
Production (Mt)	2.61	2.92	3.44	3.79	4.14	—
Comprehensive energy consumption (kgce/t)	780	729	610	549	509	500
DC power consumption (kWh/t)	—	—	310	304	309	—
Cement						
Production (Mt)	1,068.9	1,236.8	1,361.2	1,423.6	1,644.0	—
Comprehensive energy consumption[9] (kgce/t)	167	161	158	151	139	118
Power consumption (kWh/t)	92	—	91	91	90	—
Plate Glass						
Production (million weight case)	402.1	465.8	497.5	513.3	579	—
Comprehensive energy consumption (kgce/weight case)	22	19	17	16.6	16.5	15
Power consumption (kWh/t)	—	—	7.6	7.4	6.9	—
Oil Refinery						
Processed crude oil (Mt)	286.2	306.5	326.8	342.0	374.6	—
Comprehensive energy consumption[10] (kgce/t)	114	112	110	108	106	73
Power consumption (kWh/t)	—	—	60.0	60.3	60.2	—
Ethylene						

(*Continued*)

[9]Comprehensive power consumption of cement is the average value of all cement enterprises except that of building materials. In 2007, China's energy consumption of cement clinkers was 126 kgce/t, and the comprehensive power consumption was 91 kWh/t, while that of the world advanced levels were 103 kgce/t and 82 kWh/t respectively.

[10]China's power consumption for crude oil processing was 60 kWh/t in 2007. By comparing with the power efficiency of crude oil processing worldwide, energy factor consumption method is more close to reality. The energy factor reflects the depth of crude oil processing, which is calculated according to the energy consumption per unit and the configuration of processing equipment in the refinery plant. In 2005, China's energy factor consumption was 13 kgce/t, while that of the world advanced level was 10 kgce/t.

Table 8.2. (*Continued*)

Category	2005	2006	2007	2008	2009	World advanced level
Production (Mt)	7.56	9.41	10.48	9.98	10.73	—
Comprehensive energy consumption[11] (kgce/t)	1,073	1,013	1,026	1,010	976	629
Power consumption (kWh/t)	—	—	124.5	126.4	126.1	—
Sodium Ammonia						
Production (Mt)	52.20	55.93	57.87	49.95	51.34	—
Comprehensive energy consumption[12] (kgce/t)	1,700	1,700	1,702	1,661	1,591	—
Power consumption (kWh/t)	1,280	—	1,202	1,184	1,172	—
Sodium Hydroxide						
Production (Mt)	12.4	15.1	17.59	19.26	18.32	—
Comprehensive energy consumption (kgce/t)						
Diaphragm method	1,448	—	1,272	1,351	1,293	1,250
Ion-film method	1,067	—	937	922	916	910
Weighted mean value[13]	1,297	1,248	1,146	1,137	1,087	—
Power consumption (kWh/g)						
Diaphragm method	—	—	2,364	2,340	2,309	—
Ion-film method	—	—	2,213	2,201	2,186	—
Sodium Carbonate						
Production (Mt)	14.21	16	17.72	18.81	19.38	—
Comprehensive energy consumption (kgce/t)	396	402	398	355	323	310

(*Continued*)

[11]China's ethylene production mainly uses naphtha as raw materials. The world advanced level takes the value of Middle East, which mainly uses ethane as raw materials.

[12]Comprehensive energy consumption of synthetic ammonia takes the average value of the equipment of large and medium-sized enterprises using coal, oil, and gas as raw materials.

[13]Comprehensive energy consumption of sodium hydroxide is the weighted mean value of diaphragm method and ion-film method. According to the coal consumption for power generation method, China's comprehensive energy consumption of diaphragm method and ion-film method in 2007 was 1,272 kgce/t and 937 kgce/t, and was 736 kgce/t and 434 kgce/t according to the electro-thermal method, and the power consumption was 2,364 kWh/t and 2,213 kWh/t respectively.

Table 8.2. (*Continued*)

Category	2005	2006	2007	2008	2009	World advanced level
Calcium Carbide						
Production (Mt)	8.95	11.7	13.63	13.61	15.03	—
Power consumption (kWh/t)	3,450	3,666	3,465	3,440	3,395	3,000
Paper and Cardboard						
Production (Mt)	62.05	68.63	77.92	84.04	89.65	—
Comprehensive energy consumption of the entire industry[14] (kgce/t)	528	502	476	440	395	—
Power consumption of the industry (kWh/t)	745		605	578	567	—
Comprehensive energy consumption of the enterprises with pulp-making devices (kgce/t)	1,380	1,290	1,225	1,153	1,090	610
Chemical Fiber						
Production (Mt)	26.65	20.73	24.14	24.15	27.3	—
Power consumption (kWh/t)	1,396	1,207	1,086	1,094	1,091	980

[14]Of all the pulp for paper making in China in 2008, waste paper stock amounted to 60%, about 55.49 Mt, of which, 24.21 Mt were imported. The imported wood pulp was 9.52 Mt, amounting to 59% of the consumed wood pulp. 1 t wood pulp could produce 1.2 t paper, and the energy consumption for waste paper reutilization is 80% less than making paper by wood pulps.

Source: State Statistics Bureau, *China Statistical Abstract 2010*, *China Statistics Yearbook 2010*, Ministry of Industry and Information Technology; China National Coal Association, China Electrical Council, China Nonferrous Metals Industry Association, China Building Materials Industry Association, China Petroleum and Chemical Energy Conservation Association, China Paper Association, China Chemical Fiber Association, Institute of Energy Economics of Japan, *Handbook of Energy and Economic Statistics of Japan 2010*, Iron and Steel Institute of Japan, Iron and Steel Institute of South Korea, Japan Cement Association, *Journal of Japan Institute of Energy*, IEA, Energy Statistics of OECD Countries, (USA) National Mining Association.

energy-intensive products, the cumulative energy savings and CO_2 emissions reduction of the manufacturing sector were 138 Mtce and 340 Mt respectively between 2006 and 2008. During the same period, energy consumption of the above mentioned products accounted for 63.8%, 66.5%, and 64.2% energy consumption of the manufacturing sector. According to their proportions and supposing that the energy saving rate of other products is the same within 13 products, it could be calculated that cumulative energy savings and CO_2 emissions reduction of the manufacturing sector from 2006 to 2008 were 218 Mtce and 523 Mt respectively. Energy consumption of per unit produced of major energy-intensive products continued to decline in 2009. Energy savings of 13 energy-intensive products was 110 Mtce in 2009, and the overall energy savings of the manufacturing sector was 169 Mtce. CO_2 emissions reduction of 13 energy-intensive products was 267 Mt, while the overall CO_2 emissions reduction of the manufacturing sector was 411 Mt.

8.1.2 *Two pillars of performance: Technology development and structural adjustment*

(1) *Technology development*

Technology development of energy-intensive industries is essential to the achievement of energy saving in the manufacturing sector. During the *11th Five Year Plan*, technology development consists of four aspects: Displacing backward production capacity, promoting advanced technology, producing high-performance equipment and increasing investment in science and technology.

With respect to displacing backward production capacity, the coke industry had displaced 133 Mt production capacity of small coking enterprises with production under 4.3 m^3 by 2009; the cement industry had displaced 240 Mt production capacity of shaft kilns; the iron making industry had displaced 81.7 Mt production capacity of blast furnaces whose production capacity was lower than 300 m^3; the steel-making industry had displaced 59.4 Mt production capacity of small converters and electric furnaces whose production capacity was lower than 0.2 Mt; the electrolytic aluminum industry had displaced 0.8 Mt production capacity of small preroasted tanks; the iron alloy industry had displaced 5.19 Mt production capacity of <6300 kVA submerged arc furnaces; the calcium carbide industry had displaced 2.9 Mt production capacity of <6300 kVA converters.

Between 2006 and 2009, the displaced production capacity of backward technology was 110 Mtce (NDRC, 2010).

In promoting advanced technologies, penetration rate of continuous casting ratio of the steel and iron industry had increased from 82.5% in 2000 to 97.5% in 2005, and 99% in 2009. Through this technology, 200 kgce could be saved in processing 1 t steel billet, with the yield rate increasing by 12%. The penetration rate of coke dry quenching had increased from 6% in 2000 to 35% in 2005, and 70% in 2009. Through this technology, 100,000 tce could be saved in processing 1 Mt red coke. The penetration rate of blast furnace TRT had increased from 50% in 2000 to 81% in 2005, and 100% in 2009, and the generated power of 1t iron could reach 30 kWh. Production by large pre-roasted vat to the overall production in the electrolytic aluminum industry had increased from 52% in 2000 to 90% in 2009. It was observed that using >160 kV large pre-roasted tank could decrease power consumption by 9% compared with self-roasted tank. In the sodium hydroxide industry, production by ion-film method to the overall production had increased from 24.9% in 2000 to 54.7% in 2009, and its power consumption per ton sodium hydroxide was 123 kWh lower than that of diaphragm method. The proportion of NSP cement production to the overall cement production had increased from 12% in 2000 to 72.5% in 2009, and the heat consumption by NSP method was 40% lower than that of shaft kiln method;. The bulk loading rate of cement had increased from 28% in 2000 to 46.3% in 2009, thereby 3.3 million m^3 wood for making paper bags could be saved for 100 Mt bulk-loading cement, saving 4.5% broken paper bags and 2.37 Mtce energy. Proportion of float process production of the plate glass industry to the overall production had increased from 57% in 2000 to 83.4% in 2009, and the comprehensive energy consumption of float process was 16% lower than that of the vertical drawing process. Proportion of new wall materials in the heavy clay industry to the overall production had increased from 28% in 2000 to 52% in 2009, and power consumption of producing new wall materials was 40% lower than that of producing solid clay.

With respect to producing high performance equipment, the proportion of indigenously made manufacturing equipment has increased. Now, China can produce many high performance industrial devices such as 10 Mt complete equipment for large-scale refineries (import substitution rate = 90%), 300,000 t complete equipment for ethylene production (import substitution rate > 80%), complete equipment for large-scale coal chemistry production (import substitution rate = 90%), 4,000 m^3 blast furnace (import substitution rate = 95%), 1,780 mm large-scale wide strip cold rolling production line (import substitution rate = 96%), and 5,000 t/d NSP cement production line(import substitution rate > 95%). Indigenization reduces the operational cost of advanced technologies and facilitates energy-saving and emission reduction in the manufacturing sector.

With respect to science and technology investment in energy technology, in 2008, S&T expenses was RMB 30.13 billion for large and medium-sized enterprises in the iron and steel industry, RMB 8.55 billion for nonferrous metal industry, RMB 4.57 billion for building materials industry, RMB 17.66 billion for chemical industry, RMB 3.2 billion for chemical fiber industry, RMB 6.56 billion for textile and garment industry, RMB 2.56 billion for papermaking and paper products industry. Apart from the energy-intensive industries, low-energy equipment manufacturers also increased their S&T investment. In 2008, the investment was RMB 32.79 billion in the transportation equipment manufacturing, RMB 48.09 billion in communications equipment, PC and other electronic appliances manufacturing, and RMB 32.12 billion in utility equipment and special equipment manufacturing. Heavy investment in S&T is conducive to the introduction, absorption, promotion and renovation of advanced technologies by the manufacturers, and cultivates their capability in energy-saving and emission reduction.

Despite that, it is imperative to realize the gap between China and the world advanced levels with respect to indigenous renovation, R&D and designing. China does not fully possess 18 CO_2 emission reduction core technologies in iron and steel, cement and building materials industry (UNDP and RUC, 2010). New generation coking technology (e.g. SCOP21), Itmk3 iron making technology and CCS technology will need to be developed and tested in the steel and iron industry. China will need to promote: Coke dry quenching technology (China has possessed the capability of developing coke dry quenching technology, as well as the capability of processing some coke dry quenching devices, but has not possessed the ability for designing and producing high-pressure coke dry quenching equipment), waste heat and pressure recovery technology (e.g. sintering waste heat recovery technology, converter coal gas recovery technology, converter low-pressure saturated steam generation technology, coke-oven gas hydrogen making technology), steel production energy management center, coal moisture control technology, low-caloric value coal gas combustion turbine technology, injecting plastic scraps into blast furnace technology, smelting reduction technology, microwave, electric arc, heat release and heating direct steelmaking technology, advanced electric furnace, and thin strip continuous casting technology. In the cement industry, CCS and eco-cement technology (utilizing inflammable waste as alternative fuel) will need to be developed; and the key NSP cement technology (mainly automatic control device and integrated functioning), NSP simple low temperature waste heat power generation technology, and high performance grinding technology (such as advanced vertical grinding technology) will need to be promoted. Developing low-carbon technology is a challenge as well as an opportunity for China. Breakthroughs in low-carbon technology will increase its international competitiveness and reduce its employment pressure. However, it is impossible to bridge the gap with world's advanced levels in a single step. China needs enhancement of its S&T capability as well as international cooperation and mutual support for its programs.

(2) *Structural adjustment*

During the *11th Five Year Plan*, structural transformation in 2006 contributed in a positive moderate way to energy-saving in the manufacturing sector, while that in 2007 and 2008 contributed negatively. However, with progressive energy-saving programs, China's industrial structure, especially its product structure, has improved.

(1) The proportion of the heavy industry to the overall industrial value added remains high, but with a slower pace of growth. During the *11th Five Year Plan*, China's heavy industry played a major role in the national economy, and its proportion to the overall industrial value added was higher than the previous plan periods, but with a slower pace of growth. Figure 8.4 displays the variation trend of industrial value added of China's heavy and light industry between 2000 and 2008. During the *10th Five Year Plan*, the proportion of the heavy industry to the overall industrial value added increased from 62.9% in 2001 to 68.6% in 2005, increasing by 6%. During the first 3 years of the *11th Five Year Plan*, its proportion increased from 69.5% in 2006 to 71.3% in 2008, increasing by about 2%, showing a slower growth rate

(2) Product structural optimization of some manufacturing industries: During the *11th Five Year Plan*, product structure of some manufacturing industries has been optimized, with growing proportion of high value added products and low-energy products.

Taking the steel industry for an example, hot rolled steels include plate, tube and strip materials by their section shapes and usage. The

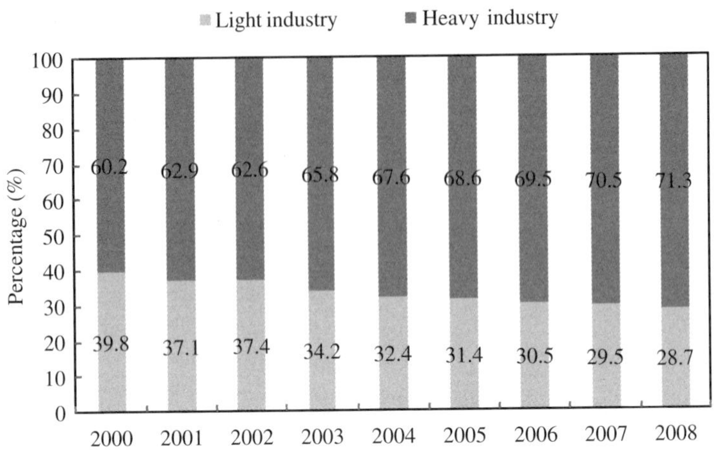

Fig. 8.4. Proportion of heavy and light industry to the industrial value added.
Source: State Bureau of Statistics.

ratio of plate, tube and strip refers to the proportion of high value added and technology-intensive plate, tube and strip materials to the total steel production, which is an important indicator for structural optimization of steel grades. During the *10th Five Year Plan*, domestic steel plate, tube and strip production was inadequate, and imported key steel grades accounted for over 80% of all imported steels. During the *11th Five Year Plan*, the proportion of domestic steel plate, tube and strip increased gradually from 49.4% in 2006 to 53.2% in 2008, increasing by about 3.6% (Fig. 8.5).

In building materials industry, the new wall materials include sintering hollow products (hollow bricks, aerated concretes, concrete blocks) and the sintering products made of industrial waste and silt, whose energy consumption of per unit produced is only about 1/4 of the traditional solid clay bricks. During the *11th Five Year Plan*, production of new wall materials increased rapidly, and its proportion to the overall production reached 56% in 2008, twice as much as that in 2000 (Fig. 8.6).

(3) Larger Enterprises: Enterprise size is an important factor affecting industrial energy performance. In China's energy-intensive industries, medium and small-sized enterprises usually adopt relatively backward production technology and raw material choices, whose energy consumption per unit produced is 30%–60% higher than that of large enterprises. During the *11th Five Year Plan*, enterprise size and equipment scale of some

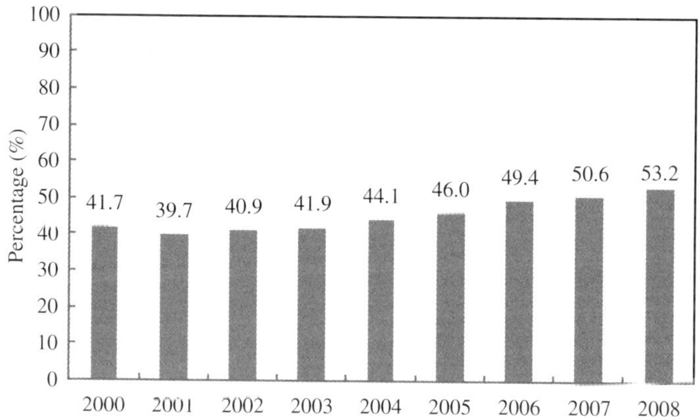

Fig. 8.5. Proportion of steel plate, tube and strip in China's steel industry.
Source: *China Steel Industry Yearbook 2009*.

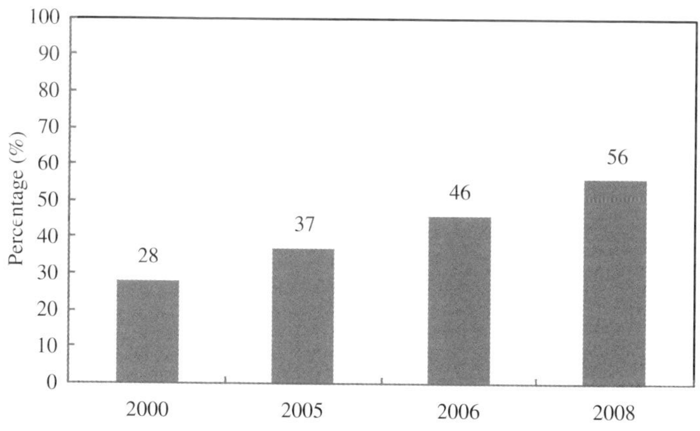

Fig. 8.6. Proportion of new wall materials to the overall production of wall materials.
Source: Qingyi, 2010.

energy-intensive enterprises became larger, contributing to their structural optimization.

In the steel industry, production capacity of the enterprises with an annual production capacity over 10 Mt accounted for 31% of the total production in 2005, and 43% in 2008. In the cement industry, there were more numbers of large NSP cement production lines; here, the production capacity of NSP clinker with over 4,000 t/d production capacity accounted for

Table 8.3. Large-scale enterprises of the steel, cement and synthetic ammonia industry.

Category		Equipment/ production capacity	Proportion to overall production capacity (%)	
			2005	2008
Steel Industry		>10 Mt/a	31	43
		5∼10 Mt/a	13	35
		1∼5 Mt/a	32	15
		<1 Mt/a	24	7
Cement Clinker	Shaft Kiln	—	55	38
	NSP	>8,000 t/d	3	1
		4,000∼8,000 t/d	12	29
		<4,000 t/d	30	32
Synthetic Ammonia		>300,000 t/a	32	39
		80,000∼300,000 t/a	25	39
		<80,000 t/a	43	22

Source: Qingyi, 2010.

Table 8.4. Scale of the float glass enterprises.

Category	2006	2007	2009
Enterprises with an annual production capacity of over 10 million weight cases	13	16	20
Proportion to the overall production capacity (%)	51.6	56	68

Source: Qingyi, 2010.

15% of the total production in 2005, and 30% in 2008. Enterprise size of the synthetic ammonia industry taking up a large proportion of the energy consumption of the chemical industry was also optimized, and production of the medium and small-sized enterprises (annual production <80,000 t/a) accounted for 43% of the total production in 2005, and 22% in 2008. See Table 8.3.

Float glass enterprises also became larger. Float glass enterprises with an annual production of over 10 million weight cases increased from 13 in 2006 to 20 in 2009, and proportion of their production capacity to the overall production increased from 52% in 2006 to 68% in 2009 (Table 8.4).

8.1.3 *Cleaner production and recycling economy*

During the *11th Five Year Plan*, cleaner production of the manufacturing sector has made some progress. For example, Shanghai Baosteel Group built the biggest clean steel plant with world's advanced technology, which adopts the smelting reduction iron making technology and has an annual pig iron production capacity of 1.5 Mt. The major equipments include one CORX iron-making furnace, two 150 t converter furnaces, one 250 mm slab caster, one 4,200 mm two-stand wide-and-heavy plate casting device. Emission of various pollutants is reduced enormously; SO_2 emission is only 4.9% of that of blast furnaces. In the cement industry, China's biggest cement enterprise CONCH group has four of the world's largest NSP cement production lines, each with a daily production capacity of over 10,000 t. Its self-developed cement kiln waste heat generation technology is world leading, which can provide 60% electricity for its own use. Its comprehensive energy consumption is 86.3 kgce/t, 24% lower than the industrial average, reaching world's advanced levels. In the sodium hydroxide industry, Shandong Lubei Chemical Co., Ltd. built a coal gangue power generation and salt-alkaline co-production system, which had become a distinguished project in salt-alkaline thermo-electricity co-production of the ion-film sodium hydroxide industry. Its energy efficiency reached 95.6%, and cost of major products reduced by 30%–50%. In addition, China has built over 60 chemical industrial parks. By sharing the raw materials, intermediate products, by-products and wastes of the enterprises, the park and the surrounding areas, energy and water consumption per RMB 10,000 produced of Shanghai chemical industrial park is 1.2 tce and 33 t respectively, about 1/2 and 1/3 of the average level.

During the *11th Five Year Plan*, recycling economy also made further progress. For example, forest-paper integration combining forest planting, management, cutting, paper pulp and papermaking could promote the development of planted forest (fast-growing and high-yield plantation), reduce wood pulp shortage, and improve the ecological environment. Compared with straw pulp, wood pulp papermaking is more favorable to water and energy conservation as well as collective pollution management. Water consumption by China's traditional papermaking method is 100–200 m^3/t, and COD emission of papermaking industry accounts for 1/3 of all industrial sectors. Forest-paper integration recycling economy system of Jiangsu Jindong Paper Co., Ltd. could produce over 1 Mt art paper annually, and its water consumption per ton paper is only 10.7 m^3, about 18% of the national standard, and its COD emission is no more than 1/6 of the national standard. In addition, refinery-chemical integration features co-production of refinery and chemical production in one enterprise, which perfectly manifests the concept of recycling economy. It characterizes in intensification, short process, safety and environmental-friendliness. All production equipments are connected with pipelines, and getting rid of storage tanks and transport vehicles, with mutual support of raw materials and high energy efficiency. All fuels are de-sulfurized and purified gas fuels, and waste heat is fully utilized, thereby resulting in high utilization rate of raw materials and energy as well as low pollutant emission. China's biggest refinery-chemical integration project was put into production on November 2009 in Quanzhou, whose production capacity is expanded from 4 Mt to 12 Mt, and its annual output of ethylene, PPE, PP and aromatic hydrocarbon is 0.8 Mt, 0.65 Mt, 0.4 Mt and 1 Mt respectively.

8.1.4 *Heavy investment in energy-saving and emission reduction*

During the *11th Five Year Plan*, the government adopted multiple modes of economic investment in energy-saving and emission reduction of the industrial sector, such as financial investment, energy-saving fund, credit incentive, tax incentive, project grant, and developing the energy service industry.

Central government's financial investment consists of financial appropriation, transfer payment and national debt fund. Between 2006 and 2010, the central finance and budgetary altogether invested RMB 211.8 billion in energy-saving and environmental protection projects. In 2006, RMB 450 million national debt funds were utilized to support major energy-saving projects, driving RMB 6.38 billion of local auxiliary fund, bank loan and funds owned by enterprises. In order to displace backward production capacity, resettle laid-off workers, compensate property losses and dispose the land, the central finance allocated RMB 3.19 billion to the underdeveloped areas in 2007.

In addition, 20 provincial governments created the energy-saving special fund of over RMB 1.3 billion yuan, which was used as energy-saving investment allowance, discount loan, energy-saving project subsidy, and for supporting key energy-saving projects, promoting energy-saving products, displacing backward production capacity, and capacity building.

In credit incentives, the China Banking Regulatory Committee, State Environmental Protection Administration and People's Bank of China jointly launched a green credit policy, which exercised credit control to the enterprises and projects violating industrial policies and environmental regulations, and checked unrestricted expansion of energy-intensive industries. In 2007, major banks released RMB 106.33 billion of loan for key energy-saving projects, RMB 3.88 billion of loan for energy-saving and emission reduction technology innovation projects, RMB 20.94 billion of loan for energy-saving and emission reduction technology renovation, and taking back RMB 3.93 billion of loan from the enterprises violating energy-saving and emission reduction policies. In 2009, central finance provided RMB 15 billion of discount loan for the enterprise's technology renovation, driving RMB 400 billion into technology renovation investment.

In tax incentives, since July 1st, 2008, VAT of power and heat generation enterprises using garbage as fuel resources has been refunded partially or completely. 50% VAT is refunded for wind, coal gangue and coal slime power generation, and for producing marsh gas with waste distillers' grains. Garbage-generated power is guaranteed to link to the grid, and has a 0.25 yuan/(kWh) subsidy based on the price of coal power; meanwhile, its VAT is exempted, and the income tax is reduced or exempted. Since January 1st, 2009, VAT has been exempted for the building materials products with over 30% waste residue, rubber power made of waste tires, retreaded tires, and reclaimed water.

In project subsidy, reward subsidy is adopted to support key energy-saving projects. 200–250 yuan/tce (200 yuan/tce for underdeveloped areas and 250 yuan/tce for developed areas) is provided to the industrial projects with an annual energy savings of over 10,000 ton standard coals. On June 6th, 2010, the Ministry of Finance allocated RMB 2 billion to the energy service companies to support their energy-saving projects. For those industrial energy-saving projects with an annual energy savings of 500–10,000 tce, central and local governments will subsidize 240 yuan/tce and 60 yuan/tce respectively.

Huge monetary investment reveals China's commitment toward energy saving and emission reduction. However, it has to be noticed that despite apparent achievement during the *11*th *Five Year Plan*, cost of energy-saving and emission reduction is still high, and the economic efficiency remains low. During the *11*th *Five Year Plan*, the central finance invested RMB 200 billion, which represented only 10%–15% of the overall investment in energy-saving (Xinhua News, 2010a). In the first four years of the *11*th *Five Year Plan*, China's energy savings and CO_2 emissions reduction was 490 Mtce and 1.13 billion ton respectively (Xinhua News, 2010b). It could be estimated that during the *11th Five Year Plan*, the cost of saving 1 tce is RMB 3,265, and cost of saving 1 t CO_2 is RMB 1,416.[16] While CDM mechanism estimates that the cost of 1 t CO_2 emission reduction is about 100 dollars in developed countries, the average cost of emission reduction in developing countries is only several to dozens of dollars.[17] Therefore, according to the full implementation cycle of CDM projects, China's cost of emission reduction during the *11th Five Year Plan* is equivalent to that of developed countries, more than 10 folds of CDM mechanism. The cost is high, and efficiency is low.

China is a developing country, therefore its priority is to increase people's income and eliminate poverty. Investments in low-carbon could impropriate resources from other fields, and reduce the investment in livelihood projects, infrastructure construction, medical health, cultural education and ecological environmental protection. With insufficient technological capability and monetary support, it is unfair and unreasonable to demand the nonaffluent Chinese people to shoulder the responsibility of realizing global and external efficiency by decreasing greenhouse gas emission. Taking radical low-carbon transformation methods in short term will incur huge cost to China. How to evaluate the potential cost of low-carbon economy should be a priority topic in future.

8.2 Energy Utilization in Energy-Intensive Industries

8.2.1 *Steel industry*

Steel industry is a basic industry in the national economy, playing a key role in the national economy and defense. From 2005 to the first half of 2008,

---

[16]In the first four years of the *11th Five Year Plan*, according to the link relative method, China's energy savings was 490 Mtce (Xinhua News, 2010), reflecting higher energy efficiency year on year, which could be treated as the newly accreted energy saving capacity. Therefore, 490 Mtce energy savings equals to 490 Mtce energy-saving capacity. Meanwhile, it also included the structural energy savings in the national industrial level and sector level, so the technological energy savings in the first four years of the *11th Five Year Plan* was lower than 490 Mtce. In addition, the estimated cost of energy-saving (3,265 yuan/tce) and emission reduction (1,416 yuan/tCO_2) is not accurate, and can only be used as a reference value.

[17]Energy savings of CDM mechanism refers to the physical energy savings instead of production value energy-saving amount. In addition, implementation cycle of a CDM project is usually about 5–7 years. Based on the implementation cycle, cost of 1 t CO_2 emission reduction of the developed countries is 500–700 dollars.

China's steel industry developed rapidly, with production of crude steel increasing from 352 Mt in 2005 to 490 Mt in 2007, increasing by 39.2%. In the latter half of 2008, growth of crude steel production slowed down due to the international financial crisis. Despite that, the crude steel production in 2008 reached 503 Mt, accounting for 38.3% of the global production; while pig iron production was 471 Mt, accounting for 48.8% of the global production. By 2008, China's steel and iron production had again ranked first in the world for 12 consecutive years, higher than the total amount of Japan, the United States, Russia, India, South Korea, Germany, and Ukraine. Economic indicators of the industry also increased rapidly, with its industrial value added from RMB 618.1 billion in 2005 to RMB 812.1 billion in 2008 (constant price in 2005), increasing by 31.3%. Average steel production per capita in 2009 was 429 kg, higher than the world average of 179 kg.

Rapid growth of production inevitably leads to higher energy consumption. In 2008, energy consumption of the steel industry was 434 Mtce, about 36% of the manufacturing sector, 15.6% of the total energy consumption; CO_2 emission was 1.31 billion ton, about 33.5% of the manufacturing sector, 20.6% of the total energy-related emission (Fig. 8.7).

Progress in energy-saving and emission reduction of the steel industry could be measured by two indicators: Energy consumption of per unit industrial value added and comprehensive energy consumption of 1.0 t steel. According to the first indicator, per unit industrial value added of the steel industry had declined from 5.73 tce/10,000 yuan in 2005 to 4.86 tce/10,000 yuan in 2008, decreasing by 15.2%; while the latter declined from 17.9 tCO_2/10,000 yuan in 2005 to 14.7 tCO_2/10,000 yuan in 2008, decreasing by 17.9% (Fig. 8.8).

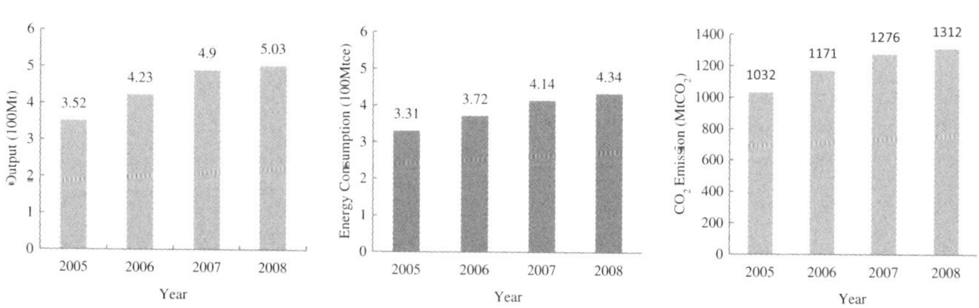

Fig. 8.7. Crude steel production, energy consumption and CO_2 emission of the steel industry.
Source: *China Energy Statistics Yearbook 2009* and *China Energy Statistics Yearbook 2010*.

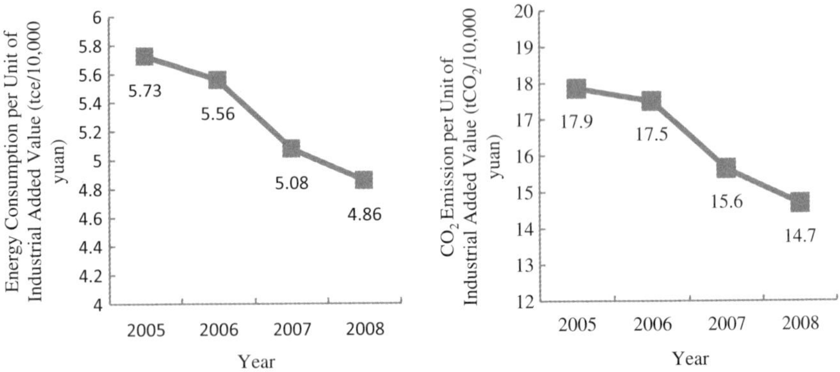

Fig. 8.8. Energy consumption and CO_2 emission of per unit industrial value added. *Source*: *China Energy Statistics Yearbook 2009* and *China Energy Statistics Yearbook 2010*.

According to the comprehensive energy consumption of 1,0 t steel, comprehensive energy consumption of the industry had declined from 1,019 kgce/t in 2005 to 973 kgce/t in 2009. The comprehensive energy consumption of large and medium-sized enterprises had declined from 760 kgce/t in 2005 to 717 kgce/t in 2009, comparable energy consumption of large and medium-sized enterprises had declined from 732 kgce/t in 2005 to 697 kgce/t in 2009 (Fig. 8.9).

Structural adjustment of steel products has made further progress. In 2008, production of steel plates for shipbuilding, pipelines for oil and gas industry, and heavy rails for railways increased by 62%, 86.1%, and 59.3% respectively. Optimization of industrial pattern and joint restructuring also made apparent progress. In 2005, production of top 10 enterprises made up 33.6% of the total production. However, with consecutive establishment of Shandong Iron and Steel Group Co., Ltd., Guangdong Iron and Steel Group Co., Ltd., and Hebei Iron and Steel Group Co., Ltd., industrial concentration level became higher, and production of top 10 enterprises made up 42.5% of the total production in 2008. Technologies and equipment of incorporated steel enterprises such as Baoshan, Anshan, Wuhan, Ma'anshan and Taiyuan Iron and Steel Plants had achieved leading levels world over.

Backward production capacity displacement in the steel industry made further progress. By 2009, iron-making industry displaced 81.7 Mt production capacity of blast furnaces whose capacity are below 300 m^3, steel-making industry displaced 59.40 Mt production capacity of small converters and electric furnaces with a production capacity of less than

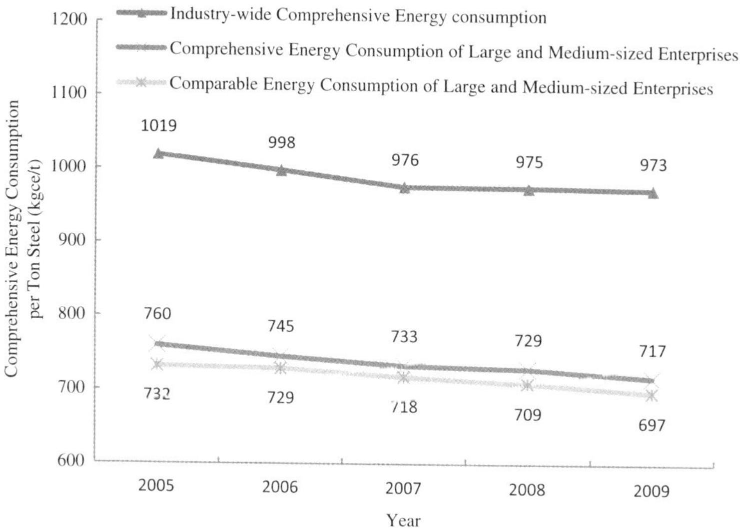

Fig. 8.9. Comprehensive energy consumption of 1.0 t steel between 2005 and 2009. *Source*: China Iron and Steel Industry Association, 2009.

200,000 t. By 2009, there remained — 45 Mt backward production capacity of small coking enterprises whose blast furnaces were below 4.3 m^3, 125 Mt production capacity of blast furnaces whose capacity were below 300 m^3, and small converters and electric furnaces whose annual production capacity were below 200,000 t (about 10% of the total production capacity). In 2010, the target is to displace 21.27 Mt, 30 Mt, and 8.25 Mt backward production capacity in the coke, iron-making and steel-making industry respectively.

Main processes of China's steel industry include coking, sintering, pelletizing, iron-making, converter, electric furnace and steel rolling (Fig. 8.10), of which, iron-making is the most energy-intensive process, amounting to 59.3% of the total energy consumption (Fig. 8.11). In 2008, energy consumption per ton steel for these processes is 120 kgce/t, 55.5 kgce/t, 30.3 kgce/t, 428 kgce/t, 5.7 kgce/t, 81.5 kgce/t, and 59.2 kgce/t respectively.

During the *11th Five Year Plan*, energy efficiency of the steel industry had improved by promoting advanced technology and processes, using large-scale equipment and reutilizing secondary energy (Table 8.5). However, there remains much potential in utilizing secondary energy for the steel industry, especially in recycling of low-temperature waste heat, flue gas sensible heat and slag sensible heat (Table 8.6).

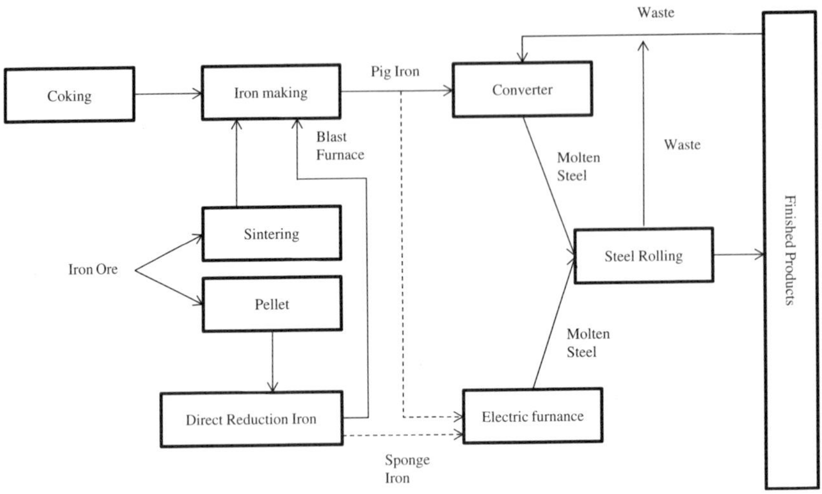

Fig. 8.10. Main processes of China's steel industry.
Note: Broken line indicates that there are two eligible technological processes, blast furnace-converter and blast furnace-electric furnace.

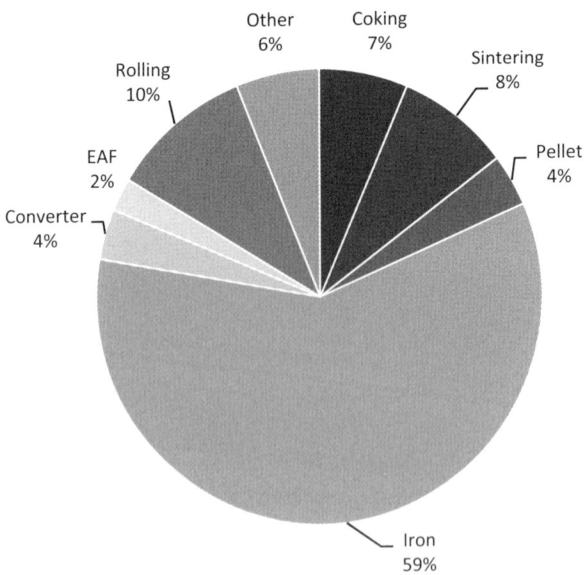

Fig. 8.11. Proportion of energy consumption of major processes to overall energy consumption.
Source: Weixing, 2008.
Note: The boxes are covering the graph (p. 127).

Table 8.5. Advanced technology, process and large-scale equipment of steel industry.

	Comprehensive energy consumption	advanced technology	Advanced process	Large-scale Equipment	Secondary energy utilization
Steel rolling (10.2%)	2008:59.2 kgce/t	heat-reserving burning (heat recovery rate > 80%, energy-saving > 30%)			
Electric furnace (2.33%)	2008:81.5 kgce/t				
Converter (3.58%)	2006:9.09 kgce/t 2008:5.74 kgce/t	negative-energy steel-making (saving 23.6 kgce per ton steel)	continuous ingot casting (saving 200 kgce per ton steel)	>120 m^3 converter	coal gas/residue heat/slag of converter
		coal gas dry purification and recovery (saving 2 kWh per ton steel)			
Iron-making (59.3%)	2005:457 kgce/t 2008:428 kgce/t	blast furnace TRT (40 kWh power generated per ton iron)		>1000 m^3 blast furnace (energy consumption lower than 19%)	coal gas/residue heat/slag of blast furnace
		coal gas dry dusting (saving 19 kWh per ton iron)			

(*Continued*)

Table 8.5. (*Continued*)

	Comprehensive Energy Consumption	Advanced Technology	Advanced Process	Large-scale Equipment	Secondary Energy Utilization
		CCPP (1 kWh power generated by per m^3 blast furnace coal gas)			
Pellet (3.73%)	backward level:49.4 kgce/t average level:30.3 kgce/t	small pellet sintering (saving 2 kgce/t)	drying grate-rotary kiln		
Sintering (8.30%)	2005: 64.8 kgce/t 2008:55.5 kgce/t	sintering machine residue heat power generation (12 kWh generated by per ton sintering ore)			flue gas sensible heat sintering residue heat
Coking (6.45%)	2005: 142 kgce/t 2008:120 kgce/t	dry quenching coking (75 kWh generated by per ton coke)	mechanical coking (saving 0.17 t/t coking coal)		coke oven flue gas red coke sensible heat

Table 8.6. Secondary energy recovery and utilization rate of steel industry.

Secondary energy	Recovery and utilization rate	Secondary energy	Recovery and utilization rate
Coke oven coal gas	97.75%	Low-temperature residue heat	<1%
Blast furnace coal gas	94%	Product sensible heat	50.4%
Converter coal gas	65 m^3/t	Flue gas sensible heat	14.92%
High-temperature residue heat	44.4%	Slag sensible heat	1.59%
Mid-temperature residue heat	30.2%		

Source: Energy Conservation and Green Power Energy Committee of China Energy Conservation Association, 2009.

8.2.2 *Building materials industry*

Building materials industry is an industrial conglomerate of considerable size, whose major products include cement, lime, plate glass, brick and tile, building (sanitary) ceramics. In China's Energy Balance Sheet, energy consumption of the building materials industry (nonmetal mineral products)\only refers to the energy that is consumed by the enterprises in the industry. However, many other industries also make building materials; for example, coal industry produced 8 billion coal gangue bricks and 19.5 Mt cement in 2008. Energy consumption of the building materials industry in the report is estimated primarily according to the production of major building materials and energy consumption per unit produced. Due to insufficient information on the industrial energy structure, CO_2 emissions could not be calculated. In 2008, building materials industry consumed 35.8 Mtce energy, second only to the steel industry (Fig. 8.12). Energy consumption of the industry made up 29.7% of the manufacturing sector, and 12.9% of China's total energy consumption.

China's unprecedented urbanization provides impetus to construction of buildings, transport facilities and infrastructure, and promotes rapid development of the building materials industry. In 2008, China's cement production was 1.42 billion ton, ranking first in the world, amounting to about 50% of the world production; building ceramics production was 6.2 billion m^2, ranking first in the world, amounting to about 40% of the world production; plate glass production was 592 million weight

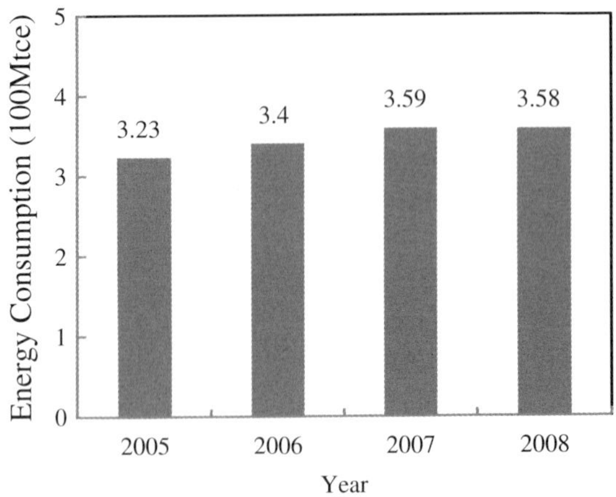

Fig. 8.12. Energy consumption of the building materials industry.
Note: Estimated according to production of major products and energy consumption per unit produced.

cases, ranking first in the world, amounting to about 48% of the world production.

Enterprises in China's building materials industry comprise several small energy-intensive enterprises in the manufacturing sector. In 2009, there were 5,500 cement enterprises above designated size with an average annual production of 300,000 t, while Japan only had 32 cement enterprises with an average annual production of 1.82 Mt. There were about 90,000 brick-tile enterprises with an average annual production of 9.7 million standard bricks, while the United States had 200 brick-tile enterprises with an average annual production of 53.7 million standard bricks. Enterprise size and industrial concentration degree affected the energy efficiency of the building materials industry. Between 2006 and 2008, the building materials industry enhanced technological innovation, displaced backward production capacity, increased industrial concentration, and accelerated structural adjustment and industrial upgrading, and the energy-saving and consumption reduction process has realized substantial achievements.

(1) *Cement industry*

Industrialization and accelerating urbanization drives cement consumption. China's cement production increased from 597 Mt in 2000 to 1.42 billion ton in 2008, increasing by 138%. From 1985 to 2008, China's cement production had ranked first in the world for 23 consecutive years, and its proportion to the world production increased from 10% to 50%. Since 2000, global cement production has grown rapidly, with a 7% growth rate, but China's growth rate is five points higher than the world average rate. In 2008, China's cement production was 8 and 17 times more than that of India (second rank) and the United States (third rank). China's production scale is much higher than other countries.

Cement industry is a typical energy-intensive industry. According to its energy consumption per unit produced and production capacity, energy consumption of cement industry was 214 Mtce in 2008, about 60% of the building materials industry, 17.8% of the manufacturing sector, and 7.7% of the primary energy consumption in China. Cement industry is also an industry with high CO_2 emission. CO_2 emission for making 1 t cement is 0.815 t in China (of which, 0.39 t is generated by burning fuel, 0.425 t is generated from decomposition of carbonate in the raw materials) (Special Column 8-1). In addition, producing 1 t cement could also produce 0.07 t CO_2 emission indirectly due to power consumption. In 2008, China's cement production was 1.42 billion ton, so China's direct CO_2 emission was 1.14 billion ton, indirect CO_2 emission was 100 Mt, accounting for 31.4% of the overall CO_2 emission of the manufacturing sector, and 19.5% of the overall energy-related CO_2 emission (Fig. 8.13).

Although overall energy consumption and CO_2 emission increases, energy consumption per industrial value added and per production had declined in 2005–2007, because the growth rate of cement production and industrial value added was higher than that of energy consumption. Energy

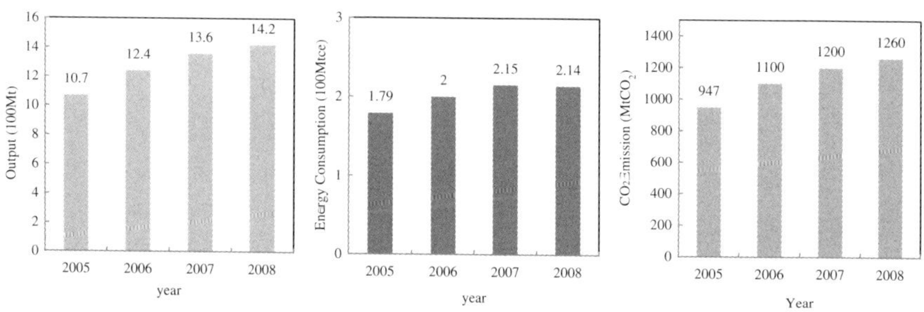

Fig. 8.13. Cement production, energy consumption and CO_2 emission of cement industry.
Source: *China Energy Statistics Yearbook 2009* and *China Energy Statistics Yearbook 2010*.

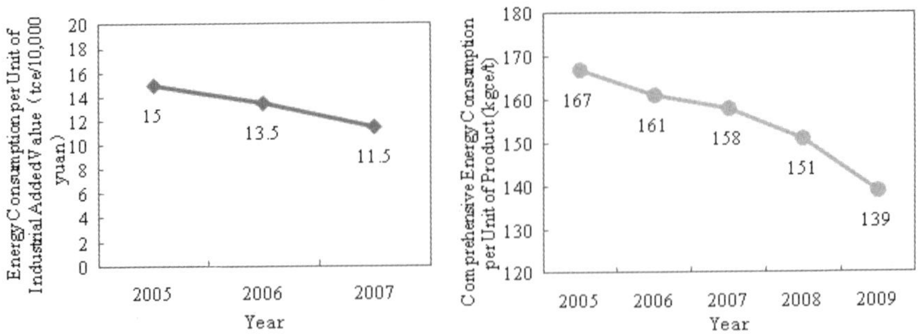

Fig. 8.14. Energy consumption per industrial value added and comprehensive energy consumption of production.
Source: Energy Research Institute of NDRC, 2010.

consumption per industrial value added declined from 15 tce/10,000 yuan in 2005 to 11.5 tce/10,000 yuan in 2007, decreasing by 23.3%. Energy consumption of per produced declined from 167 kgce/t in 2005 to 139 kgce/t in 2009, decreasing by 16.8% (Fig. 8.14).

Box 8.1. CO_2 emission in cement production.

A fundamental characteristic of greenhouse gas emissions in the cement industry is that CO_2 is emitted not only from the burning fuel, but also from the decomposition process of $CaCO_3$, a major component of the raw materials limestone, and from decomposition of $MgCO_3$. According to the analysis of cement production process, direct CO_2 emission consists of five major parts: ① calcinations of carbonates in the raw materials; ② burning of organic carbon in the raw materials; ③ burning of traditional fossil fuel; ④ burning of alternative fuel (fossil waste); ⑤ burning of noncalcinations fuel. Indirect CO_2 emission consists of two parts: ① external power generation; ② clinker production.

Table of CO_2 emission in cement production.

Source of CO_2 Emission		Parameter	Default Value
Direct CO_2 Emission			
(1) Calcinations of raw materials	(a) Clinker	Contents of CaO and MgO.	527.kg/t

(*Continued*)

Box.8.1. (*Continued*)

Source of CO_2 Emission		Parameter	Default Value
	(b) CKD	Dust amount and calcinations ratio.	calcinations ratio.=.1
(2) Organic carbon in the raw materials		Raw material/clinker ratio, content of organic carbon in raw materials.	Raw material/clinker ratio is 1.55; content of organic carbon in raw materials is 2 kg/t
(3) Traditional fossil fuel (coal, coking coal, fuel oil, natural gas)		Fuel consumption, low-calorific value of fuel, corresponding fuel emission factors.	Emission factors — in reference to the default values recommended by IPCC and CSI.
(4) Alternative burning fuel	(a) Bio-fuel	Fuel consumption, low-calorific value of fuel, corresponding fuel emission factors.	Emission factors — in reference to the default values recommended by IPCC and CSI
	(b) Alternative fossil fuel	Fuel consumption, low-calorific value of fuel, corresponding fuel emission factors.	Emission factors — in reference to the default values recommended by IPCC and CSI
(5) Non-calcinations fuel	(a) Fuel for drying raw materials	Waste heat utilization, moisture content of raw materials	0.048 t/t clinker
	(b) Transport	Fuel and power consumption of traffic vehicles	—

(*Continued*)

Box.8.1. (*Continued*)

Source of CO_2 Emission	Parameter	Default Value
Indirect CO_2 Emission		
(1) External power generation	Power consumption, CO_2 emission factor	0.847 kg/(kW.h)
(2) Production of clinker	Net clinker amount, CO_2 emission factor of clinker	862 kg/t

Source: H. Hongtao, CO_2 emission analysis and quantification in cement production, *Cement Engineering*, (1) (2009): 61–65.

By 2009, cement industry had displaced 240 Mt backward production capacity. There remains only 5 Mt backward production capacity of wet kiln in west China with low coal prices. By 2009, there was still 390 Mt backward production capacity in cement industry, and the target in 2010 is to displace 92 Mt. Displaced backward production capacity during the *11th Five Year Plan* could save over 35 Mt coal, 22.5 billion kWh electricity and reduce 3 Mt dust emission.

Cement production mainly consumes coal and electricity, i.e. heat and power consumption. Heat consumption includes drying heat consumption, calcinations heat consumption and drying heat consumption of composite materials. Power consumption includes — raw meal production, clinker calcinations, and cement grinding. NSP method is an advanced process for energy-saving and emission reduction in cement industry; other advanced technologies comprise cement kiln pure low-temperature waste heat power generation technology, high performance grinding technology, and speed regulation and energy conservation of frequency converter technology. In addition, comprehensive utilization of various wastes and increasing bulk loading rate of cement are also effective energy conservation methods.

Promoting NSP method is a requirement for energy conservation and consumption reduction, which shifts the drying and pre-heating process from the rotary kiln to the suspension pre-heater, and shifts the decomposition of raw materials from the rotary kiln to the decomposing furnace. In this way, there are only some low heat-consumptive reaction processes in the kiln, decreasing the heat load of kiln. Comprehensive energy consumption of

NSP kiln is 28% lower than that of shaft kiln, 45% lower than wet process kiln, and 53% lower than dry hollow kiln. Proportion of NSP cement in China had increased from 12% in 2000 to 72.5% in 2009.

Scale of NSP cement production line also affects energy consumption in production. Comprehensive power consumption of clinker for the production line with over 4,000 t/day production capacity is 17 kWh/t lower than that below 2,000 t/day, while comprehensive heat consumption of clinker is 19 kgce/t lower, comprehensive power consumption of cement is 15 kWh/t lower, and comprehensive energy consumption of cement is 16 kgce/t lower. Energy consumption indicators of China's production lines with over 4,000 t/day production capacity have reached world's advanced levels.

Cement kiln pure low-temperature waster heat power generation technology utilizes the residue heat of waste gas below 350°C to produce 0.8–2.5 Mpa low-pressure steam, driving the steam turbine to generate electricity. Generated electricity of per ton clinker could reach 37–45 kWh. Both the thermodynamic system and equipment of China's self-developed pure low-temperature waste heat power generation technology are mature and reliable. Its overall technological level has reached world advanced standards except that the efficiency of steam turbine is slightly lower than that of Japan. During the *11th Five Year Plan*, pure low-temperature waste heat power generation capacity has realized a geometrical growth. In 2006, 15 waste heat power generation production lines were commissioned, installing 14 machine units, with an installed capacity of 66 MW. In 2007, 86 waste heat power generation production lines were commissioned, installing 59 machine units, with an installed capacity of 571 MW. In 2008, 149 waste heat power generation production lines were commissioned, installing 106 machine units, with an installed capacity of 975 MW. In 2009, 232 waste heat power generation production lines were put into production, installing 181 machine units, with an installed capacity of 1677 MW. Between 2006 and 2010, it is estimated that altogether 60.7 Mt CO_2 emission has been reduced by waste heat utilization.

Grinding equipment is mainly used in producing raw meal and coal dust as well as cement grinding, and its power consumption makes up half of the comprehensive power consumption in cement production. Promoting efficient grinding equipment and technology is very meaningful to the power and energy conservation of the cement industry. Supposing that 60% backward ball grinders are replaced by efficient vertical mills or roller presses by 2010, 2.27 billion kWh could be saved, equaling to 813,000 tce.

There are many electric motors and fans in cement production lines, which can be utilized where frequency conversion and speed regulation technology is applicable, thereby saving 30%–40% power. At present, only 5%–7% equipments have adopted this method. Supposing that 40% equipment adopt the method by 2010, 3.8 billion kWh electricity could be saved, equal to 1.36 Mtce.

Apart from promoting advanced processes and technologies, comprehensive utilization of various types of waste is also a major method in decreasing energy consumption. There are three ways in comprehensive utilization of waste. First, waste can be used as alternative raw materials for cement, such as sewage sludge incinerated ash, papermaking sludge incinerated ash, river sludge, coal dust, blast furnace slag, calcium fluoride sludge, diatomite, coal gangue, waste plaster etc. Second, waste can be used as alternative fuels for cement, such as waste tires, waste rubbers, waste plastics, waste oils, active carbon sludge, spent bleaching clay, papermaking sludge, cokes etc. Third, cementing waste could be used as composite materials or auxiliary cementing materials in producing cement products or concrete, such as coal dust, blast furnace slag and coal gangue. China's comprehensive utilization rate of waste is higher than other countries, resulting in the lowest clinker usage in the world, and proportion of clinker in cement making had declined from 83% in 1995 to 73% in 2005 (IEA, 2007).

In addition, increasing bulk loading rate is also an effective energy-saving method. Bulk loading of 100 Mt cement could save two billion paper bags, 330 m^3 wood pulp and the energy consumption in making these bags, and could save 5 Mt cement due to broken bags and residue. China's bulk loading rate had increased from 28% in 2000 to 46% in 2008, but was much lower than the rate of 96% in the United States and Japan. Table 8.7 is a summary of the advanced technologies and processes in the cement industry.

(2) *Brick-tile industry*

There are two types of new wall materials, sintering hollow products (hollow bricks, aerated concrete blocks etc), or sintering products made of industrial waste slag (coal gangue, coal dust, shale, and slag) and river sludge (sand). Developing new wall materials is very meaningful to energy conservation. For example, replacing solid clay bricks with hollow concrete bricks could save 1/3–2/3 energy.

Production of new wall materials was 322 billion blocks (standard bricks) in 2005, and the number of sintering hollow bricks, waste slag bricks,

Table 8.7. Advanced technologies and processes of the cement industry.

Comprehensive energy consumption of cement	Major energy consumption processes	Advanced technologies	Advanced processes	Large equipment	Other energy conservation methods
	Clinker calcinations heat consumption	Pure low-temperature waste heat power generation technology (37–45kWh per ton of clinker)	NSP (28% fewer energy consumption)	Production line with the production capacity of over 4,000 t/day (16 kgce/t fewer energy consumption)	Comprehensive utilization of wastes
2005:167 kgce/t 2008:151 kgce/t	Raw meal grinding power consumption	Vertical mill equipment and technology (30% power conservation)			
	Clinker sintering power consumption	Speed regulation and energy conservation technology of frequency converter (30%–40% power conservation)			Higher rate of bulk loading
	Coal dust production power consumption				
	Cement grinding power consumption				
	Raw material and fuel drying heat consumption				
	Composite materials drying heat consumption				

coal gangue bricks and sand lime bricks were 160 billion, 140 billion, eight billion, five billion and nine billion blocks respectively. Production of new wall materials was 385 billion blocks in 2006, increasing by 19.6% year on year. Production of sintering hollow bricks in 2008 was 200 billion blocks, including 180 billion waste slag bricks. Between 2000 and 2009, proportion of new wall materials to the overall wall materials production increased from 28% to 52%. In 2008, proportion of building areas using new wall materials in the new buildings exceeded 60%. From 2000 to 2006, altogether 1.1 billion ton solid wastes have been utilized to produce new wall materials, saving 1.8 million mu of land,[18] overall saving 96.6 Mtce energy (Wang Qingyi, 2010).

In 2008, over 800 production lines with an annual production of over 30 million blocks were built, of which, nearly 300 production lines have an annual production of 60–120 million blocks (standard bricks). The production lines adopt de-airing extruders above Type-50 standard, automatic cutting, setting and loading system, as well as tunnel kiln firing process. Production of nonsintering bricks adopt hydraulic brick press and still kettle curing process. Product breeds include fired perforated bricks, hollow bricks, hollow building blocks, wall body decorative boards, paving bricks, autoclaved fly ash bricks, autoclaved sand-lime bricks etc. These processes are not only conducive to energy-saving, land conservation, waste utilization and environmental protection, but also provide better product quality and widen their applications in the building industry (China Building Materials Industry Association, 2009).

Production lines of the brick-tile industry become larger. The biggest production line of gypsum plaster board is 30 million m^2/a; size of new production lines of concrete blocks and aerated concrete are above 100,000 m^2/a; size of new production lines of sintering hollow bricks are over 30 million m^2/a, and the biggest single-line size is 80 million m^2/a.

(3) *Lime industry*

Due to sluggish international economic environment and influence of national macroeconomic regulation policies, limes for industrial use decreased widely in 2008. Except the limes for environmental use, growth rate of lime usage for all other industries in 2008 declined enormously year on year (Table 8.8). Growth rate of lime in the calcium carbide industry

[18] 1 mu = 666.7 m^2.

Table 8.8. Lime usage.

	2007		2008		Year on year (YoY) Growth Rate (%)	
Category	Quantity (Mt)	Proportion (%)	Quantity (Mt)	Proportion (%)	2007	2008
Construction and Building Materials	100	58.45	100	56.15	0	0
Steel	42.5	24.84	49	27.51	14	1.5
Calcium Carbide	12.1	7.07	11.9	6.68	21	−1.6
Nonferrous Metal	7	4.09	7	3.93	40	0
Calcium Carbonate Light	4	2.34	4.2	2.36	30	5
Other Chemical Industries	4	2.34	4	2.25	14	0
Environmental Protection	1.5	0.88	2	1.12	50	33
Total	171.1	100.00	178.1	100.00	5.2	4.1

Source: China Building Materials Industry Association, 2009.

declined from 21% in 2007 to −1.6% in 2008; the nonferrous metal industry declined from 40% in 2007 to 0% in 2008; and the calcium carbonate light industry declined from 30% in 2007 to 5% in 2008. In addition, due to technical renovation and new lime projects, supply of active lime increased. Growth rate of industrial lime consumption in the steel industry has reduced from 14% in 2007 to 1.5% in 2008. In the future, lime enterprises will mainly produce sintering lime, while active lime for steel smelting will be supplied by the steel enterprises, with ever-decreasing commodity supply.

Different from other fields, lime used for environmental protection continues to grow rapidly, especially the limes used for desulphurization in thermal power plants and for industrial waste water treatment. Further expansion of thermal power plants, technological renovation of old power plants, improvement of lime flue gas desulphurization process, and environmental protection policies issued by the government are all favorable to the expansion of lime market. In addition, pressure exerted by the national policies on energy conservation and emission reduction put more demand for lime used in acidic industrial waste water treatment (China Building Materials Industry Association, 2009).

In order to displace backward production capacity, provinces with more lime production enterprises, such as Beijing, Hebei, Shandong, Gansu and Xinjiang, shut down earth kilns and formulated regulation measures. For example, Jingxing County of Heibei Province shut down earth kilns with an overall production capacity of 3 million t/a. In 2008, altogether 20 Mt backward production capacity was displaced (China Building Materials Industry Association, 2009).

(4) *Plate glass*

Proportion of float glass production increased from 57% in 2000 to 83.4% in 2009. China's self-developed 1,100 t/d super-large float glass production line has been put into production, whose heat consumption ratio is 15%–29% lower than the 600 t/d production line. Scale of melting furnace affects the energy consumption of plate glass enormously, for example, average energy consumption of a 900 t/d melting furnace is 11%–13% lower than that of a 500 t/d melting furnace. In 2009, China's 500 t/d, 600 t/d and 700 t/d melting furnaces accounted for 75.3%, 54.48%, and 28.83% of the total melting capacity of float glass respectively. At present, mechanical and automatic equipment of China's float glass production lines has reached or caught up with the world's advanced levels. However, real energy consumption deviates greatly from the designed values during 5–12 years of continuous functioning of a kiln, and that of China is often 20%–50% higher than that of foreign countries. The fact fully reveals that there is much potential in daily energy management of China's float glass enterprises. Therefore, promoting and implementing energy management system accreditation in the building materials industry, prompting the enterprises to exercise efficient energy management and improve energy management performance has a far-reaching and practical significance.

China's enterprises with an annual plate glass production of over 10 million weight cases had increased from 13 in 2006 to 20 in 2009, whose proportion of production capacity to the overall capacity had also increased from 52% in 2006 to 68% in 2009.

Different fuels used in glass making also affects the comprehensive energy consumption of plate glass. Heat consumption of coal-burning float glass melting kiln is 73% to that of ordinary glass, and heat consumption of natural gas burning float glass melting kiln is 44% to that of ordinary glass. Average comprehensive energy consumption of float glass in Europe and the United States is 171–290 kgce/t, and the world advanced level is 132 kgce/t.

(5) *Building and sanitary ceramics*

Enterprise size of building ceramics industry has developed. At present, the average annual production of over 1,200 building ceramics enterprises exceeds 2.5 million m^2, double than that in 2000, and the largest production capacity is 100 million m^2. In addition, eliminating direct coal-fired kilns is also an important factor in improving the energy efficiency of ceramics industry. Direct coal-fired kilns have been eliminated in the ceramics industry except in the Henan Province and some other remote provinces. At present, coal water slurry and manufactured gas are used to produce low-grade ceramics products, while natural gas, liquid gas and light diesel oil are used to produce high-grade ceramics products. However, due to insufficient supply of natural gas (especially in winter), some enterprises have to be equipped with kerosene equipment. In addition, electricity used in the ceramics enterprises are supplied by the power grid, therefore many enterprises have to prepare diesel generators for emergency due to power supply shortage. Energy consumption indicators of China's large and medium-sized enterprises have reached world's advanced levels (Table 8.9).

8.2.3 *Chemistry and chemical industry*

Petroleum, chemistry and chemical industry is the mainstay industry of China's national economy, as well as an important energy industry and raw materials processing industry. Petroleum, chemistry and chemical industry consists of oil exploration, oil refinery and production of chemical products, while production of chemical products can be further divided into the chemical industry, chemical fertilizer industry, petrochemical industry, sodium carbonate industry, chlor-alkali industry, sodium carbide industry, basic organic raw materials industry, farm chemical industry, dye industry, coating industry, fine chemistry industry, rubber processing industry and new materials industry, producing about 40,000 types of products. This section mainly discusses crude oil processing and chemical products manufacturing, corresponding to the chemical raw materials industry and chemical products manufacturing in the statistics yearbook.

Based upon the absorbing inducted technologies, the chemistry and chemical industry possesses S&T innovation capability as well as a series of

Table 8.9. Energy consumption indicators of China's large and medium-sized enterprises and the world advanced levels.

Data in 2005	Items	China	World advance levels
Calcinations heat consumption	Building ceramics (kJ/kg)	2930–6279	1255–4186
	Sanitary ceramics (kJ/kg)	5023–12,558	3350–8370
Comprehensive heat consumption	Building ceramics (kgce/m^2)	2.5–15 (large and medium-sized enterprises 4.15)	0.77–6.42
	Sanitary ceramics (kgce/t)	400–1800 (large and medium-sized enterprises 1022)	238–476
Power consumption	Building ceramics (kWh/m^2)	2.5–5.5 (large and medium-sized enterprises 4.86)	2.3–5.12
	Sanitary ceramics (kWh/t/)	230–600 (large and medium-sized enterprises 301)	249–533

Source: Hongru *et al.*, 2005.

core technologies with proprietary IPR. Indigenously-developed complete technology such as large-scale ethylene cracking furnace, and extraction of PP, PE, PET and aromatic hydrocarbon have realized large-scale industrial application, and some of these technologies have been inducted by some foreign countries (Institute of Industrial Economics of CASS, 2009). Development of new products has witnessed substantial progress. For example, top-grade Changcheng lubricant of Sinopec has been successfully applied to Shenzhou spaceship, and over 85% catalysts for oil refinery and chemical production have been supplied domestically. In addition, import substitution rate of major petroleum and chemical equipment has increased substantially, of which, the import substitution rate of refinery equipment is about 90%, of 300,000 t ethylene complete equipment is over 80%. In 2008, industrial value added of the chemistry and chemical industry reached RMB 672.85 billion (constant price in 2005).

Meanwhile, the chemistry and chemical industry is also a major energy consumer in China. From 2006 to 2008, energy consumption of the chemistry and chemical industry was 195 Mtce, 213 Mtce and 220 Mtce respectively. Its energy consumption accounted for about 18% of the overall energy consumption of the manufacturing sector in 2008. The CO_2 emission of the petroleum and chemistry & chemical industry was 657 Mt in 2008, accounting for 16.7% of the overall CO_2 emission of the manufacturing sector (Fig. 8.15).

Energy consumption of per unit industrial value added of the chemistry and chemical industry declined from 4.93 tce/10,000 yuan in 2005 to 4.26 tce/10,000 yuan in 2008, decreasing by 13.6%. The CO_2 emission of per unit industrial value added of the chemistry and chemical industry declined from 10.59 tCO_2/10,000 yuan in 2005 to 8.97 tCO_2/10,000 yuan in 2008, decreasing by 19% (Fig. 8.16).

(1) *Ethylene industry*

Ethylene is one of the most-produced chemical products in the world. A series of products could be made of ethylene, such as plastics, chemical fiber and rubber, etc, which could be applied widely to all sectors of national economy. Therefore, ethylene is an important basic raw material of the petroleum, chemistry and chemical industry.

Comprehensive energy consumption of China's ethylene industry declined from 1,073 kgce/t in 2005 to 976 kgce/t in 2009, decreasing by

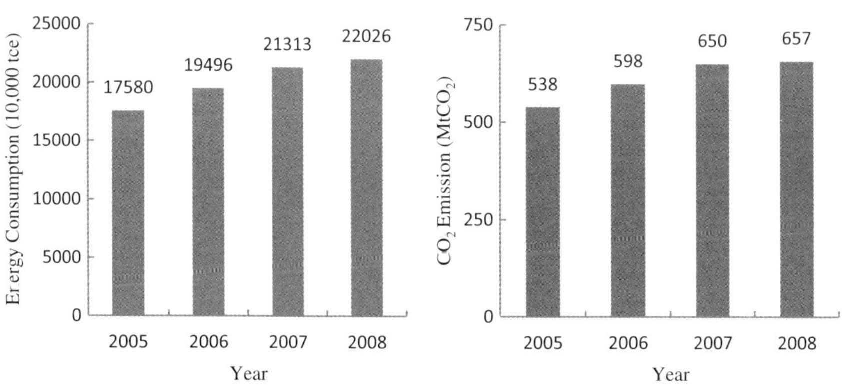

Fig. 8.15. Energy consumption and CO_2 emission of the chemical and chemistry industry.
Source: *China Energy Statistics Yearbook 2009.*

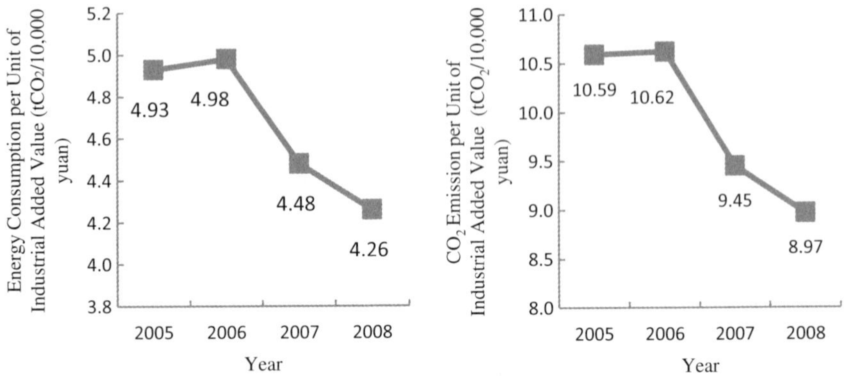

Fig. 8.16. Energy consumption and CO_2 emission per unit industrial value added of the chemical and chemistry industry.
Source: *China Energy Statistics Yearbook 2009* and *China Energy Statistics Yearbook 2010*.

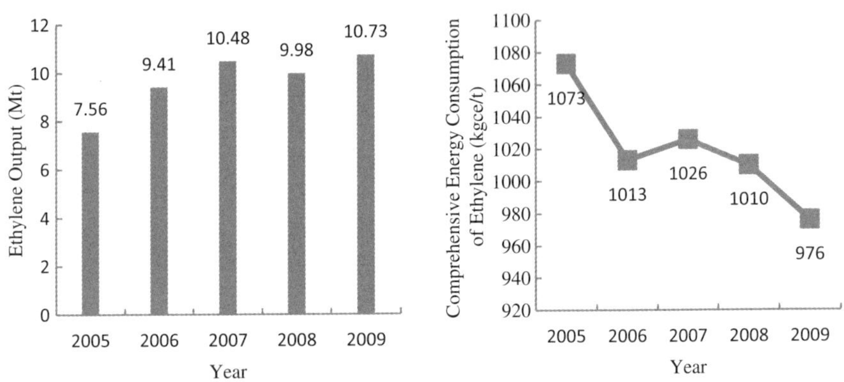

Fig. 8.17. Production and comprehensive energy consumption of ethylene.
Source: *China Energy Statistics Yearbook 2009*; Energy Conservation and Green Power Energy Committee of China Energy Conservation Association, 2009.

9% (Fig. 8.17), but still lagging far behind the level of Japan, South Korea and the Middle East. Comprehensive energy consumption of ethylene in the Middle East is 629 kgce/t.

Lower comprehensive energy consumption of China's ethylene industry is primarily attributed to the lower proportion of diesel and hydrogenated tail gas, as well as higher proportion of naphtha. Lighter cracking stock of ethylene requires lower cracking temperature. Of the cracking stock of ethylene, energy consumption of light hydrocarbon is 512–853 kgce/t, of naphtha is 853–1,365 kgce/t, and of hydrogenated tail gas and diesel is 1,365–1,706 kgce/t. In China, over 90% cracking stock of ethylene is supplied by the refinery plants, whose business orientation determines that their refinery equipments are used to produce petroleum products instead of producing naphtha as chemical raw materials. Therefore, the raw materials of China's ethylene industry tend to be heavy. On the contrary, the proportion of light hydrocarbon and raw gas is higher in foreign countries, especially in the Middle East, where the raw materials mainly consist of light and cheap raw gas.

The average scale of China's single set of ethylene equipment is reaching world's average level, and of new equipments has reached world level, i.e. 0.8–1 million t/a. China has 19 ethylene production enterprises and 22 ethylene production equipments, and the production capacity was 9.98 million t/a in 2008. Maoming Petroleum and Chemical Co., Ltd. of Sinopec has the largest production scale, producing over 1 Mt ethylene per year. The average scale of ethylene production has increased from 266,000 t/a in 2001 to 454,000 t/a in 2008, which is a little higher than the world average level of 440,000 t/a in 2004 (Table 8.10).

In addition, although China's indigenously-developed large-scale ethylene cracking furnace has been put into industrial production, its average thermal efficiency is only 90%, lower than the world average level of 94%. World-leading thermal efficiency of boilers is above 92%, while that of China is only about 88%, and the fuel energy efficiency is relatively low [CIEE (Beijing), 2009].

Table 8.10. Ethylene production scale of China and the world.

Country	Production Capacity (10,000 t/a)	Number of Equipment (set)	Average Scale (10,000 t/a)
China	998	22	45.4
The United States	2,832	41	69
Japan	751	15	50
Saudi Arabia	585	7	84
South Korea	545	11	50
World Total	11,291	254	44

Note: Data of 2008 for China, data of 2004 for other countries and the world average level.
Source: Guohui, 2005.

(2) *Sodium carbonate industry*

Sodium carbonate is one of the most important and basic raw materials of the chemical industry, which is widely used by building materials, chemical, metallurgical, daily chemicals and farm chemicals industries. At present, annual production of sodium carbonate in the world is about 47 Mt. By the end of 2008, China's sodium carbonate production capacity was about 21 Mt (Energy Research Institute of NDRC, 2010), while its production was 18.81 Mt in 2008 and 19.38 Mt in 2009. Comprehensive energy consumption of sodium carbonate declined from 396 kgce/t in 2005 to 323 kgce/t in 2009, decreasing by 18.4% (Fig. 8.18).

Advanced technologies promoted in the sodium carbonate industry include closed recovery of steam condensation water and producing alkali with transformation gas. With respect to steam condensation water closed recovery technology, it analyzes the pressure rating and sequence in the steam pipelines, deals with the problem of steam-water mixture and pipeline blockage due to different pressures by echelon pressurization and sectionalized recovery, which increases the heat efficiency by 15%–30% and the condensation water recovery rate by 10%–20%.

Since 2006, a new advanced technology, producing alkali directly with transformation gas of synthesis ammonia (associated alkali technology), has been widely adopted in the sodium carbonate industry. The new method is based upon the theory of producing alkali with low-temperature recycling, which transforms three-tower traditional alkali production method

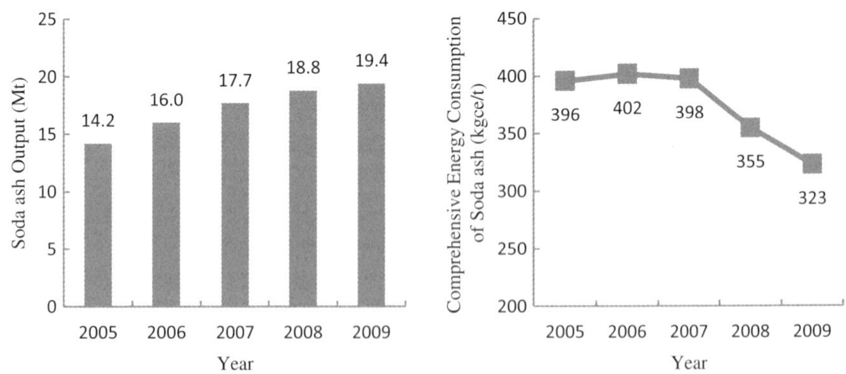

Fig. 8.18. Production and comprehensive energy consumption of sodium carbonate. *Source*: *China Statistics Yearbook 2010*; Energy Conservation and Green Power Energy Committee of China Energy Conservation Association, 2009.

Table 8.11. Technical indicators of producing alkali with transformation gas of synthesis ammonia (associated alkali technology).

Energy consumption per ton alkali (kgce/t)	Power consumption per ton alkali (kWh)	water conservation rate (kg/t)	Saved steam (kg/t)	Utilization rate of salt as raw materials (%)	Utilization rate of ammonia as raw materials (%)
273	60	15–25	70–110	99%	96%

Source: *China Statistics Yearbook 2010*; Energy Conservation and Green Power Energy Committee of China Energy Conservation Association, 2009.

into one-tower method, internal heat exchange into external heat exchange, increases the quality of bicarbonate crystallization, and realizes zero emission of waste liquid in the associated alkali system. Comprehensive energy consumption of the alkali production technology with transformation gas is 273 kgce/t, which is leading in the industry (Table 8.11). At present, penetration rate of the technology is 15%, and it is estimated that the rate would reach 50% by the end of the *11th Five Year Plan.*

(3) *Sodium hydroxide industry*

Sodium hydroxide is widely used in many industrial sectors of the national economy, such as chemical medicine, paper-making, aluminum metallurgy, wolfram metallurgy, rayon, artificial cotton and soap manufacturing. At present, production capacity and output of China's sodium hydroxide industry ranks first in the world. According to the statistics of China Chlor-Alkali Industry Association, China's production capacity of sodium hydroxide was 24.72 Mt/a by the end of 2008, of which, proportion of the production capacity of diaphragm process to the overall production capacity was 31%, that of ion-film method was 68% (Energy Research Institute of NDRC, 2010). In 2009, production of China's sodium hydroxide was 18.32 Mt.

Comprehensive energy consumption of sodium hydroxide declined from 1,297 kgce/t in 2005 to 1,087 kgce/t in 2009, decreasing by 16.2% (Fig. 8.19).

Comprehensive energy consumption of diaphragm method was 1,293 kgce/t in 2009, while the world advanced level was 1,250 kgce/t. Comprehensive energy consumption of ion-film method was 916 kgce/t in 2009 (Fig. 8.20).

Reduction of comprehensive energy consumption of sodium hydroxide in China is primarily attributed to the promotion of ion-film method.

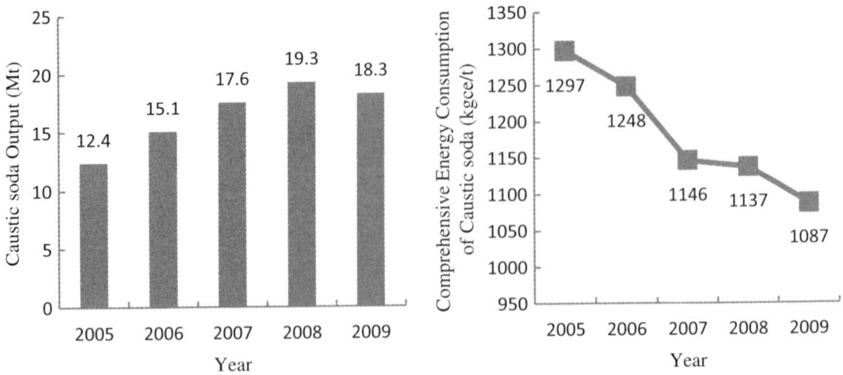

Fig. 8.19. Production and comprehensive energy consumption of sodium hydroxide.
Source: *China Statistics Yearbook 2010*; Energy Conservation and Green Power Energy Committee of China Energy Conservation Association, 2009.

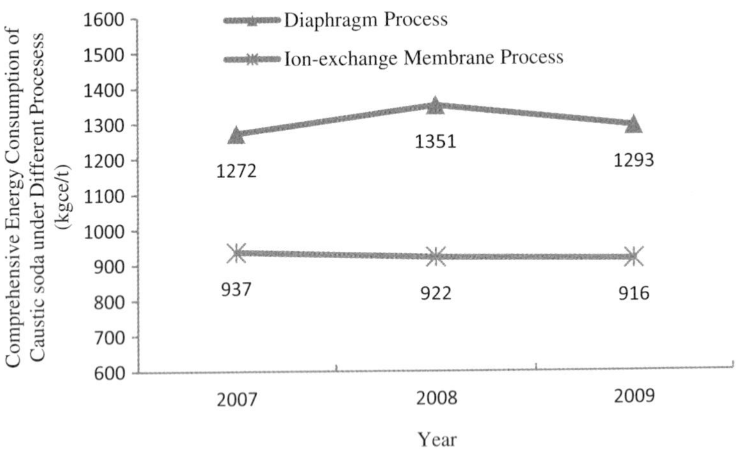

Fig. 8.20. Comprehensive energy consumption of sodium hydroxide of diaphragm and ion-film method between 2007 and 2009.
Source: Energy Conservation and Green Power Energy Committee of China Energy Conservation Association, 2009.

Electrolytic tank ion-film polar distance method deals with the problems existing in raw materials, machining parts, coating technology and assembly technology, which develops a natural circulation ion-film polar distance electrolytic tank suitable for China's enterprises. The technology could reduce DC power consumption by 100–120 kWh per ton of sodium

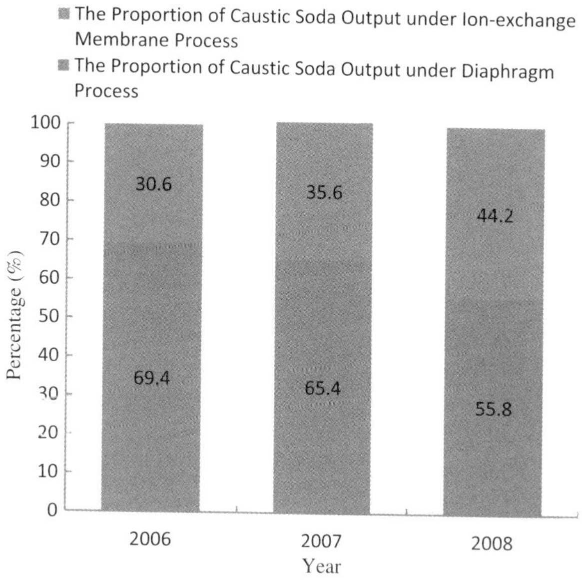

Fig. 8.21. Proportion of sodium hydroxide using diaphragm and ion-film method. *Source*: Qingyi, 2010.

hydroxide. At present, the technology has been adopted by several enterprises. Proportion of ion-film method had increased rapidly from 30.6% in 2006 to 44.2% in 2008. (Fig. 8.21).

With respect to the diaphragm method, the anode expansion and modified diagram method is designed in connection with the tank voltage of diaphragm electrolytic tanks, which could save 147 kWh/t, and has been successfully adopted by many enterprises. The average scale of China's sodium hydroxide enterprises had increased from 78,700 t/a in 2006 to 121,200 t/a in 2008. However, there is still a large gap with the average scale of foreign countries (200,000 t/a), especially the United States (720,000 t/a).

(4) *Synthesis ammonia industry*

Synthesis ammonia is a bulk chemical product primarily used to produce nitrogenous fertilizers and compound fertilizers, and has an important role in the national economy. The aim of the synthesis ammonia industry is to extract ammonia from coal, natural gas and heavy oil, and its major processes include gas making, desulphurization, gas transformation, compression, decarburization and synthesis. At present, production capacity and output of China's synthesis ammonia ranks first in the world, and the growth rate of China's production of synthesis ammonia is one of the fastest in the world. Now, China has over 550 synthesis ammonia manufacturers (Energy Research Institute of NDRC, 2010). In 2009, production of China's synthesis ammonia was 51.34 Mt.

Comprehensive energy consumption of synthesis ammonia did not change much between 2005 and 2008, decreasing from 1,700 kgce/t in 2005 to 1,591 kgce/t in 2009, decreasing by 6.4% (Fig. 8.22). World's advanced level was 1,194 kgce/t in 2005 in Western Europe, therefore, the comprehensive energy consumption of synthesis ammonia in China still lags far behind the world's advanced levels.

There are mainly two reasons for this: Coal-oriented energy structure and small production scale. According to energy structural analysis, proportion of coal used in synthesis ammonia industry was 70% in 2005 and 77% in 2008. From 2006 to 2007, out of six sets of equipment using oil as raw materials in Sinopec, four sets had been renovated to use coal as raw materials, and the proportion of coal consumption had increased from 70% in 2005 to 77.2% in 2008, proportion of natural gas remained at 20%, and proportion of oil declined from 10% to 2.8% (Fig. 8.23). All the other countries use natural gas as major raw materials. Proportion of natural gas in synthesis ammonia production is 70% for the world's average level, but is only 20% in China.

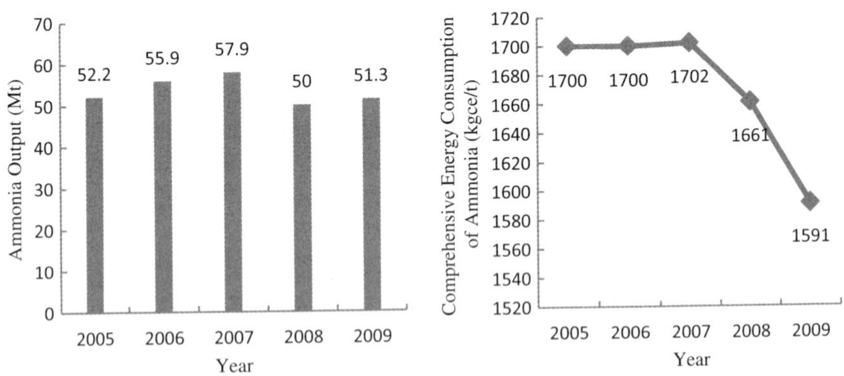

Fig. 8.22. Production and comprehensive energy consumption of synthesis ammonia between 2005 and 2009.
Source: State Statistics Bureau, 2010; China Petroleum and Chemical Industry Association.

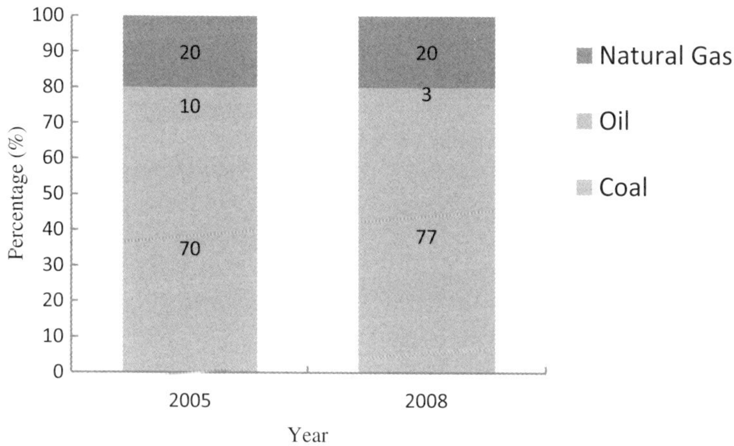

Fig. 8.23. Structure of raw materials in synthesis ammonia industry in China. *Source*: Energy Conservation and Green Power Energy Committee of China Energy Conservation Association, 2009.

In China's synthesis ammonia industry, large, medium, and small-sized enterprises coexist. Proportion of production of small enterprises (annual production lower than 80,000 t/a) declined from 43% in 2005 to 22% in 2008; proportion of production of large enterprises (annual production more than 300,000 t/a) increased from 32% in 2005 to 39% in 2008 (Fig. 8.24).

(5) *Calcium carbide industry*

Calcium carbide is an essential raw material for organic synthesis chemical industry. It can be used to produce a wide variety of organic compounds, and is widely used in the industrial, agricultural, and medicine sector. Acetylene extracted from calcium carbide can be widely used in metal welding and cutting.

China is the biggest producer of calcium carbide in the world, producing 13.61 Mt calcium carbide in 2008, accounting for over 90% of the world total production (Energy Research Institute of NDRC, 2009). In 2009, China's production of calcium carbide was 15.03 Mt.

Power consumption for China's calcium carbide industry changed from 3,450 kWh/t in 2005 to 3,395 kWh/t in 2009, decreasing by 1.6%, which is primarily attributed to the promotion of closed furnace technology (see Fig. 8.25). All open calcium carbide furnaces with severe pollution levels

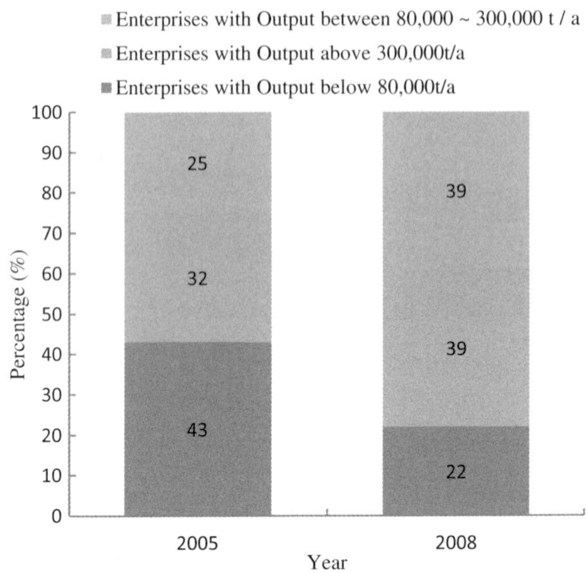

Fig. 8.24. Proportion of production scale of different synthesis ammonia enterprises in China.
Source: Qingyi, 2010.

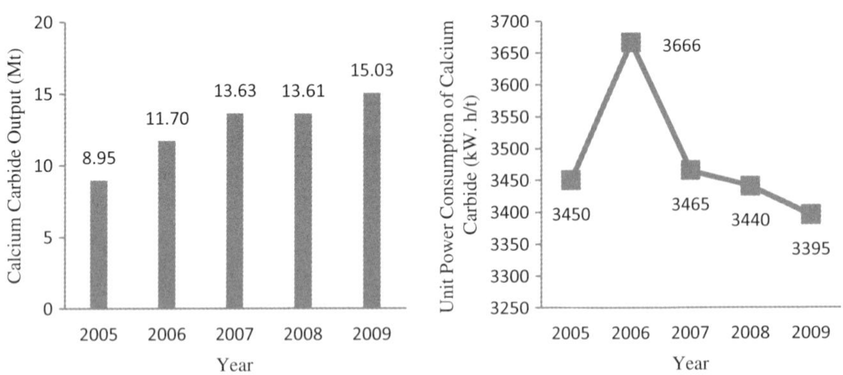

Fig. 8.25. Production and power consumption per unit of calcium carbide between 2005 and 2009.
Source: State Statistics Bureau, 2010.

have been displaced. The closed furnace could recover and reutilize its tail gas; compared with internal fired furnace, it produces fewer pollutants and has better energy conservation effect. Proportion of production capacity of closed furnaces had increased from 8% in 2006 to 20% in 2008, and possibly to 50% in 2010.

Table 8.12 summarizes the advanced technologies, processes and large equipments of chemistry and chemical industry.

8.2.4 *Nonferrous metal industry*

There are many kinds of widely-applied industrial products in the non ferrous metal industry, which plays an important role in China's economic growth, social development and defense construction. China is a major producer of nonferrous metals. By 2002, China's production of 10 nonferrous metals exceeded that of the United States, ranking first in the world. In 2008, China's production of copper, aluminum, lead and zinc accounted for 20%, 32.7%, 37.8%, and 33% of the total world production (China Energy Conservation Investment Corporation, 2009). During the *11th Five Year Plan*, nonferrous metal industry has developed rapidly, where the production of 10 nonferrous metals reached 25.5362 Mt in 2008, increasing by 56.19% compared to 2005. Due to the world financial crisis, growth rate of production slowed down in 2009, increasing only by 3.72% year on year.

Nonferrous metal industry consists of two parts, mining and dressing of nonferrous metal ores, and melting and rolling of nonferrous metal, while this report only deals with the latter part. Nonferrous metal industry is energy-intensive, and rapid growth of production inevitably brings in rapid energy consumption. In 2008, energy consumption of 10 nonferrous metals reached 57.8438 Mtce, increasing by 47.51% compared to 2005. However, growth rate of energy consumption is much lower than that of production, which is attributed to the energy conservation technology in the industry. Production, energy consumption and CO_2 emission of China's nonferrous metal industry is listed in Fig. 8.26.

Energy consumption of per unit industrial value added of the nonferrous metal industry declined from 3.58 tce/10,000 yuan in 2005 to 2.66 tce/10,000 yuan in 2008, decreasing by 25.7%; CO_2 emission of per unit industrial value added of the nonferrous metal industry declined from 8.02 tCO_2/10,000 yuan in 2005 to 5.79 tCO_2/10,000 yuan in 2005, decreasing by 27.8% (Fig. 8.27).

Table 8.12. Advanced technologies, processes and large equipments of chemistry and chemical industry.

Comprehensive energy consumption of ethylene	Advanced technology	Advanced process	Large equipment
2005:986 kgce/t 2008:942 kgce/t	Cracking oven air pre-heating energy conservation technology (saving 17 kgce per ton ethylene) Cracking oven twisted tube enhanced heat transfer technology (cracking oven with an annual production of 30,000 t could save 78.9 t fuel gas) Hydraulic cooling towers (each tower could save 440,000 kWh annually)		Average scale: 45.4 t/a
Comprehensive energy consumption of sodium carbonate 2005:396 kgce/t 2008:392 kgce/t	Heat condensation water closed recovery technology (heat efficiency increasing by 15%–30%, recovery rate increasing by 10%–20%) Transformation gas alkali production technology (comprehensive energy consumption 273 kgce/t)	Associated alkali technology Natural alkali technology	
Comprehensive energy consumption of sodium hydroxide 2005:1,297 kgce/t 2008:1,155 kgce/t	Electrolytic tank ion-film polar distance technology (saving 100–120 kWh per ton of sodium hydroxide)	Ion-film method (proportion to overall production is 44%)	Average scale:120,000 t/year

(*Continued*)

Table 8.12. (*Continued*)

Comprehensive energy consumption of ethylene	Advanced technology	Advanced process	Large equipment
	Expanded anode and modified diaphragm method (saving 147 kWh per ton of sodium hydroxide)		
Comprehensive energy consumption of synthesis ammonia 2005:1,700 kgce/t 2008:1,661 kgce/t	Ammonia synthesis loop zeolite technology (saving 27 kgce energy per ton of synthesis ammonia) Comprehensive energy conservation and renovation technology (saving 200–400 kWh per ton of synthesis ammonia)		Annual production over 300,000 t (proportion to the overall production is 39%)
Comprehensive energy consumption of calcium carbide 2005:3,450 kgce/t 2008:3,395 kgce/t		Closed furnace technology (comprehensive energy consumption decreasing by 20%-25%)	

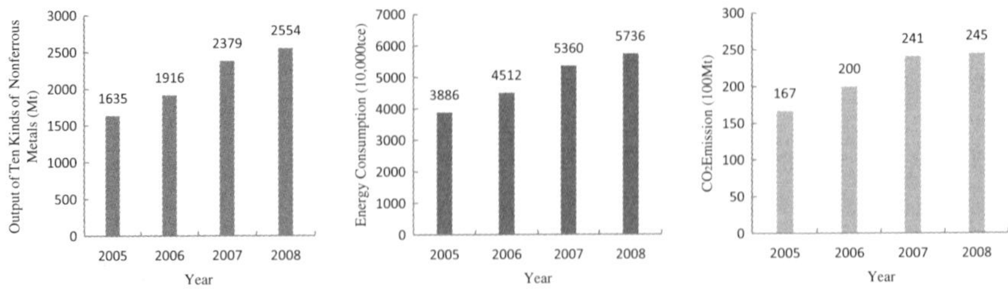

Fig. 8.26. Production, energy consumption and CO_2 emission of 10 nonferrous metals of China in 2005–2008.
Source: *China Statistics Yearbook 2010* and *China Energy Statistics Yearbook 2009*.

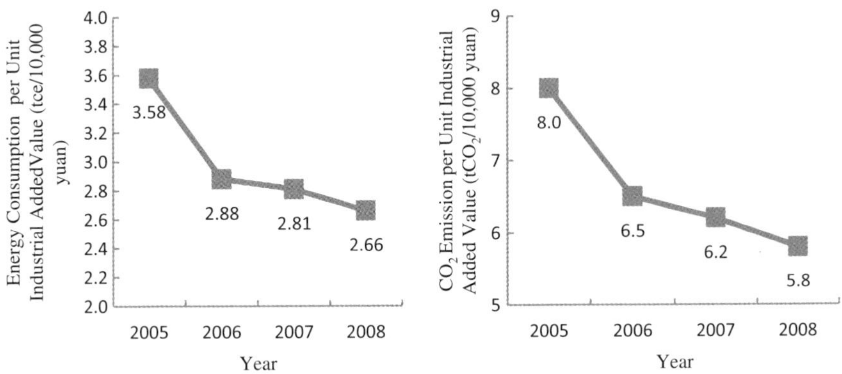

Fig. 8.27. Energy consumption and CO_2 emission per unit industrial value added of nonferrous metals industry in China in 2005–2008.
Source: *China Statistics Yearbook 2010* and *China Energy Statistics Yearbook 2009*.

During the *11th Five Year Plan*, technical indicators of the nonferrous metal industry hit an all time high, with many indicators reaching world's advanced levels. The AC power consumption of aluminum ingot declined substantially, exceeding the energy conservation objective of primary aluminum in 2010 set up by the IPAI, and reached the world advanced level. The AC power consumption reduction of electrolytic aluminum is primarily attributed to promotion of large pre-baking tank and comprehensive utilization of secondary aluminum. Large pre-baking tank over 160 kA could save 9% power than the self-baking tank. From 2000 to 2008, proportion of electrolytic aluminum produced by large pre-baking tanks over 160 kA to the overall production increased from 52% to 86%. In 2007, proportion of electrolytic aluminum produced by large pre-baking tanks over 300 kA to the overall production was 28%. Comprehensive energy consumption of copper smelting also declined rapidly, and the standards of backbone enterprises have caught up with or reached world advanced levels. With continuous technological renovation and expansion of backbone copper smelting enterprises, production proportion of these enterprises has been increasing all the time, leading to higher energy efficiency.

Increasing the utilization rate of secondary metal is an important way to reduce energy consumption of the nonferrous metal industry. Energy consumption and pollutant emission of the secondary metal is much less than that of primary metal. For example, comprehensive energy consumption of secondary copper, aluminum, lead and zinc are 18%, 45%, 27%, and 38% of the primary metals (China Energy Conservation Investment Corporation, 2009). According to the energy consumption of underground mining, opencast mining and smelting process in 2003 provided by China Nonferrous Metal Industry Association, it is calculated that the energy consumption of producing 1 ton of secondary copper, aluminum, lead and zinc is 3.33 tce, 8.20 tce, 1.36 tce, and 2.63 tce less than that of primary metals respectively. Proportion of secondary copper production to the overall production was over 50% in 2007 and 2008. Apparent reduction of comprehensive energy consumption in copper production is primarily attributed to continuous improvement of the production processes and equipments for secondary copper production. Open shaft furnace and tilting furnace not only increases the energy efficiency by 15%–30%, but also checked environmental pollution completely, and improved the sanitary conditions of the working environment. Energy consumption of producing secondary aluminum is only 3.7% to that of primary aluminum. Wide application of combined smelting furnace and proper temperature control of aluminum liquid have increased the energy efficiency by 15%–20%. Reutilization rate of secondary lead remains above 10%, while that of secondary zinc is only 2%–4% (Table 8.13).

Production of electrolytic aluminum was 7.79 Mt in 2005 and 13.17 Mt in 2008, increasing by 69.06%, but reduced to 12.89 Mt in 2009 due to the financial crisis, decreasing by 2.13% year on year. Comprehensive AC power consumption of aluminum ingot decreased from 14,680 kWh/t in 2005 to 14,171 kWh/t in 2008, decreasing by 2.43% (Fig. 8.28).

Refined copper production was 2.61 Mt in 2005 and 4.14 Mt in 2009, increasing by 45.21%. Comprehensive energy consumption of copper smelting reduced from 780 kgce/t in 2005 to 509 kgce/t in 2009, decreasing by 29.62% (Fig. 8.29).

Table 8.13. Reutilization of secondary metals.

Item	2006	2007	2008
Utilized secondary metals (Mt)	4.53	5.30	5.02
Proportion of secondary metals to overall metal production (%)	24.9	23.4	20.5
Aluminum			
Utilized secondary aluminum (Mt)	2.35	2.75	2.60
Proportion of secondary aluminum to overall aluminum production (%)	25.4	21.84	19.7
Copper			
Utilized secondary copper (Mt)	1.0	2	1.9
Proportion of secondary copper to overall copper production (%)	33.3	57	50
Lead			
Utilized secondary lead (Mt)	0.39	0.45	0.44
Proportion of secondary lead to overall lead production (%)	14.4	16.1	12.8
Zinc			
Utilized secondary zinc (Mt)	0.11	0.10	0.08
Proportion of secondary zinc to overall zinc production (%)	3.5	2.7	2.0

Source: China Nonferrous Metal Industry Association, 2009.

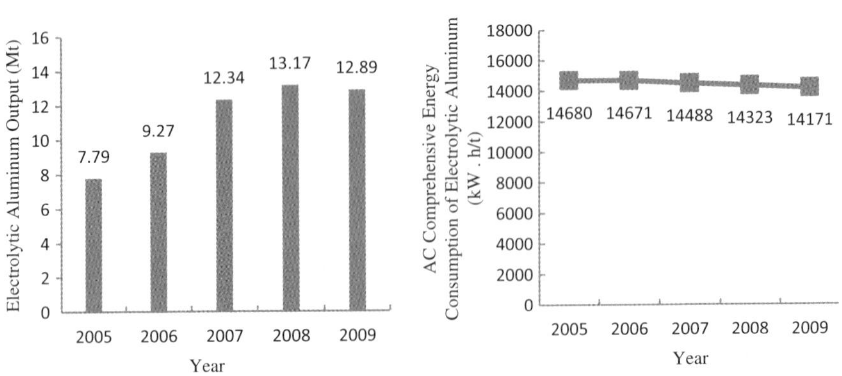

Fig. 8.28. Production and comprehensive AC energy consumption of electrolytic aluminum between 2005 and 2009.
Source: China Nonferrous Metal Industry Association, 2009; State Statistics Bureau, 2010.

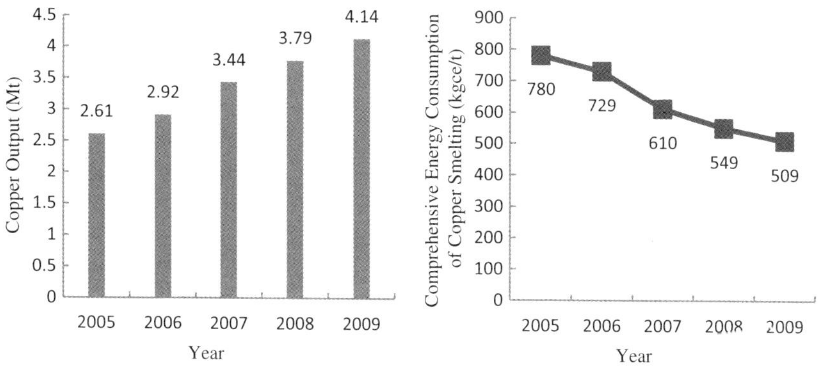

Fig. 8.29. Copper production and comprehensive energy consumption of copper smelting between 2005 and 2009.
Source: China Nonferrous Metal Industry Association, 2009; State Statistics Bureau, 2010.

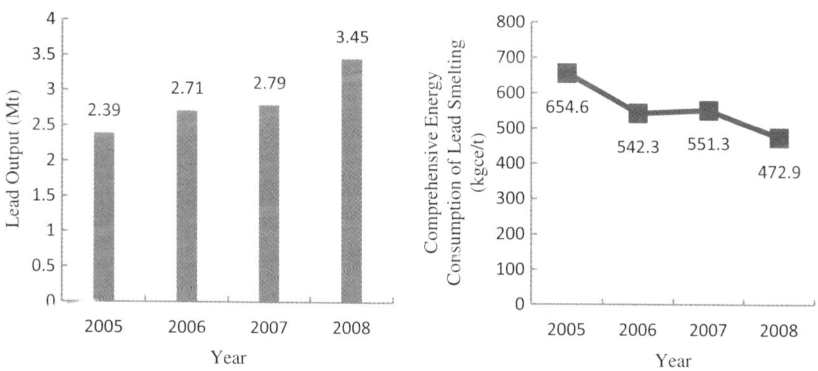

Fig. 8.30. Lead production and comprehensive energy consumption of lead smelting between 2005 and 2008.

Lead production was 239.14 Mt in 2005 and 345.18 Mt in 2008, increasing by 44.34%. Comprehensive energy consumption of lead smelting reduced from 654.6 kgce/t in 2005 to 472.9 kgce/t in 2008, decreasing by 27.76% (Fig. 8.30).

Zinc production was 277.61 Mt in 2005 and 404.23 Mt in 2008, increasing by 45.61%. Comprehensive energy consumption of electrolytic zinc reduced from 1953.1 kgce/t in 2005 to 973.9 kgce/t in 2008, decreasing by 50.14% (Fig. 8.31).

Table 8.14 summarizes the advanced technologies and processes of the nonferrous metal industry.

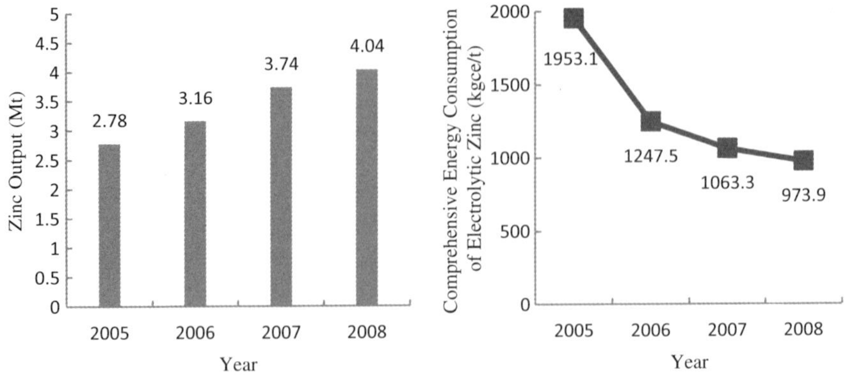

Fig. 8.31. Zinc production and comprehensive energy consumption of electrolytic zinc between 2005 and 2008.
Source: *China Statistics Yearbook 2010*; *China Nonferrous Metal Industry Yearbook 2009*.

Table 8.14. New technologies and processes of nonferrous metal industry.

AC power consumption of electrolytic aluminum	Advanced technology	Advanced process
2005:14,680 kWh/t 2008:14,171 kWh/t	Ore-dressing Bayer process in alumina production Enhanced sintering process in alumina production (saving 40% energy) New type cathode structural energy-saving technology	Large electrolytic aluminum pre-baking tank
Comprehensive energy consumption of copper smelting 2005:780 kgce/t 2008:509 kgce/t	Flash smelting technology	
Comprehensive energy consumption of lead smelting 2005:655 kgce/t 2008:473 kgce/t	ISA furnace lead smelting technology Blast furnace lead smelting technology	

(*Continued*)

Table 8.14. (*Continued*)

AC power consumption of electrolytic aluminum	Advanced technology	Advanced Process
Comprehensive power consumption of electrolytic zinc 2005:1,953 kWh/t 2008:974 kWh/t	Oxygen bottom-blowing smelting technology	Oxygen pressure leading of zinc process

8.3 Policy Evaluation of Manufacturing Sector

In order to realize the objective of decreasing energy consumption per unit GDP by 20% by 2010, the central government formulated such overall energy conservation plans as *Outline of the 11th Five Years Plan, the Decision of the State Council on Strengthening Energy Conservation Efforts* (State Council Circular, 2006, p. 28), *Reply of the State Council to the Plan of Energy Consumption Reduction Objective Per Unit GDP of All Districts During the 11th Five Year Plan* (State Council Letter, 2006, p. 94), and *Notice of the State Council on Printing and Distributing Comprehensive Working Plan of Energy Conservation and Emission Reduction* (State Council Circular, 2007, p. 15), and implemented comprehensive energy conservation policies and financial-tax policies for the industrial sector.

The comprehensive energy conservation plan stipulated explicitly that the industrial sector should focus on energy conservation of big energy consumers, i.e. the steel, nonferrous metal, coal, electric power, chemistry and chemical, and building materials industry; and highlighted that the government shall provide guidance and the enterprises shall fulfill their obligations. During the *11th Five Year Plan*, the industrial sector conducted the Top-1000 Enterprises Energy Conservation Program, the Ten Key Energy Conservation Projects and the Backward Production Capacity Elimination Program. From 2006 to 2009, these three programs saved 132 Mtce, 140 Mtce, and 110 Mtce respectively, contributing 27%, 29%, and 22% to the overall energy conservation efforts (Table 8.16). Apart from implementing administrative measures such as objective responsibility, administrative enforcement and administrative admission, the industrial sector also utilized other market mechanisms and incentive measures such

Table 8.15. Major polices for the industrial sector between 2004 and 2010.

Category of policies	List of policies	Energy conservation performance in the first 4 years of the *11th five year plan*
Comprehensive Energy Conservation Plan	Notice of NDRC on Printing and Distributing Medium and Long-Term Energy Conservation Plan (NDRC Circular, 2004, p. 2505)	132 Mtce has been saved by the top 1,000 enterprises, accounting for 27% of the total
	Outline of the 11th Five Year Plan (approved by the 4th session of 10th NPC in March 14, 2006)	
	Decision of the State Council on Strengthening Energy Conservation Efforts (State Council Circular, 2006, p. 28)	
	Reply of the State Council to the Plan of Energy Consumption Reduction Objective Per Unit GDP of All Districts During the 11th Five Years Plan (State Council Letter, 2006, p. 94)	
	Notice of the State Council on Printing and Distributing the Comprehensive Work Plan on Energy Conservation and Emission Reduction (State Council Circular, 2007, p. 15)	

(*Continued*)

Table 8.15. (*Continued*)

Category of policies	List of policies	Energy conservation performance in the first 4 years of the *11th five year plan*
Comprehensive Policy of Industrial Energy Conservation	Decisions of the State Council on Issuing and Implementing the Tentative Regulation on Accelerating Adjustment of Industrial Structures (State Council Circular, 2005, p. 40)	Eliminated 110 Mtce backward production capacity, accounting for 22% of the total
	Notice on the Implementation Plan of Top 1,000 Enterprises Energy Conservation Program (NDRC Circular, 2006, p. 571)	
	Opinions on Implementing Ten Key Energy Conservation Projects during the 11th Five Years Plan (NDRC Circular, 2006, p. 1457)	
	Notice of the State Council on Several Opinions of Accelerating Shutdown of Small Thermal Power Units (State Council Circular, 2007, p. 2)	
	Notice on Opinions of Accelerating the Implementation of EPC for Promotion of Energy Conservation Service Industry (State Council Office Circular, 2010, p. 25)	
	General Outline of China Resources Comprehensive Utilization Technology Policy (NDRC Notice, 2010, p. 14)	
Finance and Taxation Policy of Industrial Energy Conservation	Notice on Adjusting the Export Tax Rebate Rate of Certain Goods and Increasing the List of Prohibited Goods for Processing Trade (Ministry of Finance, 2006, p. 139)	Ten Key projects saved 140 Mtce, accounting for 29% of the total

(*Continued*)

Table 8.15. (*Continued*)

Category of Policies	List of Policies	Energy Conservation Performance in the First 4 Years of the *11th Five Year Plan*
	Tentative Regulations on Management of Financial Incentive Fund for Energy Conservation Technology Renovation (Ministry of Finance, 2007, p. 558) Tentative Regulations on Management of Central Finance Incentive Fund for Eliminating Backward Production Capacity (Ministry of Finance, 2007, p. 873) List of Preferential Enterprise Income Tax on Resources Comprehensive Utilization (2008 Edition) (MOF, 2008, p. 117) Tentative Regulations on Management of EPC Finance Incentive Fund (MOF, 2010, p. 249) Circular on Handling Issues Concerning Preferential Power Price for Energy-Intensive Enterprises (NDRC, 2010, p. 978) Policies and accomplished energy savings of the Top-1,000 Enterprises Energy Conservation Program Policies and accomplished energy savings of the Backward Production Capacity Elimination Program Policies and accomplished energy savings of the Ten Key Energy Conservation Project	

Source: NDRC, 2010.

Table 8.16. Performance of industrial energy conservation action.

Energy conservation action	Saved energy in 2006–2009* (100 Mtce)	Proportion to the overall saved energy 490 Mtce† (%)
Top-1,000 Enterprises Energy Conservation Program	1.32	27%
Ten Key Energy Conservation Project	1.4	29%
Backward Production Capacity Elimination Program	1.1	22%

Note: There are much overlapping in the Top-1,000 Enterprises Energy Conservation Program and Ten Key Energy Conservation Project, which cannot be added up twice.
Source: *NDRC, 2010; †Xinhua News, 2010b.

as tax preference, discriminatory electric price and Energy Performance Contracting (EPC), and enforced binding indicators for energy conservation and consumption reduction. Policy evaluation of the section mainly consists of four parts: Top-1,000 Enterprises Energy Conservation Program, Ten Key Energy Conservation Project, Backward Production Capacity Elimination Project, and EPC.

8.3.1 *Top-1,000 enterprises energy conservation program*

Top-1,000 Enterprises Energy Conservation Program aims to improve the energy efficiency of top 1,000 energy-intensive enterprises in nine key energy-intensive sectors, which is an important policy in decreasing energy consumption per unit GDP by 20% during the *11th Five Year Plan*. The program learns from the practice of international energy agreements (voluntary agreement of negotiation agreement), i.e. the government and industrial enterprises sign agreements and lay down energy conservation objectives according to the obligations enumerated in the agreements. Before its initiation, Shandong Economic and Trade Committee had conducted voluntary agreement pilot projects in Jinan Iron & Steel Group Corporation and Laiyang Iron and Steel Group Corporation, which accomplished the objectives successfully and become a model for the Top-1,000 Enterprises Energy Conservation Program.

In 2006, the NDRC, the Office of National Energy Leading Group, the State Statistics Bureau, the General Administration of Quality Supervision, the State-owned Enterprises Supervision and Administration Commission jointly formulated the *Implementation Plan for Top-1,000 Enterprises Energy Conservation Program* (NDRC Circular, 2006, p. 571), and began to implement the Top-1000 Enterprises Energy Conservation Program. The implementation plan appealed top 1,000 energy-intensive enterprises to improve their energy efficiency substantially, whose energy consumption per unit of major products shall reach domestic advanced levels, and that of some enterprises shall reach world's advanced level or industry's advanced levels. The implementation plan also stipulated clearly that 100 Mtce shall be saved from 2006 to 2010.

Top 1,000 enterprises refer to the enterprises with an annual energy consumption of over 180,000 tce in the nine energy-intensive sectors of steel, nonferrous metal, coal, power, petroleum and petrochemical, chemistry and chemical, building materials, paper-making and textile industry. There were 1,008 enterprises in 2004. In 2006, there were 19 additional enterprises as well as 29 withdrawing enterprises due to close-down or merger, so altogether 998 enterprises signed the letters of responsibility for energy conservation. In 2007, three enterprises went bankrupt, 15 enterprises were merged with other enterprises, 26 enterprises were closed down and energy consumption of one enterprise underwent major changes, leaving only 953 enterprises for examination and appraisal. The number was 953 in 2008, and 901 in 2009 (Fig. 8.32).

Overall energy consumption of top 1,000 enterprises was 670 Mtce in 2004, accounting for 33% of the overall energy consumption in China, and 47% of the industrial sector. Average comprehensive energy consumption of top 1,000 enterprises in 2004 was 670,000 tce, while that of the building materials was 350,000 tce, of petroleum and petrochemical industry was 1.07 Mtce, of steel industry was 1 Mtce, and of other industries was lower than 670,000 tce.

The plan requests top 1,000 enterprises to set up energy conservation management offices and lay down energy conservation objectives, create an energy utilization report system, conduct energy auditing and energy conservation training, establish energy conservation incentive mechanism, and increase energy conservation investment. The top 1,000 enterprises report energy utilization progress directly to the State Statistics Bureau every three months via network, which ensures the accuracy and reliability of the data, alleviating the burden of both parties. The State Statistics Bureau disseminates information of average and overall energy consumption of each industry instead of each enterprise. Of the top 1,000 enterprises, 95% established specialized or part-time energy management offices and arranged

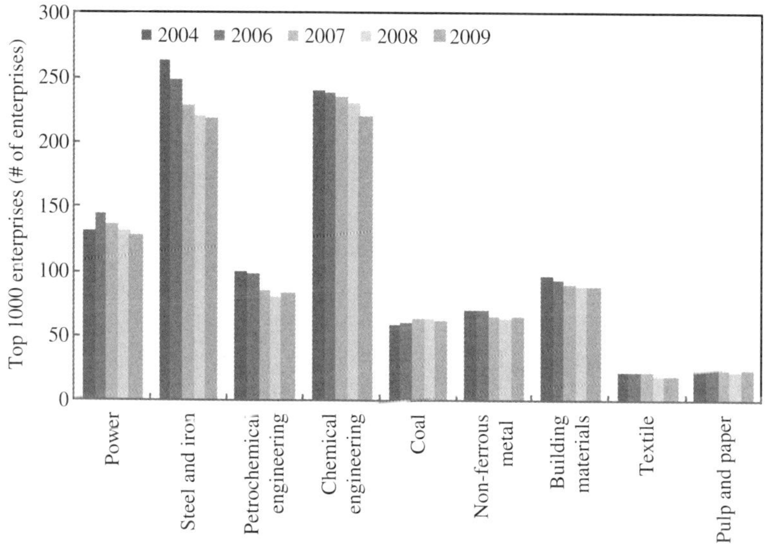

Fig. 8.32. Top 1,000 enterprises.
Source: NDRC and State Statistics Bureau, 2007; 2008; 2009; 2010.

management personnel. Enterprise energy management mechanism includes setting up energy procurement management system, energy utilization management system, and quota examination system of various production procedures.

With regard to signing letters of responsibility on energy conservation objectives, altogether 998 enterprises signed the letters of responsibility in 2006, of which, 15 state-owned enterprises signed letters of responsibility directly with the NDRC, while other enterprises signed letters with local governments. It is relevant to mention that the enterprises under state-owned enterprises also need to sign letters of responsibility with provincial governments. At the governmental level, local governments also signed letters of responsibility with the NDRC to shoulder energy conservation responsibilities. For example, the NDRC and Shanghai government signed a letter of responsibility, while the latter also signed letters of responsibility with 14 enterprises under state-owned enterprises and nonstate-owned enterprises to lay down energy conservation objectives for each enterprise.

The letter of responsibility consists of energy conservation objectives as well as assessment and evaluation. Energy conservation objective stipulates the energy savings of enterprises, while assessment and evaluation lays down regulations on assessment targets, contents as well as reward and punishment. Energy conservation objectives are measured by integrated evaluation method, and the results consist of four levels — over-fulfilled, fulfilled, basically fulfilled and unfulfilled. The results will be factored into assessment of the central government to provincial governments and provincial officials. Enterprises and areas with unfulfilled energy conservation objectives shall be deprived of the opportunity to compete for the year-end reward and titles of honor. Inside fully state-owned or state-hold enterprises with unfulfilled energy conservation objectives, the responsible persons cannot be promoted while assessing the annual performance. The examination and assessment system prompts the government officials to supervise and assist the local enterprises to fulfill their energy conservation objectives.

In laying down energy conservation objectives, the NDRC takes the energy consumption in 2004 as the benchmark standard and lays down energy conservation objectives of 2010. The NDRC also takes into account the basic conditions of each enterprise, such as industrial natures and overall technological levels, into consideration. However, detailed energy conservation potentials of each enterprise or industry are not evaluated due to limited time. According to the experience worldwide, the work of laying down objectives should start 3–4 years prior to the initiation of the program (LBNL, 2010). However, Top 1000 Enterprises Energy Conservation Program was designed to act in concert with the *11th Five Year Plan*, these norms could not be adopted due to limited timeline and large number of participating enterprises.

Energy auditing planning for top 1000 enterprises began in the fourth quarter of 2006. In the first half of 2007, energy conservation authorities of each provincial government organized experts to conduct collective auditing to the energy reports of enterprises, and requested them to amend the unqualified reports. Energy auditing planning for all enterprises was completed during the first half of 2007. The enterprises could conduct energy auditing by themselves or employ a third party. The third party is usually an engineering and design entity instead of a national accreditation body. At present, 60%–70% enterprises have chosen to conduct energy auditing themselves.

Since implementation of the Top 1,000 Enterprises Energy Conservation Program, the cumulative energy savings from 2006 to 2008 was 20 Mtce, 58 Mtce, 106 Mtce, and 132 Mtce respectively. Of the energy consumption of top 1,000 enterprises in 2006, coal consumption accounted for 36.10%,

21.30% for crude oil, 12.97% for electricity, 8.91% for coke, 1.91% for steam, 1.48% for natural gas and 17.33% for other energies. Supposing that the energy structure of top 1,000 enterprises remains the same during 2006–2009, cumulative CO_2 emission reduction would reach 396 Mt by 2009. Figure 8.33 is the cumulative energy conservation and CO_2 emission reduction of each industry of the top 1,000 enterprises from 2006 to 2009.

From 2006 to 2009, the top five provinces of overall energy consumption were Shanxi, Shandong, Liaoning, Hebei, and Jiangsu province, whose cumulative energy savings were 59.41 Mtce, accounting for 45% of the overall energy savings of top 1,000 enterprises (132 Mtce), and whose CO_2 emissions reduction was 181 Mt (Fig. 8.34).

In general, under the Top 1000 Enterprises Energy Conservation Program, altogether 132 Mtce energy was saved from 2006 to 2009, accounting for 27% of the overall energy savings of 490 Mtce in the first four years of the *11th Five Year Plan.*

8.3.2 *Ten key energy conservation projects*

The Ten Key Energy Conservation Projects is an important element of the *Medium and Long Term Energy Conservation Plan* (NDRC Circular, 2004, p. 2505) and has been included in the *Outline of the 11th Five Years Plan.* In 2006, the NDRC, the Ministry of Science and Technology, the Ministry of Finance, the Ministry of Construction, the General Administration of Quality Supervision, the Inspection and Quarantine, the State Environmental Protection Administration, the Government Offices Administration of the State Council, the Government Offices Administration of the Central Committee of CPC jointly formulated the *Implementation Plan for Ten Key Energy Conservation Projects During the 11th Five Year Plan* (NDRC Circular, 2006, p. 1457). Of the ten key energy conservation projects (Special Column 8.2), boiler (kiln) renovation of the coal-fired industry, regional heat and power cogeneration, waste heat and pressure utilization project, oil conservation and alternative project, electric motor system energy conservation project, and energy system optimization project are key energy conservation projects of the industrial sector, and will be discussed in the section. Apart from above-mentioned projects, there are also the building energy conservation project, the green lights project, the governmental agency energy conservation project, and the energy conservation monitoring and technical service system construction project.

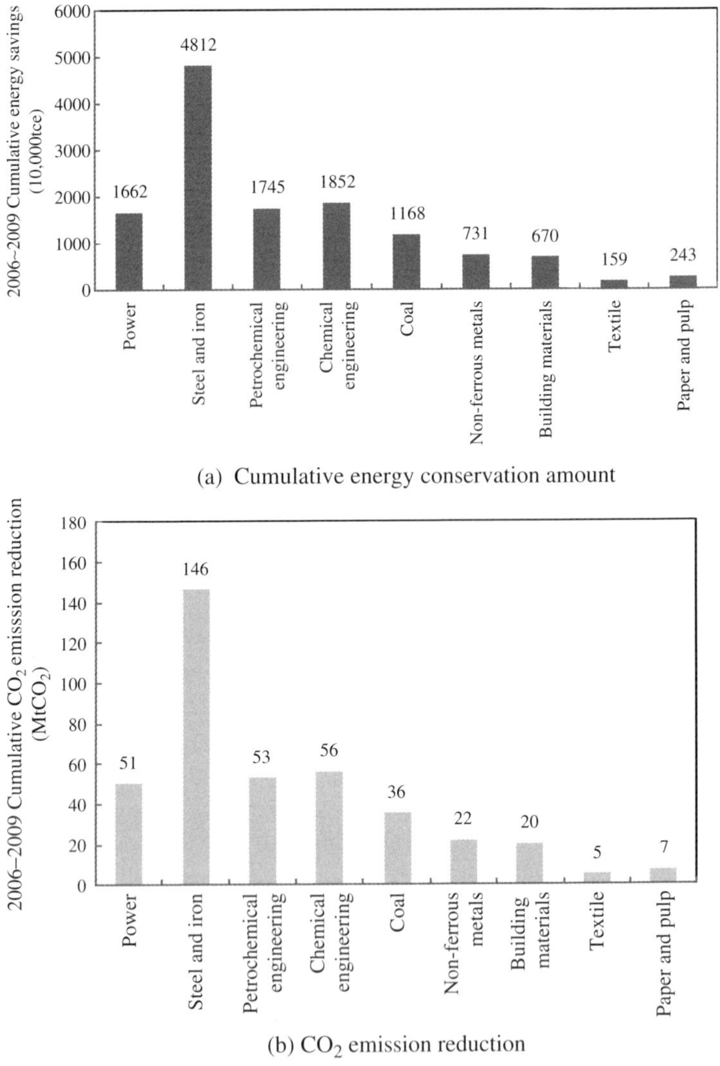

Fig. 8.33. Cumulative energy conservation and CO_2 emission reduction of each industry of top 1,000 enterprises from 2006 to 2009.
Source: NDRC and State Statistics Bureau, 2007; 2008; 2009; 2010.

The objective of the Ten Key Energy Conservation Projects is to create an energy savings of 240 Mtce during the *11th Five Year Plan*. From 2006 to 2009, 140 Mtce energy savings have been achieved (NDRC, 2010).

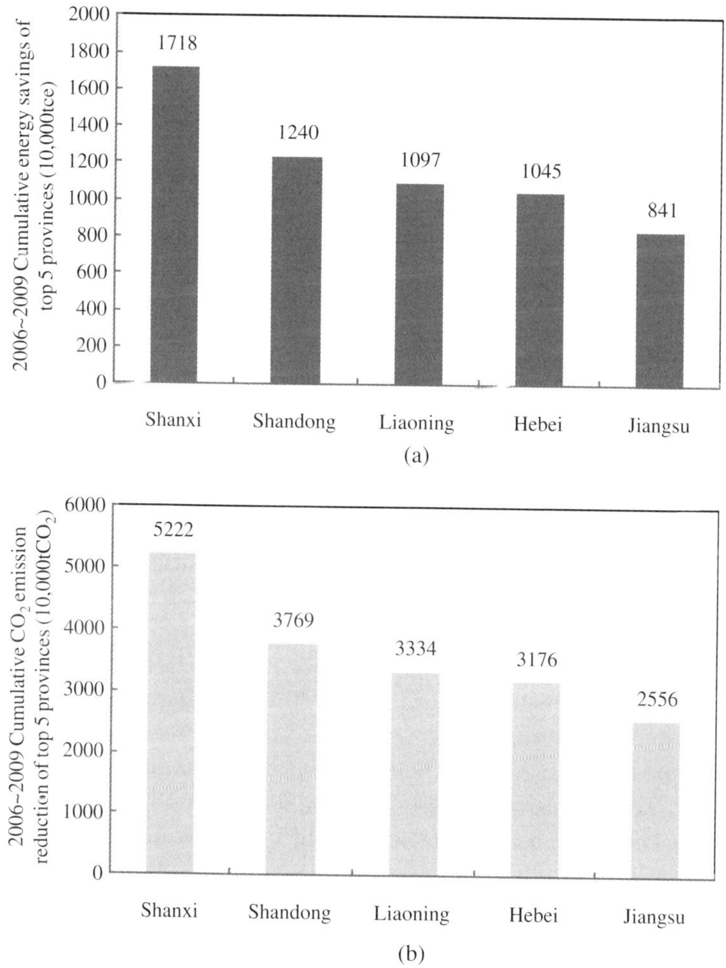

Fig. 8.34. Cumulative energy savings and CO_2 emission reduction of top five energy consuming provinces from 2006 to 2009.
Source: NDRC and State Statistics Bureau, 2007; 2008; 2009; 2010.

The essence of Ten Key Energy Conservation Projects is to encourage the enterprises to conduct energy conservation technology renovations and enhance their energy conservation and emission reduction capabilities by increasing government investment. The implementation plan demands accelerating the development of energy conservation technologies, especially supporting the pilot projects with independent IPR, and enhance the independent innovation capability. On Aug. 2007, the Ministry of Finance and the NDRC issued and distributed the *Tentative Regulations on Management of Financial Incentive Fund for Energy Conservation Technology Renovation* (MOF, 2007, p. 558) (*Regulations* in short term), which stipulated that the central finance shall allocate necessary initiation funds, and provide proper support and reward to top 10 energy conservation projects by method of "reward subsidy". The regulations also stipulated that, the central finance will offer financial incentives to the boiler (kiln) renovation of the coal-fired industry, waste heat and pressure utilization project, oil conservation and alternative project, electric motor system energy conservation project, and energy system optimization project. The details are as following: For the enterprises with an energy savings of over 10,000 tce, 200 yuan/tce, and 250 yuan/tce will be subsidized respectively for the East Region and Central-West Region. From 2006 to 2009, the central finance altogether allocated RMB 28.76 billion in supporting the Ten Key Energy Conservation Projects.

There are many overlaps between the Ten Key Energy Conservation Projects and Top 1000 Enterprises Energy Conservation Program, i.e., a fairly amount of key projects are implemented by the top 1,000 enterprises. With respect to setting up energy conservation objectives, incentive standards, energy conservation auditing and objective assessment, both programs act in concert. Further, in setting up energy conservation objectives of 10 key projects, various development and reform committees, economic and trade committees, enterprises and public institutions will negotiate in formulating the implementation plans of local key energy conservation projects and annual implementation schedules, and coordinate with each other for their implementation. Incentive standards and assessment methods have been illustrated in the previous section.

With regard to objective assessment, implementation, and performance of ten key projects will be included in the annual work assessment system of relevant governments, enterprises and public institutions. In addition, various levels of governments and enterprises are mainly responsible for the completion of energy conservation objectives, while the industry associations also have to shoulder the responsibility of providing more guidance to energy conservation projects.

Assessment of energy conservation amount is conducted in two phases, by 28 governmental agencies designated by the Ministry of Finance. During the first phase, enterprises report the project's energy savings to the municipal governments before implementing energy conservation technological renovation, and the latter submits the report to the provincial governments after preliminary assessment for comprehensive assessment. Then, the provincial governments will submit the report to the NDRC, which will be assessed by 28 governmental agencies. The Ministry of Finance will allocate 60% incentive fund to the enterprises with approved energy conservation projects before initiation of the project. During the second phase, governmental agencies will conduct another assessment of the energy savings after project completion, and will allocate another 40% incentive fund to the projects with accomplished energy savings, and allocate a certain amount of fund in proportion to the finished energy savings of the projects failing the objectives. In order to ensure transparency and justice of the assessment procedure, 28 governmental agencies are prohibited to assess the enterprises located in the same provinces with them. By 2010, over 3,000 energy conservation projects have completed energy conservation reward assessment, and the assessment is still going on.

In general, the central finance altogether allocated RMB 28.76 billion to support the Ten Key Energy Conservation Projects from 2006 to 2009, creating an energy-saving capacity of 140 Mtce, contributing 29% to the overall energy conservation of 490 Mtce in the first four years of the *11th Five Year Plan.*

Special Column 8.2. Ten key energy conservation projects.

(1) Boiler (kiln) renovation project of the coal-fired industry: renovate low-efficiency industrial boilers, build regional collective coal distribution and processing centers for boilers, displace backward industrial kilns, and conduct comprehensive energy conservation renovation to existing industrial kilns.

(2) District heat and power cogeneration project: build heat and power cogeneration systems primarily for the heating system and for industrial use; distributed heat and power cogeneration and heat-power-chilled water cogeneration system; and pilot thermo-electric plant with comprehensive utilization of low-caloric value fuel and straw.

(3) Waste heat and pressure utilization project: in the energy-intensive industries such as steel, building materials as well as chemistry and chemical industry, renovate and build waste heat and pressure utilization equipment and devices, such as the pure low-temperature waste heat generation systems, pressure difference power generation,

and reutilization of by-product inflammable gas and low-caloric value gas.

(4) Oil conservation and alternative project: in the power, petroleum and petrochemical, building materials, chemistry and chemical industry, as well as traffic and transportation industry, implement oil conservation and alternative renovation, develop coal and LPG products, alcohol ether fuel and biochemical diesel oil.

(5) Electric motor energy conservation project: Renovate low-efficiency motors, regulate and adjust the large and medium asynchronous motors, and conduct energy-saving renovation to the driven equipment of the motor systems.

(6) Energy system optimization project: Conduct energy conservation renovation to the oil refineries, ethylene, synthesis ammonia and steel industries.

(7) Buildings energy conservation project: Implement 50% energy conservation standards for all new buildings, and 65% for the new buildings in four municipalities directly under the central government, and in northern severe cold and cold areas, and conduct strict all-round supervision. Build pilot projects of low and super-low energy consumption buildings with integrated recyclable energy, conduct city-level pilot renovation of existing living and public buildings, promote the industrialization of new wall materials and energy-saving building materials.

(8) Green lights project: Renovate production lines of electricity-saving lamps with enhanced product quality, reduced cost of production and more indigenous renovation capability; promote the application of high-efficiency lighting products.

(9) Governmental agency energy-saving project: Energy-saving renovation of existing buildings and comprehensive electricity efficiency renovation, energy conservation assessment of new buildings and all-round monitoring, promote governmental procurement of energy-saving products.

(10) Energy conservation monitoring and technical service system project: Renovate the energy-saving monitoring devices and equipment of provincial energy-saving monitoring (inspection) centers, conduct energy assessment to key energy-intensive enterprises.

Note: Compiled from the website materials of the NDRC.

8.3.3 *Elimination of backward production capacity*

Elimination of backward production capacity is very meaningful to optimizing industrial structure, increasing the overall industrial level and realizing transformation of economic growth mode. Eliminating backward production capacity of energy-intensive industries is a major energy-saving activity and a focus of industrial structural optimization during the *11th Five Year Plan*. With the joint functions of the *Decision of the State Council on Issuing and Implementing the Tentative Regulation on Accelerating Adjustment of Industrial Structures* (State Council Circular, 2005, p. 40), *Notice of the State Council on Several Opinions of Accelerating Shutdown of Small Thermal Power Units* (State Council Circular, 2007, p. 2), *Notice on Adjusting the Export Tax Rebate Rate of Certain Goods and Increasing the List of Prohibited Goods for Processing Trade* (MOF, 2006, p. 139), *Tentative Regulations on Management of Central Finance Incentive Fund for Eliminating Backward Production Capacity* (MOF, 2007, p. 873) and *Circular on Handling Issues Concerning Preferential Power Price for Energy-Intensive Enterprises* (NDRC, 2010, p. 978), altogether 110 Mtce energy savings has been realized during the first four years of the *11th Five Year Plan* (NDRC, 2010) (Table 8.17).

The elimination indicators for six industries of iron-making, steelmaking, cement, electrolytic aluminum, glass and paper-making issued by the Ministry of Industry and Information Technology in 2010 are much higher than that stipulated in the *Notice on Further Expanding Work Efforts to Ensure the Achievement of Energy Conservation and Emission Reduction in the 11th Five Year Plan* (State Council Circular, 2010, p. 12). The elimination objective for papermaking industry is 4.32 Mt, increasing by 715% compared to 530,000 t in the *Notice*. It highlights the urgency for economic structure adjustment, as well as the central government's determination to ensure the fulfillment of emission reduction objectives in the *11th Five Year Plan*.

Elimination of backward production capacity is primarily conducted by taking governmental administrative measures. For the enterprises with excessive consumption or whose production capacity does not comply with relevant industrial policies, administrative compulsory measures will be taken to enforce dismantling or production suspension. There are two kinds of shut-down enterprises: For the dismantled or suspended legal enterprises with necessary approval and examination proceedings but whose production capacity does not comply with relevant industrial policies, the government

Table 8.17. Energy savings of backward production capacity elimination between 2006 and 2009 and elimination objectives in 2010.

Industry	Contents	Objectives of the 11th five years plan	Cumulative eliminated production capacity between 2006 and 2009	Existing backward production in 2009	Elimination objectives in 2010
Coal	Production capacity of small enterprises declined from 1,080 Mt in 2005 to 700 Mt in 2010	380 Mt	311 Mt		200 Mt
Coke	Small coking enterprises with the blast furnaces of coking chambers smaller than 4.3 m^3	80 Mt	133 Mt	45 Mt	21.27 Mt
Electricity	Replacing small thermal power units with large ones	50 GW	60.06 million kW	Unit < 100,000 kW 14.66 million kW	10 million kW
Iron Smelting	Blast furnaces smaller than 300 m^3	100 Mt	81.7 Mt	125 Mt	30 Mt
Steel Smelting	Small converters and electric furnaces with an annual production of less than 200,000 t	55 Mt	59.4 Mt	10% of the production capacity	8.25 Mt
Cement	Equal amount of cement clinker with shaft kiln	250 Mt	240 Mt	390 Mt	91.55 Mt
Glass	Backward plate glass	30 million weight cases	25 million weight cases	80 million weight cases	6.48 million weight cases

(Continued)

Table 8.17. (*Continued*)

Industry	Contents	Objectives of the 11th five years plan	Cumulative eliminated production capacity between 2006 and 2009	Existing backward production in 2009	Elimination objectives in 2010
Electrolytic Aluminum	Small pre-baking tank	650,000 t	800,000 t	1 Mt	339,000 t
Calcium Carbide	Calcium carbide production capacity of furnaces lower than 6,300 kVA	2 Mt	2.9 Mt	27% of production capacity	718,000 t
Metal Alloy	Submerged arc furnace lower than 6,300 kVA	4 Mt	5.19 Mt	10% of production capacity	1.44 Mt
Papermaking	Grass pulp production equipment lower than 34,000 t per year; chemical pulping production line lower than 17,000 t per year; paper mills using waste paper as raw materials whose annual production is lower than 10,000 t and with unqualified emission standards	6.5 Mt	6 Mt	—	4.32 Mt

Note: Of the eliminated production capacity of coking industry, 50 Mt are primitive and modified coking enterprises, while 82.5 Mt are small coking enterprises.

Source: Ministry of Industry and Information Technology State Statistics Bureau, China Coal Industry Association, China Coking Industry Association, China Iron and Steel Industry Association, China Building Materials Industry Association, China Calcium Carbide Industry Association, China Paper Association, *China Environmental News*, 2010-6-9; *China Chemical Industry News*, 2010-4-8.

will offer some compensation; for the enterprises without legal approval procedures, the enterprises will be banned to operate and the owners will suffer further fines and penalties. Central finance's incentive fund and local finance's special compensation funds are used to compensate for the losses suffered by the local governments, investors and employees during shutdown process. For the districts, cities or counties whose local financial income decreased due to large-scale shutdown actions, financial transfer payment will be increased. For the districts, cities or counties with insufficient shutdown actions, their accountability will be pursued and the financial transfer payment will reduce.

Market measures are supplementary to the administrative measures, one of which is to apply discriminatory electricity price to energy-intensive enterprises. In June 2006, enterprises of six energy-intensive industries of electrolytic aluminum, metal alloy, calcium carbide, sodium hydroxide, cement and steel industry were divided into four categories of elimination, restriction, permission and encouragement, and were applied with discriminatory electricity prices. The NDRC issued the *Opinions on Completing the Discriminatory Electricity Price Policy* (State Council Circular, 2006, p. 77), which adjusted the electricity price for elimination-category enterprises from 0.05 yuan/(kWh) to 0.20 yuan/(kWh); and adjusted the electricity price for restriction-category enterprises from 0.02 yuan/(kWh) to 0.05 yuan/(kWh). The *Circular on Handling Issues Concerning Preferential Electricity Price for Energy Intensive Enterprises* (NDRC Price, 2010, p. 978) in 2010 made further adjustment to the electricity price differences. The *Circular* stipulated that discriminatory electricity price continues to be applied to eight industries of electrolytic aluminum, metal alloy, calcium carbide, sodium hydroxide, cement, steel, yellow phosphorous and zinc smelting industry, and the electricity price of elimination-category enterprises will be increased from 0.20 yuan/(kWh) to 0.30 yuan/(kWh). The *Circular* also stipulated that price standards could be further increased for the elimination-category and restriction-category enterprises according to local requirements. The State Electricity Regulatory Commission reported in 2008 that, about 2,000 energy-intensive enterprises had been shut down or converted to other enterprises since implementation of discriminatory electricity price, and over 70% were elimination-category enterprises (Northern Star Power Grid, 2008).

Elimination of backward production capacity has realized certain achievements, but not without difficulties. There are mainly four problems: Economic interests, vested interests of local governments, incomplete compensation mechanism and over-emphasis on administrative measures. First, the basic reason for existence of backward production capacity in market economy is that the backward enterprises are still profitable. According to cost analysis, although the cost of raw material and energy consumption of the backward enterprises is higher than that of the advanced enterprises, they are still profitable in market competition due to small overall investment, low labor cost, less environment-related cost and meager social cost. At present, the backward production elimination mechanism has not realized internalization of external costs, giving cost advantage to the backward enterprises, so they do not have to shut down. Second, recognition of local governments differs from the central government. Many local governments are interested in pursuing maximum GDP growth, when eliminating backward production capacity affects local GDP growth and financial income, some local governments will give silent consent and even protect the backward production capacity. Third, current efforts puts more emphasis on responsibilities and obligations but with insufficient compensation for local government, enterprise owners and employees; therefore, it faces resistance from local stakeholders. Various levels of governments shall create a special fund for elimination of backward production capacity and explore market compensation mechanisms for various industries, so as to deal with the economic and social problems on personnel resettlement, capital settlement and GDP loss. Lastly, too much emphasis is put on the restrictive and obligatory functions of administrative measures, with insufficient emphasis on incentive measures. Although administrative measures could check the development of backward production capacity in short term, they cannot allocate resources reasonably and establish a long-term mechanism for backward production capacity elimination.

In general, eliminating backward production capacity is central to the transformation of industrial structure and economic growth mode, and is an opportunity for enhancing industrial levels and creating an awareness of scientific development. From 2006 to 2009, 110 Mtce energy conservation capacity had been realized due to backward production elimination, contributing 22% to the overall capacity of 490 Mtce in the first four years of the *11th Five Year Plan*.

8.3.4 *Energy performance contracting (EPC)*

There are many technically applicable and economically reasonable projects in the industrial energy conservation sector, which face financing bottlenecks due to lack of necessary economic means in the market. Energy service companies are usually small in scale with insufficient self-owned assets, and cannot obtain bank loan due to lack of mortgage and guaranty. Apart from financing problems, the energy service industry is also faced with other difficulties, such as incomplete taxation policies, absence of supportive policies, unclear budgetary expenditure, and different accounting practices. Under such circumstances, implementing EPC is an effective measure in promoting energy conservation and emission reduction within the market mechanism.

With the support of International Finance Corporation (IFC) and the Global Environmental Facility (GEF), China Energy Conservation Promotion Program was implemented in 1998. Three demonstration energy service companies were established in Beijing, Shandong and Liaoning; the Energy Conservation Service Industry Committee was also established, and an EPC mechanism was introduced into China. There were mainly four roles for the IFC in China Energy Conservation Promotion Program. First, with respect to project loan, the IFC provides mortgage for energy service companies, utilizes GEF grants to channel in the funding from banks, and shares the investment risks of projects. Second, with respect to financial service, IFC provides technical assistance to banks to enhance their professional ability in dealing with energy efficiency projects. Third, with respect to marketing, IFC becomes partner to the banks as well as the energy service companies, and serves as their bridges and facilitators. Fourth, with respect to sharing benefits of savings, IFC provides technical guidance to the final energy consumer enterprises. By China's Energy Conservation Promotion Program, IFC created a sustainable and reproducible business mode in the energy conservation sector by enhancing the abilities of market players such as the financial institutions, energy service companies and terminal energy consumption enterprises.

In recent years, EPC has developed rapidly in China, and the operational mode has evolved from simple shared savings to guaranty and trusteeship. From 2003 to 2009, investment in EPC projects increased from RMB 850 million to RMB 28 billion, and the annual energy conservation capacity increased from 560,000 tce to 13.5 Mtce. In 2009, there were 502 energy service companies and over 4,000 energy-saving projects, and the employees in the industry had increased from 65,000 by the end of 2008 to 113,000, increasing by 74%.

In April 2010, the State Council issued the *Opinions for Accelerating the Implementation of EPC for Promotion of Energy Conservation Service*

Industry (State Council Office Circular, 2010, p. 25), which is an important policy for creating and cultivating a beneficial market environment. First, the opinions included EPC projects into the budgetary investment and special funds for energy-saving emission reduction of the central finance, providing subsidies and rewards; second, it implemented preferential taxation policies, exempting or reducing the business tax, value-added tax and income tax of the energy service companies having EPC projects. Third, it completed the accounting system, made clear the expenditure items of energy service costs for the governmental institutions, public institutions and enterprises which implemented energy-saving renovation projects by utilizing EPC method. Fourth, it improved the financial services, encouraging the financial institutions such as banks to provide financial service for projects of energy service companies, and permitting the service companies to apply for mortgage loan with the fixed assets of EPC projects.

In June 2010, the Ministry of Finance and the NDRC jointly formulated the *Tentative Regulations on Management of EPC Finance Incentive Fund* (MOF, 2010, p. 249), which stipulated that the central finance would allocate RMB 2 billion in 2010 to support the energy service companies in taking energy-saving renovation through EPCs. The policy provided rewards to the industrial energy-saving renovation projects with an annual energy savings of 500 tce (inclusive)–10,000 tce, of which, central finance reward is 240 yuan/tce while provincial reward is no less than 60 yuan/tce. Meanwhile, provincial financial and energy management authorities are subjects of implementation for EPC, which are responsible for assessing the projects and contracts, defining the annual energy savings of projects, and monitoring and examining the implementation and its effects.

Both policies cover a wide range of projects with great efforts, providing direct subsidy to energy-saving efforts and energy service companies. According to the standards, subsidies account for 30% of the standard coal price (RMB 960). Subsidies of some provincial finances are very heavy, for example, Shanghai municipal government subsidizes 360 yuan/tce, that is to say, altogether 600 yuan/tce is subsidized. Therefore, the subsidies account for 25%–30% of the investment in energy-saving projects.

Meanwhile, the threshold for energy-saving subsidies is very low, the industrial energy-saving projects with an annual energy savings of 500–10,000 ton can be subsidized. It varies greatly with the Ten Key Energy Conservation Projects and Top 1,000 Enterprises Energy Conservation Program, which only provide subsidies to the big projects with annual energy savings of 10,000 ton. In addition, the subsidies are given according to energy savings instead of equipment acquisition. Therefore, only those energy-saving projects with real effects could be benefited, thereby encouraging the healthy development of industry. Lastly, the subsidies are provided directly to the energy service companies, helping them exploring market recognition and increasing profitability.

According to estimation of China Energy Conservation Service Industry Committee, gross output value of the industry would be RMB 80 billion in 2010, increasing by 36% year on year. Total market scale would be about RMB 400 billion in future, so there is a huge space for future development. Both the *Opinions* and *Regulations* are new policies enforced in 2010, so there is no statistics for effect evaluation. However, we could analyze the possible existing problems from the perspective of system and welfare.

From the perspective of system, heavy subsidies and low thresholds would generate more energy service companies, thereby accelerating market development and competition. On the other hand, many small-scale companies might only have the ability of renovating specific projects instead of the entire system. While based upon the characteristics of industrial energy conservation, optimization of the entire system is the vital part. Calculations of the IEA reveal that energy savings of system optimization (such as heat-power cogeneration, secondary energy recovery and reutilization) is 1.1–1.2 times of that derived from single process optimization (IEA, 2007). Making up the defects within the market mechanism is a problem during the development of EPC.

Moreover, energy price will affect the enthusiasm of terminal energy-consuming enterprises in adopting EPC. Low energy prices and low costs of energy consumption would lead to unobvious energy conservation benefits; furthermore, the costs of energy conservation renovation cannot be covered. Under such circumstances, if the terminal energy consumers do not adopt energy conservation renovation, the energy service industry will have no client.

From the perspective of welfare, when the service companies and enterprises are sharing the energy-saving benefits, the former might give a low expected energy savings to have a larger proportion, while the latter might take evasive measures or do not operate according to contractual regulations so that the energy savings is below the expected objectives. Or, energy savings of the service companies could not satisfy the terminal energy-consuming enterprises, who might give up EPC and invest a lump sum of money in buying energy-saving products. Whatever the case, unreasonable behaviors during the game playing of energy service companies and energy-consuming enterprises might sabotage the social welfare. Therefore, governments or independent third-party agencies set up by the governments need to conduct contractual auditing, monitor the implementation of contracts, provide accurate measurement for energy-saving effects, and improve the information transparency of both parties.

In a nutshell, industrial energy-saving policies during the *11th Five Year Plan* feature in focusing on key problems, passing down pressures, cultivating capabilities and completing mechanisms. Focusing on key problems means grasping key industries and key enterprises. Energy-intensive industries are major industrial energy consumers and constitute a breakthrough point for industrial energy conservation. Industrial energy conservation shall be implemented on key energy-consuming units, which are usually large enterprises with competitive edges in money, technology and influence. Grasping key enterprises is an important mechanism in realizing highly-effective energy conservation and emission reduction. In addition, China's unitary administrative system determines that the government could utilize its resources to pass down the energy conservation pressure for implementation in a short time. Energy conservation program of the *11th Five Year Plan* is faced with the difficulties of high objectives, short time limit and heavy tasks. Therefore, administrative measures are the best way to realize energy conservation in a short time when the production elements market has just taken shape. In addition, the government invests directly and heavily in energy conservation and emission reduction during the *11th Five Year Plan*, which aims to provide fund incentives to enterprises to improve their equipment, technology and competitiveness. Of course, the energy conservation mechanism dominated by administrative measures needs to be completed. China's government is changing its means of intervention in economy management, and improving the energy conservation market mechanism is becoming a new trend. Developing the market with the guidance of the government will become a new channel for realizing highly-effective industrial energy conservation.

8.4 System Analysis

During the *11th Five Year Plan*, energy conservation of the industrial sector primarily features in governmental guidance. In order to realize the restrictive objective of decreasing energy consumption of per unit GDP by 20% by 2010, the industrial sector conducted the Top 1,000 Enterprises Energy Conservation Program, the Ten Key Energy Conservation Projects and the

Elimination of Backward Production Capacity Program with the lead of the government, which achieved remarkable results. Two issues will be analyzed in the section: Why governmental guidance is adopted for industrial energy conservation; and how can it achieve remarkable results?

8.4.1 *Reasons for governmental guidance*

Chinese government has played a major role in promoting economic development. Different from the laissez-faire model of some western free-market countries, in China's model of governance, politics is intermingled with economic structure and not above it. Since the establishment of the PRC, industrial sector has been a focus for China's economic development and governmental intervention. Before the reform and opening up, the government allocated economic resources in the industrial sector by administrative measures, exerted direct intervention in its development, participated in industrial investment and production directly, and promoted its rapid development. Since the reform and opening up, the central government delegated its powers to local governments. By controlling their promotion opportunities and financial transfer payment, it prompts local officials to utilize some key resources, such as administrative approval, land expropriation, loan warranty and preferential policies, to develop the industrial economy. The central and local governments had the power to allocate the resources within industries and enterprises, so it was very applicable for the governments to give pressure and incentives in guiding industrial energy conservation. This is an important institutional background for guiding the energy conservation process of industrial enterprises by the government during the *11th Five Year Plan*.

In addition, China's market is not well developed yet. Chinese government has exerted control of various extents to energy prices, which cannot reflect their scarcity, therefore the market cannot play a major role in energy conservation management. Moreover, it is very urgent to realize the indicators for energy consumption of per RMB 10,000 GDP during the *11th Five Year Plan*, while completing the market mechanism and implementing a market-oriented energy conservation system is unable to realize the objectives timely, the government could utilize the responsibility system and up-down mobilization to put the energy conservation tasks into practice. In this way, governmental guidance becomes the major mode of industrial energy conservation, and the administrative intervention measures and incentive measures become the primary driving force for industrial energy conservation.

8.4.2 *Analysis of industrial energy conservation governance mechanism*

Enterprises are the subject of implementation for industrial energy conservation. By taking effective energy conservation measures, the government fulfills its purpose of improving energy efficiency and reducing energy consumption. There are mainly two reasons for the effectiveness of government-oriented industrial energy conservation process. First, the administrative measures such as the target responsibility system and the incentive and guidance measures play a restrictive and encouraging role in the energy conservation of industrial enterprises; second, with the restrictions and encouragement of the government, the enterprises respond positively to the energy conservation policies and take appropriate activities taking account of the cost-benefit analysis, political achievement and future space for development.

(1) *Target responsibility system between the government and enterprises*

Administrative measures of the government are an important method for placing energy-saving pressures on industrial enterprises. Various governments exercise control over the energy-saving activities of industrial enterprises by way of target responsibility system, administrative compulsion as well as administrative permission. Target responsibility system during the *11th Five Year Plan* plays a major role in energy conservation, which is an important cause for the excellent performance of industrial energy conservation, and is a major cause for more compulsive measures in the last year of the *11th Five Year Plan* compared to adoption of both incentive and restrictive measures during the first four years.

Target responsibility system for energy conservation consists of two ways, i.e., target responsibility system between the superior and inferior governments, and between governments and enterprises. The target of reducing energy consumption per unit of GDP by 20% is legally binding, and the central government breaks down the target and allocated it to each province (including autonomous region and municipalities directly under central government) by the target responsibility system. The target is set up by the central government; therefore, the target responsibility system between higher and lower governments forms the foundation for letters of responsibility system between governments and enterprises.

During the first four years of the *11th Five Year Plan*, target responsibility system was mainly implemented through the Top 1,000 Enterprises Energy Conservation Program and local key enterprises energy conservation program. Local governments included the energy-intensive enterprises into the target responsibility system, increasing the number of enterprises signing letters of responsibility by leaps and bounds. Top 1,000 Enterprises Energy Conservation Program is the biggest energy-saving project of target responsibility system between the government and industrial enterprises. Based upon the program, each province, autonomous region and municipality directly under central government defined the key energy-intensive enterprises under its jurisdiction, and took concrete measures by signing letters of responsibility with relevant enterprises, passing down energy conservation tasks, and laying down restrictive requirements to the energy-savings of the enterprises. Target control of key enterprises also reveals the administrative feature of focusing on big ones and relaxing the control over small ones.

With energy-saving target responsibility, energy conservation and consumption reduction indicators have a veto power with administrative nature, and the central government stipulated clearly that its completion relates directly to the assessment of local officials, giving it a strong restrictive nature over local governments, urging them to pay attention to energy conservation and passing down the pressure to key energy-intensive enterprises. For the enterprises, completion of energy conservation indicators not only relates to annual awards, but to the approval of other projects and acquisition of energy conservation fund. Therefore, the enterprises are also faced with incentive as well as restrictive measures.

By 2010, it has become more difficult to accomplish the energy conservation tasks. Energy consumption per unit GDP had declined 15.6% in the first four years of the *11th Five Year Plan*, lower than the target of 20%. However, energy consumption per unit GDP actually increased in the first half of 2010 year on year, giving much pressure to local governments. This condition led to major changes in the energy consumption mechanism, some local governments even exercised power restriction and production restriction measures to urge the industrial enterprises in saving energy. For example, in September 2010, Wu'an city in Hebei province, a city with many steel and coking enterprises, issued a notice to 18 steel plants within its jurisdiction, demanding shutdown of all blast furnaces and steel rolling production lines within designated time limit, and some local coking enterprises and iron ore dressing enterprises were also demanded to stop production. Tangshan municipal government issued the *Notice on Control of Products of High Energy Consumption and Pollution Industries*, which formulated a production control plan for the steel and coking enterprises, and prepared to shut down power supply to the enterprises exceeding planned production. For local governments, taking compulsory measures is the most direct and effective method in realizing energy-saving indicators. Difficult situation and urgency prompts some local governments to take the compulsory measures. Therefore, it is evident that target responsibility system and administrative measures exercise practical and effective restrictions over the industrial enterprises.

(2) *Encouragement and guidance of the government to enterprises*

Enterprises are the major force for energy conservation, and the energy-saving targets enforced by governmental agencies on energy-intensive enterprises need to be realized by the enterprises with improved capabilities. Apart from giving pressure, governmental agencies also take incentive measures to urge the enterprises in taking energy-saving measures positively, such as financial reward, preferential taxation policies and lower energy price.

Technical renovation is a major method for energy conservation. In order to encourage the enterprises to improve energy efficiency by technical renovation, the central government allocated special fund to reward the projects with an annual energy savings of over 10,000 tce. The scope of reward includes the boiler (kiln) renovation of coal-fired industries, waste heat and pressure utilization, oil conservation and alternative, electric motor system energy conservation and system optimization of Ten Key Energy Conservation Projects. In addition, it also provides a lump sum of reward to the EPC projects with an annual energy savings of 5,000–10,000 tce. For the shared-saving EPC projects meeting the requirements of national standards, the energy service companies could apply for financial reward to the provincial financial departments and competent energy conservation authorities. For the backward production capacity without technical renovation potentials, the central government creates special funds to offer reward to the under-developed areas through special transfer payment, the amount of which is decided by the scale of displaced production capacity and reward standards. It covers 13 industries, including electricity, iron smelting, steel smelting, electrolytic aluminum, metal alloy, calcium carbide, coke, cement, glass, paper-making, alcohol, MSG and citric acid.

Preferential taxation policies means to increase the enterprises' profits by tax rate changes, thereby reducing their energy consumption. Taxation policies on resources comprehensive utilization are the major taxation measures during the *11th Five Year Plan*. Resources comprehensive utilization of the manufacturing sector includes waste heat and pressure utilization and improvement of recyclable materials reutilization rate. Income tax of the enterprises adopting resources comprehensive utilization could be exempted or reduced for a certain time limit, giving them a higher margin of profit in reducing energy consumption, serving the purpose of pursuing more profits, thereby encouraging them to improve the comprehensive utilization rate of resources.

Energy-intensive enterprises of China's manufacturing sector mainly use coal, oil, electricity and natural gas as energy. Coal's price is mainly decided by the market. With energy tax imposition and increasing transportation cost, coal's price rises sharply, giving much pressure to the enterprises using coal as the primary energy. Since 2009, price-forming mechanism of oil products has improved. With price hike in the world crude oil market, price of China's oil products also witnessed several rises. In June 2010, China raised the factory price of natural gas, got rid of the double-track price system, and expanded the fluctuation range, leading to appropriate allocation of natural gas resources. In November 2009, the price of power was raised all across the country. Since June 2010, price markup standard of restriction-category enterprises of eight industries including electrolytic aluminum and metal alloy increased from 0.05 yuan/kWh to 0.10 yuan/kWh, of elimination-category enterprises increased from 0.20 yuan/kWh to 0.30 yuan/kWh. General rise of energy prices forces the enterprises to reduce its consumption by increasing investment in technical renovation.

(3) *Analysis of enterprises' energy-saving motivation*

(1) Cost-benefit analysis prompts the enterprises to save energy and reduce consumption. Enterprises are the subject of markets, and responding to market price is their motive force for energy conservation. From the perspective of cost-benefit analysis, technical renovation, energy structure adjustment and increasing resources comprehensive utilization rate are three methods for energy conservation.

In recent years, rise of energy price is much higher than that of industrial products, and the pressure of rising cost is the primary motive force for them to take energy conservation measures. Statistics show that in 2006, fuel price increased by 71.2% compared to 2001, while that of metallurgical products increased by 41.8%, of chemical products increased by 12%, of nonmetal products increased by only 1.4% (Liu Chang *et al.*, 2008). Products of steel, nonferrous metal, building materials, as well as chemistry and chemical industry facing serious competition are very sensitive to cost rises and margin of profits. The enterprises need to take effective energy-saving measures to reduce cost and survive the competition. Meanwhile, they also need to adjust the product structures and increase the product value added to acquire higher margin of profits. Energy conservation techniques and management are major methods to reduce energy consumption. Enterprises need to conduct technical renovation to existing production equipment, improve the internal management, and assess each process and raw materials. Energy consumption is the major component of production cost assessment of each department, and each enterprise will determine whether the energy consumption could be assessed separately according to practical conditions. Two-grade and Three-grade measurement and statistics system is the foundation of internal assessment. Therefore, more standardized organization system within an enterprise is more conducive to its energy conservation. For these enterprises, energy conservation is a spontaneous behavior under market pressure. Coal price upsurge and lower cost of domestically-made energy conservation equipment shortens the payback period of technical renovation investment in energy conservation. Reward fund of the central and local finance further reduces the cost of enterprises. Adoption of energy conservation equipment plays an important role in increasing the average energy consumption level of various industries.

Price of different energies respond to difference on their scarcity, and price difference stimulates adjustment and transformation of the enterprises' energy structure. Although rise of overall energy price is not much, large price rise differences are formed due to different price regulation levels. Contrary to the rising coal price, energy-intensive enterprises enjoy preferential electricity prices. In 2003, many provinces formulated preferential power price measures for energy-intensive enterprises by way of linkage power generation and supply, power supply by agreement, direct supply to large users and bidding into the grid. Shortly after the enactment of export tax rebate policy in 2008, the international financial crisis erupted, and some energy-intensive enterprises such as electrolytic aluminum were deeply impacted. In order to promote local development, some provinces (autonomous regions, municipalities directly under central government) provided various preferential power prices to local energy-intensive enterprises, prompting some enterprises to transform their production process and increase the proportion of electricity in energy consumption. Energy efficiency of power is relatively high, and a larger portion of electric power in the energy structure is favorable to the improvement of the overall energy efficiency.

Increasing resource comprehensive utilization rate can reduce energy consumption and bring more benefits to the enterprises. For the manufacturing sector enterprises such as the nonferrous metal, steel making and cement industry, waste heat and pressure utilization, waste gas recovery and better recyclable resources utilization rate could reduce the energy consumption per produced and the production cost, as well as enjoy preferential taxation.

(2) Administrative pressure forces the enterprise leaders paying more attention to energy conservation. Of the energy-intensive enterprises, state-owned enterprises only constitute a small portion, but with larger size, and the proportion of their industrial value added is fairly large in the energy-intensive industry (Fig. 8.35). Realization of their energy conservation targets directly affects the overall energy consumption of the industry. Governmental agencies prompt the energy-intensive enterprises to take

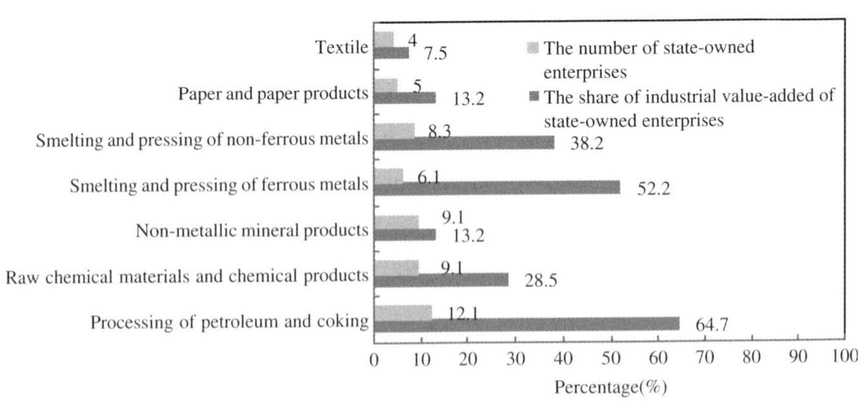

Fig. 8.35. Proportion of state-owned enterprises in energy-intensive industries. *Source*: State Statistics Bureau, 2006.

effective measures by controlling their leadership appointment and space for development. Leaders of state-owned enterprises are faced with a veto power and the restriction of some detailed assessment indicators, which directly affects their personal income. Performance evaluation to leaders of state-owned enterprises affects their choice of strategies by incentive and restrictive mechanisms, and encourages them to allocate enough resources for the smooth functioning of energy conservation process. Taking an enterprise under Sinopec group for example, Sinopec will assess the energy consumption per produced of the enterprise, which will face economic punishment in exceeding the standards, i.e., deducting 10% performance-related pay of the entire enterprise. In addition, 5% performance-related pay of the leaders will be deducted. The evaluation method draws high attention of the leading group, which divides the indicators and relevant reward and punishment mechanism into each team, so as to realize the overall energy conservation target by internal evaluation. For the enterprises failing the targets in 2009, the NDRC stipulated clearly that "putting forward renovation measures within a month after announcing the evaluation results and reporting to the competent provincial energy conservation authorities; disqualified for annual awards, titles of honor, and supportive policies such as national inspection exemption." In addition, various provincial governments also put forward similar restrictive measures to the energy conservation targets of state-owned enterprises. Taking Hebei province for example, Hebei provincial development and reform committee announced the key enterprises for energy conservation and emission reduction in 2007, covering coking, chemical fertilizer, oil, natural gas, coal and power industry. The provincial government also stipulated that the corporate representatives of the state-owned enterprises failing the designated targets would be removed immediately, and the private enterprises have to shut down for renovation (Yang Yisong, 2007). Shandong province conducted evaluation to 13 indicators (including energy conservation amount and energy auditing) of 103 key energy-intensive enterprises in 2007, with five enterprises failing the evaluation, which would be disqualified for award and titles of honor within a certain time limit; and the members of leading group of state-owned and state holding enterprises which failed energy conservation targets cannot have evaluation rewards (Liu Baosen, 2007). Performance evaluation is an important tool in measuring the political performance of enterprise leaders, and including the energy conservation targets into evaluation will have a big influence on the political promotion of enterprise leaders. On the one hand, the pressure prompts the enterprise leaders to give more priority to energy conservation and invest in more personnel and resources for realization of energy conservation targets; on the other hand, failing energy conservation targets will affect adversely to their development and indirectly to their economic performance. Double pressure that administrative measures bring to the leadership and enterprise business performance prompts the implementation of energy conservation efforts within the enterprises.

(3) Investing in energy conservation for more space of development. Investing in future development space is another important impetus for the enterprises to conduct energy conservation activities. Top 1,000 enterprises or key energy-intensive enterprises of each province failing energy conservation targets will face external restrictions on expansion. For the top 1,000 enterprises failing energy conservation evaluation standards, the governmental agencies will suspend examination and approval of new energy-intensive investment projects and new land for industrial usage. Shandong province exerted more strict controls over the failing enterprises with respect to approval of new projects, land usage approval and financial credit (Liu Baosen, 2007). On the contrary, investment in energy conservation will be encouraged by governmental agencies. For example, Hebei province provided more preferential policies for enterprises' development in energy conservation and technical renovation, and gave priority support to new, expansion and renovation projects related to the energy conservation and emission reduction of key enterprises with respect to project approval, land acquisition and credit (Yisong, 2007). Energy-intensive enterprises are monitored and restricted by local governments in their production elements and space of development, and the enterprises usually need to maintain a good cooperation with governmental agencies. If the enterprises behave badly in energy conservation targets, it will affect the energy conservation evaluation conducted by the central government to local governments, and affect the ability of local governments in acquiring resources. Therefore, the enterprises will try their best to meet the regulations and standards formulated by governmental departments. The enterprises with good business performance and space of development usually adopt technical renovation measures positively to fulfill energy conservation targets. On the contrary, the enterprises with difficulties will balance the cost of sustenance and shutdown, and will decide whether to eliminate backward production capacity and quit the market. If the financial reward and other compensations provided by the governmental agencies are fairly sufficient and could help the enterprise in dealing with real difficulties, the enterprises usually choose to eliminate the backward capacity. Enterprise are restricted in many aspects

by the government, therefore investing in energy conservation for future development is a reasonable choice for them.

8.4.3 *Existing problems*

Industrial energy conservation guided by the government has achieved good results, but not without important problems. First, there is a gap between energy conservation targets and the practical capabilities of local governments and industrial enterprises. Formulation and decomposition of energy consumption reduction indicators per unit GDP does not fully take local realities into consideration, forcing some places to take extreme measures such as power cuts to fulfill the targets, injuring normal social and economic development, as well as brining economic losses to relevant industrial enterprises. Second, there is a gap between requirements of energy conservation and support of energy-saving mechanism under the guidance of the government. Government-oriented industrial energy conservation is suitable to the common management mode in China's unitary system, is favorable to realizing energy conservation targets in short time. However, it depends too much on administrative measures, leading to insufficient functioning of the market system, thereby leading to higher cost of industrial energy conservation. Lastly, there is a difference between the motive power of the central government and the enterprises. Central government plays a major role in industrial energy conservation during the *11*th *Five Year Plan*, providing important incentives and restrictions to local governments and industrial enterprises from bringing forth energy conservation tasks to passing down auxiliary policies. Compared to the central government, some local governments and industrial enterprises lack the momentum for energy conservation, playing a passive role in energy conservation treatment and taking measures just for completing the energy conservation targets. Industrial energy conservation has not become a spontaneous action for the enterprises. Therefore, much remains to be done in the independent implementation of the industrial energy conservation system.

Appendix: List of Industry Energy-Saving Policies

Energy conservation policies such as comprehensive energy conservation planning, integrated policies and finance and taxation policies play their role in industrial energy conservation during the *11*th *Five Years Plan*, which will be introduced in the following paragraphs.

1. *National comprehensive energy conservation planning*

(1) Notice of the NDRC on *Printing and Distributing Medium and Long-Term Energy Conservation Plan* (NDRC Circular, 2004, p. 2505)

In November 2004, based upon the total energy production and consumption as well as energy consumption per unit GDP before 2002, the NDRC formulated the medium and long-term energy conservation plan, which covers the *11th Five Years Plan* and before 2020. The plan lays down the energy conservation target by 2010 (2.25 tce per RMB 10,000 GDP[19]) and the priorities, and brings forward the target by 2020 (1.54 tce per RMB 10,000 GDP). Special energy conservation plan is an important element of China's medium and long-term energy planning, a guidance document for China's medium and long-term energy conservation and a basis for implementation of energy-saving projects. The plan also puts forward key energy-saving areas and projects for the first time (i.e., Ten Key Energy Conservation Projects).

(2) *Eleventh Five-Year Plan for National Economic and Social Development of the PRC*

In March 14th, 2006, the fourth meeting of 10th NPC approved the *11th Five Year Plan*, which is an important guidance document for fulfilling the responsibilities of economic regulation, market monitoring, social management and public service by the government. It stipulated clearly the target of "reducing energy consumption per unit GDP by 20%" and the requirement of "grasping the energy conservation of steel, nonferrous metal, coal, power, chemical and chemistry, and building materials industry and energy-intensive enterprises".

(3) *Decision of the State Council on Strengthening Energy Conservation Efforts* (State Council Circular, 2006, p. 28)

In August 2006, the State Council brought forward a series of policy proposals to fulfill the objectives of the *11th Five Year Plan*, which clearly stipulated to "establish energy conservation target responsibility system as well as assessment and evaluation system", "enhance industrial energy conservation", and pointed out particularly "emphasizing on steel, nonferrous, coal, power, petrochemical, chemical and chemistry and building materials industry, and on energy conservation of the enterprises with an annual energy consumption of over 10,000 t standard coal; implement Top 1,000 Enterprises Energy Conservation Program; prompt the enterprises to adjust

[19] In this place, energy intensity is calculated by the equivalence method.

product structure positively; accelerate energy conservation technical renovation and reduce energy consumption". It also stressed that "relevant departments as well as local governments and competent authorities should implement Ten Key Energy Conservation Projects stipulated in *the 11th Five Year Plan* seriously".

(4) *Reply of the State Council to the Plan of Energy Consumption Reduction Objective Per Unit GDP of All Districts During the 11th Five Year Plan* (State Council Letter, 2006, p. 94)

During the *11th Five Year Plan*, energy consumption per unit GDP declined from 12.2 tce/1,000 yuan in 2005 to 9.8 tce/1,000 yuan in 2010, decreasing by 20%. The State Council replied in September 2006 that the *Plan for Energy Consumption Reduction Indicators Per Unit GDP of Various Places During the 11th Five Year Plan* (NDRC, 2006, p. 1816) was approved in principle. Energy consumption per unit GDP is a legally binding restrictive indicator, and each province (district and city) is requested to include the indicator into its comprehensive evaluation of social and economic development, performance evaluation and political performance evaluation, and is decomposed and implemented by various cities (prefectures), districts, industries and key enterprises.

(5) *Notice of the State Council on Printing and Distributing the Comprehensive Work Plan on Energy Conservation and Emission Reduction* (State Council Circular, 2007, p. 15)

In order to fulfill the energy conservation targets of the *11th Five Year Plan*, the State Council issued the *Notice of the State Council on Printing and Distributing the Comprehensive Work Plan on Energy Conservation and Emission Reduction* in May 2007, which stipulated that the government's guidance role and subject responsibility of enterprises shall be brought into play. The notice restressed the energy conservation targets, and indicated to "accelerate the elimination of backward production capacity" and "accelerate the implementation of Top Ten Energy Conservation Projects".

2. *Comprehensive industrial energy conservation policy*

(1) *Decisions of the State Council on Issuing and Implementing the Tentative Regulation on Accelerating Adjustment of Industrial Structures* (State Council Circular, 2005, p. 40)

In December 2005, according to the requirements of national macroeconomic development and industrial structural adjustment, the NDRC issued the *Guidance Catalog for Industrial Structural Adjustment* (2005 edition) (State Council Circular, 2005, p. 40), which divided the industries into three categories of encouragement, restriction and elimination, which in turn served as the basis for enterprises enjoying preferential power prices and taxation policies. In 2007, the State Council issued the *Notice of the State Council on Accelerating the Structure Adjustment of Industries with Surplus Production Capacity*, which also encouraged the development of high-tech industries. In addition, the NDRC issued the following documents, the *Notice on Problems in Advancing the Adjustment of Industrial Structures*, the *Notice on Preventing Unrestricted Expansion of Energy-Intensive Industries*, and *Opinions on Strengthening Fixed Asset Investment Regulation and Exercising Strict Control over New Projects.* Meanwhile, the NDRC also formulated a series of guidance opinions, industrial policies and admission conditions for 10 energy-intensive industries of steel, oil refinery, ethylene, cement, aluminum, coal, coking, calcium carbide, power and textile.

In December 2009, in order to further promote energy conservation process and eliminate backward production capacity and energy-intensive equipment, the Ministry of Industry and Information Technology issued the *Elimination Catalog of High Energy-Consuming and Backward Electrical and Mechanical Equipments (Products)* (First Batch) (MIIT, 2009, p. 67), which included nine categories of 272 kinds of equipments (products). It requested the enterprises to change into highly-efficient energy conservation equipments (products), and various levels of energy conservation monitoring agencies should enhance monitoring and inspection of the equipments (products) listed in the *Catalog*.

(2) *Notice on the Implementation Plan for Top 1000 Enterprises Energy Conservation Program* (NDRC Circular, 2006, p. 571)

In April 2006, based upon the *Decision of the State Council on Strengthening Energy Conservation Efforts* (State Council Circular, 2006, p. 28), the NDRC, the State Energy Leading Group Office, the State Statistics Bureau, the General Administration of Quality Supervision, Inspection and Quarantine, and the SASAC decided to conduct Top 1,000 Enterprises Energy Conservation Program in key energy-intensive enterprises, and formulated the *Implementation Plan for Top 1,000 Enterprises Energy Conservation Program.* Top 1,000 enterprises refer to 1,008 enterprises above designated scale and with independent accounting whose comprehensive energy consumption was over 180,000 tce in 2004. The enterprises cover nine key energy-intensive industries of steel, nonferrous metal, coal, power, petrochemical, chemistry and chemical, building materials, textile, and paper-making. By method of signing letters of responsibility for energy conservation targets with the enterprises, about 100 Mtce is expected to be saved.

(3) *Notice of the State Council on Several Opinions of Accelerating Shutdown of Small Thermal Power Units* (State Council Circular, 2007, p. 2)

In January 2007, the State Council issued a notice which demanded that "the objectives of closing small thermal power units shall be broken down into each province (district, city), and signing letters of responsibility for closing small thermal power units with provincial governments and state-owned large power companies". The notice encouraged all provinces and enterprises to shut down small thermal power units and build large power units, which also stipulated that the power units that were included in the shutdown plan of each province during the *11th Five Year Plan* and shut down in due course could enjoy preferential power generation indicators within certain time limit (no more than three years), or obtain a certain amount of economic compensation by transferring part of its capacity to large power units. Power generation indicators and extent of period decrease with the delayed time of shutdown.

(4) *Opinions on Implementing Ten Key Energy Conservation Projects during the 11th Five Year Plan* (NDRC Circular, 2006, p. 1457)

The *Opinions on Implementing Ten Key Energy Conservation Projects during the 11th Five Years Plan* was compiled by the NDRC, the MOST, the MOF, the MOC, the AQSIQ, the SEPA, the Government Offices Administration of State Council, and the Government Offices Administration of the Central Committee of CPC, and was issued formally in July 2007. It made detailed arrangement of the Ten Key Energy Conservation Projects, consisting of guiding theory, principles and objectives, contents of implementation, and safeguard mechanism. Each key project is introduced from the aspects of status quo and problems, main contents, auxiliary measures, and implementation.

(5) *Notice on Opinions of Accelerating the Implementation of EPC for Promotion of Energy Conservation Service Industry* (State Council Office Circular, 2010, p. 25)

In order to bring market mechanism into play and enhance policy support and guidance, the State Council issued the *Notice on Opinions of*

Accelerating the Implementation of EPC for Promotion of Energy Conservation Service Industry in April 2010, which provided monetary subsidy or reward to EPC companies, exempted or reduced their taxation, adopted proper preferential taxation policies, completed relevant accounting system, and improved the financial service.

(6) *General Outline of China Resources Comprehensive Utilization Technology Policy* (NDRC Notice, 2010, p. 14)

In July 2010, the NDRC, the MOST, the MIIT, the Ministry of Land and Resources, the MOHURD and the Ministry of Commerce jointly issued the *General Outline of China Resources Comprehensive Utilization Technology Policy*, which primarily aimed to "accelerating the development, demonstration and promotion of comprehensive resources utilization technology, guiding the flow of social funds, providing technical support for relevant enterprises in utilizing comprehensive resources utilization technology, and improving the overall level of China's comprehensive resources utilization technology". In the outline, "waste gas, waste heat and waste pressure utilization technology" played an important guiding and promoting role for industrial energy conservation.

3. *Financial and taxation policies*

(1) *Notice on Adjusting the Export Tax Rebate Rate of Certain Goods and Increasing the List of Prohibited Goods for Processing Trade* (Ministry of Finance, 2006, p. 139)

In order to reduce the pressure on domestic energy consumption due to exported energy-intensive products, the MOF, the NDRC, the MOC, the General Administration of Customs, and the State Administration of Taxation jointly issued the *Notice* in September 2006, which adjusted the export tax rebate rate of some exported commodities.

(2) *Tentative Regulations on Management of Financial Incentive Fund for Energy Conservation Technology Renovation* (Ministry of Finance, 2007, p. 558)

The *Regulation* was issued by the MOF and the NDRC in Aug. 2007. In order to ensure the practical energy conservation effects of technical renovation projects, the Ministry of Finance adopted the method of linking monetary reward to energy savings, offering reward to the enterprises realizing energy conservation objectives. The regulation also stipulated that, the central finance will offer financial incentives to the boiler (kiln) renovation of the coal-fired industry, waste heat and pressure utilization project, oil conservation and alternative project, electric motor system energy conservation project, and energy system optimization project. The details are as following: for the enterprises with an energy saving of over 10,000 tce, 200 yuan/tce and 250 yuan/tce subsidy will be provided respectively for the East Region and Central-West Region. The Ministry of Finance will allocate 60% of the incentive fund to the enterprises with approved energy conservation projects before inception of the project, while the remaining part will be allocated properly after energy-saving assessment conducted after project completion.

(3) *Tentative Regulations on Management of Central Finance Incentive Fund for Eliminating Backward Production Capacity* (Ministry of Finance, 2007, p. 873)

In December 2007, the Ministry of Finance issued the *Tentative Regulations on Management of Central Finance Incentive Fund for Eliminating Backward Production Capacity*, which stipulated that the central finance would create incentive funds, and encourage the underdeveloped areas to eliminate backward production capacity by special payment transfer. Local governments shall be responsible for the elimination of backward production capacity, and the central government will offer appropriate reward according to the scale of eliminated production capacity, and the reward will be managed by the local governments according to practical conditions. It covers 13 industries, including electricity, iron smelting, steel smelting, electrolytic aluminum, metal alloy, calcium carbide, coke, cement, glass, paper-making, alcohol, MSG, and citric acid.

(4) *List of Preferential Enterprise Income Tax on Resources Comprehensive Utilization* (2008 Edition) (Ministry of Finance, 2008, p. 117)

The Enterprise Income Tax Law and its implementation rules stipulated that for the enterprises using prescribed resources in the *List of Preferential Enterprise Income Tax on Resources Comprehensive Utilization* as their major raw materials to produce unrestricted or un-forbidden products complying with relevant national and industrial standards, only 90% gross revenue will be recorded for income taxation. The *List* also made regulations on comprehensively-utilized resources, products and technology standards.

(5) *Circular on Handling Issues Concerning Preferential Power Price for Energy-Intensive Enterprises* (NDRC Price, 2010, p. 978)

In May 2010, the NDRC issued the *Circular on Handling Issues Concerning Preferential Power Price for Energy-Intensive Enterprises*, which stipulated to suspend the preferential power price for energy-intensive enterprises (including polycrystalline silicon), enhance implementation efforts for discriminatory power prices, and that local governments shall further enhance the price markup standards for elimination-category and restriction-category enterprises based upon the price adjustment in 2006. And for those enterprises whose energy consumption level per unit produced exceeds the energy consumption (power consumption) standards prescribed by the central government or local governments, they will be inflicted with punitive power prices.

(6) *Tentative Regulations on Management of EPC Finance Incentive Fund* (Ministry of Finance, 2010, p. 249)

In June 2010, according to the *Notice on Opinions of Accelerating the Implementation of EPC for Promotion of Energy Conservation Service Industry* (State Council Office Circular, 2010, p. 25), the central finance allocated incentive fund to support EPC and development of energy service industry. In order to regulate management of financial fund and increase its utilization efficiency, the Ministry of Finance issued the *Tentative Regulations on Management of EPC Finance Incentive Fund*, which stipulated that the Ministry of Finance would provide a lump sum of monetary reward to EPC projects according to annual energy savings and prescribed standards. The monetary reward will be provided by the central finance and provincial finance departments, of which, the central finance reward is 240 yuan/tce while the provincial reward is no less than 60 yuan/tce.

CHAPTER 9

Building

9.1 Low Carbon Performance in the Building Sector

In the life cycle of buildings, building energy consumption and CO_2 emissions consists of two categories, one is related to the process of construction,[1] and the other is related to daily operation of buildings.[2] According to the international practices, building energy consumption usually refers to the latter one, while the former one is classified into the industrial energy consumption. In addition, buildings are built for either industrial or civil use. Energy consumption of buildings for industrial use is mainly related to the manufacturing process and is classified into industrial energy consumption too. Therefore, building energy consumption in the report refers to the operational energy consumption of civil buildings.

According to international convention, building energy consumption can be further divided into residential part and commercial part. However, in China, Due to the urban–rural duality and climate differences,[3] building energy consumption is divided into four categories in

[1]It refers to the energy consumption in production, transportation, and utilization of building materials, components, and equipments, i.e., the energy consumption in building materials industry, cargo transportation industry, and construction industry. It is notable that energy use of this part accounts for 20%–30% of the country's total energy consumption, due to the penetrating urbanization.

[2]It refers to the energy consumption of lighting, heating, air-conditioning, and other energy carriers in buildings. Energy is constantly needed to maintain certain level of indoor environment and fulfill required services, as long as the building is occupied.

[3]Historically, district heating systems of city- and community-level are commonly applied in buildings of North China, due to its severe cold weather there in winters. Consequent energy consumption for such purposes is dramatic and is featured with unique characteristics, in comparison with energy use for other end-use purposes in buildings, which

the report as: ① energy consumption of district heating system in North China's urban areas[4]; ② energy consumption in urban residential buildings (but excluding category ①); ③ energy consumption in commercial buildings (but excluding category ①) and; ④ energy consumption in rural residential buildings.[5]

Energy types for daily operation in buildings usually include commodity types, such as electricity, heat, coal, gas and oil, and noncommodity types, such as biomass and solar energy and so on. It focuses, in this report, mainly on consumption of commodity energy, due to difficulties in collecting data of noncommercial energy. In addition, electricity is converted to standard coal equivalent according to national plant coal consumption of the current year, while other types of commodity energy are converted to standard coal equivalent by their low calorific value. Factors for CO_2 emission calculation are listed in Chapter 4. Explanation for appeared noncommodity energy use will be footnoted properly.

9.1.1 *Growth rate of CO_2 emission in China's building sector declines apparently, with lower energy intensity than that of developed countries*

(1) *Total energy consumption and CO_2 emission*

Calculation[6] in the report reveals that, in 2008, commodity energy use in China's building sector total to 540 Mtce, accounting for 18.4% of China's

---

usually varies greatly by households or by persons. For better understanding, it is considered separately in this report. Energy use for household heating facilities, especially in South China, is in accordance with specified individual or household requirements, and then taken into account together with energy use of other end-usages in residential and commercial buildings, respectively.

[4] For better analysis, North China refers to the provinces, autonomous regions and municipalities as: Beijing, Tianjin, Hebei, Shanxi, Inner Mongolia, Liaoning, Heilongjiang, Shandong, Henan, Shannxi, Gansu, Qinghai, Ningxia, and Xinjiang.

[5] Due to lack of relevant statistics, it is assumed in the report that all rural buildings are residential ones. To specify, schools, healthcare centers, retail shops, and administrative offices in the rural area are considered as residential buildings.

[6] Methodology:(a) energy consumption of urban residential buildings (excluding energy use for district heating system): Is the aggregate of total coal, coke, coke gas, other gases, kerosene, LPG, natural gas, electricity, and 5% diesel oil of urban residential consumption according to the China Energy Balance Sheet in *China Energy Statistics Yearbook*; (b) energy consumption of rural residential buildings is the aggregate of total coal, coke, coke gas, other gases, kerosene, LPG, natural gas, electricity, and 5% diesel oil of rural residential consumption according to the China Energy Balance Sheet

total energy consumption; CO_2 emission reached 1,260 Mtce, sharing 19.8% of China's total. According to the research on China's rural energy consumption conducted by Tsinghua University in 2006–2007, there are about 100 Million tons of standard coal consumed in rural areas, yet out reach of the statistic system of China. Taking this part into account, total commodity energy consumption in China's building sector would be 640 Mtce in 2008, which accounts for 22.2% of China's total and lead to about 1.53 billion tons of CO_2 emissions.

Energy consumption and CO_2 emissions in China's building sector has been increasing in recent years. Energy consumption in 2008 was 118 Mtce more than that in 2005, increasing by 30%, with an annual growth rate of 8.6%. During the same period, CO_2 emissions increased by 2.53 billion tons, or 25%, with an annual growth rate of 7.8%. However, average growth rate of CO_2 emissions in China's building sector from 2006–2008 was slower than that during the 10th *Five Year Plan* (*FYP*) (9.1%). See Fig. 9.1.

Compared to other countries, China shares about 13% in the total CO_2 emissions by the building sector of the whole world, in 2006. See Fig. 9.2.

Total CO_2 emission in the building sector equals to the CO_2 emission per unit floor area multiplying the total floor area. CO_2 emission per unit floor area is determined by way of consuming energy in buildings. Total floor area reflects the size of a certain building stock. In addition, CO_2 emission factors of different energy types also significantly affect total CO_2 emissions.

(2) *Floor area*

China's urban population has witnessed a rapid growth with continuous urbanization, and higher living standards lead to higher demand for larger floor area per capita. Growth of population and floor area per capita results in rapid growth of total floor area in China's urban areas. See Fig. 9.3 and Fig. 9.4.

---

in *China Energy Statistics Yearbook*; (c) energy consumption of commercial buildings excluding district heating system is the aggregate of total coal, coke, coke gas, other gases, kerosene, LPG, natural gas, electricity, and 65% diesel oil of the wholesale and retail industry and catering industry according to the China Energy Balance Sheet in *China Energy Statistics Yearbook*; (d) energy consumption of district heating system in North China's urban areas is calculated according to the results of *2009 Annual Report of R&D of China's Building Energy Conservation*.

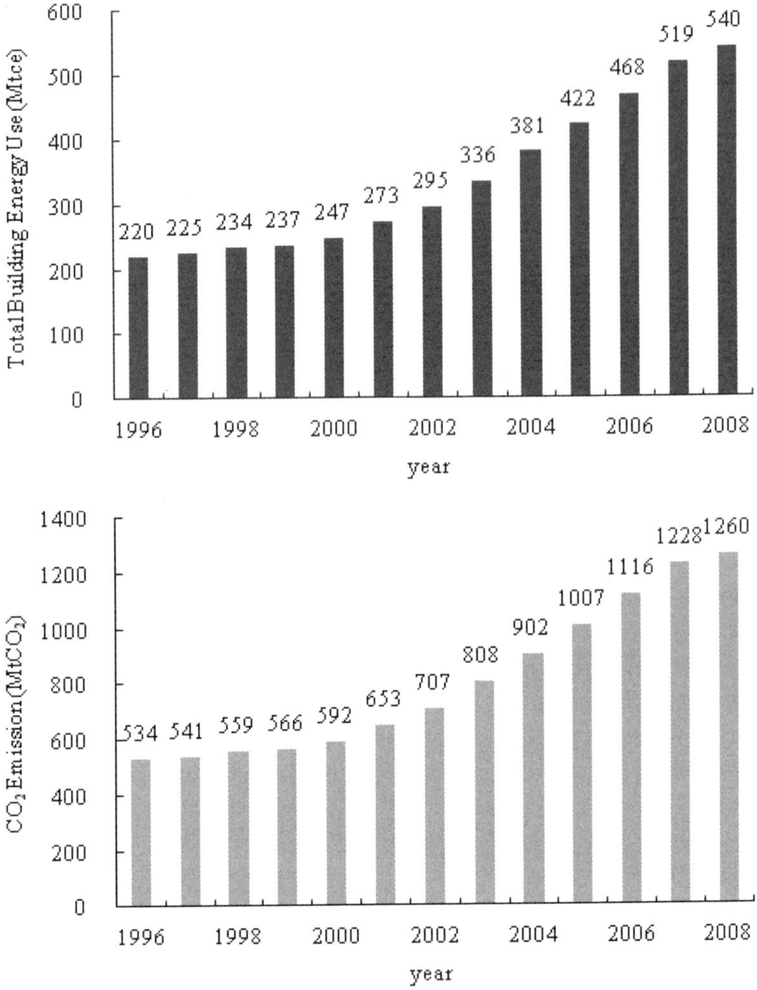

Fig. 9.1. Total energy consumption and CO_2 emissions in China's building sector, from 1996 to 2008.

(1) China's urbanization began with the reform and opening up in 1978, after which, urban population kept increasing and construction work started.
(2) By 1996, China's urbanization ratio reached 30%, entering the stage of rapid urbanization and increasing by 1%–1.5% annually. By 2008, the urbanization level reached 46%, with urban population increasing

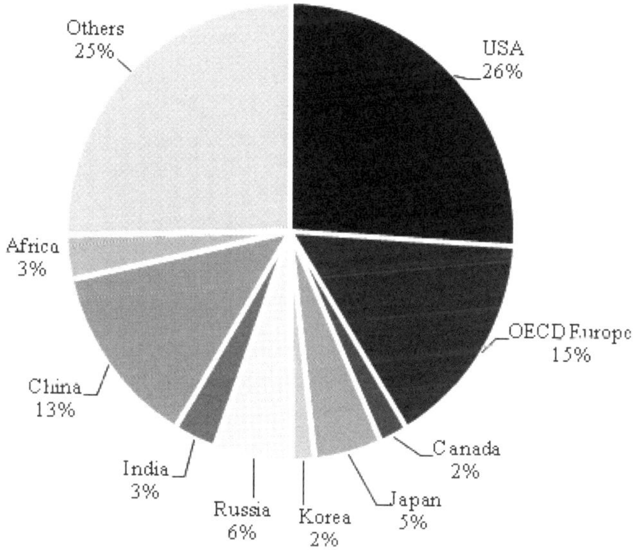

Fig. 9.2. Total CO_2 emissions in world building sector in 2006.
Data Source: ① World Bank, *World Development Indicators database* (World Bank, Washington, 2007); ② State Statistics Bureau, *China Statistics Yearbook 2009* (China Statistics Press, Beijing, 2009); ③ The United State Department of Energy, *2009 Buildings Energy Data Book* (D&R International, Ltd., USA, 2009); ④ Natural Resources Canada, *2006 Energy Use Data Handbook* (Energy Publications Office of Energy Efficiency, Canada, 2007); ⑤ The Energy Data and Modeling Center, *2009 Handbook of Energy & Economic Statistics in Japan* (The Energy Conservation Centre, Japan, 2009); ⑥ Ministry of Commerce, Industry and Energy and Korea Energy Economics Institute, *Yearbook of Energy Statistics 2007* (Korean Energy Economics Institute, Seoul, 2007) ⑦ Energy Information Administration, *International Energy Outlook 2009* (EIA Publications, USA, 2009).

from 370 million to 610 million. Meanwhile, floor area per capita had increased continuously, and the growth rate of buildings was much faster than that of urban population.

(3) China's urban building has developed rapidly since 2000, and the floor areas of urban buildings increased from 7.7 billion m^2 in 2000 to almost 15 billion m^2 in 2004, and 19.5 billion m^2 in 2008, increasing by 1.5 folds in eight years.

(3) *Energy consumption and CO_2 emission intensity*

Figure 9.5 shows the energy consumption and CO_2 emission intensity in China's building sector from 1996 to 2008. In 2008, energy consumption

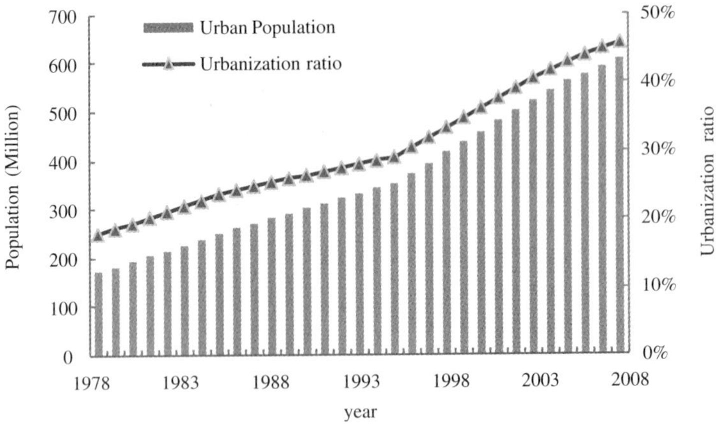

Fig. 9.3. China's urbanization level and urban population from 1978 to 2008.
Source: State Statistics Bureau, *China Statistics Yearbook 2009* (China Statistics Press, Beijing, 2009).

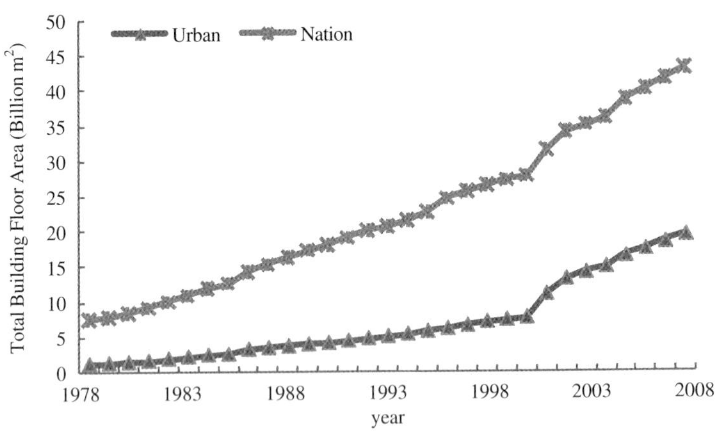

Fig. 9.4. Total floor area of China's buildings and urban buildings from 1978 to 2008.
Note: Floor areas in 2007 and 2008 are estimated values.
Source: State Statistics Bureau, *China Statistics Yearbook 2009* (China Statistics Press, Beijing, 2009).

intensity in China's building sector was 12.8 kgce/m^2, CO_2 emission intensity was 29.3 kgCO_2/m^2.

Energy consumption intensity in 2008 increased by 16.0% compared to 2005, with an annual growth rate of 5.1%, CO_2 emission intensity in 2008 increased by 12.2%, with an annual rate of 3.9%. Growth rate of energy

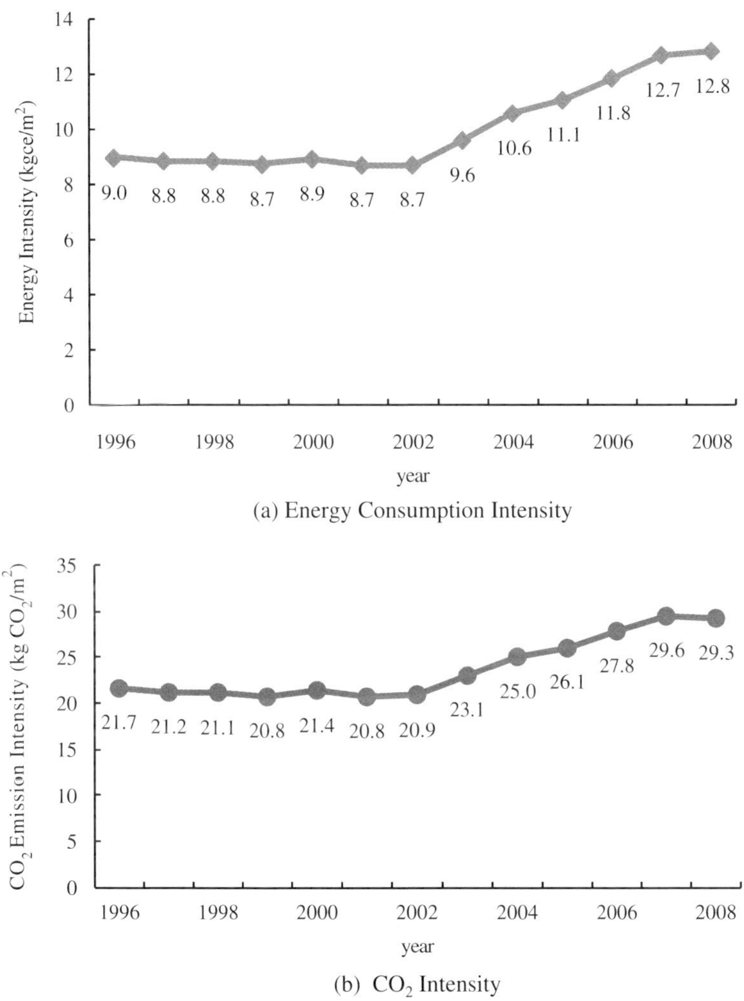

Fig. 9.5. Energy consumption and CO_2 emission intensity in China's building sector from 1996 to 2008.
Note: Floor areas in 2007 and 2008 are estimated values.
Source: State Statistics Bureau, *China Statistics Yearbook 2009*, China Statistics Press, Beijing, 2009.

consumption intensity was much lower than that of energy consumption and CO_2 emission.

Average growth rate of CO_2 emission per unit floor area from 2006 to 2008 was slightly lower than that during the 10th *FYP* (annual growth rate is 4.06%).

By comparing the energy consumption intensity in China's building sector with that of developed countries, it can be discovered that:

(1) China's average CO_2 emission per capita was 849 kgCO_2/person in 2006, lower than the world average level (1,326 kgCO_2/person), about 1/9 of the USA, 1/4 of Japan, and 1/5 of OECD Europe. See Fig. 9.6.
(2) CO_2 emission per unit floor area was 27.9 kgCO_2/m^2 in 2006, less than 1/3 of the USA and lower than that of developed countries. See Fig. 9.7.

The huge difference derives from different ways of consuming energy by Chinese and foreign residents (Zhang, 2009). It needs to be noticed that, on one hand, the difference does not necessarily mean corresponding difference in the resident's satisfaction to their livelihood. Research shows that there is no direct relationship between energy consumption and satisfaction of residents, and there is no apparent difference in the satisfaction rate of two sample groups whose energy consumption differs by 10 folds (China Urban

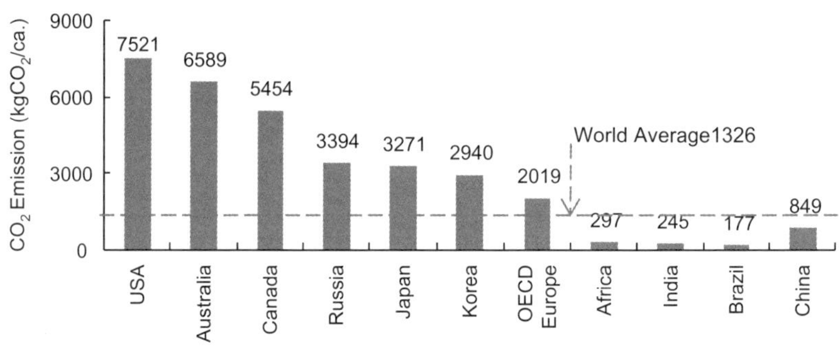

Fig. 9.6. CO_2 emission per capita in the building sector of different countries in 2006.
Source: The same with Fig. 9.2.

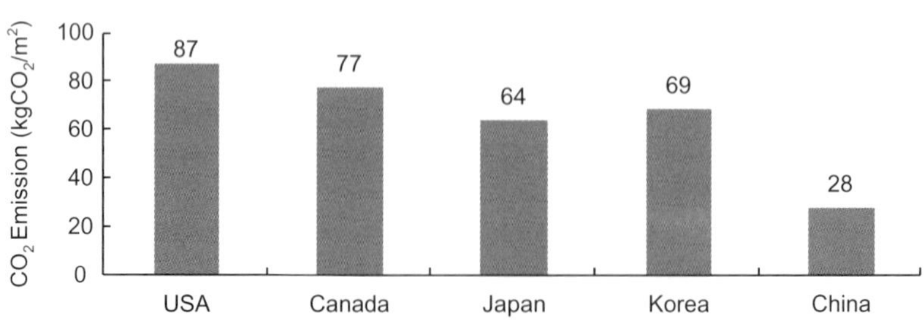

Fig. 9.7. CO_2 emission per unit floor area in the building sector of different countries in 2006.

Energy Consumption and Energy Conservation Policy Research Group, 2010). On the other hand, research shows that the satisfaction is in exponential relationship with energy consumption. When the satisfaction reaches a certain level, a little improvement in satisfaction would cost enormous energy consumption. Under current energy background, it does not pay to do so. (Building Energy Research Center of Tsinghua University, 2009).

(4) *Energy Structure*

From 1996 to 2008, coal and electricity dominate in the final energy pan in buildings, see Fig. 9.8. Proportion of coal consumption had declined from 75% in 1996 to 56% in 2008, but total coal consumption had increased from 120 Mtce to 215 Mtce. Power consumption had increased most rapidly from 186 billion kWh in 1996 to 730 billion kWh in 2008, increasing by three folds; and its proportion had increased from 32% to 44%.

With respect to noncommodity energy, primary biomass energy (such as straw and firewood) is widely used in China, mainly in rural areas. According to statistics, biomass energy consumption in China's rural areas was 252 Mtce in 2008 (Wang, 2009), almost half of China's total commodity energy consumption in the building sector.

(5) *Four categories of building energy consumption*

Total CO_2 emissions and CO_2 intensity for four categories of building energy consumption are listed in Fig. 9.9 and Fig. 9.10.

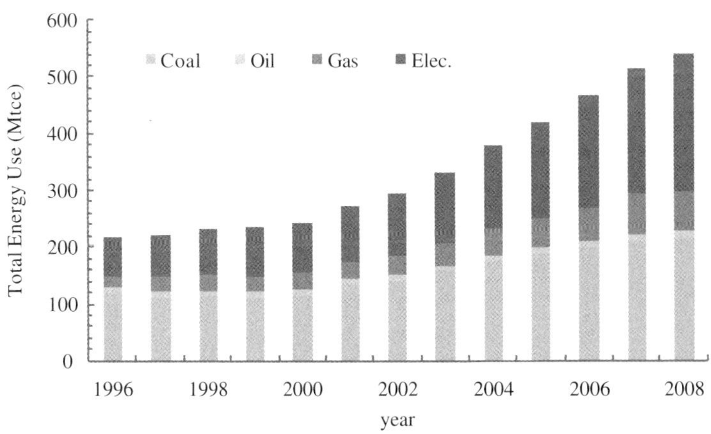

Fig. 9.8. Energy structure in China's building sector from 1996 to 2008.

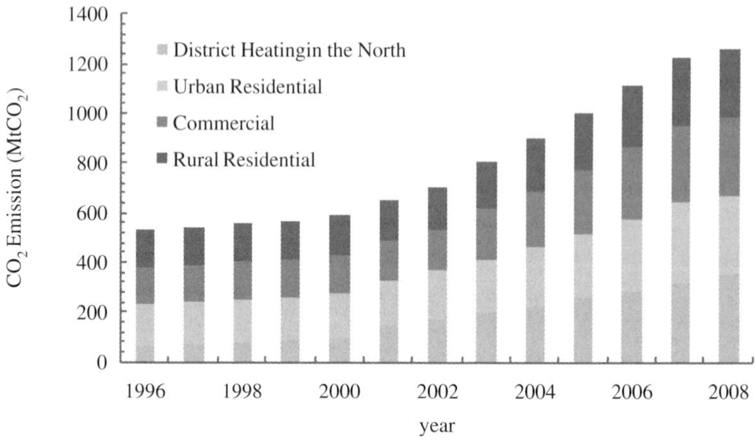

Fig. 9.9. CO_2 emissions of various categories of building energy consumption from 1996 to 2008.

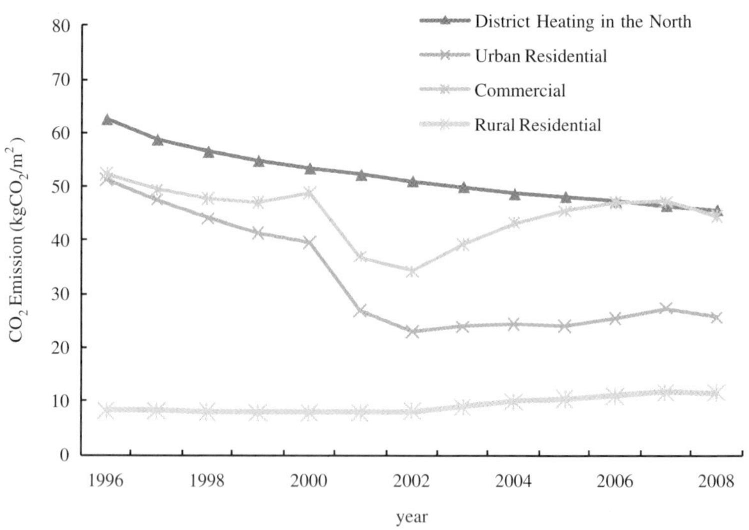

Fig. 9.10. CO_2 emission per unit floor area of various categories of building energy consumption from 1996 to 2008.

Total CO_2 emission for district heating system in North China's urban areas has increased most rapidly (mainly due to fast growth of building floor areas that apply district heating system), accounting for 28% of total CO_2 emission in the building sector in 2008.

CO_2 emission per unit floor area for district heating system in North China's urban areas has decreased continuously, while that for commercial buildings has increased most rapidly in recent years, which had caught up with the category of district heating system in North China's urban areas by 2008, about 44.6 $kgCO_2/m^2$.

There is significant difference of CO_2 emissions between urban and rural areas. CO_2 emission of rural buildings is much lower than that of urban residential buildings. By taking energy consumption of heating systems into consideration, CO_2 emission of rural residential buildings only accounts for about 1/3 of urban residential buildings.

Characteristics, developing trend, low-carbon performance, and major influencing factors of energy consumption and CO_2 emission for four categories will be analyzed in the following paragraphs.

9.1.2 *Better technology structure and enhanced envelopes contributing to declining energy intensity for district heating system in northern cities*

At present, most buildings in Northern cities are installed with district heating system, while the remaining buildings adopt various household heating facilities. Technologies for producing heat for district heating system include combined heating and power (CHP) systems and boilers. Natural gas boilers are widely utilized in Beijing city, while coal boilers are more commonly used in other northern cities.

(1) *Total energy consumption and CO_2 emission*

Energy consumption of district heating system in northern cities was 133 Mtce in 2008, accounting for 25% of the total energy consumption in the building sector (Building Energy Conservation Research Center of Tsinghua University, 2010); CO_2 emission was 353 Mt, accounting for 28% of the total CO_2 emission. In recent years, total energy consumption and CO_2 emission has increased dramatically. Energy consumption in 2008 was 35.91 Mtce more than that in 2005, increasing by 37%, with an annual growth rate of 11.1%. Meanwhile, CO_2 emission in 2008 was 94.12 Mt more than that in 2005, increasing by 36%, with an annual growth rate of 10.9%. See Fig. 9.11. Notably, from 1996 to 2008, energy consumption and CO_2 emission of the district heating system in Northern cities increased by more than four times.

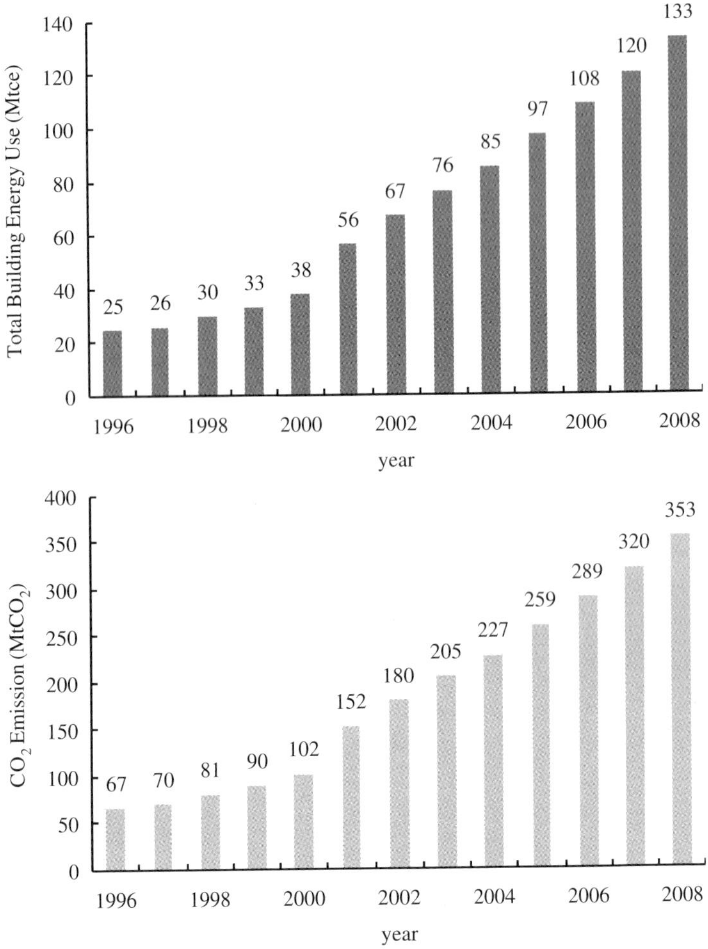

Fig. 9.11. Total energy consumption and CO_2 emission of the district heating system in Northern cities from 1996 to 2008.

(2) *Floor area*

From 2005 to 2008, the floor area with district heating system increased from 5.4 billion m^2 to 7.7 billion m^2, increasing by 42.5%. See Fig. 9.12. From 1996 to 2008, the floor area with district heating system increased more than six folds. Proportion of floor areas with district heating system to the total floor area in Northern cities has been increasing.

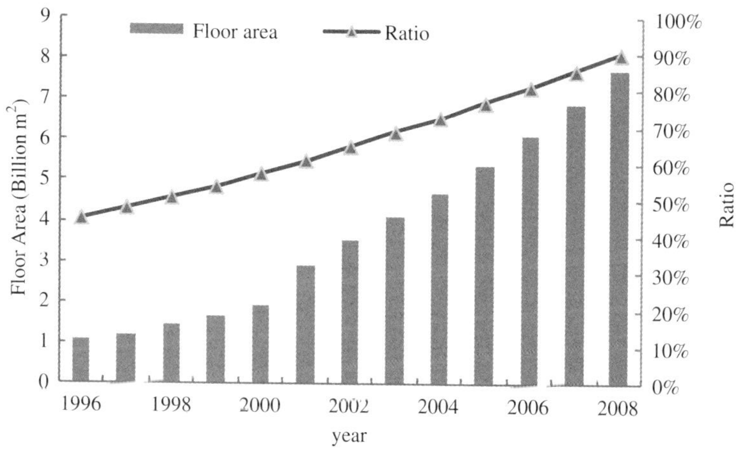

Fig. 9.12. Floor areas with district heating system and its proportion to total floor area in Northern cities from 1996 to 2008.

(3) *Energy consumption and CO_2 emission intensity*

Declining CO_2 emission per unit floor area and energy consumption intensity marked prominent low-carbon performance of district heating system in northern cities. See Fig. 9.13. CO_2 emission per unit floor area in 2008 was 45.7 $kgCO_2/m^2$, decreasing by 5.1% compared to 2005 (48.2 $kgCO_2/m^2$); decreasing by 27% compared to 1996 (62.6 $kgCO_2/m^2$). Reduction of energy consumption per unit floor area slows down the growth of total heating-related energy consumption and CO_2 emission in Northern cities. In 2008, proportion of district heating system in northern cities to the total building energy consumption was about 25%, which indicates that enhancing energy efficiency of heating could significantly contribute to slowing down growth of total energy consumption in China's building sector.

In 2008, average energy consumption per unit floor area of the district heating system in Northern cities was slightly lower than the heating-related energy consumption of South Korea in 2005 (17.6 $kgce/m^2$) (Korean Energy Economics Institute, 2005), similar to that of Japan's Hokkaido in 2005 (15.3 $kgce/m^2$) (Yutaka *et al.*, 2005).

Building insulation and efficiency of technologies for heat production are two major factors affecting heating-related energy consumption. The former one is a key factor that determines the heat demand of buildings, while the latter one reveals technological capability for fulfilling that heat

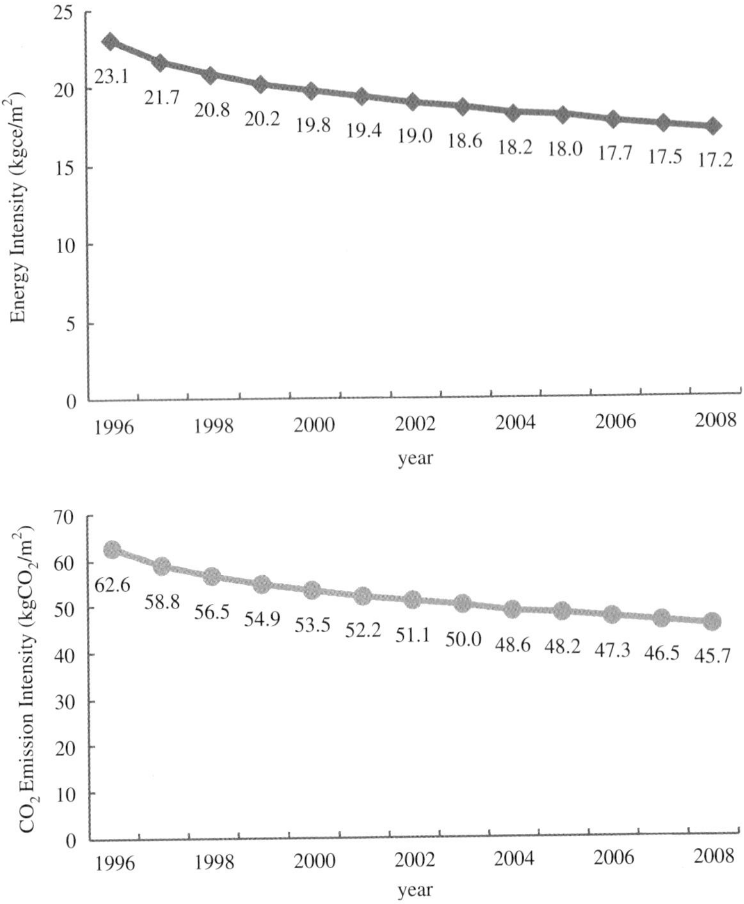

Fig. 9.13. Energy consumption and CO_2 emission intensity of the district heating system in Northern cities from 1996 to 2008.

demand. Both factors contribute to the declining building energy consumption intensity and CO_2 emission intensity in Northern cities.

(4) *Building insulation*

At present, design standards for building energy efficiency in China mainly focus on improving the insulation performance of envelope, and are implemented substantially in recent years In 2000, only 180 million m^2 floor areas in China complied with the design standards, which is less than 0.6% of the total building stock; by the end of 2009, the number reached 4.08 billion m^2

(Ministry of Construction, 2010, p. 45), and 2.4 billion m^2 were in Northern cities, accounting for 28% of the total floor areas in Northern cities.

Better insulation of newly-built buildings contributed much to the reduction of average heat demand of buildings in northern cities from 106 kWh/m^2 in 1996 to 92 kWh/m^2 in 2008 (Yang, 2010), decreasing by 1/8.

(5) *Heat source efficiency*

Coal is still the major energy source for district heating system in northern cities. However, energy efficiency of different coal-fired technologies varies greatly. For example, thermal efficiency of coal could reach 148%[7] with CHP system; while that of small coal boilers is usually lower than 50%. In general, thermal efficiency of CHP system is higher than that of boilers, while that of gas-fired boilers is higher than coal boilers, and that of large coal boilers is higher than smaller ones.

From 1996 to 2008, proportion of floor areas using high-efficiency technologies, especially CHP systems to the total floor areas with district heating system had increased steadily. See Fig. 9.14. With structural

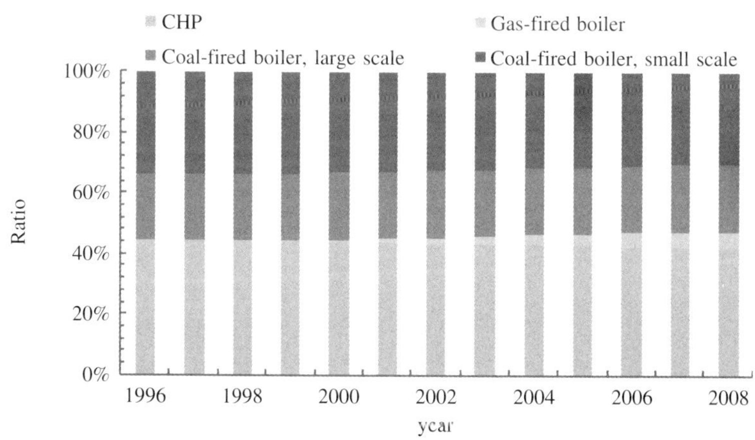

Fig. 9.14. Proportion of floor areas of various district heating systems in Northern cities 1996–2008.

[7]For large-scale CHP system, 1 MJ coal could produce 0.3 MJ power and 0.4 MJ heat. In high-efficiency thermal power plants [supposing 300 gce/(kWh)], producing 0.3 MJ power needs 0.73 MJ coal, which equals that producing 0.4 MJ heat needs 0.27 MJ coal, so the heat generation efficiency is 148% (Building Energy Conservation Research Center of Tsinghua University, 2009).

optimization of technologies, efficiency of each technology also improved due to technical and management enhancement. Both contributed to the improvement of comprehensive efficiency for heat production in district heating systems, which increased from 54% in 1996 to 65% in 2008 (Yang, 2010), increasing by 1/5.

9.1.3 *Energy structural optimization and changing life styles lead to slight increase of energy consumption intensity in urban residential buildings*

Energy consumption of urban residential buildings (excluding energy use of district heating system) refers to energy use for purposes of household heating, cooling, lighting, household appliances, cooking, and domestic hot water and so on. Energy types utilized in urban residential buildings include electricity, coal, LNG, town gas, (including natural gas and coal gas) and so on. Solar energy use for water heating or other purposes in household is not taken into account since that only commodity energy is dealt with within this section.

(1) *Total energy consumption and CO_2 emission*

Commodity energy consumption of urban residential buildings (excluding energy use of district heating system) was 152 Mtce in 2008, accounting for 28% of the total energy consumption in the building sector; CO_2 emission was 316 Mt, accounting for 25% of the total in the building sector. In recent years, total energy consumption and CO_2 emission has increased continuously. Energy consumption in 2008 was 31.64 Mtce more than that in 2005, increasing by 26%, with an annual growth rate of 8.1%. Meanwhile, CO_2 emission in 2008 was 57.16 Mt more than that in 2005, increasing by 22%, with an annual growth rate of 6.9%. See Fig. 9.15. From 2006 to 2008, average growth rate of the total building CO_2 emission was slower than that during the 10th *FYP* (annual growth rate was 8.14%).

(2) *Floor area*

China's economy is developing rapidly and continuously, and the housing conditions of urban residents have improved. As a consequence, the floor areas of urban residential buildings increased rapidly (Fig. 9.16). From 1996 to 2008, urban population increased from 373 million to 607 million, the

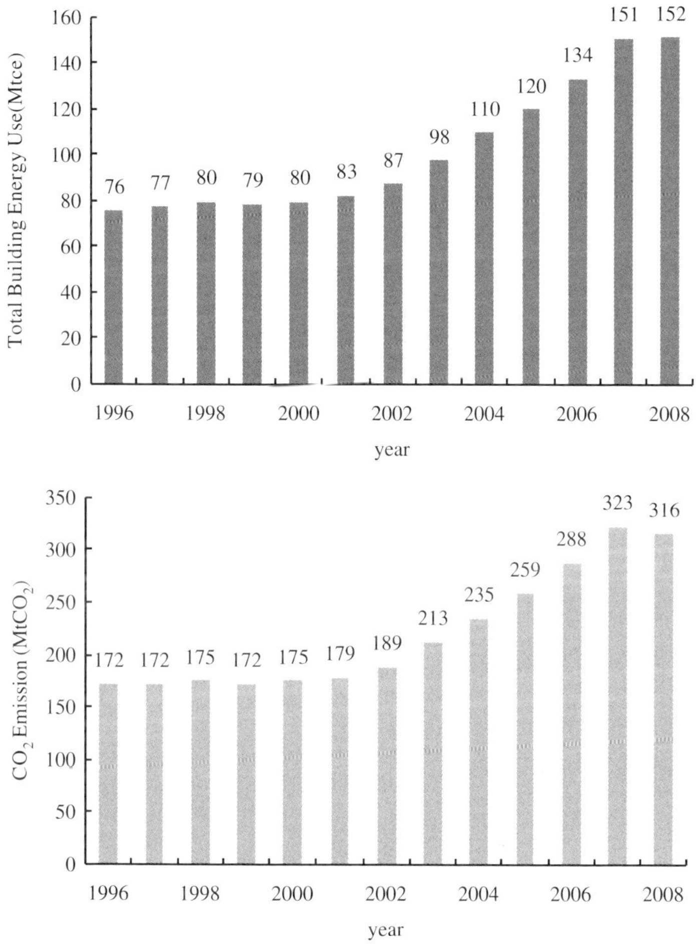

Fig. 9.15. Total energy consumption and CO_2 emission of China's urban residential buildings from 1996 to 2008.

floor area per capita increased from $9.0\,\text{m}^2$ to $20.8\,\text{m}^2$, and the total floor area of urban residential buildings increased by 2.6 times.

(3) *Energy consumption and CO_2 emission intensity*

Energy consumption and CO_2 emission per unit floor area of urban residential buildings witnessed a fall and rise. See Fig. 9.17.

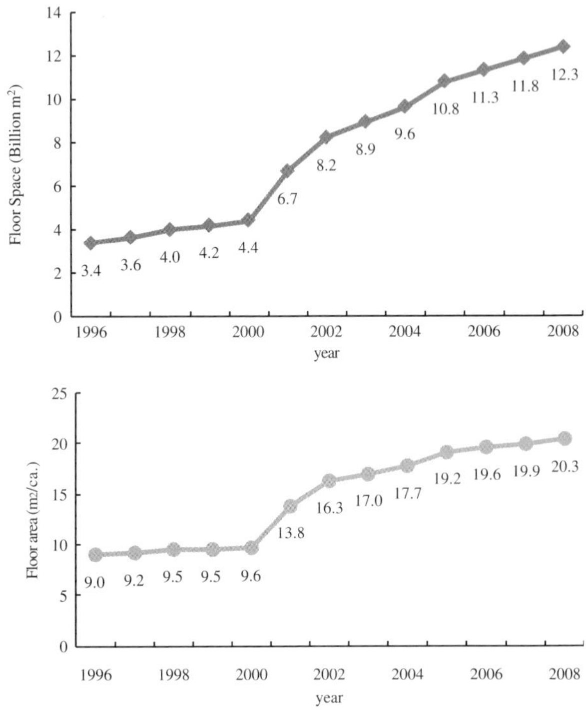

Fig. 9.16. Floor areas of China's urban residential buildings from 1996 to 2008.

(1) From 1996 to 2002, energy consumption per unit floor area decreased from 22.6 kgce/m^2 to 11.0 kgce/m^2, decreasing by over 50%. It is largely due to the expanding use of district heating system in Northern China, and resulting decrease of coal consumption for household heating in families. See Fig. 9.18.
(2) From 2002 to 2007, energy consumption per unit floor area increased to 12.8 kgce/m^2, lower than that of Japan (22 kgce/m^2) (The Energy Data and Modeling Center, 2009), less than half of the USA (27 kgce/m^2) (The United States Department of Energy, 2009).
(3) Energy Consumption intensity in 2008 was slightly lower than that in 2007.
(4) CO_2 emission intensity changes along with changing energy consumption per unit floor area.

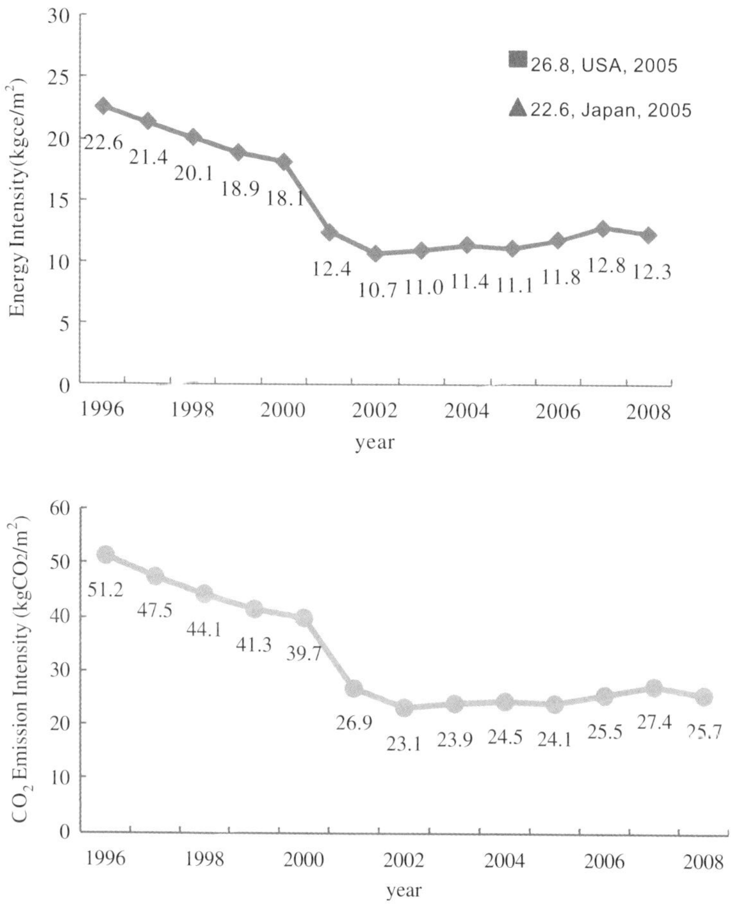

Fig. 9.17. Energy consumption and CO_2 emission intensity of China's urban residential buildings from 1996 to 2008.

(4) *Energy structure*

Energy structural change in urban residential buildings is listed in Fig. 9.19. Share of coal decreased rapidly, which explained well why growth rate of CO_2 emission in urban residential buildings was lower than that of the whole building sector. Proportion of electricity, natural gas, and LNG to the total energy consumption witnessed a substantial growth, and the real consumption of all energy has increased by more than two folds.

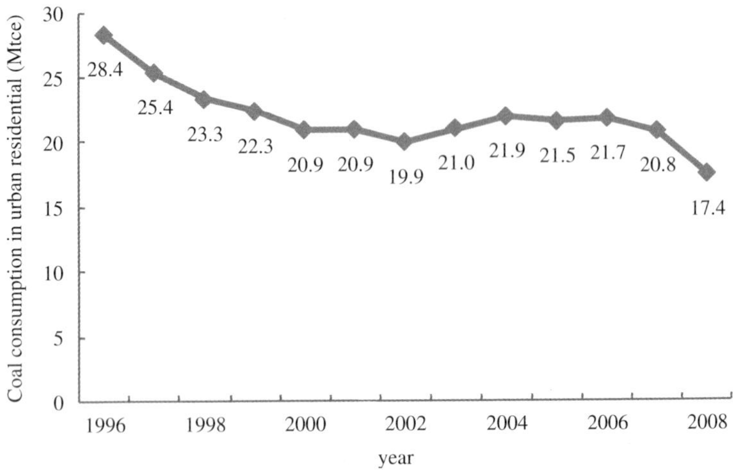

Fig. 9.18. Coal consumption in China's urban residential buildings from 1996 to 2008.

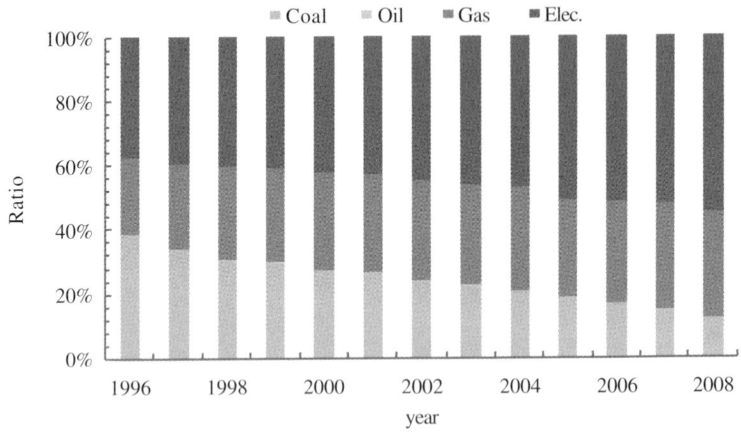

Fig. 9.19. Energy consumption (excluding district heating system) of urban residential buildings from 1996 to 2008.

(5) *Lifestyles are determinants of residential energy consumption*

As stated previously, enormous gap, in terms of energy consumption per capita and per unit floor area in residential sector, was observed between China and developed countries, such as USA and Japan. Similar energy gaps, too, were witnessed between Chinese residents, according to an investigation by Tsinghua University on energy consumption of nearly

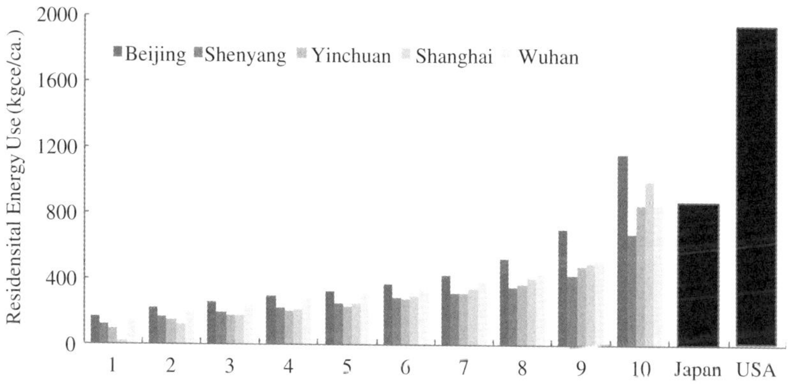

Fig. 9.20. Building energy consumption per capita of five cities in China in 2008 and comparison with the USA and Japan.
Note: Energy consumption data of the USA and Japan is that of 2005.
Source: The same with Fig. 9.2.

10,000 residents in five cities in 2009 (Zhang, 2010). Researchers sequenced all samples by energy consumption per capita and divided the samples equally into 10 groups, with the purpose of investigating the energy consumption per capita of each group (Fig. 9.20). Results reveal that there is a huge difference between these groups. Energy consumption per capita of Group 10 is higher than that of Japan's average level, and is over 10 times than that of Group 1.

Further analysis reveals that the difference is closely related to the lifestyles of different sample groups[8] (Zhang, 2010). Lifestyles of Group 10 and Group 1 differ enormously. However, the primary difference lies more in operation of energy carriers than ownership of them. For example, samples in Group 10 take much more showers, apply cooling devices day and night, and are more used to energy-intense appliances such as electric dryers and so on.

Considering the hiking economic development and enhancing living standards, it is predictable that household energy consumption will continue

[8]Lifestyle refers to the subjective and objective behaviors of an individual in conducting building energy consumption in his daily life, which shall include preferences and requirements for specific activities in buildings, selection of equipment and tools used within the activity, operation manner and setup of the selected equipment and tools, and evaluation after the activity took place. Different combinations of above choices constitute different lifestyles.

to increase. It would load heavy burden on domestic energy supply, if lifestyles of majority citizens turn to a Group-10 style, or even a Japan- or USA-style.

9.1.4 *Rising proportion of large-scale commercial buildings lead to continuous increase of the total energy consumption in commercial sector*

In commercial buildings, apart from hot water and steam used in the district heating system (and will not be discussed in this section), electricity dominates in the energy pan, and is mainly used for cooling, lighting, elevators, office equipment, ventilation, water supply and sewage, and other equipments. Yet, coal, natural gas, and LNG are also used for cooking, hot water producing, and in a few buildings, for heating and cooling, too.

(1) *Total energy consumption and CO_2 emission*

Energy consumption of commercial buildings (excluding energy use for district heating system, which is as in 9.1.2) was 141 Mtce in 2008, accounting for 26% of the total energy consumption in the building sector; CO_2 emission was 318 $MtCO_2$, accounting for 26% of the total emission in the building sector. In recent years, total energy consumption and CO_2 emission has increased. Energy consumption in 2008 was 29.16 Mtce more than that in 2005, increasing by 26%, with an annual growth rate of 8.0%. Meanwhile, CO_2 emission in 2008 was 59.17 Mt more than that in 2005, increasing by 23%, with an annual growth rate of 7.1%. See Fig. 9.21. Growth rate of total CO_2 emission of commercial buildings in the first three years of the 11th *FYP* is slower than that in the 10th *FYP* (annual growth rate was 10.3%).

(2) *Floor area*

From 1996 to 2008, total floor area of commercial buildings increased from 2.8 billion m^2 to 7.1 billion m^2, increasing by 1.5 folds, and the floor area per capita increased from 7.4 m^2/capita to 11.7 m^2/capita, increasing by over 50% (Fig. 9.22).

(3) *Energy consumption and CO_2 emission intensity*

Energy consumption and CO_2 emission per unit floor area also witnessed a fall and rise. See Fig. 9.23.

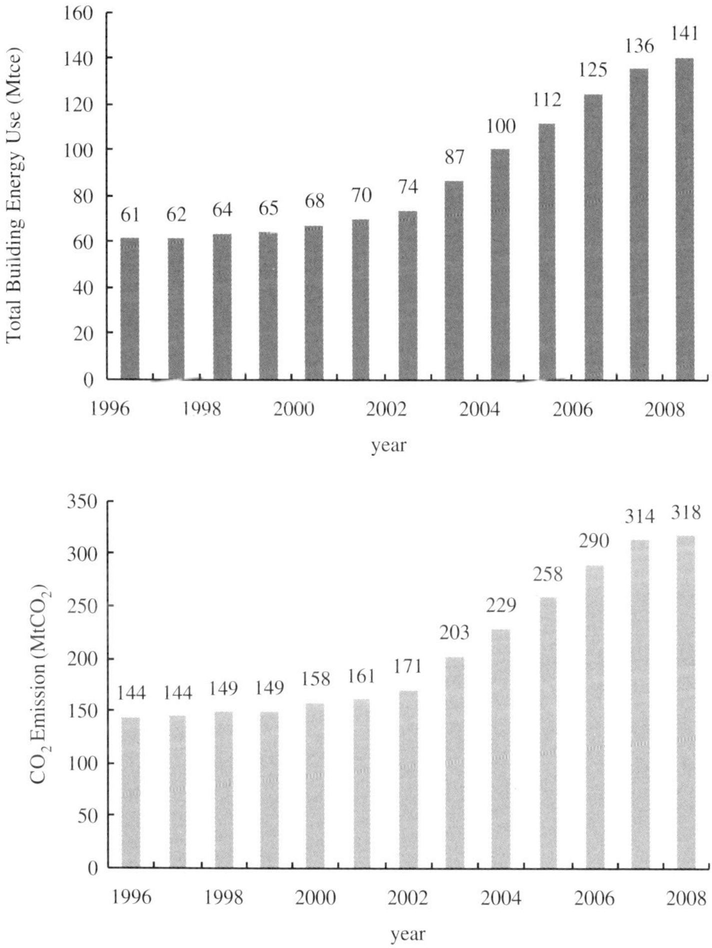

Fig. 9.21. Total energy consumption and CO_2 emission in China's commercial buildings from 1996 to 2008.

(1) From 1996 to 2002, energy consumption per unit floor area decreased from 22.3 kgce/m^2 to 14.8 kgce/m^2, decreasing by more than 40%.[9]

(2) From 2002 to 2007, energy consumption per unit floor area increased to 20.4 kgce/m^2, much lower than that of Japan (70 kgce/m^2) (The Energy

[9]It is due to fast expanding use of district heating system in Northern China, and resulting decrease of coal consumption for heating systems in single buildings.

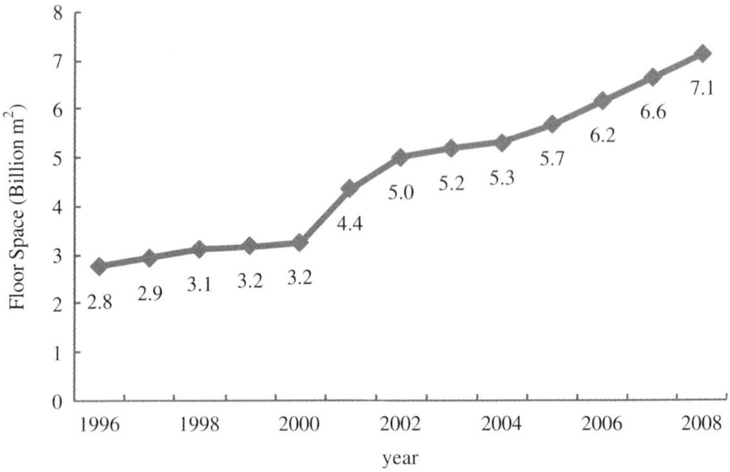

Fig. 9.22. Floor areas of China's commercial buildings from 1996 to 2008.

Data and Modeling Center, 2009) and of the USA (78 kgce/m^2) (The United States Department of Energy, 2009).

(3) Energy consumption intensity in 2008 was slightly lower than that in 2007.

(4) *Energy structure*

Fuel mix of energy consumption in commercial buildings is listed in Fig. 9.24, which mainly features in apparent lowering of coal-share and hiking electricity-share.

(5) *Energy consumption for commercial buildings of different scales*

For commercial buildings, energy consumption is largely determined by building sizes. Generally, larger buildings have higher energy consumption per unit floor area. Accordingly, China's commercial buildings can be divided into large-scale and ordinary commercial buildings. As is defined, building volume of the former is usually over 20,000 m^2 per building, and elevators, central air-conditioning system, and mechanical ventilation systems are usually installed. As a result, the energy consumption per unit floor area is 2–5 times more than that of the latter (Building Energy Conservation Research Center of Tsinghua University, 2010).

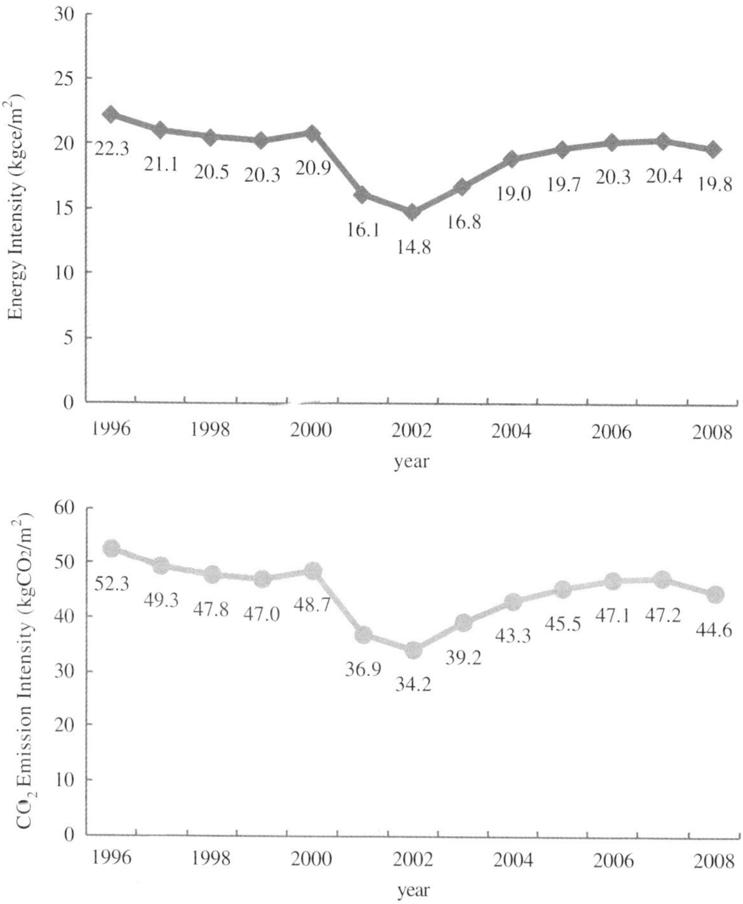

Fig. 9.23. Energy consumption and CO_2 emission intensity of China's commercial buildings (excluding district heating system) from 1996 to 2008.

In recent years, under the expanding thinking of "being geared to international standards" and "Western noble life styles" in urban China, construction of large-scale commercial buildings becomes unrestricted and its floor area in total increased dramatically. Meanwhile, many existing commercial buildings are retrofitted by introducing luxurious Western-style systems and equipments without necessary considerations for localization, leading to rapid growth of energy consumption. From 1996 to 2008, floor areas of large commercial buildings increased from 80 million m^2 to

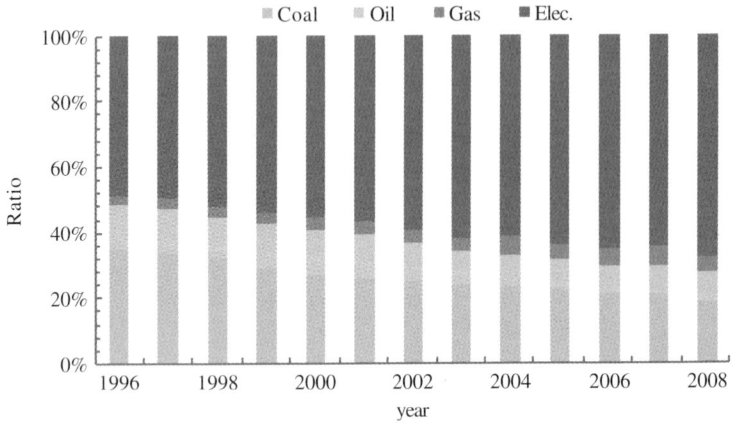

Fig. 9.24. Fuel mix of energy consumption (excluding district heating system) in commercial buildings from 1996 to 2008.

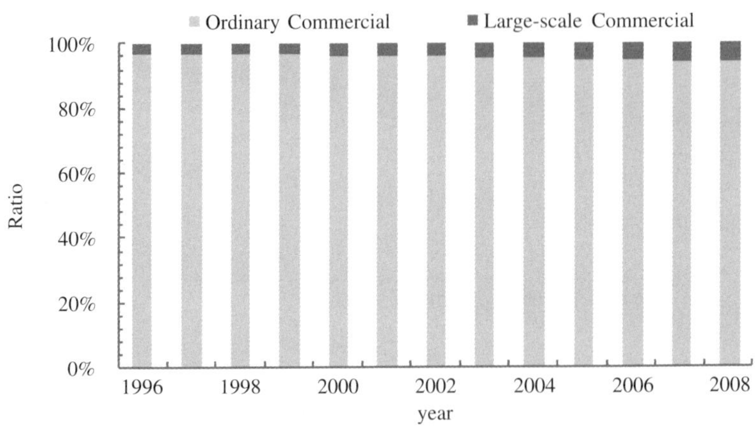

Fig. 9.25. Proportion of ordinary and large-scale commercial buildings from 1996 to 2008.

430 million m^2, increasing by more than four times, and its proportion to total commercial building stock has increased continuously (Building Energy Conservation Research Center of Tsinghua University, 2009). See Fig. 9.25. In 2008, proportion of large commercial buildings accounted for 6% of the total floor areas of commercial buildings (Yang, 2010), but their power consumption nearly accounted for 20% (Building Energy Conservation Research Center of Tsinghua University, 2010).

Since 1990s, growing energy intensity of commercial buildings (excluding energy use for district heating system) was due to structural increase of large-scale commercial buildings.

9.1.5 *Lowering proportion of renewable energy, and growth rate of rural energy consumption exceeding that of building floor arcas in rural areas*

In rural residential buildings, energy is mainly used for heating, cooking, lighting and household appliances. Apart from major commodity energy sources such as coal, LNG, and electricity, there are many primary bio-mass energy sources such as straw and firewood, as well as some renewable energy sources such as methane and solar-heated domestic hot water. To specify, these primary bio-mass energy and renewable energy sources are primarily used for heating in winter and daily cooking.

(1) *Total energy consumption and CO_2 emission*

Commodity energy consumption of rural residential buildings was 115 Mtce in 2008, accounting for 21% of the total energy consumption in the building sector; CO_2 emission was 273 Mt, accounting for 22% of the total in the building sector. Total energy consumption and CO_2 emission experienced successive increase in recent years. Energy consumption in 2008 was 21.50 Mtce more than that in 2005, increasing by 23%, with an annual growth rate of 7.2%. Meanwhile, CO_2 emission in 2008 was 42.80 Mt more than that in 2005, increasing by 18.6%, with an annual growth rate of 5.9%. See Fig. 9.26. Average growth rate of CO_2 emission during the 11th *FYP* was slower than that in the 10th *FYP* (during which the annual growth rate was 7.9%).

(2) *Floor area*

Due to fast urbanization in last decade, rural population decreased from 850 million to 720 million, while the floor areas of rural residential buildings increased from 18.5 billion m^2 to 23.4 billion m^2, increasing by 1/3. See Fig. 9.27.

(3) *Energy consumption and CO_2 emission intensity*

In 2008, energy consumption and CO_2 emission per unit floor area of rural residential buildings were 4.86 kgce/person and 11.6 kgCO^2/person respectively. See Fig. 9.28.

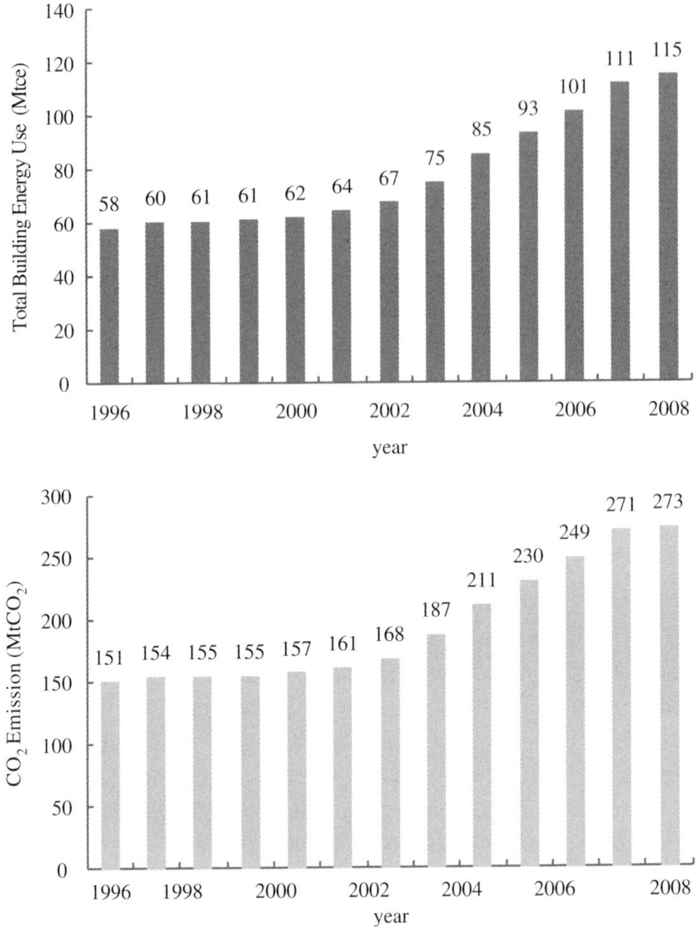

Fig. 9.26. Total energy consumption and CO_2 emission in China's rural residential buildings from 1996 to 2008.

Energy consumption and CO_2 emission per unit floor area of rural residential buildings has increased in recent years. From 2005 to 2008, energy consumption per unit floor area increased by 15.7%, with an annual growth rate of 5.0%; CO_2 emission per unit floor area increased by 15.8%, with an annual growth rate of 5.0%. Average growth rate of CO_2 emission during the 11th *FYP* is slower than that in the 10th *FYP* (annual growth rate was 5.8%).

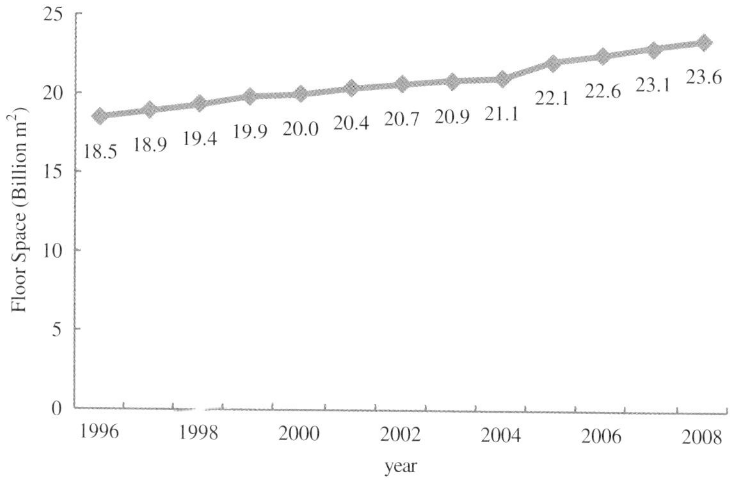

Fig. 9.27. Floor areas of China's rural residential buildings from 1996 to 2008.

Comparison of residential energy consumption (including energy use for district heating[10]) between rural and urban area revealed significant energy gap between them. The energy consumption per unit floor area in rural residential buildings was only 1/4 of the counterpart in rural area. In addition, enhancing living standards further fuels up continuous increase of energy consumption and CO_2 emission per unit floor area in rural residential buildings. From 1996 to 2008, energy consumption per unit floor area of rural residential buildings, and CO_2 emission increased by 50% and 40% respectively. It should be noted that there is still considerable potential for further improvement of rural living standards, considering the existing gaps between rural and urban area, and the new-countryside construction project undertaking and deepening within China by the central government, which would naturally and inevitably increase energy demand in rural areas.

[10]Supposing that the energy consumption per unit floor area for urban residential buildings and commercial buildings with district heating system are the same, urban residential buildings in North China was 5.40 billion m^2 in 2008, heating-related energy consumption per unit of floor area 17.2 $kgce/m^2$, therefore, total building energy consumption of urban residential buildings in North China was 92.88 Mtce. Therefore, total energy consumption of China's urban residential buildings was 245 Mtce in 2008, and the average energy consumption per unit floor area was 19.8 $kgce/m^2$.

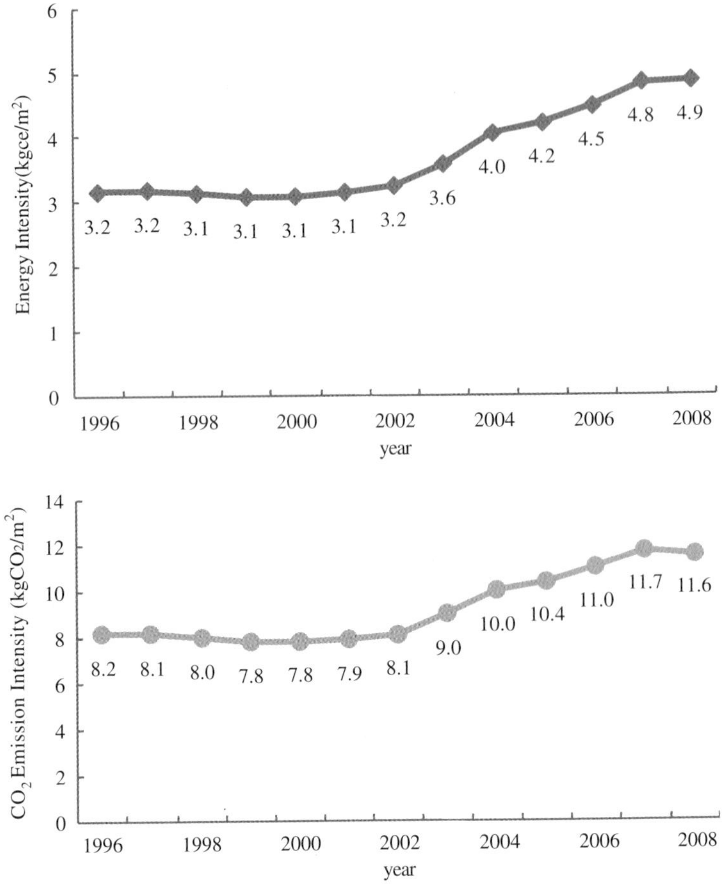

Fig. 9.28. Energy consumption and CO_2 emission intensity in China's rural residential buildings from 1996 to 2008.

(4) *Energy structure*

Total consumption of commodity energy sources has increased greatly, and total coal consumption also increased, but its proportion to the former has declined to some extent, which is conducive to the slowdown of CO_2 emission. Meanwhile, proportion of electricity to total energy consumption has been increasing with facilitated and expanding use of electricity. See Fig. 9.29.

Proportion of bio-mass energy such as straw and firewood has been declining. See Table 9.1. According to *China Rural Statistics Yearbook 1985*,

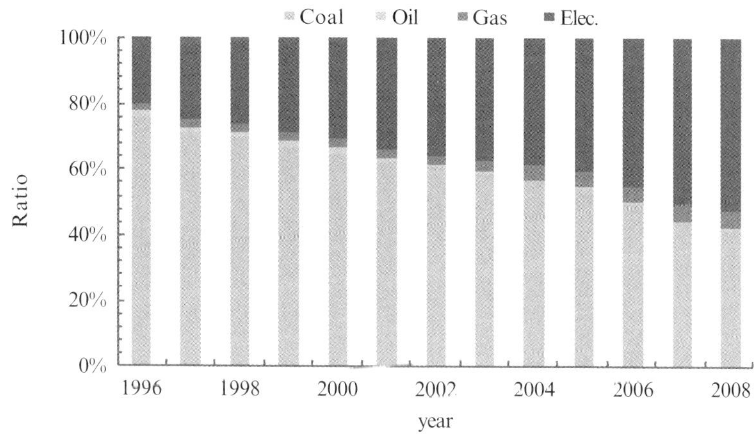

Fig. 9.29. Fuel mix of commercial energy sources (excluding bio-mass energy) in rural residential buildings from 1996 to 2008.

proportion of bio-mass energy such as straw and firewood was 80% in 1980s. However, with enlarging living space and increasing income, rural residents could afford more commodity energy, so the proportion of bio-mass energy has declined continuously to 60% by 1998.

Promoting renewable energy in rural areas has always been a focus in the course of China's rural energy conservation. In recent years, utilization of renewable energy such as solar water heater and methane has been increasing, which plays an important role in meeting the requirement of improving living standards and fulfilling rising energy demand in rural area. See Table 9.1.

9.1.6 *Summary*

Energy consumption in the building sector has increased largely in China, along with fast urbanization in last decade. In 2008, commodity energy consumption in China's building sector was 540 Mtce, accounting for 18.4% of the total energy consumption in China; CO_2 emission was 1.26 billion tons of CO_2, accounting for 19.8% of the total CO_2 emission within the country. With respect to total building energy consumption, China ranks second in the world, just behind the USA, while the energy consumption per capita and per unit floor area of China is much lower than that of developed countries.

Table 9.1. Consumption of bio-mass energy sources and proportion to total energy consumption in rural areas.

Category		Unit	1980	1995	2000	2005	2006	2007
Commercial energy		100 Mtce	—	1.30	1.62	1.93	2.01	2.11
Primary bio-mass energy	Straw	100 Mtce	—	1.51	1.23	1.60	1.78	1.60
	Firewood	100 Mtce	—	1.00	0.81	1.03	0.97	0.93
	Total	100 Mtce	—	2.51	2.04	2.63	2.75	2.53
Marsh gas tank	Year-end households	10,000 households	369	517	764	1,807	2,175	2,623
	Gas production	100 million m^2	—	14.7	25.9	70.6	81.9	98.8
Solar water heater	Floor area	10,000 m^2	—	449	1,108	3,206	3,944	4,286
Proportion of primary bio-mass energy		%	80	63	56	58	58	55

Note: Commercial energy consumption in rural areas quoted in the table is 100 Mtce more than that calculated according to the statistics yearbook.

Source: ① Department of Education, Science and Technology of Ministry of Agriculture, *Rural Energy Statistics Yearbook 2008* (China Agriculture Press, Beijing, 2008). ② Department of Education, Science and Technology of Ministry of Agriculture, *Rural Energy Statistics Yearbook 1997* (China Agriculture Press, Beijing, 1997). ③ Q. Wang, *China Sustainable Energy Project Reference Material 2009* (Energy Foundation, Beijing, 2010).

From 1996 to 2008, China's building energy consumption had increased along with fast urbanization, enlarging building stock and enhancing livings of citizens in China, and has still considerable potential to further increase. During the first three years of the 11th *FYP*, growth rate of building energy consumption and CO_2 emission was lower than that during the 10th *FYP*, mainly due to actions undertaking in China for energy conservation considerations. With respect to four categories of building energy consumption, the following could be concluded:

(1) Low-carbon performance of district heating system in Northern cities is most prominent, and CO_2 emission and energy consumption intensity per unit floor area has declined continuously. CO_2 emission per unit floor area in 2008 was 5.1% less than that in 2005, 27% less than that in 1996. The floor areas with district heating system increased by 42.5% from 2005 to 2008, and by more than six times from 1996 to 2008, and proportion of the total floor area with district heating system in Northern urban area had also increased. Reduction of heating-energy consumption per unit floor area slows down the growth rate of the total energy consumption and CO_2 emission for district heating in Northern urban area.
(2) For CO_2 emission per unit floor area of urban residential buildings (excluding energy use for district heating system), a marked decrease of over 50% was observed during the period of 1996–2002, which was due to the rapid expansion of district heating system, the resulting decrease of household heating facilities and consequent dramatic decrease of coal consumption in urban households. The unit area CO_2 emission went up steadily from 2002 to 2007 and decreased a little bit in 2008.
(3) From 1996 to 2002, the unit area CO_2 emission for commercial buildings (excluding district heating system) decreased near 40%, which could also be attributed to less coal consumption in buildings due to expanding use of district heating system. The unit area CO_2 emission went up steadily from 2002 to 2007 and decreased a little bit in 2008.
(4) Rural residential buildings ranked the lowest among four categories, in terms of unit area CO_2 emission. The value in 2008 is only 1/3 and 1/4 of that in urban residential and commercial buildings, respectively. Enhancing living standards and corresponding increase potential of energy demand due to existing gaps, were the chief attribution to continuous increase of energy consumption and CO_2 emission per unit floor area in rural residential buildings.

9.2 Evaluation of Policies in the Building Sector

9.2.1 *Better developed policy system leads to enormous energy-saving capacity*

With respect to the four categories of building energy consumption, China's policy system in building energy efficiency consists of three levels: ① Clauses in laws and regulations; ② major national plans; ③ a series of policies (usually, several policies share the same objective, which are named as key policy clusters in the report) based upon above two aspects. Policy clusters consist of building standards, administrative measures, as well as economic incentive measures. These measures support with each other for effective implementation of building laws, regulations and standards.

Major polices of the first two levels and corresponding categories of building energy consumption are listed in Fig. 9.30.

Laws and regulations

Relevant provisions of existing laws and regulations, mainly the *Renewable Energy Law* (2006), the *Energy Conservation Law* (2008), the *Administration Regulations for Public Institutions* (State Council Order: 531) and the *Energy Conservation Regulations for Civil Buildings* (State Council Order: 530), encouraged the application of energy efficiency building materials and equipments in buildings, which mainly relates to two major factors affecting building's operational energy consumption, i.e., envelope as well as internal systems, and equipment. These laws and regulations also stipulated the procedures for implementation and evaluation of the energy-saving measures. For example, *Administration Regulations for Public Institutions* mandated that the construction unit should specify the contents of building energy efficiency in its project completion report submitted to competent construction administrative departments for record keeping; while the *Energy Conservation Law* further stipulated that real estate enterprises should provide detailed description of energy efficiency measures and warranty period of building insulation works in written form; and the *Energy Conservation Regulations for Civil Buildings* further stipulated that the real estate enterprises should provide energy consumption indicators apart from above information.

National Planning.

Since the beginning of the 8th *FYP* in 1990, the targets in various national plans have become increasingly higher, with more quantitative and accurate

<table>
<tr><th>Category</th><th>Rural energy consumption</th><th>Energy consumption for district heating system in North urban China</th><th>Energy consumption for commercial buildings excl. district heating</th><th>Energy consumption for urban residential buildings excl. district heating</th></tr>
<tr><td rowspan="2">State Law</td><td colspan="4">Energy Conservation Law (2008)</td></tr>
<tr><td>Renewable Energy Law (2006)</td><td></td><td></td><td></td></tr>
<tr><td rowspan="2">Government Regulation</td><td></td><td colspan="3">Administration Regulations for Public Institutions (2008)</td></tr>
<tr><td></td><td></td><td>Energy Conservation Regulations for Commercial Buildings (2008)</td><td></td></tr>
<tr><td rowspan="8">National Planning</td><td colspan="4">Long and Medium-Term Energy Conservation Plan (2004)</td></tr>
<tr><td colspan="3">Notice of the State Council about Recent Key Works of Building Conservation-oriented Society (2005)</td><td></td></tr>
<tr><td></td><td colspan="3">Outline of 11th Five Years Plan for Building sector (2006)</td></tr>
<tr><td colspan="3">11th Five Years Plan for Energy Development (2007)</td><td></td></tr>
<tr><td></td><td colspan="3">Notice of the State Council on Printing and Distributing the Comprehensive Work Plan on Energy Conservation and Emission Reduction (2007)</td></tr>
<tr><td>Renewable Energy Development Plan (2007)</td><td></td><td>Renewable Energy Development Plan (2007)</td><td></td></tr>
<tr><td></td><td colspan="2">Work Plan for Energy Conservation and Emission Production in 2009 (2009)</td><td></td></tr>
<tr><td></td><td></td><td colspan="2">Long and Medium-Term Energy Conservation Plan. (2009)</td></tr>
</table>

Fig. 9.30. Laws, regulations, and national plans for energy conservation policy system in China's building sector.

indicators. Taking building envelope for example: 8th *FYP* proposed for wall materials retrofit; 9th *FYP* laid down the objective of 30%[11] by phases

[11]New residential buildings with district heating system: First phase (before 1996), reducing 30% energy consumption based upon the universal design standards for energy consumption in 1980–1981; second phase (1996–2004), reducing 30% energy consumption based upon the standards of the first phase; third phase (after 2005), reducing 30% energy consumption based upon the standards of the second phase.

for energy efficiency of new buildings as well as retrofitting existing buildings; the *Outline of* 10th *FYP for Building Energy Conservation by the Ministry of Construction* stipulated that the energy efficiency objective for new buildings should be 50%[12]; the *Long and Medium-Term Energy Conservation Plan* issued in 2004 changed the objective to 65%[13] for some districts, and put forward different objectives for cities of different scales[14]. The *Outline of* 11th *FYP for Building Sector* issued in 2006 stipulated that 2.15 billion m^2 energy-efficiency new buildings should be built during the 11th *FYP*, that the energy efficiency objective for all new buildings should be 60%,[15] and the objective of energy efficiency retrofit to existing buildings should be the same with that stipulated in the *Long and Medium-Term Energy Conservation Plan. Notice of the State Council on Printing and Distributing the Comprehensive Work Plan on Energy Conservation and Emission Reduction* (State Council Circular, 2007, p. 15) issued by the State Council in 2007 further stipulated to conduct energy efficiency retrofit to 150 million m^2 building floor areas during the 11th *FYP*. Detailed planning for different categories of building energy consumption will be discussed in the following paragraphs.

Key policy clusters for different categories of building energy consumption.

As is disclosed in Fig. 9.31, 4–6 key policy clusters could be formed according to different categories of building energy consumption. With the support of key policy clusters, China's building sector realized a huge energy conservation capacity during the 11th *FYP*. As is revealed in Fig. 9.32, altogether 77.16 Mtce energy saving capacity had been realized (energy saving capacity does not mean real energy savings, but is the difference between

---

[12]Design standards for reducing 50% energy conservation are applied to all new buildings with district heating system in Northern cities with severe cold and cold temperature. The standards should be implemented before 2001 for large and medium-sized cities, before the end of 2003 for small cities, and before the end of 2005 for all districts and counties.

[13]For a few large cities such as Beijing and Tianjin.

[14]By 2010, for energy efficiency renovation of existing residential and commercial buildings, 25% of which should be renovated in large cities, 15% in medium-sized cities and 10% in small cities.

[15]By 2010, energy efficiency of new residential buildings should be over 60%, water-saving rate should increase by over 20%, and consumption of nonrenewable energy should decrease by 10%. Design standards for 50% energy conservation are applied to new buildings, and 65% for new buildings in some large cities and areas with severe cold and cold temperature.

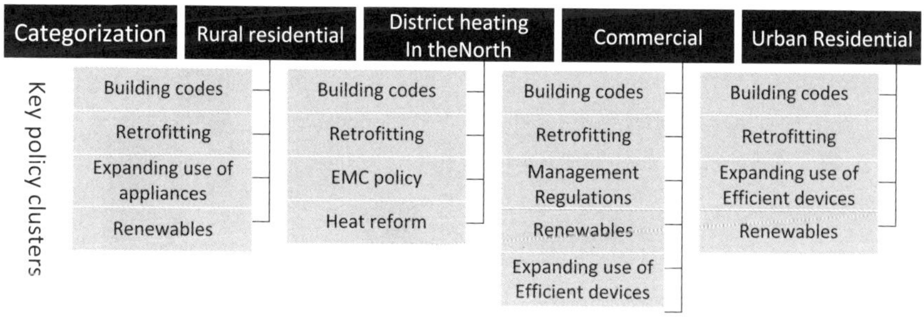

Fig. 9.31. Key policy clusters for different categories of building energy consumption in China's building sector.

real energy consumption and energy consumption under a business-as-usual (BAU) scenario[16]) between 2006 and 2008, of which, energy consumption of district heating system in North China had realized 37.72 Mtce energy saving capacity, almost half of the total.

This section explains detailed disaggregation of the energy saving capacity, and conducts evaluation to relevant key policy clusters, which make planning and designing according to the characteristics of different categories of building energy consumption, serving the purpose of energy efficiency. Therefore, characteristics and influencing factors of different categories of building energy consumption should be known first.

Actually, building energy consumption is not only relevant to energy efficiency and kind of energy resource. Due that buildings are designed to provide services to building users, building carbon emission is mainly affected by human behaviors. There are two aspects: First is the management and operation of buildings, for example, operating season and period of the central air-conditioning system in commercial buildings determines its running time; second is the service demand of the occupants, which is primarily decided by their living habits. For example, preference to a certain degree of indoor temperature determines the running time of air-conditioners and lighting devices. Therefore, carbon emission in the building sector is determined by five aspects of energy efficiency, structure, management, demand,

[16]Reference energy consumption is based upon the data of energy consumption intensity in 2005, supposing that the growth rate of energy consumption intensity is the same with the average growth rate between 2001 and 2005.

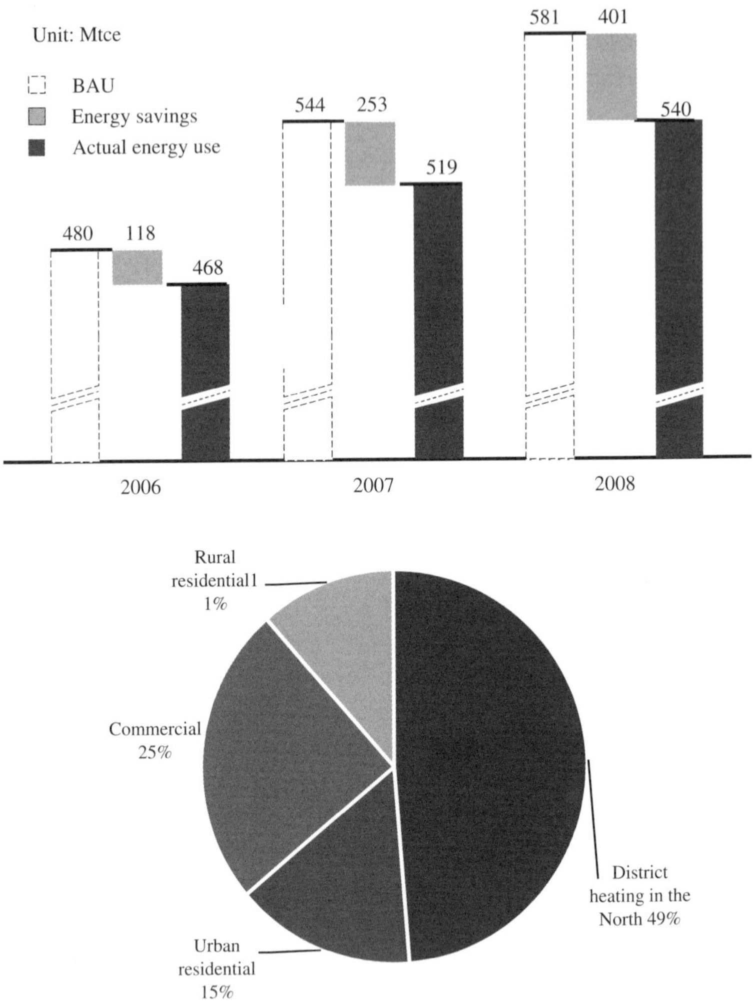

Fig. 9.32. Energy saving capacities and disaggregation from 2006 and 2008.
Note: BAU energy is based upon the data of energy consumption intensity in 2005, supposing that the growth rate of energy consumption intensity is the same with the average growth rate between 2001 and 2005.

and building floor area. Details of building energy consumption of these aspects and their relationships are listed in Table 9.2.

Key policy clusters of four categories of building energy consumption will be introduced and their energy-saving capacity will be analyzed separately in the following paragraphs.

Table 9.2. Transformation of influence factors for four categories of building energy consumption and carbon emission.

Influencing factors	Energy efficiency		Structure		Management and operation		Demand		Floor area	
	Actual status	Influence	Actual status	Influence	Actual status	Influence	Actual status	Influence	Actual status	Influence
Energy consumption of district heating system in North China	Higher heat source efficiency	Decrease of energy consumption	Higher proportion of heat supply by CHP system	Decrease of energy consumption	Longer heating period	Increase of energy consumption	Higher indoor temperature	Increase of energy consumption	Bigger overall floor areas	More energy consumption
							Open window more often			
	Less pipeline heat loss Better building insulation level									
Energy consumption of urban residential buildings	Higher energy efficiency of household devices	Decrease of energy consumption	More solar water heaters	Decrease of energy consumption	—	—	Change of life patterns and more service demand	Increase of energy consumption	More air-conditioning and lighting consumption	Increase of energy consumption
Energy consumption for commercial buildings	higher efficiency of energy consumption devices	Decrease of energy consumption	Higher proportion of large-scale commercial buildings	Increase of energy consumption	More district heating systems	Increase of energy consumption	Better environment and more office functions	Increase of energy consumption	More heating, air-conditioning, lighting, and office equipment consumption	Increase of energy consumption

(*Continued*)

Table 9.2. (*Continued*)

Influencing factors	Energy efficiency		Structure		Management and operation		Demand		Floor area	
	Actual status	Influence	Actual status	Influence	Actual status	Influence	Actual status	Influence	Actual status	Influence
Energy consumption for rural residential buildings	More energy efficiency lamps, energy efficiency kitchen ranges, etc.	Decrease of energy consumption	Lower proportion of bio-mass energy	Increase of energy consumption	—	—	Better living standards, and more service demand	Increase of energy consumption	Energy consumption is more relevant to number of households and residents rather than, floor areas (excluding heating)	—

9.2.2 *Policy evaluation for energy consumption of district heating system in northern cities*

Key policy clusters related to the energy consumption of district heating system in northern cities consist of energy efficiency in new buildings, retrofit of existing buildings, EMC, and heating reform. Composition of key policy clusters is listed in Fig. 9.33. According to the intensity level of existing policies, heating reform is the key point to energy efficiency of urban buildings. Contents, implementation, and estimated energy-saving potential of various key policy clusters will be briefly introduced in the following paragraphs.

(1) *Criteria for new buildings and retrofit of existing buildings*

Energy consumption of district heating system in northern cities is the combination of that of residential and commercial buildings in North China's urban areas. Criteria and policies for new buildings and retrofit of existing buildings will be introduced in detail in Sec. 9.2.3 and 9.2.4.

By 2009, altogether 4.08 billion m^2 floor areas of energy efficiency buildings had been built in China, including 109.49 million m^2 floor areas of retrofitted buildings in 15 Northern provinces and municipalities. Among these, 2.4 billion m^2 floor areas of energy efficiency buildings had been realized in northern cities, accounting for 27% of the total floor areas in Northern cities.[17] Compared to the heat demand of building in 1980, about 16.08 Mtce[18] energy-saving capacity has been realized annually (energy-saving capacity does not equal to real energy savings), which means that about 11.64 Mtce[19] energy-saving capacity has been realized annually since 2005.

[17]Due to lack of statistics, floor areas of new-builds in northern cities of each year/floor areas of new-builds each year = floor areas of new energy efficiency buildings in Northern cities of each year/floor areas of new energy efficiency buildings of each year. Other data is collected from *China Statistics Yearbook 2009* and *Circular on Energy Conservation Inspection of 2009 Energy Conservation Emission Reduction Supervision and Inspection in China's Building sector* (MOC, 2010, p. 45).
[18]According to the 50% standard, it could be estimated that about 58 kWh/m^2 is reduced for heating-related energy consumption of new buildings on the basis of 1980, therefore, multiplying the number with 2.4 billion m^2, an energy-saving capacity of 130.7 billion kWh (equal to 16.08 Mtce) of heat could be realized.
[19]There were about 700 million m^2 energy efficiency buildings in 2005, realizing an energy-saving capacity of about 4.44 Mtce.

<table>
<tr><th rowspan="3">Categories</th><th colspan="5">District heating system in northern cities</th></tr>
<tr><th rowspan="2">New buildings</th><th rowspan="2">Retrofitting</th><th colspan="2">Heating reform</th><th rowspan="2">EMC</th></tr>
<tr><th>Heat charging system</th><th>Heat source structural adjustment</th></tr>
<tr><td rowspan="2">Construction Criteria</td><td colspan="2">Design Standard for Energy Efficiency of Residential Buildings (Heating Region), 1996</td><td></td><td></td><td></td></tr>
<tr><td colspan="2">Design Standard for Energy Efficiency of Public Buildings, 2005</td><td></td><td></td><td></td></tr>
<tr><td rowspan="2"></td><td colspan="2">Code for Acceptance of Energy Efficiency Buildings, 2007</td><td></td><td></td><td></td></tr>
<tr><td colspan="2">Technical Guideline for Energy Efficiency Evaluation and Labeling of Civil Buildings(Tentative), 2008</td><td>Technical Guideline for Heat Metering, 2008</td><td></td><td>General Technical Rules for EMC, 2010</td></tr>
<tr><td>Management Measures</td><td colspan="2">Building energy efficiency inspection, 2004</td><td colspan="2">Guiding Opinions for Pilot Works of Heat Supply System Innovation in Cities, 2003</td><td></td></tr>
<tr><td></td><td></td><td></td><td colspan="2">Opinions for Further Promoting Heat Supply System Innovation in Cities, 2005</td><td></td></tr>
<tr><td></td><td></td><td colspan="2">Opinions for Implementation of Heat Metering, 2006</td><td></td><td></td></tr>
<tr><td></td><td></td><td></td><td>Tentative Regulations on Heating Price Management of Urban Areas, 2007</td><td></td><td></td></tr>
<tr><td></td><td></td><td colspan="2">Opinions for Implementation of Heat Metering and Energy Efficiency Renovation in North China with District Heating Systems, 2008</td><td></td><td></td></tr>
</table>

Fig. 9.33. Policy chart for northern cities with district heating system.

	Tentative Regulations on Energy Evaluation and Labeling of Civil Buildings, 2008		Regulations on Heat Metering of Civil Buildings, 2008		
	Ordnance on Energy Conservation of Civil Buildings, 2008		Opinions on Further Promoting Heat Metering Innovation, 2010		Opinions on Accelerating Development of EMC and Energy Service Industry, 2010
		Notice of Enhancing Work Efforts for Completing the Task of Heat Metering and Energy Conservation Renovation to Existing Residential Buildings in North China, 2010			
Incentive Policies			Notice on Taxation of Heat Supply Enterprises, 2004		
			Notice on Further Implementing Preferential Policies on Taxation of Heat Supply Enterprises, 2006		
		Tentative Regulations on Management of Incentive Funds for Heat Metering and Energy Conservation Renovation of Existing Residential Buildings in North China with District Heating System, 2007			Tentative Regulations on Management of Incentive Funds of EMC Projects, 2009

Fig. 9.33. *(Continued)*

(2) *EMC*

Relevant EMC policies will be enumerated in Sec. 9.2.4. Now, we will mainly introduce the development of EMC system for the district heating system in northern cities and its estimated energy-saving capacity.

At present, EMC mode has been widely adopted by building energy efficiency projects, which collects heating charges from heat users by contracting existing boilers, such as supply system management, energy efficiency retrofit as well as equipment, and pipeline network maintenance. For example, by way of energy trust service, a Beijing EMC company has conducted successful management to over 120 heat supply projects for governmental agencies, public institutions and enterprises as well as residential communities, covering 18 million m^2, realizing an output value of RMB 300 million.

It is estimated that by the end of 2010, 30 million m^2 of existing buildings would be renovated by EMC, 20 million m^2 of which would be commercial buildings, while the remaining 10 million m^2 would be residential buildings (Lin and Hao, 2010). And 220,000 tce of energy-saving capacity[20] could be realized each year.

(3) *Heat supply system innovation*

By retrofitting existing heat charging system and adopting market methods, China's heating reform aims to reduce wastage of terminal heat users, enhance pipeline energy efficiency of heat supply enterprises, share the costs of energy efficiency renovation, and promote energy efficiency renovation. Therefore, policies on heating reform consist of two aspects: ① Administrative orders, defining the key role of heat metering and charging system in heat supply system renovation, and furthering the heat supply system renovation involving clients and heat supply enterprises; ② incentive policies for better implementation of the foregoing reform.

On one hand, China formulated various administrative orders to further the reform of heat supply system. On July 21st, 2003, the MOC, the NDRC, the MOF, the Ministry of Personnel, the Ministry of Labor and Social Security, the State Administration of Taxation, and the State Environmental Protection Administration jointly issued and distributed the *Guiding*

[20] According to the 50% standard, it could be estimated that about 58 kWh/m^3 is reduced for heating-related energy consumption of renovated buildings on the basis of 1980, therefore, multiplying the number with 30 million m^2, an energy-saving capacity of 1.8 billion kWh (equal to 220,000 tce) of heat could be realized.

Opinions for Pilot Works of Heat Supply System Innovation in Cities, China's first document on heat supply system renovation, which stipulated clearly to "promote heat metering and charging system steadily, and promote energy conservation of both parties". After two years of experiment, eight ministries and commissions including the Ministry of Construction issued and distributed the *Opinions for Further Promoting Heat Supply System Innovation in Cities* in December 6th, 2005, which stipulated that heat charging system innovation is the key point of heat supply system reform.

With respect to heat metering and charging reform, the MOC issued and distributed the *Opinions for Implementation of Heat Metering System* in June 28th, 2006, which stipulated that governmental agencies and commercial buildings shall take the lead in implementing heat metering system, as well as in establishing and perfecting the heat metering and charging system. On June 3rd, 2007, the NDRC and the MOC issued and distributed the *Tentative Regulations on Management of Urban Heat Supply Price*, which was a reference document for completing the heat supply price system and regulating heat price management. On February 2nd, 2008, four ministries and commissions including the Ministry of Construction distributed the *Opinions on Further Promoting Heat Metering Innovation Works*, which formulated provisions on different aspects of heat metering innovation, and defined the responsibilities and provisions of reward and punishment for government authorities and heat supply enterprises. On May 21st, 2008, the MHURD distributed the *Opinions for Implementation of Heat Metering and Energy Efficiency Renovation in North China with District Heating Systems*; and in June 1st, 2010, the MHURD and the MOF jointly issued the *Notice of Enhancing Work Efforts for Completing the Task of Heat Metering and Energy Conservation Renovation to Existing Residential Buildings in North China*, to supervise the implementation process of the reform.

On the other hand, in order to stimulate the enthusiasm of heat supply enterprises, the MOF and the State Administration of Taxation formulated a series of economic incentive policies, for example the *Notice on Taxation of Heat Supply Enterprises* on February 5th, 2004, and the *Notice on Further Implementation of Preferential Policies on Taxation of Heat Supply Enterprises* in November 27th, 2006, which provided tax reduction and exemption to heat supply enterprises. In order to fully motivate the enthusiasm of local governments, the MOF issued the *Tentative Regulation on Management of Incentive Funds for Heat Metering and Energy Conservation Renovation of Existing Residential Buildings in North China with*

District Heating System on December 20th, 2007, which formulated regulations on management of central-allocated incentive funds for heat metering and energy efficiency renovation of existing residential buildings in North China's heating regions. The incentive funds cover three aspects including retrofitting of building envelope, renovation of indoor heat metering and temperature regulation, as well as renovation of heat source and heat pipeline networks, which pays attention to renovation capacity and energy-saving effects.

Considering from implementation — ① the heat source efficiency. Heat source efficiency of various district heating systems in Northern cities has been improved by technical advancement and renovation. ② Heat source structure. CHP systems and large coal boilers have taken place of small coal boilers, which has improved the coal utilization rate enormously. See 9.14. ③ Pipeline network loss. There are fewer heat escapes and leakages, reducing heat loss in the heating system.

China started compulsory meter installation in 2006 to realize heat charging by real heat consumption. This prompted the users to reduce heat consumption and conduct envelope renovation to reduce heat transfer; meter installation also prompted the heat supply enterprises to improve the operation level of pipeline networks and efficiency of heat source, so as to reduce the cost of heat production and supply. In this way, meter installation can be connected directly or indirectly to energy conservation and emission reduction of the district heating system in Northern cities. By the end of the heating period in 2009, the metered areas had been doubling each year. As in Fig. 9.34, floor areas with heat meters reached 360 million m^2 in four years, accounting for 4.5% of the total heating area in northern cities; and 150 million m^2 of which were also heat charging areas, saving about 370 Mtce[21] energy each year. The promotion of heat metering and charging is relatively stronger in large cities. For example, Tianjin's floor areas with

[21]According to statistics, heat loss of heat supply networks in Beijing's district heating system is about 5–10 kWh/m^2, and uneven end heat loss is 25 kWh/m^2; by improving pipeline management and reducing energy consumption of residents by incentive market measures, heat loss of heat supply networks could be controlled within 8 kWh/m^2, while uneven end heat loss could be controlled within 7 kWh/m^2 (Building Energy Conservation Research Center of Tsinghua University, 2007). Although heat metering and charging systems vary with each other, 10%–20% energy-saving rate could be realized generally (Construction Technology, 2009; Liu Lanbin *et al.*, 2009), equal to 10 kWh/m^2; then 150 million m^2 floor areas could realize an energy-saving capacity of about 3 billion kWh (equal to 370,000 tce).

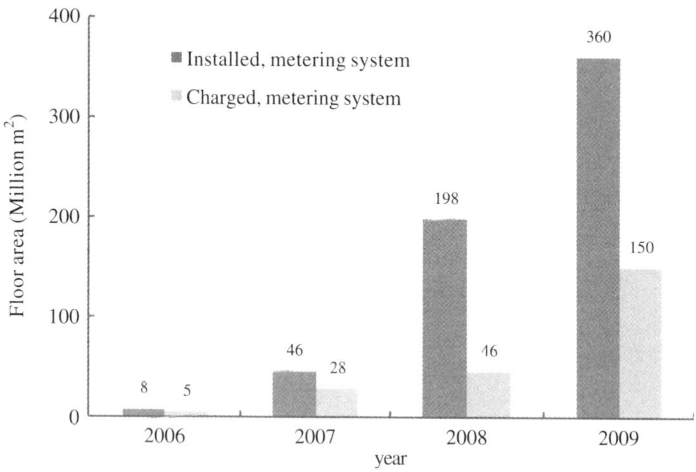

Fig. 9.34. Heat metering and charging floor area during the 11th FYP.
Source: the MHURD, 2010. Six Promising Aspects and Prospects for China's Building Energy Efficiency—Speech of Deputy Minister Qiu Baoxing in the 6th International Conference of Green and Energy-Efficiency Buildings and New Technologies and Products Expo. http://www.mohurd.gov.cn/ldjh/jsbfld/201004/t20100408-200306.htm (accessed on 27 August 2010).

district heating system was 190 million m^2 (State Statistics Bureau, 2009) in 2009. By 2009, about 13.42 million m^2 existing buildings had realized heat metering, accounting for 7% of the total.

Therefore, we can obtain the energy-saving capacity in 2007 and 2008 due to relevant policies on district heating system in northern cities (Fig. 9.35). With higher indoor temperature, longer heating period, and continuous growth of total floor areas due to better living standards, the estimated energy consumption in northern cities is about 132 Mtce; while with energy conservation of new buildings, renovation of existing buildings, heating reform and EMC, about 11.5 Mtce energy saving capacity could be realized, and the real energy consumption increased by 23.50 Mtce compared to 2005.

Therefore, heating reform is related to the key policy clusters for district heating system in Northern cities, which has a high energy-saving potential. Proportion of high-efficiency heat supply source is increasing; however, heat metering floor area is less than 5% of the total area with district heating system in Northern cities. Therefore, its implementation will be analyzed in Sec. 9.3.2, so as to discover the root cause for slow progress of heating reform.

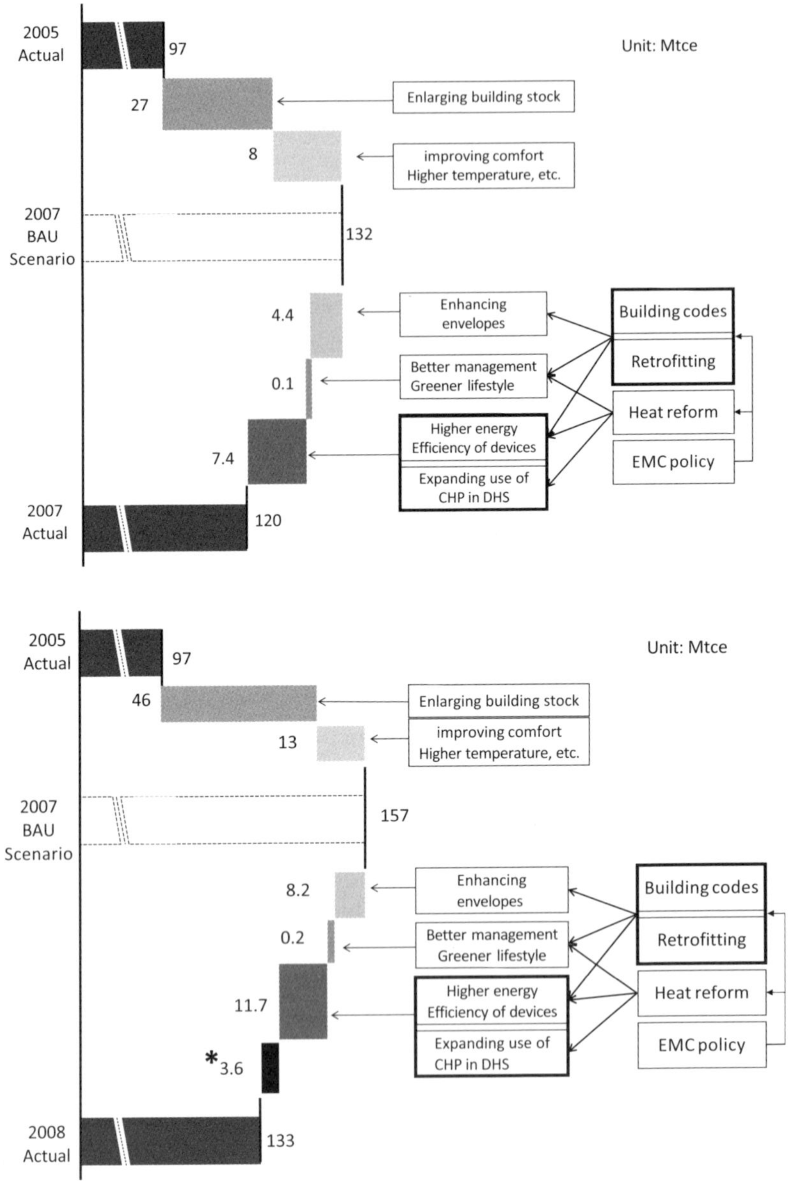

Fig. 9.35. Energy-saving capacity deriving from low-carbon policies for district heating system in Northern cities in 2007 and 2008 (unit: 10,000 tce).
*This part of energy-saving capacity cannot be explained by policy. There might be many reasons for this, for example, statistical error or life patterns irrelevant to policies which cannot be calculated and verified due to lack of statistical data.

9.2.3 *Policy evaluation for energy consumption of urban residential buildings*

Key policy clusters related to energy consumption of urban residential buildings aims at four aspects as: Energy conservation of new buildings, renovation of existing buildings, energy efficiency appliances and lighting, and renewable energy reutilization. Composition of key policy clusters is listed in Fig. 9.36. In addition, policies on energy efficiency appliances play a more significant role in the course of energy conservation in residential sector, considering its impact on energy use. Contents, implementation, and estimated energy-saving potential of various key policy clusters will be briefly introduced in the following paragraphs.

Categories	Energy Conservation of Urban Residential Buildings (Excluding District Heating System)			
	New buildings	Retrofitting	Energy efficiency appliances and lighting	Renewable energy
Construction Criteria	Design Standards for Energy Conservation of Residential Buildings (Heating Region), 1996		Energy efficiency standards for household appliances and lighting devices	Technical Regulations for PV application in Civil Buildings, 2008.10
	Design Standards for Energy Conservation of Residential Buildings in Hot-summer Cold-winter Areas, 2001			
	Design Standards for Energy Conservation of Residential Buildings in Hot-summer Warm-winter Areas, 2003			
	Evaluation Standards for Green Buildings, 2006			
Management Measures	Specialized Inspection on Building Energy Efficiency, since 2004		Energy-saving Products Helping People Project, 200 9.5	Implementation Plan for Promoting Solar Thermal Utilization, 2007.4
			Green Lighting Project, 2004.11	Implementation Opinions for Accelerating BIPV Application, 2009.3

Fig. 9.36. Policy chart for energy consumption in urban residential buildings (excluding district heating system).

Incentive Policies			Tentative Regulations on Subsidy Management for Promoting High-efficiency Lighting Products, 2007.12	Tentative Regulations on Subsidy Management for BIPV Buildings, 2009.3
			Tentative Regulations on Subsidy Management for Promoting High-efficiency Energy-saving Products, 2009.5	Implementation Plan for Urban Demonstrations of Renewable Energy Building Application, 2009.7
			Regulations on Promoting High-efficiency and Energy-saving Room Air Conditioners by Energy-saving Products Helping People Project, 2009.5	

Fig. 9.36. (*Continued*)

(1) *Codes for new buildings and renovation of existing buildings*

Policy systems for energy consumption of newly-built urban residential buildings consist of two parts: First, a series of building codes at both national and local level; second, a series of governmental regulations for better implementation of above standards.

To specify, building codes are aiming at: ① Enhancing envelopes of buildings to reduce heating and cooling demand; ② encourage more use of energy carriers of higher energy efficiencies in buildings, as well as better operation of them; ③ promoting passive designs in buildings, which could reduce ventilation and lighting demand in much lower price. China's current national standards (Table 9.3) have taken various weather conditions fully

Table 9.3. Compulsory building design standards and energy-saving targets in China.

Standards			Energy-saving Targets				
Buildings	Climate Zone	Standards	District Heating System	De-centralized Heating System	Air-conditioning	Lighting	Others
Residential Buildings	Cold	JGJ26-95	50% lower to 1980		—	—	—
	Hot-summer cold-winter	JGJ134-2001	—	50% lower than before		—	—
	Hot-summer warm-winter	JGJ75-2003	—	50% lower than reference buildings		—	—
Commercial Buildings	Cold	GB50189-2005		50% lower than before			—
	Hot-summer cold-winter						—
	Hot-summer warm-winter						—

into consideration, and it has formulated various compulsory regulations on major end-usages in buildings.

Major actions the government takes for implementation include: ① Formulate energy-saving targets, including targets for implementation of new buildings standards, and for renovation of existing buildings according to effective standards[22]; ② enforce compulsory implementation of the standards, and define responsibilities of each evolved body; ③ conduct energy-saving monitoring and inspection; ④ encourage labeling of green buildings, and deepen the work of energy conservation in building sector.

As above, due to national and historical conditions, over 70% urban buildings in North China utilize district heating system, which is energy-intensive, featured as different from other end-usages in civil buildings, and consequently, needs to be considered separately. Therefore, the following paragraphs will introduce the contents of standards and their implementation related to energy consumption of urban residential buildings (excluding energy use for district heating systems), while that of other end-usages will be introduced in later sections.

Design standards for residential buildings go through three stages of energy efficiency, i.e., 30%, 50%, and 65%.[23] China's first building design standard is the *Design Standards for Energy Efficiency of Civil Buildings* (*Heating Residential Buildings*) (JGJ26-86) issued by the MURCEP in 1986, which brought forward that heating energy consumption of new residential buildings should reduce 30% on the basis of energy use by buildings built in 1980–1981. Later, China successively issued the *Design Standards for Energy Conservation of Residential Buildings* (*Heating Residential Buildings*) (JGJ26-95), the *Design Standards for Energy Conservation of Residential Buildings in Hot-summer Cold-winter Areas* (JGJ134-2001, J116-2001) and the *Design Standards for Energy Conservation of Residential Buildings in Hot-summer Warm-winter Areas* (JGJ75-2003, J275-2003), which brought forward that energy consumption for cooling and heating should reduce 50%, compared with those built in the 1980s; which

[22]Renovation of buildings refers to renovating the envelopes of existing buildings that did not meet with the requirements in local building codes, to better their energy performance to required level (Yong *et al.*, 2007).

[23]30%, 50%, and 65% standards: Heating energy consumption (unit: kgce/(m^2a) of new residential buildings in cold and severe cold areas should reduce 30%, 50%, and 65% on the basis of energy use by typical buildings built in 1980–1981; that is to say, saving corresponding quantity of coal for heating purpose.

not only formulated provisions on insulation level, sun-shading measures, heating and air-conditioning measures, and equipment performance of residential buildings, but also referred to a few regulations on other end-usages such as ventilation and lighting. Meanwhile, the *Outline of the* 11th *Five Years Plan for Building Sector* (MOC, 2006, p. 53) encouraged that "metropolitans in cold and severe-cold areas should adopt the energy efficiency design standards of 65% for new buildings as long as conditions allows". For example, Beijing city issued the *Design Standards for Energy Conservation of Residential Buildings in Beijing City* (DBJ01-602-2004) in 2004, and Heilongjiang province issued the *Design Standards for 65% Energy Conservation of Residential Buildings in Heilongjiang Province* (DB23/1270-2008) in 2008.

Standards for construction now in China are the *Regulations on Construction Quality Acceptance of Building Energy Efficiency Project* (GB50411-2007).

The government formulated detailed objectives for implementation of such standards. With respect to energy efficiency of new buildings, the MOC issued the *Outline of the* 11th *FYP for Building Sector* in March 2006, which stipulated that total floor area of new buildings conforming to building codes shall reach 2.15 billion m^2 at the end of 11th *FYP*. With respect to renovation of existing buildings, the *Medium and Long Term Energy Conservation Plan* issued in 2004 stipulated that the proportion of renovated buildings shall reach 25% for large cities, 15% for middle cities and 10% for small cities. The *Notice on Further Expanding Work Efforts to Ensure the Achievement of Energy Conservation and Emission Reduction in the* 11th *FYP* (State Council Circular, 2010, p. 12) stipulated that by the end of 2010, "50 million m^2 residential buildings in northern areas with district heating system shall be renovated in accordance with local building codes and applied with heat metering and charging system, so as to meet the target of renovating 150 million m^2 set by State Council for the 11th *FYP*".

In order to implement the standards and fulfill the targets, the government issued the *Regulation on Energy Efficiency Management of Civil Buildings* (MOC Order, 2005, p. 143, nullified) and the *Ordnance on Energy Efficiency of Civil Building* (State Council Order, 2008, p. 530), which stipulated the responsibilities of and punitive measures for construction administrative departments at all levels and market players in the process of designing, construction, acceptance, marketing, and operation. In addition, the State Council issued the *Comprehensive Working Plan of Energy*

Conservation and Emission Reduction in May 2007, which stipulated to implement specialized evaluation on building energy efficiency; to ensure that construction allowance and relevant documentations will not be issued until its design and construction are in accordance with building codes; and to bring forward further compulsory implementations of building codes.

Since 2004, the MOC conducted nation-wide specialized inspection on building energy efficiency to inspect the implementation of building codes during designing, construction, acceptance, and marketing process of buildings. Notably, the implementation of building codes for new buildings and renovation of existing buildings are focus. Meanwhile, in order to act in concert with the special inspection, competent construction administrative departments of some localities implemented closure monitoring for designing, construction and acceptance of buildings such as Tangshan city's "closure management procedure for energy efficiency buildings".

In order to promote building energy efficiency and enhance building performance, China has always been committed to the establishment of building labeling system, and promotion of the concept of green buildings. The pioneer projects were the scarce green building projects in 1990s, and China's first *Technical Guidelines for Green Buildings* was formulated by the national scientific and technological project during the 10th FYP. By the end of 2003, the research team of Tsinghua University, the China Academy of Building Research and Beijing Institute of Architectural Design formulated the GOBAS, the first evaluation and assessment system for green buildings in China, which laid down solid foundations for building labeling system. In early 2006, the MOC and the MOST jointly issued the *Evaluation Standards for Green Buildings* (GB/T50378-2006), China's first regulation on standards of green buildings, which divided the green buildings into three levels, i.e., one-star, two-star, and three-star green buildings. The standards are not compulsory, which will be evaluated by competent governmental authorities, and its promotion primarily relies on official training. Later, the MOC and the MOST issued a series of policies, such as the *Regulations of Management of Green Building Evaluation Labeling* (MOC and MOST Letter, 2007, p. 206), the *Technical Manual on Evaluation of Green Building* (MOC and MOST Letter, 2007, p. 205), the *Supplementary Explanation to Technical Manual on Evaluation of Green Building (Operation)* (MOC and MOST Letter: [2009] 235) and the *Supplementary Explanation to Technical Manual on Evaluation of Green Building (Planning and Designing)* (MOC and MOST Letter, 2008, p. 113), so as to provide guidance to the green buildings evaluation and labeling projects as well as green buildings designing evaluation and labeling projects. By then, China's building energy efficiency labeling system has been established preliminarily.

With regard to implementation of building codes: According to the statistics issued by the MOC in February 2005, only 180 million m^2 floor areas of China's urban and rural buildings met with requirements in building codes by the end of 2000, accounting for 0.6% of the total building stock. As is revealed in Table 9.4, after nearly 10 years of hard work, over 90% (and 960 million m^2 of) new buildings were built in accordance with the 50% code in 2009. So the total floor area of buildings conforming to building codes reach 4.08 billion m^2 by the end of 2009, accounting for 20% of the total building stock. In addition, there is 109.49 million m^2 renovated buildings in 15 Northern provinces and municipalities (including residential and commercial buildings, new buildings, and renovated existing buildings).[24] And 2.4 billion m^2 of those buildings are residential ones, accounting for 19% of the total floor areas of urban residential buildings.[25] Compared to 1980, about 9.12 Mtce[26] energy-saving capacity was achieved in China, which actually refers to reduction of heating demand in buildings and consequently does not equals to direct energy use for heating in

[24] *Circular on Energy Conservation Inspection of 2009 Energy Conservation Emission Reduction Supervision and Inspection in China's Building Sector* (MOC and MOST, 2010, p. 45).

[25] Due to lack of statistics, it is assumed that the floor areas of new residential buildings of each year/the floor areas of new buildings of each year = the floor areas of new energy efficiency residential buildings of each year/the floor areas of new energy efficiency buildings of each year. Other data is collected from *China Statistics Yearbook 2009* and *Circular on Energy Conservation Inspection of 2009 Energy Conservation Emission Reduction Supervision and Inspection in China's Building Sector* (MOC and MOST, 2010, p. 45).

[26] In 1996, heat demand of district heating system in northern cities was 116 kWh/m^2, and supposing that the heat demand in 1980 was similar to that in 1996 due to lack of building standards before 1996. According to the 50% energy efficiency standard, heat demand of now buildings should reduce by 50%, i.e., about 58 kWh/m^2. In addition, 1.4 billion m^2 of 2.4 billion m^2 energy efficiency buildings are in northern cities (the floor areas of new residential buildings in northern cities of each year/the floor areas of new residential buildings of each year = the floor areas of new energy efficiency residential buildings in Northern cities of each year/the floor areas of new energy efficiency residential buildings of each year. Other data is collected from *China Statistics Yearbook 2009* and *Circular on Energy Conservation Inspection of 2009 Energy Conservation Emission Reduction Supervision and Inspection in China's Building Sector* (MOC and MOST, 2010, p. 45), producing about 80 billion kWh heat saving capacity (equal to 9.12 Mtce).

Table 9.4. Implementation rate of new energy efficiency buildings 2005–2009.

Year	Total floor areas (100 million m^2)	New buildings (100 million m^2)	Implementation rate in designing stage (%)	Implementation rate in construction stage (%)	New energy efficiency buildings (100 million m^2)	Cumulative floor areas of energy efficiency buildings (100 million m^2)	Proportion to existing urban buildings (%)
2005	164.5	15.4					
2006	174.5	10.0	96	54			
2007	186.0	11.5	97	71		21.2	11.7
2008							
2009			99	90	9.6	40.8	21.7

Source: ① *Report on Energy Conservation Inspection of 2009 Energy Conservation Emission Reduction Supervision and Inspection in China's Building Sector* (MOC and MOST, 2007, p. 15); ② *Circular on Energy Conservation Inspection of 2009 Energy Conservation Emission Reduction Supervision and Inspection in China's Building Sector* (MOC and MOST, 2010, p. 45); ③ State Statistics Bureau, 2009; ④ *China Statistics Yearbook 2009* (China Statistics Press, Beijing, 2009).

buildings, as well as 1.47 Mtce for decentralized heating and cooling system in residential buildings.[27]

With regard to building labeling, by October 1st, 2010, 46 projects had been publicly notified by the MOC (MHURD, 2010), of which, 21 were residential ones, covering a total floor area over 3 million m^2; and seven were three-star projects. Green building labeling system has been conducted extensively.

(2) *Renewable energy utilization*

Renewable energy utilization in residential buildings mainly consists of solar collectors as well as a few PV power generation pilot projects. Relevant policies will be introduced in Sec. 9.2.5. Results of policy implementation and estimated energy savings will be discussed in the following paragraphs.

Solar heat collectors are mainly used in residential buildings. By the end of 2009, total area of solar heat collectors had reached 145 million m^2 (Energy Foundation, 2007), of which, 95 million m^2 were in urban areas,[28] producing an annual energy-saving capacity of 3.06 Mtce.[29]

(3) *Promotion of energy efficiency products*

Promoting use of high-efficiency household appliances and energy-efficient lamps (which refers mainly to fluorescent lamps here in China) is an important method to reduce energy consumption in residential buildings. Relevant work was divided into three aspects.

[27] According to statistics, average power consumption for household cooling was 2.1 kWh/m^2 in China in 2005 (Zhang *et al.*, 2010). According to the 50% energy efficiency standard, power consumption of air-conditioners in new buildings should reduce by 50%, about 1.1 kWh/m^2. Therefore, it could be estimated that 2.4 billion m^2 energy efficiency buildings could realize 2.7 billion kWh energy-saving capacity (equal to 1.02 Mtce). In addition, there are about 100 million m^2 decentralized heating areas in northern cities, with heat demand per unit floor area decreasing by 50% to 58 kWh/m^2, about 5.8 billion kWh heat saving capacity could be achieved (equal to 450,000 tce).

[28] *Rural Energy Statistics Yearbook 2008* revealed that solar heat collector area in rural areas was 42.86 million m^2 in 2007. It is estimated it would reach 50 million m^2 in 2009, and it is estimated that solar heat collector area for urban residential buildings is about 95 million m^2.

[29] Calculation of energy-saving capacity: Supposing 300 showers will be taken per household per year, each shower consuming 60 L water at a temperature of 25°C, then the heat demand per household would be 523 kWh/a. The heat is supplied by solar collectors, with each household having a collector area of 2 m^2, then 95 million m^2 could produce 24.9 billion kWh heat energy-saving capacity, equal to about 3.06 Mtce.

First, promoting energy-efficient household appliances at the national level. The *Medium and Long-Term Energy Conservation Plan* formulated in 2004 stipulated to implement energy efficiency standards and labeling system, promote energy efficiency refrigerators, air-conditioners, TVs, washing machines, computers, as well as other household and office appliances, reduce standby energy consumption and regulate the market of energy efficient products; implement energy efficiency standards on lighting products, promote CFLs and increase its utilization rate.

Second, formulating relevant standards for energy efficient products, regulate relevant market and raise the grade of labeling system. For example, the *Energy Efficiency Grades and Minimum Allowable Values of Energy Efficiency for Room Air Conditioners* (GB12021.3) stipulated that the minimum allowable value of energy efficiency for air-conditioners in market would be raised from grade V (COP = 2.6) to Grade 2 (COP = 3.2). On May 1, 2009, the *Energy Efficiency Grades and Minimum Allowable Values of Energy Efficiency for Refrigerators* (GB12021.2) was enacted formally, which raised the standard for Grade I refrigerator-freezer by 27%, and increased the minimum allowable value of energy efficiency by 20%.

Third, conducting pilot projects, formulating relevant incentive policies and administrative measures, and promoting energy-efficiency products. At present, there are two key projects: Green Lighting Project, which is among the Ten Key Energy Conservation Projects brought forward by the *Notice of the State Council on Printing and Distributing the Comprehensive Work Plan on Energy Conservation and Emission Reduction* (State Council Circular, 2007, p. 15); and Energy-efficient Products Benefiting People Project, which is according to *Notice of the State Council on Printing and Distributing the 2009 Work Plan on Energy Conservation and Emission Reduction* (State Council Circular, 2009, p. 48).

There are sorts of incentive policies, too. ①With regard to lighting, in December 2007, the MOF and the NDRC issued the *Tentative Regulations on Subsidy Management for Promoting High-efficiency Lighting Products* (MOF, 2007, p. 1027), which stipulated that the central finance will provide 30% or 50% subsidy to energy-efficiency lighting products of the bid-winning enterprises, benefiting bulk users as well as urban and rural residents;②with regard to household appliances, the MOF issued the *Tentative Regulations on Subsidy Management for Promoting High-efficiency Energy-saving Products* (MOF: [2007] 1027), which stipulated that the central finance will provide subsidy to enterprises which shall sell energy-efficiency products to the consumers at subsidized prices. In May 2009, the MOF and the NDRC formulated the *Notice of Conducting Energy-saving Products Benefiting People Project* (MOF, 2009, p. 213), which stipulated to promote 10 kinds of household appliances (including air-conditioners) with Grade I–II energy efficiency between 2009 and 2012; and issued the *Notice on Printing and Distributing Regulations on High-efficiency and Energy-saving Room Air Conditioner by Energy-saving Products Helping People Project* (MOF, 2009, p. 214), which stipulated that from June 1st, 2009, the central finance will provide a subsidy of RMB 500–850/set to Grade I air-conditioners, and RMB 300–650/set to Grade II air-conditioners.

With support of above policies, energy efficient products are used widely in China. From 2008 to 2009, the central finance provided subsidy for 180 million energy efficient lamps, including 120 million lamps in 2009, saving 3.8 billion kWh (about 1.20 Mtce)[30] each year. On the other hand, many energy-efficient lamps used in China now have not been subsidized. According to statistics of China Association of Lighting Industry, China produced 3.7 billion energy efficiency lamps in 2009, 65% of which were exported. Supposing that the remaining 1.2 billion lamps for domestic demand were put to use,[31] then 25.3 billion kWh (8.02 Mtce) could be saved.

According to statistics of the Ministry of Finance, by the end of April 2010, altogether over RMB 7 billion financial subsidies had been provided under the Energy-saving Products People Benefiting Project. By May 2010, market share of energy-efficient air-conditioners had reached over 80%. According to incomplete statistics, over 15 million energy-efficiency air-conditioners were sold in 2009 (Xinhua News, 2009), which can save about 360 million kWh (110,000 tce)[32] annually.

---

[30]Supposing that the average power of energy efficiency lamps is 11 W, which can be used to replace 40 W incandescent lamps, and 2 hours of usage each day, then it can be calculated that 3.8 billion kWh (equal to 1.2 Mtce) could be saved for 180 million lamps.

[31]Actually, apart from those lamps in stock, energy efficiency lamps are also used for lighting of nonresidential buildings, such as road lighting, landscape lighting, and lighting of commercial buildings. Due to lack of relevant statistics, it is difficult to differentiate.

[32]According to statistics, average power consumption of residential air-conditioners was 2.1 kWh/m^2 in 2005 (Zhang *et al.*, 2010). There are 177 million households in urban areas, with each household possessing one air-conditioner; 3.1 persons in each household, average floor areas in urban areas is 20.1 m^2 per capita, then average floor areas per household is 64 m^2 (State Statistics Bureau, 2009), and average power consumption for air-conditioners per household is 135 kWh/(household/a). According to the *Energy Efficiency Grades and Minimum Allowable Values of Energy Efficiency for Air Conditioners* (GB12021.3), with 15% higher energy-efficiency ratio for air-conditioners, each household could save 21 kWh/a, then 15 million air-conditioners could save about 360 million kWh (110,000 tce).

Therefore, we can obtain the energy-saving capacity in 2007 and 2008 deriving from related policies on energy consumption of urban residential buildings (Fig. 9.37). Due to changing life patterns of urban residents, higher demand for energy as well as increasing floor areas of residential buildings, it could be estimated that energy consumption of urban residential buildings (excluding energy use for district heating system) in 2007 would be 35 Mtce more than that in 2005. While due to energy management of new buildings, renovation of existing buildings, promotion of energy efficiency appliances and lighting devices, as well as utilization of renewable energy, about 4 Mtce energy-saving capacity could be realized. Therefore, the real energy consumption in 2007 would be 31 Mtce more than that in 2005.

It can be discovered that rebound effect of rising energy demand due to changing lifestyles overwhelms the energy-saving benefits by technological enhancement. Therefore, carbon emission in urban residential buildings is still growing considerably.

9.2.4 *Policy evaluation for energy consumption of commercial buildings*

Key policy clusters for energy consumption of commercial buildings mainly relate to building codes, renovation of existing buildings, better operational management, promotion of energy efficient products, expanding use of renewable energy, and EMC subsidies policies. See Fig. 9.38. On the one hand, from the perspective of policy concentration level, strict implementation of standards for new buildings and deepening energy efficiency renovation of commercial buildings have been two pillars for energy efficiency policy system of commercial buildings. On the other hand, administrative measures and policies on operational management are the most numerous, which reveals that the importance of operational management to energy consumption of commercial buildings is more known to common people. In addition, there are more policies on energy consumption equipments, technologies, and service systems for commercial buildings. Photovoltaic technology integrated with commercial buildings becomes hot in China recently. Moreover, China is making great efforts in enlarging the ESCO (Energy Service Company) industry in commercial building sector, due to its successful development in developed countries. Now, contents, implementation, and estimated energy-savings of key policy clusters will be introduced below.

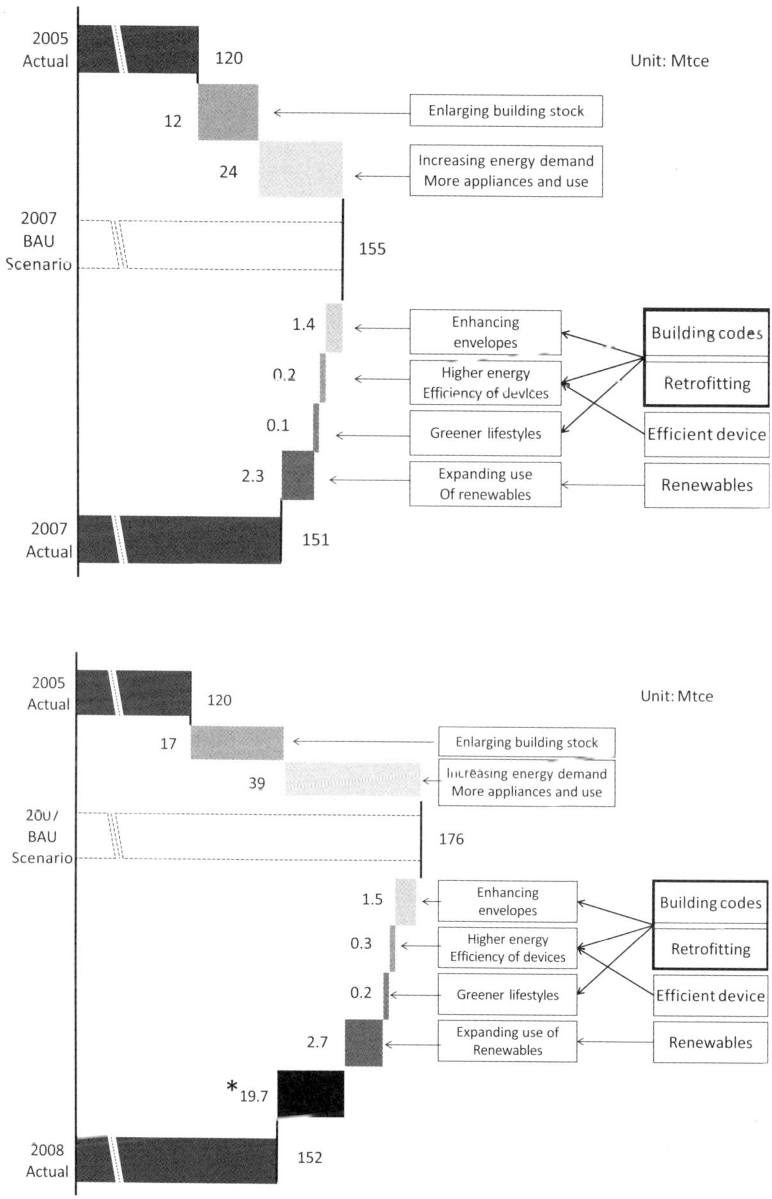

Fig. 9.37. Energy-saving capacity deriving from low-carbon policies for urban residential buildings excluding district heating system in 2007 and 2008 (unit: 10,000 tce).
*This part of energy-saving capacity cannot be explained by policy. There might be many reasons for this, for example, statistical error or life patterns irrelevant to policies which cannot be calculated and verified due to lack of statistical data.

Categories	Commercial buildings					
	New buildings	Retrofitting	Energy management	Renewable energy utilization	Promotion of energy-efficient products	ESCO
Codes and Standards	Design Standards for Energy Conservation of Commercial Buildings, 2005					
	Standards for Evaluation of Green Buildings, 2006					
	Testing and Evaluation System for Energy Efficiency of Civil Buildings, 2007		Guidelines for Energy Audit on Governmental Office Buildings and Large Commercial Buildings, 2007			
			Standards for Data Collection of Energy Consumption of Civil Buildings, 2008			General Technical Rules for EMC, 2010
Administrative Measures	Inspection on building energy efficiency, 2004				Government Procurement List of Energy-efficient Products, 2010	Opinions on Accelerating EMC and Promoting ESCO Industry, 2010
	Guidance Opinions on Development of Energy-saving and Land-saving Residential and Commercial Buildings, 2005					
	Several Opinions on Strengthening Engineering Management of Large Commercial Buildings, 2007					
	Management Measures on Evaluation and Labeling of Building Energy Efficiency, 2007		Notice on Strictly Implementing the Air-conditioning Temperature Control Standards for Commercial Buildings, 2007			
	Notice on Temporarily Implementing the Civil Building Energy Efficiency Evaluation and Labeling System, 2007					
		Implementation Opinions on Strengthening the Energy-saving Management of Governmental Office Buildings and Large Commercial Buildings, 2007				
			Implementation Plan for Energy efficiency Supervision System in Governmental Office Buildings and Large Commercial Buildings, 2010			
			Statistic and reporting system for energy consumption and efficiency of civil buildings, 2010			
Incentives		Tentative Regulations on Management of Energy-saving Fund for Governmental Office Buildings and Large Commercial Buildings, 2007		Tentative Regulations on Management of Financial Subsidies for BIPV, 2009		Tentative Regulations on Management of Financial Subsidies for EMC, 2009

Fig. 9.38. Chart of key policy clusters for commercial buildings.

(1) *Standards for new buildings and renovation of existing buildings*

Policies on standards for new buildings and renovation of existing buildings have been introduced in detail in Sec. 9.2.3. Following are brief summary on standards specified for commercial buildings and their corresponding energy savings.

There are two stages of design standards for commercial buildings. *Energy Conservation Design Standards on Building Envelop and Air Conditioning for Tourist Hotels* (GB50189-93) in 1993 only applies to a limited scope of commercial buildings. However, due to increasing imbalance between power supply and demand as well as hiking energy consumption for commercial buildings in 2004, the Ministry of Construction formulated the *Design Standards for Energy Conservation of Commercial Buildings* (GB50189-2005) in 2005, also called 50% energy efficiency design standards[33]; and the former standard in 1993 was nullified.

The government also formulated a series of administrative measures to ensure proper implementation of above standards, which have been introduced in Sec. 9.2.2. It is worth to mention that, commercial buildings are the focus of renovation work stipulated during the 11th *FYP*. The *Medium and Long Term Energy Conservation Plan* formulated in 2004 stipulated that "renovation of governmental buildings and the heating, air-conditioning, and lighting systems shall be the focus of the 11th *FYP" and that* 20% of the total floor areas of governmental buildings shall be renovated according to the energy efficiency design standards."

The result was encouraging. By 2009, the total floor areas of energy efficiency buildings were 4.08 billion m^2 in China, including 109.49 million m^2 floor areas of renovated buildings in 15 Northern cities. To specify, 1.6 billion m^2 are of commercial buildings,[34] accounting for 22% of the total commercial buildings. Compared with the energy consumption of commercial buildings before taking the energy efficiency measures, altogether

[33]GB50189-2005 stipulated that total energy consumption of heating, ventilating, air-conditioning, and lighting system in the commercial buildings conforming to design standards should decrease by 50% than before.

[34]Due to lack of statistics, the floor areas of new commercial buildings of each year/the floor areas of new buildings of each year = the floor areas of new energy efficiency commercial buildings of each year/the floor areas of new energy efficiency buildings of each year. Other data is collected from *China Statistics Yearbook 2009* and *Circular on Energy Conservation Inspection of 2009 Energy Conservation Emission Reduction Supervision and Inspection in China's Building Sector* (MOC and MOST, 2010, p. 45).

6.96 Mtce[35] of heating-related energy consumption could be saved annually (which refers to heating demand of building itself, and has been included in Sec. 9.1), and 9.99 Mtce[36] for other end-usages such as the cooling, ventilating, lighting and so on.

With regard to commercial building labeling, by October 1st, 2010, 46 projects had been notified publicly as green building designing evaluation and labeling projects by the MHURD, 25 of which were commercial ones, covering a total floor area of over 1.2 million m^2, and 16 were three-star projects. Extensive efforts are made to promote the green building labeling system.

(2) *Renewable energy utilization and promotion of energy-efficient products*

Conventional energy carriers in commercial buildings have been included in 10 categories of energy efficiency products promoted by the Energy-saving Products Benefiting People Project as well as the energy efficiency lamps promoted by the Green Light Projects. Relevant policies and implementation has been discussed in Sec. 9.2.3.

BIPV is observed as the major way of renewable energy utilization in commercial buildings. Relevant policies have already been enumerated in Sec 9.2.5. It will mainly introduce the effect of above policies and estimated energy-savings in following paragraph.

[35]According to the 50% standard, heating demand of new buildings will decrease by 58 kWh/m^2 on the basis of 1980. In addition, out of 1.6 billion m^2 energy-saving commercial buildings, about 1.1 billion m^2 locate in North China. (the floor areas of new commercial buildings in North China of each year/the floor areas of new commercial buildings in China of each year = the floor areas of new energy efficiency commercial buildings in North China of each year/the floor areas of new energy efficiency commercial buildings in China of each year. Other data is collected from *China Statistics Yearbook 2009* and *Circular on Energy Conservation Inspection of 2009 Energy Conservation Emission Reduction Supervision and Inspection in China's Building Sector* [MOC and MOST, 2010, p. 45]), saving about 60 billion kWh heat energy (equal to 6.96 Mtce).

[36]According to statistics, average primary power consumption of air-conditioning, ventilating and lighting system in commercial buildings was 12 kgce/m^2 in China in 2005 (Zhang *et al.*, 2010). According to 50% energy-saving standard, power consumption of air-conditioners, ventilation and lighting system in new buildings should reduce by 50%, about 6 kgce/m^2. Therefore, it can be estimated that 1.6 billion m^2 energy efficiency commercial buildings could save 9.99 Mtce.

In fact, BIPV is majorly used in commercial buildings in China. By the end of 2009, overall installed capacity of PV buildings had reached 59.6 MWp (Wp is the peak power of solar cells) in China (CSIA, 2010), saving about 90 million kWh (equal to 30,000 tce)[37] each year. It should be noted that, in cities with high building intensity, sunshine area for PV panels to produce power is very limited, and cannot replace much building energy consumption, which reduces very much the attention building owners willing to attach. In addition, its operation and maintenance is very inconvenient, therefore its energy-saving effect is very limited.

(3) *EMC*

Energy service system consists of two parts. First, ESCOs offer EMC service, that is to say, offering consultation, inspection, designing, financing, renovation, and operation and management service to building users for reducing building energy consumption. Establishing EMC system aims to prompt the building users to conduct energy renovation and optimal management by making use of market mechanism, with the purpose of reducing the energy consumption level of buildings. On the other hand, the society and governments offer energy-saving education to ordinary citizens, encouraging them to save energy in their daily life. Now, related policies on EMC system will be introduced in the following paragraphs.

China has always been supportive to EMC development. The *Decisions on Strengthening Energy Conservation Works* (State Council, 2006, p. 28) issued by the State Council in 2006 stipulated clearly to "accelerate promotion of EMC system and technical energy efficiency renovation of enterprises." Introduction of EMC system can be traced back to 1998, when Chinese government, the World Bank, and the GEF jointly conducted an international cooperation program "WB-GEF-China Energy Conservation Project", to promote EMC development in China. Under the *aegis* of the project, EMCA was created in 2005 to serve as a bridge for Chinese government and energy service enterprises, with the purpose of promoting sustainable development of China's energy service industry, providing support

[37]Generating capacity of PV power generation system is 1.56 kWh/(Wp.a) in 2010 (Li Junfeng *et al.*, 2010a). Then, it can be estimated that 59.6 MWp PV panels could produce 90 million kWh in 2009. Average coal consumption for power generation is 0.317 kgce/(kWh) in 2009, then it can be estimated that 59.6 MWp PV panels could save 30,000 tce annually.

to China's energy service enterprises and finally realizing industrialization of the energy service sector.

However, EMC mainly focuses on the industrial sector at present, the energy service industry in the building sector is in its initial stage, and the energy service system has not been complete. Therefore, China is in urgent need of establishing a sound energy service market in the building sector. Now, policies related to EMC and energy service system mainly consist of the following three aspects.

First, the government issues relevant management measures in the form of administrative orders to establish EMC system and create a sound energy service market in the building sector. The *Notice of the State Council on Printing and Distributing the Comprehensive Work Plan on Energy Conservation and Emission Reduction* (State Council Circular, 2007, p. 15) stipulated to "cultivate the energy service market, accelerate EMC development, and support the specialized energy service companies in priority for conducting diagnosis, designing, financing, renovation and management to enterprises, party and governmental office buildings, public facilities and schools". In August 1st, 2008, the State Council issued the *Regulations of Energy Conservation of Public Institutions* (State Council Order, p. 531), which stated clearly that "public institutions could delegate the energy service agencies to conduct energy efficiency diagnosis, designing, financing, renovation and management by way of EMC". The *Medium and Long Term Energy Conservation Plan* in 2009 further encouraged specialized energy service companies to conduct energy efficiency renovation to medium and small-sized enterprises and public institutions by way of EMC.

Second is to issue relevant standards to regulate the industrial development. In Aug. 9th, 2010, the Standardization Administration of China issued the General *Technical Rules on EMC* (GB/T 24915-2010) that would be implemented by January 1st, 2011. It stipulated the terms, definitions, types of contracts, technical requirements and reference contracts of EMC, and reduced the contractual expenses of negotiation, transaction and implementation in EMC projects.

Third is to formulate economic incentive policies to boost EMC development in the building sector. In March 2010, the State Council circulated the *Notice on Opinions of Accelerating the Implementation of EMC for Promotion of Energy Conservation Service Industry* (State Council Office Circular, 2010, p. 25), which stipulated to support the development of energy service industry by subsidy, taxation, accounting system and financial services. June 3rd, 2010, the MOF and the NDRC issued the *Tentative*

Regulations on Management of Financial Subsidies for EMC (MOF, 2010, p. 249), which stipulated that central finance and provincial finance provides 240 yuan/tce and 60 yuan/tce respectively for EMC. The enterprises with a registered capital of over RMB 5 million could apply for the subsidy, which is mainly used to support shared-savings contracts (see Sec. 3.3). Meanwhile, local governments also formulated relevant policies. For example, in September 9th, 2010, Sichuan provincial government formulated the *Implementation Opinions on Accelerating EMC and Promoting Energy Service Industry* (Sichuan Government, 2010, p. 80), which defined the objective of EMC promotion in Sichuan; stipulated that in implementing energy efficiency renovation, the public institutions, the owners, real property managers or long-term renters of large commercial buildings should take EMC methods in priority; and stipulated to offer credit and taxation support, such as temporarily exempting business tax and exempting value-added tax etc.

After more than 10 years of development, EMC has developed into a promising energy service industry, and EMC has become a widely-applied model in China's energy service industry. According to the statistics of EMCA, by the end of 2009, 502 EMC companies had adopted EMC (China Energy Conservation Association, 2010), and EMC project investment reached RMB 28 billion in 2009, increasing by 67.37% year on year, producing an overall output value of RMB 58 billion, saving 13.5 Mtce annually (the State Council, 2010), which reveals the rapid growth of energy service industry and its enormous potential. Energy service industry in the building sector has been expanding, and its overall investment was RMB 550 million in 2007, while it is estimated that the building energy service market would reach RMB 2.34 billion in 2010 (SHENYIN and WANGUO Securities, 2008), saving 2.10 Mtce annually.[38] On the other hand, concentration level of the industry is not high. By October 2010, China had listed three companies promoting building energy efficiency by EMC, whose mid-year reports revealed that by 2009, altogether 26 EMC projects had realized benefit sharing.

---

[38]Statistics on the promising building energy service industry are lacking. The calculation method: Energy savings = EMC investment in the building sector/total EMC investment × total EMC energy savings. Actually, cost of building energy efficiency is much lower than that of industrial energy efficiency. Therefore, with same investment, energy savings in the building sector would be more than that in industrial sector.

(4) *Operational management of commercial buildings*

Scientific and reasonable operational plan for building energy consumption system is very important to reduce energy use in commercial buildings. Pre-set energy efficiency effects could be realized only by establishing sound energy management measures, and supplementing them with auxiliary building energy consumption statistics, energy auditing and energy consumption quota management. With regard to China's policy system (See Fig. 9.38), it has most policies on operational management of commercial buildings, which consists of the following aspects.

First is to put forward objectives for energy efficiency of commercial buildings by laws, regulations and national planning. The *Regulations on Energy Conservation of Civil Buildings* (State Council Order, p. 530) and the *Regulations on Energy Conservation of Public Institutions* (State Council Order, p. 531) stipulated that governmental office buildings and large commercial buildings should realize energy metering and conduct energy auditing; local governments above county level should impose power consumption limits to local commercial buildings; and an energy-saving objective of 20% should be realized during the *11*th *FYP*. The *Comprehensive Working Plan for Energy Conservation and Emission Reduction* issued in 2007 stipulated to establish energy consumption statistics, energy auditing and energy consumption quota system for large commercial buildings in 25 demonstrative provinces and municipalities, saving 12.5 Mtce.

Second is to formulate relevant standards and regulations, regulate the operational management of building energy consumption system and the technical essentials of auxiliary energy consumption statistics system. The MOC issued the *Regulations on Operational Management of Air-conditioning and Ventilation System* (GB50365-2005) in 2005 to enhance operational management of commercial buildings. Meanwhile, it also issued the *Guideline Principle for Energy Auditing of Governmental Office Buildings and Large Commercial Buildings*, the *Regulations on Establishment, Acceptance and Operational Management of Energy Consumption Monitoring System for Governmental Office Buildings and Large Commercial Buildings*, the *Technical Guideline for Designing and Installation of Separate Metering System of the Energy Consumption Monitoring System for Governmental Office Buildings and Large Commercial Building*, the *Technical Guideline for Data Collection of Separate Energy Consumption*, the *Technical Guideline for Data Transmission of Separate Energy Consumption*, and the *Technical Guideline for Establishment* and *Maintenance of*

Data Center, to provide guidance for energy auditing of large commercial buildings.

Third is to formulate relevant management measures and incentive policies to promote the development of energy-saving supervision system, and to lay a solid foundation for energy-saving renovation of existing buildings, establishment of energy consumption quota system and above-quota surcharge system. Actually, the MOC had conducted civil buildings energy consumption statistical analysis in 23 cities in 2007. In order to support energy consumption statistics, the MOF issued and distributed the *Tentative Regulations on Management of Financial Incentive Fund for Energy Conservation of Governmental Office Buildings and Large Commercial Buildings* (Ministry of Finance, 2007, p. 558), which provided funds for establishing the building energy efficiency supervision system, including the fund for creating the building energy consumption supervision platform, as well as conducting building energy consumption statistics, building energy auditing and building energy efficiency public notification. In 2009, the MHURD and the SSB jointly issued the *Statistical Report System for Energy Consumption and Conservation Information of Civil Buildings*, and issued the *Notice on Printing and Distributing the Statistical Report system for Energy Consumption and Conservation Information of Civil Buildings* (MOC, 2010, p. 31), which stipulated to conduct energy consumption statistical analysis to all buildings in provinces and municipalities in China.

According to the implementation process, the major effort is still focusing on the creation of energy efficiency supervision system for governmental office buildings and large commercial buildings. By the end of 2009, altogether 29,359 governmental office buildings and large commercial buildings had completed energy statistics, 2,647 were defined as key energy-consuming buildings, 2,175 had completed energy auditing, energy consumption of 2,441 buildings were notified publicly, and energy consumption of 758 buildings were monitored dynamically.[39] And it is estimated that about 5.35 Mtce could be saved each year.[40] In 2007, the MOC organized

---

[39] *Circular on Energy Conservation Inspection of 2009 Energy Conservation Emission Reduction Supervision and Inspection in China's Building Sector* (MOC and MOST, 2010, p. 45).

[40] Calculation method: About 30,000 large commercial buildings, supposing that each covering an area of 20,000 m^2; energy consumption per unit floor area of large commercial buildings is about 44 $kgce/m^2$, and after energy auditing and energy efficiency renovation, about 20% energy could be saved. Then, it can be estimated that 5.35 Mtce could be saved.

competent construction authorities in 24 provinces and municipalities to conduct energy auditing and energy efficiency public notification for governmental office buildings and large commercial buildings. The first batch of pilot cities such as Beijing, Tianjin, and Shenzhen have completed the construction of platforms, and it is estimated that by the end of 2010, dynamic energy consumption supervision platforms for 465,[41] 260,[42] and 317[43] large commercial buildings would be established respectively; the second batch of pilot cities in Jiangsu, Chongqing, and Inner Mongolia have initiated the platform construction process, and it is estimated that the projects will complete by the end of 2010; the third batch of pilot cities in Hunan, Liaoning, and Wuhan have completed the preparation of implementation plans, and are going to initiate the construction process. In general, construction of energy consumption supervision platform goes well and the precious experiences will benefit the operational management and energy consumption statistics of commercial buildings enormously.

Therefore, we can obtain the energy-saving capacity deriving from policies relevant to energy consumption of commercial buildings in 2007 and 2008 (see Fig. 9.39). Due to changing life patterns, more commercial buildings and higher proportion of large commercial buildings, it can be estimated that energy consumption of commercial buildings in 2007 would be 30 Mtce more than that in 2005. While due to energy management of new buildings, renovation of existing buildings, operational management, renewable energy utilization, promotion of energy efficiency products and energy service system, about 7 Mtce could be saved, so the real energy consumption in 2007 would be 23.70 Mtce more than that in 2005.

EMC policies just being implemented do not form the largest energy-saving capacity, but are related to other key policy clusters for commercial buildings, whose promotion and implementation are conducive for further

[41]By the end of 2009, platforms for 150 large commercial buildings have been established. According to the *Notice on Enhancing the Energy Conservation Management of Governmental Office Buildings and Large Commercial Buildings* issued by the MHURD in July 18th, 2010, an additional 315 platforms will be established in 2010.

[42]By the end of 2009, platforms for 95 large commercial buildings have been established. According to the *Notice on Enhancing the Energy Conservation Management of Governmental Office Buildings and Large Commercial Buildings* issued by the MHURD, an additional 165 platforms will be established in 2010.

[43]By the end of the first half of 2009, platforms for 67 large commercial buildings have been established. According to the *Notice on Enhancing the Energy Conservation Management of Governmental Office Buildings and Large Commercial Buildings* issued by the MHURD, an additional 250 platforms will be established in 2010.

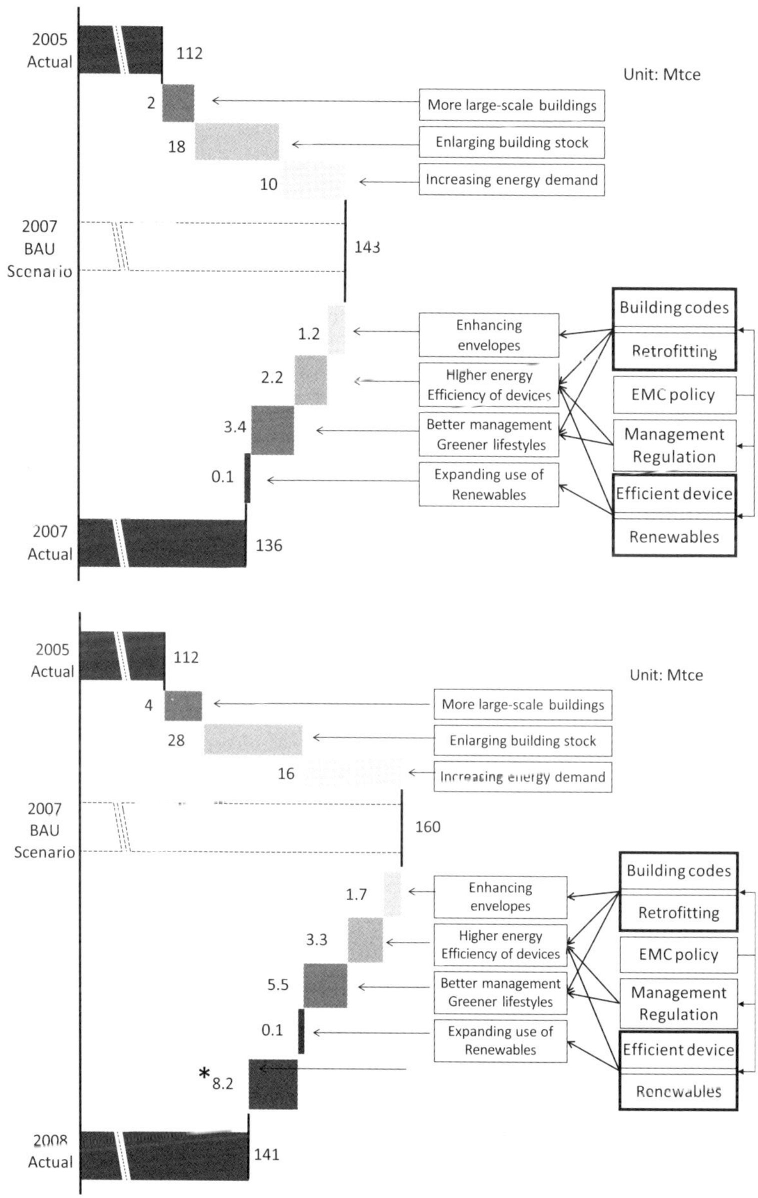

Fig. 9.39. Energy-saving capacity deriving from low-carbon policies for commercial buildings (excluding district heating system) in 2007 and 2008 (unit: 10,000 tce).
*This part of energy-saving capacity cannot be explained by policy. There might be many reasons for this, for example, statistical error or life patterns irrelevant to policies which cannot be calculated and verified due to lack of statistical data.

implementation of energy policies for commercial buildings. Detailed analysis is listed in Sec. 9.3.3.

9.2.5 *Policy evaluation for energy consumption in rural areas*

Key policy clusters on rural energy consumption mainly concentrate on building codes, renovation of existing buildings, project on prevailing household appliances in countryside area, and renewable energy utilization, consisting of industrial regulations and standards, administrative measures and economic incentive measures. See Fig. 9.40. At present, renewable energy, especially solar power utilization has been the focus of energy-saving policies in rural areas. The contents, implementation and estimated energy savings of the key policy clusters will be introduced in the following paragraphs.

(1) *New building standards and renovation of existing buildings*

The MHURD issued the *Technical Guideline on Energy Conservation of Rural Residential Buildings in Cold and Severe Cold Areas* (Tentative) in June 2009, which is the first systematic standard for rural buildings. It consists of ① regulations on thermal performance of the building envelop in building new buildings and renovating existing buildings; ② regulations on heating and ventilation system, including stove-brick "Kang" technology for heating and natural ventilation designing; ③ regulations on renewable energy utilization, especially the solar thermal energy, including passive sun house and solar water heater etc. Statistics are lacking due that the standards are newly formulated.

Viewed from history, rural residential buildings in China have distinctive local features due to local weather divergences, for example, heated brick bed (or "Kang") in North China and cave dwellings in Western regions. With the development of rural economy and technical advancement, China develops some new technologies for rural residential buildings including insulation technology for envelop structure, passive sun house, new cave dwellings and Chinese-style "Kang" technology. Rural buildings have distinctive local features and are diversified, and it is difficult to conform to a universal standard. Therefore, local governments should play a positive role in demonstrating and promoting energy-saving technologies for rural residential buildings.

Categories	Rural Residential Buildings			
	New buildings	Retrofitting	Energy efficiency products	Renewable energy utilization
Building Standards	Technical Guideline for Energy Conservation of Rural Residential Buildings in Cold and Severe Cold Areas (Tentative), 2009, 2005			
			Household Appliances to Countryside Labeling	Technical Regulations on PV Application in Civil Buildings, 2008
Administrative Measures	Promotion of applicable technologies		Notice on Promoting Household Appliances to Countryside Project in China, 2008	Implementation Plan for Promoting Solar Thermal Power, 2007
				Opinions on Accelerating the Comprehensive Utilization of Straws of Agricultural Products, 2008
				Construction Plan for Marsh Gas Project in China's Rural Areas (2006–2010), 2007
Incentive Policies			Implementation Plan for New Subsidized Products in Household Appliances to Countryside Project, 2006	Tentative Regulations on Management of Special Funds for Renewable Energy Utilization in Buildings, 2006
				Regulations on Management of Funds for Marsh Gas Project in Rural Areas, 2007
				Tentative Regulations on Management of Reward Funds for Bio-energy and Biochemical Products, 2007
				Tentative Regulations on Management of Subsidies for Straw Utilization, 2008
				Notice on the Implementation Plan of Accelerating the Utilization of Renewable Energy in Rural Buildings, 2010
				Amendment to the Renewable Energy Law proposed to create a renewable energy development fund, 2010

Fig. 9.40. Policy chart for energy consumption in rural residential buildings.

(2) *Renewable energy utilization*

The central government advocates application of renewable energy technologies in buildings, so as to reduce the consumption of fossil energy by increasing equipment energy efficiency and proportion of renewable energy. At present, China's renewable energy utilization in buildings mainly consists of solar power and bio-mass energy. Solar power utilization consists of solar-thermal and solar-electric methods. Bio-mass energy utilization consists of in-situ combustion and making marsh gas and solid fuel.

Major tasks of Chinese government in renewable energy utilization include: ① advocate the application of renewable energy in buildings, especially in integrated buildings, and formulate objectives for development and application of renewable energy in buildings;② formulate standards and regulations, administrative measures and incentive policies, and initiate BIPV pilot projects to promote the application of PV technology and products.

The *Renewable Energy Law* is China's basic law on the utilization of renewable energy, which stipulated clearly that the government encourages installation and utilization of solar water heaters, solar heating and cooling system, and PV power generation system by social units and individuals; encourages utilizing renewable energy sources in rural areas to meet the energy demand by rural residents, improve the production and living standards of rural areas, protect ecological environment and further promote ecological construction, increase the income of rural residents, and promote rural economic and social development. In 2008, the State Council issued the *Regulations on Energy Conservation of Civil Buildings* (State Council Order, p. 530), which re-stressed that the government encourages and supports adopting renewable energy such as solar power and earth heat in constructing new buildings and renovating existing buildings, and encourages local governments and departments with favorable conditions to support installation and utilization of solar systems such as solar water heaters, solar lighting systems, solar heating systems and solar heating/cooling systems by social units and individuals.

The government also formulated corresponding national plans for developing and promoting various renewable energy utilization technologies based upon above-mentioned laws and regulations. The *Medium and Long-Term Plan for Renewable Energy* in 2007 brought forward that by 2010, commercial buildings should realize the target of 50,000 kW PV power generation. In the same year, the 11th *Five Years Plan for Energy* proposed that by 2010, 40 million households in villages and towns would use marsh gas, marsh gas production in rural areas would reach 16 billion m^3, and there would be 50 million solar water heaters and 1 million solar kitchen ranges. Local governments also responded positively to the national plans and formulated local plans. In January 5th, 2010, the *Guiding Opinions on Accelerating Solar Power Utilization and Promoting Solar Power Industrial Development of Beijing* (Beijing, 2009, p. 43) brought forward that by 2012, solar collector area would reach 7 million m^2, and the solar power generation system would have a generating capacity of 70 MW.

The government issued many standards, administrative policies and incentive polices to achieve above-mentioned targets, and initiated pilot projects to promote its application.

First is to formulate relevant standards and regulations to enhance capability building and accelerate the application of relevant technologies. Taking solar power for an example, the MHURD initiated the compilation and amendment process for *Technical Regulations on Application Technologies of PV System in Civil Buildings*, the *Drawings for Design and Installation of PV System in Civil Buildings* and the *PV Components in Civil Buildings*, which will offer guidance to designing, installation and acceptance of BIPV.

Second is to formulate regulative policies in the form of administrative orders. In March 2007, the MOA formulated the *Construction Plan for Marsh Gas Project in Rural Areas* (2006–2010) (MOA Plan, 2007, p. 7), which has been approved by the NDRC (NDRC, 2007, p. 66). The plan stressed on the necessity and feasibility of developing marsh gas projects in rural areas.[44] In April 2007, the NDRC issued the *Implementation Plan for Promoting Solar Thermal Utilization in China*, which stipulated clearly that China would formulate a compulsory policy for installation of solar water heaters. In May 2007, the NDRC and the MOC issued the *Notice on Accelerating the Application and Promotion of Solar Water Heaters* (NDRC, 2007, p. 1031), which stressed the importance of promoting solar water heaters, and supported the technical progress and industrial development of solar water heaters.

[44]Actually, the plan also covers the contents and planning objectives, regional pattern and construction scale, investment evaluation, benefit analysis and guarantee measures, which aims to deal with the demand for living energy, guarantee the achievement of returning agricultural land to forestry, change agricultural growth pattern, increase the income of farmers, improve rural living and production conditions, and promote the construction of new villages.

Third is to formulate economic incentive policies. In May 2006, the MOF issued the *Tentative Regulations on Management of Special Funds for Renewable Energy Utilization* (MOF, 2006, p. 237), which stipulated to offer monetary support in the form of free subsidy and loan discount to renewable energy projects in buildings, especially "the renewable energy utilization projects on living energy in rural and pastoral areas". At the same year, the MOF issued the *Tentative Regulations on Management of Special Funds for Renewable Energy Utilization in Buildings* (MOF, 2006, p. 460), which proposed to offer financial subsidies to key BIPV projects by unit floor areas, such as supply of hot domestic water, heat supply and air-cooling, PV conversion and lighting. In the following two years, the MOF also issued the *Regulations on Management of Funds for Marsh Gas Project in Rural Areas* (MOF, 2007, p. 434), *Tentative Regulations on Management of Reward Funds for Bio-energy and Biochemical Products* (MOF, 2007, p. 7), *Tentative Regulations on Management of Subsidies for Straw Utilization* (MOF, 2008, p. 735), which proposed to provide subsidy for equipment acquisition and installation of rural marsh gas projects and straw solid fuel projects directly. In July 2009, the MOF and the MHURD jointly issued the *Notice on the Implementation Plan of Accelerating the Utilization of Renewable Energy in Buildings in Rural Areas* (MOF, 2009, p. 306), which selected the rural primary and middle schools, residential buildings and health centers as major targets for subsidization, providing 60 yuan/m^2 for buildings applying earth heat pumps, 15 yuan/m^2 for buildings using BIPV technology, and providing 60% newly-accreted investment to household sun-houses. These measures have promoted the utilization of nonfossil energy such as bio-mass energy and solar power in rural areas, and helped in substituting for and saving fossil energy.

With the advancement of PV technology as well as development of foreign and domestic market, PV technology has become a focus in national economic incentive policies. In March 23rd, 2009, the MOF and the MHURD jointly issued the *Implementation Opinions for Accelerating the Application of BIPV* (MOF, 2009, p. 128), which supported the application of pilot PV buildings, proposed to carry out the Solar Roof Plan, and promoted solar roofs and PV walls in large and medium-sized cities.[45] The plan also supported off-grid power generation in rural areas and transmitting electricity to each household. Meanwhile, the MOF also issued the

[45] At present, solar PV buildings are mainly commercial buildings, with almost no urban residential buildings.

Tentative Regulations on Management of Financial Subsidies to Application of BIPV (MOF, 2009, p. 129), which proposed to offer a maximum of 20 yuan/Wp to qualified PV buildings. In July 2009, the MOF and the MHURD jointly issued the *Notice on Implementing Golden Sun Demonstrative Projects* (MOF, 2009, p. 397) and *Tentative Regulations on Management of Financial Subsidies to Golden Sun Demonstrative Projects*, which reiterated the policies on PV buildings in document [2009]128 issued by the MOF.

Local governments responded positively and formulated relevant local policies to support the utilization of renewable energy. In 2010, Beijing city initiated the Golden Sun Project, which stipulated that Beijing government would offer 200 yuan/m^2 to first 1 million m^2 solar collector areas in commodity buildings before December 31st, 2012; offer 1 yuan/kWh/a for 3 years to the commercial buildings with solar roof PV power generation systems; provide a subsidy equal to 30% renovation costs to rural residential buildings with solar heating systems. Xingtai, Hebei province also formulated a series of preferential policies, for example, exempting 50% urban maintenance expenses, providing RMB 500 for each household, and offering a subsidy equal to 10% equipment cost for the social units and households utilizing solar power; and county (city, district) finance would offer RMB 300–500 to each household of the demonstrative new villages utilizing BIPV technology.

There are also certain problems during policy implementation, for example, the fund does not stably and consistently financed, the amount and scope of its application was not as fixed as policies incentives, and cannot provide long-term and stable support to the development of renewable energy. For example, the special fund budget for renewable energy was RMB 8.5 billion in 2008, while the real amount was only RMB 3.9 billion[46] (Beijing Oriental Environment Institute *et al.*, 2009). Therefore, China formulated the *Renewable Energy Law* (Amendment) in April 1st, 2010, which stipulated clearly to create a renewable energy development fund to deal with the problem of money shortage for the development of renewable energy. It stipulated clearly that "the central finance will create a renewable energy development fund, whose money sources include the special fund allocated by the central finance as well as the extra income from

[46]Part of the fund is not available due that some supportive policies have not been formulated.

electricity charges generated with renewable energy. For the renewable energy utilization projects in the guidance list for the development of renewable energy industry and with qualified credit conditions, financial agencies could provide discount loans, and the government also provides preferential taxation policies."

Fourth is to initiate renewable energy demonstrative projects. In July 2009, the MOF and the MHURD jointly issued the *Notice on Printing and Distributing the Implementation Plan for Urban Demonstration of Application of Renewable Energy in Buildings* (MOF, 2009, p. 305), to accelerate the development of new energy industry as well as energy-saving and environmental-friendly industry, and promote wide application of renewable energy in urban buildings. The notice stipulated that the central finance supported the application of solar water heaters, earth heat pumps, solar heating and cooling, and combination of solar power and earth heat pumps in urban buildings. With respect to auxiliary measures, the MOF formulated the *Tentative Regulations on Management of Special Funds for Renewable Energy Utilization in Buildings* (MOF, 2006, p. 460) and the *Regulations on Evaluation of Renewable Energy Utilization in Buildings* (MOF, 2006, p. 459).

In general, current situations are:

Policies on utilization of bio-mass energy mainly focus on rural areas. By the end of 2007, 24.87 million households utilized marsh gas, producing 9.9 billion m^3 methane gas annually, realizing an energy-saving capacity of 9.45 Mtce,[47] which was 3.81 Mtce more than that in 2005.[48] By the end of 2009, 30.50 million households (Xinhua News, 2009) utilized marsh gas, realizing 11 Mtce energy-saving capacity annually, which was 5.35 Mtce more than that in 2005.

China's solar power utilization mainly focuses on residential buildings. By the end of 2009, China's solar collector area reached 145 million m^2 (Energy Foundation, 2007), 50 million m^2 was in rural areas,[49] realizing

---

[47] *Rural Energy Statistics Yearbook 2008* revealed that in 2007, China's marsh gas production was 9.9 billion m^3, heat generation capacity is 5000–6000 kcal (1 kcal = 4185.85 J). Supposing it is 500 kcal, then energy-saving capacity of 9.9 billion m^3 marsh gas is 9.45 Mtce.

[48] *Rural Energy Statistics Yearbook 2008* revealed that in 2005, China's marsh gas production was 7.1 billion m^3, equal to 5.65 Mtce.

[49] *Rural Energy Statistics Yearbook 2008* revealed that in 2007, solar collector areas in rural areas were 42.86 million m^2, 4 million m^2 was increased each year from 2005 to 2007, then it can be estimated that the number would be 50 million m^2 in 2009.

570,000 tce energy-saving capacity on the basis of 2005[50]; 1.63 million solar kitchen ranges[51] were utilized, realizing 180,000 tce energy-saving capacity on the basis of 2005[52]; 16.13 million m^2 sun houses[53] were utilized, realizing 10,000 tce energy-saving capacity on the basis of 2005[54]; electrification in rural areas realizes 95.6 MWp PV power generation (CSIA, 2010), realizing 160 million kWh energy-saving capacity each year (equal to 50,000 tce).[55]

BIPV was mainly used in urban commercial buildings and little in rural areas. Therefore, the report will not calculate the energy-saving capacity of BIPV in rural areas.

(3) *Household appliances to countryside*

In order to promote the construction of new socialist rural areas, build a harmonious society and put into practice of the Household Appliances

[50]Calculation of energy-saving capacity: supposing 300 showers will be taken per household per year, each shower consuming 60L water of 25°C, then heat demand per household would be 523 kWh/a. The heat is supplied by solar collectors, with each household having 2 m^2 solar collector area, then 50 million m^2 could produce 13 billion kWh heat energy-saving capacity, equal to about 1.60 Mtce. It is estimated that 32.06 million m^2 heat collector areas in 2005 could save 1.03 Mtce, and then about 570,000 tce energy-saving capacity was realized in 2009 on the basis of 2005.

[51]There were 1.12 million solar kitchen ranges in rural areas in 2007, 250,000 solar kitchen ranges were increased each year from 2005 to2007 (*Source*: Department of Education, Science and Technology, Ministry of Agriculture, 2008; Rural Energy Statistics Yearbook 2008 (China Agriculture Press, Beijing, 2008)), then it can be estimated that the number would be 1.63 million in 2009.

[52]In 2006, 4.65 Mt LNG was utilized in China's rural areas, with 41.05 million households using LNG as cooking energy (Construction Energy Conservation Research Center of Tsinghua University, 2008), then each household consumes 194 kgce each year for cooking. Solar kitchen ranges could meet the demand for cooking energy in areas with sufficient sunshine, 940,000 solar kitchen ranges are increased on the basis of 2005 (690,000), thereby producing 180,000 tce energy-saving capacity.

[53]*Rural Energy Statistics Yearbook 2008* revealed that in 2007, there were 13.14 million m^2 sun houses in rural areas, 750,000 m^2 were increased each year from 2005 to 2007, then it can be estimated that the number would be 16.13 million m^2 in 2009.

[54]China's average energy consumption is 4.9 kgce/m^2 (Construction Energy Conservation Research Center of Tsinghua University, 2008); supposing sun houses with affluent sunshine could meet 40% energy demand, then 16.13 million m^2 sun houses could realize 30,000 tce energy-saving capacity, 10,000 tce more than that in 2005 of 13.14 million m^2 sun houses.

[55]Generating capacity of PV power generation system is 1.56 kWh (Wp.a) in 2010 (Li, 2010), then it can be estimate that 95.6 MWp PV panels could generate 160 million kWh in 2009. Average coal consumption for power generation is 0.317 kgce/(kWh) in 2009, then it can be estimated that 95.6 MWp PV panels could realize 50,000 tce energy-saving capacity annually.

to Countryside Project initiated by the State Council, the MOF and the Ministry of Commerce proposed to provide financial subsidies for the project after repeated research and analysis, so as to expand domestic demand, improve people's living standards and accelerate the construction of new socialist rural areas.

The policy has had three stages, i.e., experiment stage, promotion stage and nationwide promotion stage. ① Experiment stage: The *Notice of Ministry of Finance on Conducting Pilot Test of Household Appliances to Countryside Project* (MOF, 2007, p. 710) stipulated that from December 2007, Household Appliances to Countryside Project would be experimented in Shandong, Henan and Sichuan province and Qingdao city, providing 13% direct financial subsidy to colored TVs, fridges (freezers) and cell phones, and each household could only buy one piece of each category of household appliances. ② Promotion stage: The *Notice of Ministry of Finance and Ministry of Commerce on Printing and Distributing the Work Plan for Promoting Household Appliances to Countryside* (MOC, 2008, p. 426) stipulated that while continuously conducting the project in above three provinces and one city, 14 provinces (autonomous regions, municipalities directly under central control) and cities specifically designated in the national plan with sound marketing and after-sales service networks and more motivation, including Inner Mongolia, Liaoning, Dalian, Heilongjiang, Anhui, Hubei, Hunan, Guangxi, Chongqing and Shaanxi would be included into the project, and the implementation period is four years. ③ Nationwide enforcement stage: In 2008, the MOF, the Ministry of Commerce and the MOST issued the *Notice on Promoting Household Appliances to Countryside Project in China* (MOF, 2008, p. 862), which stipulated that from February 1, 2009, the project would be promoted in the entire country apart from the 14 provinces and cities. Later, The *Notice on Enhancing the Implementation Efforts for Household Appliances to Countryside Project* (MOF, 2009, p. 48) increased the categories of products to TVs, washing machines, fridges (freezers), cell phones, air-conditioners, computers, water heaters, microwave ovens, electromagnetic ovens, electric bicycles, and each household could buy two pieces of each category of products.

The basic requirement for these household appliances is that they should be energy efficient and environment-friendly. According to investigation in rural markets, energy efficiency grade of fridges and refrigerators should be Grade II of new standard, equal to A+ and A of EU standards; of automatic washing machine should be above Grade II, of twin tube washing machine should be above grade III, which are 2–3 grades higher than the average market level.

After several years of practice, these products have become the farmers' first choice, and helped in improving their living standards. Information management system for household appliances to countryside of the Ministry of Commerce revealed (MOC, 2010) that by December 31st, 2009, altogether 37.6798 million pieces of household appliances have been sold; by the end of the first three quarters of 2010, 52.573 million pieces of household appliances have been sold.

However, higher energy efficiency does not necessarily mean lower energy consumption. With more energy carriers in homes, lifestyles of rural residents and consequent energy demand would rise considerably, which might lead to significant increase in real energy consumption of rural residential buildings.

Therefore, we can obtain the energy-saving capacity for rural areas in 2007 and 2008 deriving from relevant policies (Fig. 9.41). Due to changing lifestyles in rural areas, lower proportion of bio-mass energy and more buildings, it can be estimated that energy consumption of rural residential buildings in 2007 would be 21 Mtce more than that in 2005. While due to four key policy clusters on energy efficiency new-builds, renovation of existing buildings, household appliances to countryside project and renewable energy utilization, 2.7 Mtce could be realized, so the real energy consumption in 2007 would be 18 Mtce more than that in 2005.

With popularity of energy-efficient products and renewable energy utilization, an energy-saving capacity of 2.7 Mtce was achieved in 2007. However, energy demand due to higher living standards, especially the increasing demand for commodity energy, far exceeds the growth rate of renewable energy utilization. Therefore, carbon emission in rural areas has been increasing. Though the growth rate of carbon emission due to energy consumption in rural areas has reduced to 9%, it is still much higher than the growth rate of rural residential buildings (1.6%), which reveals that the function of existing policy clusters in decreasing rural carbon emission cannot match the increasing demand for commercial energy in rural areas.

9.2.6 *Summary*

The Outline of the 11th FYP for Building Sector (MOC, 2006, p. 53) stipulated that 101 Mtce should be cumulatively saved during the *11th FYP*, including 61.5 Mtce deriving from energy-efficient new buildings.

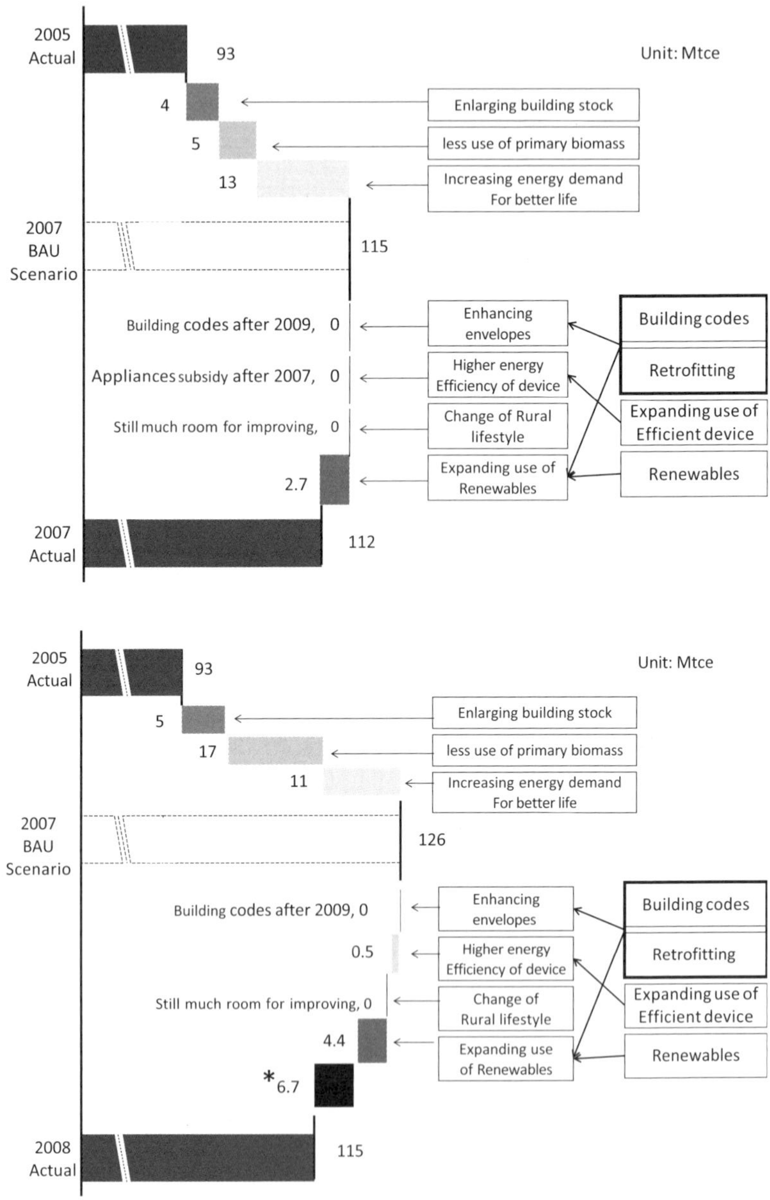

Fig. 9.41. Energy-saving capacity deriving from low-carbon policies for rural residential buildings in 2007 and 2008 (unit: 10,000 tce).
*This part of energy-saving capacity cannot be explained by policy. There might be many reasons for this, for example, statistical error or life patterns irrelevant to policies which cannot be calculated and verified due to lack of statistical data.

1) During the first three years (2006–2008) of the *11th FYP*, energy efficiency efforts in China's building sector developed steadily, all policies had been put into practice, producing 77.16 Mtce energy-saving capacity, and completing 75% of the total energy-saving target.
2) According to the analysis to energy-saving capacity of different policies, the completed energy savings could be decomposed just like Table 9.5 and Fig. 9.42. Energy-saving capacity deriving from new energy efficiency standards for new buildings is 31 Mtce,[56] accounting for 40% of the total capacity; and the energy-saving capacity of new buildings would reach 44.51 Mtce when added up with the energy-saving capacity deriving from other policies,[57] accounting for 70% of the total energy-saving objective.

Therefore, building energy efficiency during the *11th FYP* has progressed smoothly.

9.3 Analysis on Policy Mechanism in Building Sector

9.3.1 *Analysis on environment of building energy policies in the context of urbanization*

As stated in above two sections, major factors for carbon emission in the building sector include size of building stock, lifestyles of people, and efficiency of energy carriers in buildings. As is observed, China's building stock kept increasing in recent years, along with changing lifestyles of citizens and enhancing efficiency of energy carriers. This section will analyze the development of above three major elements and their effect to building energy consumption and carbon emission.

[56] In the report, "energy-saving standards for new buildings" differs with "energy savings of new buildings" in that the latter concept is the combination of the former concept as well as the energy savings due to higher energy efficiency of household appliances, EMC and renewable energy utilization.

[57] It includes: 6.42 Mtce energy-saving capacity for new buildings due to heat supply system innovation, 3.04 Mtce due to renewable energy utilization, 3.08 Mtce for EMC and 970,000 tce for energy efficient products.

Table 9.5. Energy-saving capacity deriving from building energy efficiency policies from 2005 to 2009.

	Different Categories of Buildings-related Energy Consumption				Key Policy Clusters				
Year	District heating system in North China	Urban residential buildings	Commercial buildings	Rural residential buildings	Standards for new buildings	Heating reform	Renewable energy	EMC	Energy efficiency products
2006	582	304	176	118	483	363	284	29	19
2007	1,184	392	681	270	1032	759	500	206	29
2008	2,006	458	1,052	494	1,585	1,204	715	410	96
Sub-total	**3,772**	**1,153**	**1,908**	**882**	**3,100**	**2,326**	**1,500**	**645**	**145**
2009	—	—	1,727	—	—	—	941	723	173

Note: Sub-total line does not include the data of 2009. "–" means that the energy-saving capacity in 2009 could not be calculated due to lack of data.

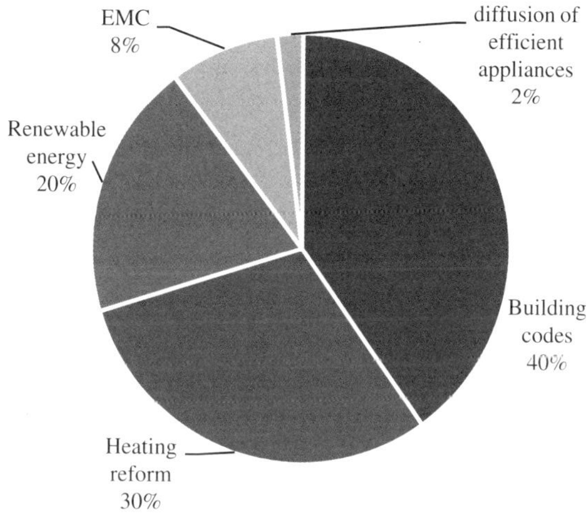

Fig. 9.42. Decomposition of energy-saving capacity in the building sector from 2006 to 2008 by Key Cluster Policies.

(1) *Penetrating building stock due to rapid urbanization results in hiking energy use in buildings*

China's urbanization rate exceeded 30% in 1995, and the growth rate was kept over 1% afterward. As a result, over 10 million farmers were migrating into cities each year, leading to sharp increase of demand for housing, education, medical health and other public facilities and so on. Yet notably, the growth rate of urban building stock exceeded the population growth rate. Urban population in 2008 was 60% more than that in 1996, but the total floor areas in urban areas increased by more than two folds, which was regarded as among the most significant attributions for hiking building energy consumption in China. To specify, the energy consumption intensity per unit floor area in 2008 was 1.4 times more than that in 1996, while the overall building energy consumption in 2008 was 2.4 times more than that in 1996.

Table 9.6 is a comparison of residential floor areas per capita between China and developed countries in 2005. According to the table, China's residential floor area per capita is lower than that of developed countries. The *Outline of the 11th Five Years Plan for Building Sector* (MOC, 2006, p. 53) stipulated to maintain a steady growth of building stock in both urban and rural areas, optimize the supply of residential buildings, and, as working focus, provide enough residential buildings for middle-and-low

Table 9.6. Comparison of residential floor areas per capita between china and developed countries in 2005 (unit: m^2/person).

Item	China's average	China's urban areas	China's rural areas	USA	Germany	French	Japan	South Korea
Floor area per capita	25	20	29	63	42	50	34	28

Source: ① SSB, *China Statistics Yearbook 2006* (China Statistics Press, Beijing, 2006); ② Q. Wang, energy data in *China Sustainable Energy Project Reference Material 2009* (Energy Foundation, Beijing, 2009).

income residents. According to the *11th FYP*, 500–600 million m^2 new urban residential buildings would be built each year. Yet, from 2005 to 2008, altogether 3 billion m^2 urban buildings are built, including 1.6 billion m^2 residential ones. Under such high growth rate, China's total urban floor areas would increase by 10 billion m^2 by 2020, reaching 30 billion m^2, about 1.5 times more than that in 2008, and the urban floor area per capita would be 36 m^2 per capita, which is still lower than that of developed countries.

Therefore, China's rapid urbanization will last for a certain period of time, which would further increase energy demand in the building sector. Even if the energy intensity remains the same, the overall building energy consumption would increase accordingly if the total building stock doubled.

According to a scenario research (Yang, 2010), the total floor area in 2030 under two scenarios of "Business as usual (BAU)" and "strict control" would be 61.5 billion m^2 and 46.8 billion m^2 respectively (see Table 9.7). Even if the energy intensity remains at a low level, energy consumption of the "BAU" scenario would be 157 Mtce more than that of the "strict control" scenario, which is about 1/3 of the overall building energy consumption in 2008.

Besides operation energy consumption in buildings, burdens would be load upon energy use for the construction sector and complexities of land-use in cities as the building stock enlarges. It is estimated (Li and Jiang, 2006) that between 1998 and 2003, energy consumption in building materials production and in the construction process was about 1/4 of the overall energy consumption in China. Therefore, exercising a reasonable control over the total floor areas, especially the growth of urban building stock, is critical for creating an energy-saving society and realizing the objective of energy conservation and emission reduction.

Table 9.7. Different scenarios for building floor areas in 2030.

Scenarios	Status	Urban residential floor area per capita (m^2/person)	Rural residential floor area per capita (m^2/person)	Commercial floor area per capita (m^2/person)
Present	29 m^2/person, lower than developed countries	18	31	9
BAU	Urban floor area per capita equal to Japan (45 m^2/person, 2004), 35 m^2/person for rural areas; lower than that of Europe and the USA in 2004	30	35	15
Strict control	Higher than present state, lower than that of developed countries in 2004	20	30	12

(2) *Pursuing luxurious lifestyles is a major cause for growing building energy consumption in last decade*

By comparison, it can be noticed that China's building energy consumption per capita is much lower than that of developed countries. According to the difference in ways of energy consuming: ①At present, China's urban buildings can basically meet the demand of urban residents under current lifestyles, while rural buildings could not. ②Different from foreign countries (Zhang *et al.*, 2009), Chinese prefer utilizing the natural environment in prior to meet their daily requirement, such as natural lighting and drying clothes by sunlight and winds; while Westerners choose mechanic methods over natural-prior ones to get services needed within a building, such as strict control over ventilation volume by mechanic ventilation systems, constant temperature and humidity by sophisticated air-conditioning system, and drying clothes with electric dryers.

And lifestyle of Chinese keeps changing rapidly and gradually, with glorious economic success they achieved and largely facilitated cultural communications by advancing technologies with the left world taking place. On one hand, positively, living conditions are enhanced to some extent, especially in rural households and some of urban families such as longer time for heating and cooling, better domestic hot water supply and more convenient usage of household appliances. At present, China's energy consumption in urban households is much lower than that in developed countries, and improving the residents' living standards is an important objective for building a harmonious society. Therefore, changing life styles is inevitable and acceptable. But on the other hand, some residents took a Western ways of life over traditional ones. They pursue a "high-class and luxurious" life style, living a "fashionable and world-class" life, resulting in much higher energy use. And the number of such citizens kept increasing along with burgeoning economy of China, which has become a major cause for growth of building energy consumption in China's urban areas, and has to be dealt with directly.

Fast changing of lifestyles are taking place now within China, which inevitably brings growth of building energy consumption and carbon emission. Relevant research (Yang, 2010) conducted a scenario analysis, considering potential distributions of different lifestyles (see Table 9.8). According to the research, when the proportion of population group of high energy consumption rises, energy consumption of urban residential buildings would increase by 100 Mtce, which is about 2/3 of the energy consumption of urban residential buildings in 2008.

At present, developed countries also pay great attention to the impact of life styles on building energy consumption, and advocate their people to

Table 9.8. Scenarios for proportion of population groups in urban residential buildings (excluding district heating system) (unit: %).

	Population Groups		
Serial	Low energy consumption	Middle energy consumption	High energy consumption
Scenario I	20	60	20
Scenario II	20	30	50

Note: Population group of "low energy consumption" refers to those maintaining a natural-oriented way of life, which assures basically comfort and consumes as much energy as the average level of China in 2008; population group of "middle energy consumption" refers to those improving living standard moderately by technological advancement, and thus using slightly more energy on the basis of 2008; population group of "high energy consumption" refers to those taking Western ways of life and using as much energy as Japanese do in 2006.

change life styles for saving energy. For example, LOHAS became popular in developed countries, which calls for natural way of living. For instance, it encourages people replacing mechanic ventilation with natural one, and reduces use of air-conditioning systems by rising room temperature slightly in summers and lowering it in winters. These concepts coincide with Chinese traditional ways of life. Advocating natural and harmonious development, promoting energy-saving life styles, maintaining Chinese traditional living habits and propensity to energy-saving behaviors, and avoiding transforming into an energy-intensive society are important measures for energy conservation and emission reduction of buildings.

(3) Focusing on energy efficiency and cultural atmosphere, dealing with the relationships between technical progress and lifestyles properly

On one hand, technical progress improves energy efficiency of energy carriers, which is conducive to less energy use in buildings. Adopting proper technologies according to local conditions in buildings are important energy-saving measures. On the other hand, culture plays a covert role in changing people's life styles, affecting their acceptance and application of technology. Moreover, it is a fact that proper applications of certain technologies are strictly constrained according to features of buildings and local geographic conditions, as well as lifestyles of residents in buildings. In a word, people decide what kind of life they prefer and then choose different technologies to fulfill requirements of their preferred life in buildings they live in.

China is a large country, covering several climate zones. As a result, building types and lifestyles of different localities and nationalities vary greatly with developed countries. According to practice, some technologies of developed counties might not be applicable to China. For example, by using household central air-conditioning system people could maintain constant temperature and humidity and ventilation volume in quite an efficient way. As a result, it is popular among USA people who want such an isolated indoor environment. But when one want keep more frequent contact with the outdoor environment, just like most Chinese preferring, the system becomes less efficient and not flexible to the demand. Turning to a split-up type of air-conditioner would benefit both in comfort and energy, which, as is in China, is only about 1/5–1/3 of the centralized systems in USA. If the household central air-conditioning system is introduced unrestrictedly into China, it will first force Chinese people to live a life that the technology requires and then increase the energy consumption enormously.

On one hand, technology is serving the people. Energy-saving technologies should be developed by taking local residents' demand into consideration. On the other hand, technology is changing our life, and the key problem is how to respond. From the perspective of energy conservation, developing harmonious life styles and introducing them to people by cultural promotion and technical influence is the long-term strategy for building an energy-saving society as well as reducing energy consumption and CO_2 emission.

Taking China's national characteristics into consideration, under the prerequisite that building energy consumption stabilizes or increases only slightly, the best way for building energy efficiency might be, adopting proper building energy efficiency policies, developing suitable technical methods, encouraging a harmonious life style, improving the interior environmental quality of buildings, and improving people's living standards.

9.3.2 *Analysis on heating reform policy*

Heating reform policy refers to: (1) "stop free heat supply and realize commercialization and monetization of it", which was almost done. People accepted to pay for heating supply and were able to recognize how they received subsidies as they using heating in winters. (2) "Applying heat metering and charging system gradually, and promote energy-saving renovation of existing urban residential buildings and heating system". There are some problems with respect to the second aspect. First, disputes emerge on the metering and charging system, which obstacles to significant extent its implementation. In fact, heating fee was charged according to size of flats people live in, instead of the heat people really used for heating purposes. Second, heat supply enterprises are fighting for heat sources. Third, people are not motivated for renovation of their buildings.

"Applying heat metering and charging system gradually" indicates transformation of charging methods from the one according to floor areas people live in, to the one according to amount of heat people use for heating. Yet, Charging by amount of heat people use for heating, or by real heat consumption for short, is still vague and somewhat perplexing. Water and electricity consumption of each household is separate and not subject to the interference of other households. However, considering the heat transmission features, heat metering can only be measured building by building, and then divided into each household according to a certain proportion. There are two ways of charge division. First is to divide into each household by floor areas. Second is the so-called metering by household, i.e., dividing heat charges of the entire building into two parts, one part is divided into each household according to its proportion to the total floor area, taking into consideration of the heat transmission among households, another part is divided into each household according to its proportion to the overall heat supply, considering the difference of heat consumption among households. It is worth to mention that although the first method also divides charges by floor areas, it is not the same with existing method of charging by floor areas, which charges by a fixed heat price per unit floor area and does not reflect the heat-insulating performance of the buildings. While the heat price per unit floor area of the first method is variable, it is related to the heat consumption of the entire building, and reflects its heat-insulating performance. The second method also differs from the two-fold heat price system. In the two-fold heat price system, heat price by fixed floor areas could not reflect the heat-insulating performance of different buildings, and the relations between the combination of two-fold heat prices of each household and heat prices of the whole building are ambiguous.

The biggest misunderstanding in heat supply system reform is that "charging by floor areas" is wasteful and not good for energy conservation, so it needs to be replaced by "charging by real heat consumption". "Charging by real heat consumption" encourages energy conservation, reduces energy consumption and increases energy efficiency, but we cannot say "charging by floor areas" leads to waste or inefficiency. For example, heat metering by household could reduce energy consumption, but requires meter installation, maintenance, meter reading and price counting. If the expenses of above additional items are higher than the benefit of saved energy, then it is uneconomical, and "charging by real heat consumption" is wasteful.

Efficiency of the heat metering and charging system attributes to energy-saving benefits as well as costs, whose change reflects the change of restrictive conditions. Whether a select system is efficient or not is subject to certain restrictive conditions, whose change could bring change to existing systems. Essence of heat supply metering system reform is that changing restrictive conditions leads to the necessity of changing heat supply system. Causes for adopting a certain charging method have to be made clear before tapping into its restrictive conditions.

As stated above, essence of heat metering and charging system is the division of heat charges. There are three kinds of charge division method

Table 9.9. Compatible conditions for three heat charge division methods.

Item	No metering	Metering by building	Metering by household
Charging method	Fixed heat price	Dividing heat charges of the building by floor areas of each household	Dividing heat charges of the building by floor areas and heat consumption of each household
Contents of metering	Household floor areas	Number of buildings and household floor areas	Number of buildings, household floor areas and heat volume
Metering expenses	Nil	low (installing, maintaining and reading the building heat meters)	high (installing, maintaining and reading building and household-based heat meters)
Compatible conditions	Same thermal conditions of building envelop, identical room temperature control, same vacancy rate	Different thermal conditions of building envelop, identical room temperature control, same vacancy rate	Different thermal conditions of building envelop, different room temperature control, different vacancy rate

(Table 9.9), first is to divide the heat charge of the entire heating area by floor areas; second is to conduct heat metering by building, and then divide the heat charges by floor areas; third is to conduct heat metering by household, that is to say, divide the heat charges by floor areas as well as by real heat consumption. It is apparent that the first method has the least metering expense, the second method needs to conduct heat metering to the entire building, and the third method needs to conduct heat metering by household, incurring large metering expenses (meter installation, maintenance, meter reading, and price counting). More detailed metering had more encouragement to energy-saving behavior of the users.

For example, when a heat supply enterprise supplies heat to a residential community, the compatible conditions for the first charging method is that the heat demand of building envelops in the community is similar, and heat supply per unit floor area for different households is also similar. There is a fixed heat price for per square meters, and it only needs to charge by floor areas of each household. The compatible conditions for the second method is that the heat demand of building envelops in the community is different, but the demand for heating temperature of each household is similar. Then, the total heat charges should be divided by building and then the heat charges of a building should be divided into each household by its floor areas. The compatible conditions for the third method is that heat demand per unit floor area of different households varies greatly, including different heat demand due to different household locations, different room temperatures and different vacancy rates. Then, the total heat charge of the entire community should be divided by building according to building-based metering, which in turn should be divided into two parts: One part is divided into each household by its proportion to total floor area of the building, while the other part is divided into each household by its proportion to total heat consumption of the building (Table 9.9). Then, why is the necessity for reforming the heat supply charging system?

The widely adopted "charging by floor areas" system has a fixed heat price set up by the government. In June 3rd, 2007, the NDRC and the MOC jointly issued the *Notice on Printing and Distributing the Tentative Regulations on Urban Heat Supply Prices* (NDRC, 2007, p. 1195), which stipulated that "heat price is fixed or guided by the government in principle, which should be decided by competent pricing departments of provincial (district, city) governments or by delegated municipal and county governments". For example, in Beijing the standard for natural gas-fired heat supply is 30 yuan/m^2, for coal-fired heat supply is 24 yuan/m^2. However, with higher implementation rate of building energy consumption standards, the difference between the heat demands of new building envelops and that of old ones becomes larger. When the Ministry of Construction conducted energy-saving inspection to new buildings in 2004, the implementation rate for large cities such as Beijing and Tianjin was only about 30%–40%; but with urbanization and more new buildings, by 2009, the rate has reached 90% for Beijing and Tianjin. Collecting identical heat charges per unit floor area from residential buildings saving 50%–65% energy and from energy-intensive buildings would bring into the following four problems. First, heat supply enterprises will fight for new energy-efficient residential communities and give a cold shoulder to old ones, and their competition changes from vying for higher efficiency to fighting for good communities. Second, with the same heat price, the local governments have to provide heat supply subsidy to old residential communities, but could not collect the energy-saving benefits from the heat supply enterprises supplying heat to energy-saving residential communities. Third, the users or owners are not motivated to conduct energy-saving renovation to building envelops due that their investment brings no direct economic gains. Lastly, it provides misleading information on heating cost to the users.[58]

Heat metering and charging by building is the basis for heat metering by household. At present, heat metering by household has some problems, the adaptability, accuracy and endurance of various heat meters needs further improvement, and corresponding regulations and standards for meter installation, maintenance, reading and price counting need to be formulated. China's heat metering by household is lagging far behind Germany. The essence or starting point for heat metering and charging regulations of Germany is that the accreted expenses of heat metering and charging should not be larger than the energy-saving benefits. One of the most important difficulties for China's metering by household is that the energy-saving benefits cannot match the metering expenses. Therefore, metering by household should be advanced gradually. With more complete standards and regulations, lower metering expenses, and higher price of coal and natural gas, it would become easier to carry out metering by household.

It should be noted that metering by building and metering by household deals with different problems. Metering by building could promote energy-saving renovation of building envelops, while metering by household, despite it could also serve above purpose, also aims to meet the requirements for different room temperatures by different users. However, China's energy source per capita is very limited, so such requirements might not be advocated. The other purpose for metering by households is to differentiate the energy consumption of different households, and another purpose is to advocate energy conservation, such as reducing heat supply to vacant buildings, which is meaningful to those new residential communities with higher vacancy rate.

"Metering by heat consumption" vigorously promoted at present mainly adopts two-fold heat price system, which collects heat charges by household,

---

[58]In a community which was originally in the municipal district heating network, an EPC heat supply company promised that after heat supply renovation, heat price/m^2 would be reduced to over RMB 10, about half of the municipal heat price 30 yuan/m^2, and the residents felt satisfied. However, the real cost of heat supply was only more than RMB 10 presupposed that the community supplied heat by itself and without EPC service. This revealed that for some energy-saving residential buildings, government-guided heat price differs greatly with real heat supply cost.

i.e., paying part of the heat charges by household floor areas (18 yuan/m^2 for Beijing), and paying rest of the heat charges by heat consumption of the household (44.4 yuan/GJ). The purpose is to motivate the users' energy-saving enthusiasm by heat metering by household. While on the other hand, considering heat transmission among households, it also needs to collect part of the heat charges by floor areas.

However, two-fold heat price system has several problems. First is that the fixed price for metering by floor areas could not reflect the energy-saving performance of buildings; second is that heat charges paid by the households in one building by two-fold heat price system usually differ from the total heat charges of the building by metering by building. Therefore, two fold heat price system could only be applied to buildings instead of households. In order to avoid the effect of heat transmission and promote energy-saving behaviors of the users, it is suitable to divide the heat charges by building to all households in a building, i.e., dividing the charges by floor areas and by heat consumption, which is also a two-fold system. However, it is different from the previous one.

Linking heat charges of a building to its heat consumption is the best way to promote energy efficiency. There is no heat transmission within buildings; therefore, charging by building is the essence of heat supply charging system reform. At present, charging by household is the focus in the heat metering and charging system, which ignores the essence of heat metering, and is the key problem for heat supply charging system reform.

At present, all household-based heat meters (including heat meters, heat allocation meters and on-off heat meters) need the support of building-based heat meters. And it is very convenient to charge by building and then divide the heat charges by floor areas and heat consumption. However, it is seldom adopted. The reasons might be that charging by floor areas in existing two-fold heat price system enjoys cross subsidies, charging by floor areas with the same price is favorable to nonenergy-saving buildings, whose residents are mainly low-income population groups. Therefore, charging by floor areas or by existing two-fold heat price system can be treated as a subsidizing method to low income groups.

With rapid development of the real estate industry, value of the old residential communities in downtown area has been increasing, so is the wealth of the residents, which might reduce the resistance of low-income groups to heat supply price system reform. Meanwhile, outward appearance of the old buildings could be renovated, creating a larger space for appreciation, and giving more momentum to energy-saving renovation.

Another important problem in charging by heat metering relates to the formation mechanism of heat charges, i.e., what kinds of operational expenses could be included into the heat charges and how to measure them.

Germany's heat metering system started early, and its relevant laws and regulations are very complicated in transforming from charging by floor areas to charging by heat consumption. However, Germany's charging by floor areas is to make out the operational expenses of a building, and then dividing the operational expenses into each household by floor areas. Therefore, the operational expenses are variable, which changes with heat supply volume and energy price. Germany's heat supply regulations center on two aspects, heat charge division and regulations on heat supply operational expenses. Now, China pays attention to heat metering regulations, but overlooks the regulations on heat supply operational expenses.

Germany's regulations on heat supply operational expenses clarify what expenses could be counted as operational expenses. Of course, the most important element is energy cost, which changes with price fluctuation. For example, stored energy is bought in installments, the amount and price of each installment is different, then the energy cost will be decided by FIFO principle. Germany's residential communities mainly utilize oil burning boilers, and the price of burning oils is geared to international standards and is fluctuating. Therefore, there are detailed regulations as to the operational expenses of heat supply. China's heat supply energy price is regulated by the government. For example, before commercialization of coal market, coal price was fixed by the government, and the price of natural gas is still fixed by the government right now. Therefore, China's regulations on operational expenses are not as detailed as that of Germany. In recent years, coal price fluctuates greatly due to commercialization, while natural gas price changes slightly. It can be predicted that heat supply costs of coal-fired heat supply enterprises is more volatile than that of natural gas-fired enterprises. With increasingly higher coal price, government-guided heat price of coal-fired heat supply enterprises is more flexible.

China's charging by floor areas system is to multiply the floor area with a fixed heat price, that is to say, the heat charge for a household is fixed. Germany's charging by floor areas system is to divide the heat charges of the entire building to each household by floor areas, that is to say, the heat charge for a household is variable. Changing heat charge reflects the change of heat supply operational expenses. In the preliminary stage of heat supply system reform, energy price was regulated by the government, and the thermal conditions of residential building envelops were almost identical, then it was convenient and feasible to charge by a fixed heat price. However, with coal market reform and fluctuating coal price, heat supply costs also changes day to day. Therefore, a fixed heat price could not reflect the fluctuating heat energy price, and could not reflect the operational performance of heat supply enterprises.

To sum up, there are mainly two problems in China's district heating system:

(1) Existing heat price does not take into account the operation fee for heat supply, and thus could not adapt to the fluctuating energy price, and could not motivate the heat supply enterprises to improve the networks and their operation scheme.
(2) Existing two-fold heat pricing system measures heat consumption by households, which could not reflect well envelop levels of different buildings. In addition, its cost is relatively high. As a result, it could not motivate terminal heat users effectively to renovate their systems and envelopes.

9.3.3 *Analysis on EPC policy in the building sector*

EPC system covers every aspect of building renovation and routine energy consumption; therefore its function will be affected inevitably by relevant policies. In this sense, EPC system has a clear objective, and cooperates with other policies for commercial buildings.

Thus, analyzing EPC system is favorable to appreciate the policy system for commercial buildings in depth.

Operational process of EPC system is listed in Fig. 9.43. After confirming project implementation in the project feasibility analysis phase, a qualified third party will conduct energy auditing to the energy status of the consumption unit, including its energy consumption as well as energy-consuming devices of and in the buildings. Based upon above information, the energy service company will bring forward a comprehensive energy renovation plan to realize certain energy savings. Then, the energy service company shall sign a contract with the client, and begin to carry out the renovation plan. After completion, a qualified third party will conduct energy auditing to the renovated buildings, verifying the real energy savings after renovation against the estimated energy consumption of the current

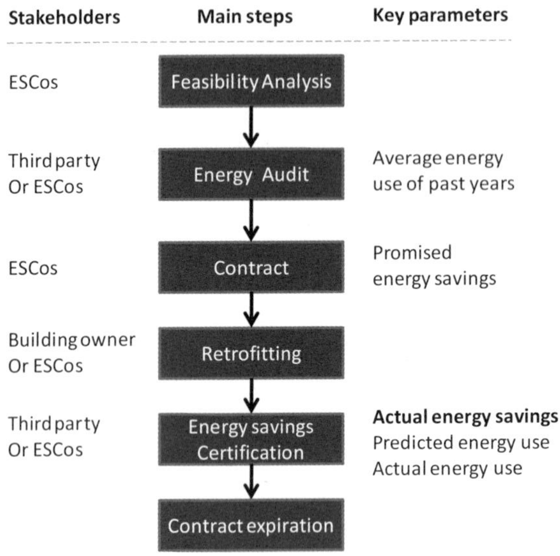

Fig. 9.43. Main process of EPC.

year, and obtain the real energy savings of the current year (real energy savings of the current year = estimated energy consumption — real energy consumption). By comparing the real energy savings with the estimated energy savings in the contract, the client pays to the energy service company till expiration of the contract.

According to Fig. 9.43, it can be find out that ① EPC centers on contractual implementation; ② real energy savings (estimated energy consumption — real energy consumption) is the key for incurred expenses during EPC, which will be inspected and obtained by a qualified third party; ③ incurred expenses during project implementation consist of the consultation expenses of each phase, equipment renovation expenses, and consultation fees for optimal management plan; ④ subsidies for expenses would be decided by relevant national policies. Now, the four aspects will be analyzed in the following paragraphs.

(1) *Shared savings contract is more adaptable to fluctuating building energy consumption*

There are mainly three business models for EPC, i.e., guaranteed savings contract, shared savings contract and energy cost trusteeship contract. Due that estimated energy consumption differs from real energy consumption, three types of contracts have different compatible conditions in the building sector.

Comparison of three business models is in Table 9.10 and Table 9.11. Uncontrollable factors such as weather and personnel have enormous effect on building energy consumption, leading to very volatile changes. Therefore, securing a fixed amount of energy savings is more risky than in the industrial sector, while shared savings contract is conducive to reduce risks of both parties, and is easier to be promoted. According to the *Tentative Regulations on Management of EPC Finance Incentive Fund* (MOF, 2010, p. 249), China's financial subsidy mainly focuses on shared savings contract. From this perspective, existing policies are favorable to the development of EPC in the building sector.

(2) *Lacking energy-saving auditing standards leads to indefinite contractual energy savings*

China's *General Technical Rules on EPC* makes detailed regulations on EPC procedures, but energy savings auditing standards in the building sector is still lacking. Usually, the estimated as well as the real energy consumption of the current year is audited by the energy service companies or by both the companies and the clients. However, this might lead to two serious problems. First is that it is difficult for both parties to agree on real energy savings, especially for commercial buildings. Through energy-saving renovation and optimization offered by the energy service companies, energy consumption of the centralized systems such as air-conditioning, heating, lighting, and hot water system could be reduced effectively; while for the decentralized equipments such as computers, copiers, printers, and water dispensers, energy-saving behaviors could have an obvious effect. In addition, the primary energy consumption for commercial buildings is air-conditioning and heating, which is subject to the change of weather conditions. Without relevant standards, it is difficult to calculate the estimated energy consumption of the current year, and both parties might have different opinions for energy savings, leading to project delay and adverse conditions e to the development of energy service industry. Second is that the energy service companies and the clients might negotiate on energy savings to obtain more national subsidies, which is a waste of limited national resources.

Table 9.10. Comparison of EPC models.

Types of contracts	Main obligations	Implementation		
		Real energy savings $R < E$	Real energy savings $R = E$	Real energy savings $R > E$
Guaranteed savings contract	In realizing estimated energy savings (E), the client pays bill (B) to the company, and energy-saving cost (C) is afforded by the client	Client affords $B + C$, gets E; company realizes $(E - R)$, gets B	Client affords $B + C$, gets E; company gets B	Client affords $D + B + C$, gets R; company realizes $D + B$, while $D = (R - E)$
Shared savings contract	In realizing E, company shares proportion (P), while C is afforded by the company	Client gets EP, company affords $(E - R)P + C$, gets R(1-P)	Client gets EP, company affords C, gets E(1-P)	Client gets RP, company affords C, gets R(1-P)
Outsourcing of energy system management	Client pays B to company (equivalent to E), company affords C	Client pays B, gets E; company affords $C + (E - R)$, gets B	Client pays B, gets E; company affords C, gets B	Client pays B, gets E; company affords C, gets $B + (R - E)$

Note: "company" refers to the energy service company.
Source: W. Yong, L. Chanqing, L. Yingzong *et al.*, *Research on the Innovation of Building Energy Conservation Management System in China* (China Architecture and Building Press, Beijing, 2007).

Table 9.11. Compatible conditions of major EPC models.

Contract	Guaranteed savings	Shared savings	Energy cost trusteeship
Renovation cost affording party	Client	Energy service company	Energy service company
Client's risk	Large	Small*	Nil
Company's risk	Small	Small	All
Adaptable conditions	Steady energy consumption	Fluctuating energy consumption	Both

*From the perspective of clients, their only return of investment is the energy savings, so they have a large risk; however, their investment is decided by the estimated gains, which reduced their risk enormously.

Therefore, formulating relevant standards enables the energy service industry developing rapidly and healthily in the building sector. Meanwhile, it needs to formulate an auxiliary qualification certification system. When the clients and service companies have disputes, a qualified third party could be authorized to deal with these problems.

(3) *Existing expense system could not weigh up the service value of energy service companies correctly*

Expenses incurred during project implementation usually consist of equipment replacement expenses and consultation fees for optimal management plans.

From the perspective of energy consumption, many cases of energy efficiency renovation of commercial building revealed that, energy savings of equipment replacement is less than that of optimal management plans. While from the perspective of invested fixed asset, cost of equipment replacement is much higher than that of optimal management plans. Therefore, paying consultation fees to optimal management plans constitute a more cost-effective option.

However, from perspective of the clients, equipment replacement brings more obvious changes than the optimal management plans. On the other hand, these plans link closely with the clients' daily behaviors, making the clients feel that the energy savings are the result of their own behaviors instead of deriving from the consultations of energy service companies.

This might lead to two consequences. First, the clients only care about the energy-saving effect of equipment replacement paid by themselves;

second, the energy service companies might increase the proportion of equipment replacement budget to energy-saving effect so that the clients might accept it more easily. In addition, if there are relatively more equipment replacement expenses in the project, it is more convenient to conduct project auditing and financing, and is easier to get tax credit discount. Therefore, equipment replacement expenses usually take a major part. These phenomena inevitably lead to an illusion that "energy savings derive from equipment replacement rather than higher operational management levels", resulting in unreasonable expansion of energy-saving equipment market and collusion of energy service companies with equipment suppliers. While the cost-effective energy-saving operational management, which is the most valuable service that could be provided by the energy service companies, does not have its rightful place.

(4) *High threshold for the subsidy policy, which is unsuitable to the decentralized building energy service industry*

The Tentative Regulations on Management of Financial Subsidies of EPC projects stipulated that only those companies with a registered capital of RMB 5 million could apply for financial subsidies, which is suitable for the scale economy such as the industrial sector, but not for building sector, especially for the commercial buildings.

Buildings are products with their own features, whose structures, systems, equipment and operation vary with each other, leading to different energy-consuming characteristics. Therefore, energy-saving renovation of different buildings should also have different priorities. To some extent, the relationship between energy service companies and commercial buildings owners resembles that of doctors and patients in that the doctors should prescribe medicine to the patients according to their specific symptoms. From this perspective, energy consumption of commercial buildings differs from that of industrial enterprises in that it is difficult to realize scale economy and modularity. Taking the enormous numbers of commercial buildings into consideration, the energy service industry in the building sector should mainly consist of small service companies with their own characteristics. Then, for most energy service companies in the building sector, the threshold of RMB 5 million makes them unable to obtain the financial subsidies. In addition, EPC projects put up a high demand for the financing capability and financing channels of the energy service companies, while most small companies usually "do not possess sufficient circulating funds, unable to provide enough credit guarantee or mortgage to the banks, therefore it is very difficult for them to apply for a large sum of loans from the banks" (Chen and Chen, 2009). This financing obstacle is unfavorable to the fast and healthy development of the energy service industry in the building sector.

It also needs to be noticed that too low threshold in the energy service industry also leads to problems (Zhang *et al.*, 2009), especially in that some high-risk energy service companies take great efforts to conceal the risks from the clients, leading to misunderstanding and unnecessary losses.

Therefore, apart from providing subsidies according to energy savings, the government could also consider establishing a special energy-saving fund to provide guarantee to the project financing of energy service companies, reduce the financing difficulties in the preliminary stage of EPC projects (Zhou, 2009), and encourage their development in the building sector. Relevant analysis has introduced in Sec. 6.2.1.

To sum up, although the government has formulated subsidy policies to offer support to the shared savings contracts in the building sector and support the development of energy service companies, the energy service industry still develops slowly. This is primarily due to lack of relevant energy auditing standards and third-party energy auditing institutes and systems. Therefore, energy savings, the key parameter for contract execution, could not be verified by competent authorities. On one hand, energy savings in EPC is not operable, leading to contractual disputes. On the other hand, the energy-saving effects of equipment replacement and optimal management could not be differentiated, the clients could not get a clear idea of the return of investment, and consultation of the energy service companies could not be evaluated properly, which might lead to collusion of the companies and equipment suppliers as well as ineffective project financing. In addition, too high subsidy threshold does not adapt to the characteristics of the building energy service industry, and it is difficult for the service companies to raise finance.

Therefore, it is imperative to formulate energy savings auditing standards, so as to create a benign market environment, and promote the effective and healthy development of the energy service industry in the building sector.

(Special thanks to Cai Hong Wu of Beijing Sanbaili Energy Conservation Co., Ltd., who helped a lot during compilation of this book.)

CHAPTER 10

Transportation

10.1 Performance of Energy Consumption and Carbon Emission in Transportation Sector

The production activity in transportation sector, the tertiary industry in national economy that covers industries engaged in transporting passengers and freight through railways, highways, waterways, civil aviation and pipes, is to realize the displacement of people and objects. In the year 2008, the energy consumption in transportation sector is 290 Mtce, accounting for 10.4% of total energy consumption and 15.9% of end-use energy consumption. The emission of CO_2 is 630 million tons, accounting for 9.9% of the total end-use emission.

Energy consumption and carbon emissions in transportation sector rose respectively by 14.1% and 13.3% in the first three years of the 11th Five Year Plan (FYP) period, which is closely related to the growing passenger and freight traffic turnover volume and structural changes of transport mode. However, the growth rate of energy consumption and carbon emissions in transportation sector was remarkably slowed down during this period. During the 11th FYP period, only the energy consumption per transport-unit[1] in water transport, diesel locomotive in railway and gasoline freight vehicle in highway was increased, while the energy consumption per transport-unit in other modes was decreased or kept unchanged comparing with the 2005 level. Thus significant energy savings were brought during the first three years of the 11th FYP, in which energy saved cumulatively by railway and aviation was respectively 5.83 Mtce and 2.1 Mtce; fuel consumption per hundred kilometres of light passenger vehicles of road traffic, influenced by fuel economy policy, constantly fell off, saving 1.24 Mtce.

[1]The energy consumption per transport-unit in this report refers to unit turnover traffic volume energy consumption.

10.1.1 *Though energy consumption and carbon emission were raised in transportation sector, the growth rate was slowed down*

Figure 10.1 shows that the energy consumption in transportation sector has been continuously increasing with an average annual growth rate from 2000

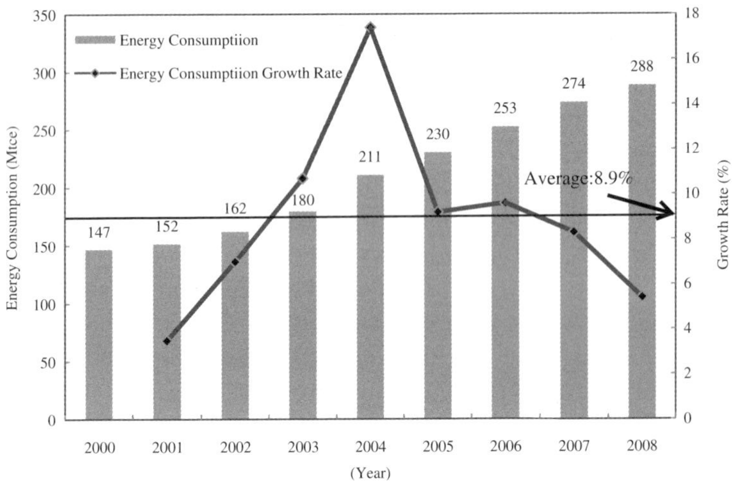

Fig. 10.1. Total energy consumption and its growth rate in transportation sector. *Note*: The definition and algorism of total energy consumption can be seen in footnote.[2] The growth rate is link relative ratio, i.e., using the previous year as the base year. *Data Source*: Energy Statistics of National Bureau of Statistics, *China's Energy Statistics Yearbook 2009* (China Statistical Publishing House, Beijing, 2010).

[2]Calculation method: China's energy balance sheets in China's Energy Statistics Yearbook classified transportation, storage and post and telecommunications into transportation sector. (A) in all sectors, manufacture, construction (except for personal consumption), 95% of gasoline consumption and 35% of diesel oil consumption are classified into transportation sector; all gasoline consumption and 95% of diesel consumption in construction are classified into transportation sector; all gasoline consumption and 25% of diesel consumption in agriculture are classified into transportation sector; besides, 15% of electricity consumption in transportation sector is classified into construction sector. (B)The total terminal energy consumption is calculated by electrothermal equivalent of fuels in transportation sectors; (C) The summed transportation fuels include: coal, gas, coal total, coke, other gas, crude oil, gasoline, kerosene, diesel, fuel oil, liquefied petroleum, natural gas, heat, electricity and others. (D) carbon consumption in transportation does not include fuel charge of external boats and machines or oversea fuel charge of Chinese boats and machines.

to 2008 being 8.9%, and the total consumption reached to about 290 Mtce in 2008, which increased by 25.2% and 96.6% comparing with the year of 2005 and 2000, respectively.

The energy consumption in China's transport sector continues to grow, but the growth rate has been slowing down since 2004 and the trend was particularly evident during the 11th FYP period, as shown in Fig. 10.1.

The overall trend of total emission in transport sector in China is similar to that of energy consumption. As shown in Fig. 10.2, from 2000 to 2008, carbon emissions were increased year by year with an average annual growth rate of 8.6%, reaching to 630 million tons in 2008. Carbon emissions in transportation in 2008 were increased by 24.2% comparing with the year of 2005 and by 93.2% comparing with the year of 2000. However, as China adopted a series of energy saving policies and measures, the growth rate of carbon emission was slowed down, especially significant since the 11th FYP. The rapid development of China's economy, improvement of people's living standards and the development of motorization process lead to the sharp increase in carbon emissions. While urbanization and urban modernization are advancing, to advocate low-carbon and energy-saving mode of

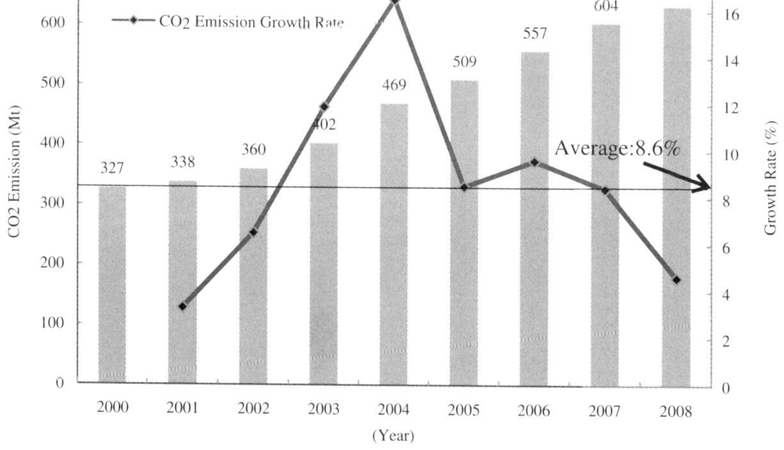

Fig. 10.2. CO_2 emission and its growth rate in transportation sector.
Note: Carbon emission is obtained by fuels (including power) consumed by transportation sector multiplying by carbon emission factor. The growth rate is link relative ratio.
Data Source: Energy Statistics of National Bureau of Statistics, *China's Energy Statistics Yearbook 2009* (China Statistical Publishing House, Beijing, 2010).

transport, such as Green Road, Green commuting, and Green living, will exert a significant influence on the future.

10.1.2 *Energy consumption intensity in transportation sector is reduced and energy saving effect is significant*

(1) *The growth rate of energy consumption and carbon emission in transportation is lower than that of production in transportation*

Though carbon consumption and carbon emission in transportation are increased during the 11th FYP period, the growth rate is slowed down, even slower than the growth rate of passenger and freight turnover volume of traffic. As shown in Fig. 10.3, carbon consumption and emission were increased respectively by an average annual growth rate of 7.8% and 7.5% from 2005 to 2008, while passenger and freight turnover volume of traffic were grown respectively by 10.0% and 23.9% during the same period. Besides, the growth rate of carbon consumption and emission in 2008 were

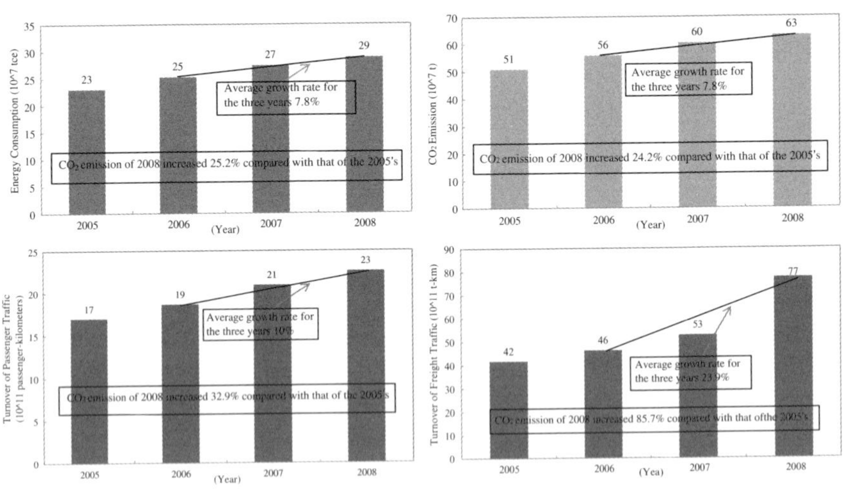

Fig. 10.3. Energy consumption, carbon emission and turnover volume of passenger and freight traffic in communications and transportation from 2005 to 2008.
Data Source: Energy Statistics of National Bureau of Statistics, *China's Energy Statistics Yearbook 2009* (China Statistical Publishing House, Beijing, 2010); China communications and transportation association, *China Transportation Yearbook 2006–2009* (China Transportation Yearbook Clubs, Beijing, 2006–2009).

respectively 25.2% and 24.2% comparing with that of 2005, while the passenger and freight traffic turnover were increased respectively by 32.9% and 85.7%. All these indicate that the actual energy consumption by unit production capacity was decreased during the 11th FYP period.

(2) *The energy consumption intensity of most modes in transportation is lowered*

The rapid growth of passenger and freight volume leads directly to the growth in energy consumption and carbon emission in transportation sector. However, it is found in a detailed study of transportation modes that only energy consumption of waterways,[3] diesel locomotive in railway and gasoline freight vehicle in highway was increased, while the energy consumption per transport-unit in all other modes was decreased or kept unchanged comparing with the 2005 level.

(1) Railway: Railway, characterized by energy saving and environmental friendliness, is a transportation means which China strives to develop. Calculations show that for the same generated transport amount, railway locomotive produces the least energy consumption[4] (Table 10.1). The railway transportation mileage is increased from 75,400 km in 2005 to 79,700 km in 2008 with an average annual increase of 1433 km. China started to adjust energy consumption structure of the transport industry during the 10th FYP, and one of the policies is to replace steam traction by diesel and electric traction in railway. Steam locomotives were completely forced out of railway transport in 2003. This move significantly improved the railway in running speed, economic benefit and social benefits. In 2009, China's railway electrification rate reached to 41.7% (Ministry of Railways of PRC, 2009), and the railway energy consumption structure was improved significantly.

During the 10th FYP period (2001–2005), energy saved in railway mainly comes from restructuring locomotive. With internal combustion and electric locomotives replacing steam locomotive, it is made more difficult to save energy in railway during the 11th FYP period. Nevertheless, by the

[3]In 2007, good momentum of water transportation raised its tariffs, which in turn brings a low fuel costs relative to income. This led to the shipping industry increasing horsepower and improving the speed of the transportation and production. The ship fuel consumption is positively correlated with the speed cube, leading to a rising fuel consumption per transport-unit of waterway transport.

[4]The unit locomotive consumption is not the comprehensive railway consumption.

Table 10.1. Energy consumption intensity of all transportation means (measurement: tce/(10^6 tkm)).

Transportation modes	Transportation means	2005	2008*
Railway	electric locomotive	1.37	1.36
	combustion motor	3.58	3.63
Highway	gasoline bus	141.8	140.7**
	diesel bus	143.7	138.7**
	gasoline freight vehicle	85.9	89.2**
	diesel truck	78.0	78.0**
Waterway	boats	10.0	17.1**
Aviation	airplanes	488.5	453.7

*Ministry of Transport embarked on a special national highway and waterway traffic investigation in 2008. Since statistical coverage has been changed greatly, the data is not suitable for historical comparison.
**Refers to data in 2007. There is no quantitative analysis for tunnels in this report since this data is not available.
Data Source: China communications and transportation association, *China Transportation Yearbook 2006–2009* (China Transportation Yearbook Clubs, Beijing, 2006–2009).

end of 2009, the comprehensive energy consumption for unit transportation workload has been steadily declined in railway transport sector (without holding joint venture), presenting a decrease of 18.2% (Fig. 10.4) comparing with that in the late 10th FYP Period (2005), thus bringing evident energy saving effect to railway transport. Energy saved in 2006, 2007, and 2008 was respectively 970 thousand tons, 2.1 million tons, and 2.76 Mtce, cumulatively 5.83 Mtce (Fig. 10.5).

Although the consumption per transport-unit of diesel locomotive was increased comparing with the level during the 10th FYP period, the consumption per transport-unit of electric locomotives in 2009 was reduced by 3.6% comparing with the year of 2005 and by 5% comparing with the year of 2000 (Table 10.2). Seen from turnover volume of freight and passengers traffic, although highways still remains to be the main mode of transport at present, as the railway network is being expanded and the speed of railway transportation is being further improved, railway is expected to shunt highway and air traffic in the future; from the perspective of internal railway restructuring, with the improvement of railway electrification rate, railways' potential to reduce emission is huge.

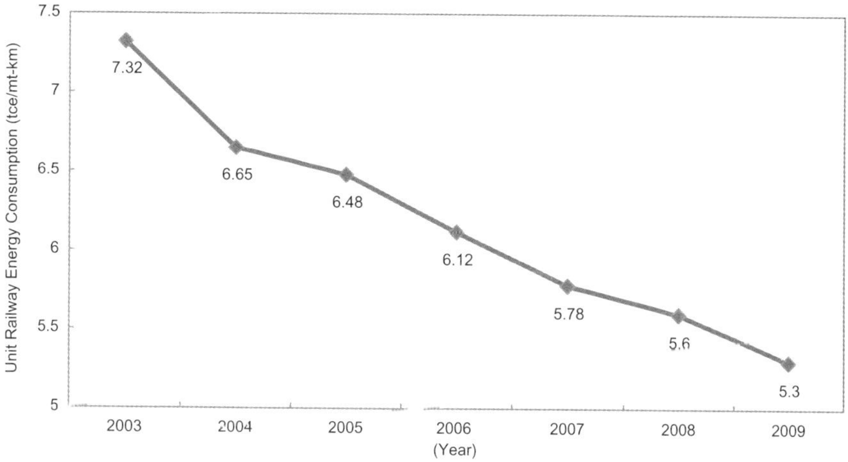

Fig. 10.4. Railway consumption per transport-unit.

Note: Holding joint venture is excluded.

Data Source: M.O.R. Statistics Center of P.R.C, 2009 Railway Statistical Bulletin, (2010), http://www.china-mor.gov.cn/zwgk/gongbao2009.html (accessed on 28 October 2010).

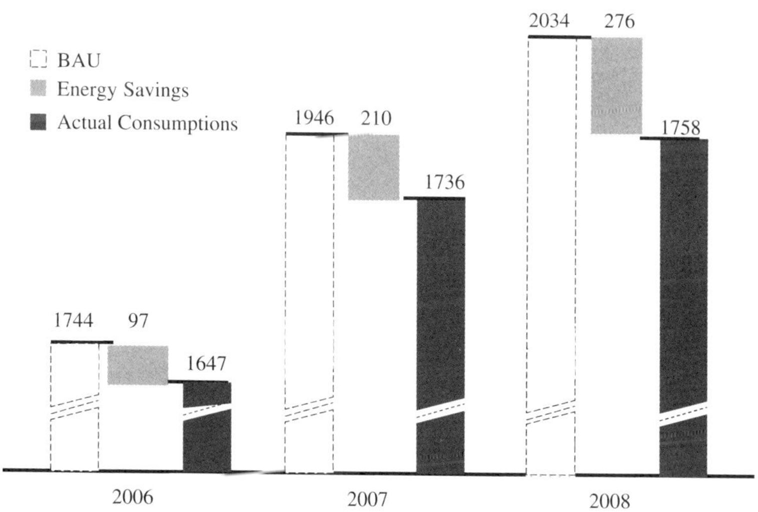

Fig. 10.5. Railway energy savings.

Note: The consumption per transport-unit in 2005 is used as the energy consumption benchmark; actual energy consumption in each year is calculated using the actual consumption per transport-unit; constant ratio energy saving is then obtained by their difference. Holding joint venture is excluded from this data.

Data Source: M.O.R. Statistics Center of P.R.C, 2009 Railway Statistical Bulletin, (2010), http://www.china-mor.gov.cn/zwgk/gongbao2009.html (accessed on 28 October 2010).

Table 10.2. Consumption per transport-unit of diesel locomotive and electric locomotive.

Railway locomotive	Measurement	2000	2001	2002	2003	2004	2005	2006	2007	2008	2009
Electric locomotive	kW·h/(106 tkm)	11 320	11 310	11 080	11 000	11 120	11 180	11 000	10 950	11 060	10 790
	tce/(106 tkm)	1.39	1.39	1.36	1.35	1.37	1.37	1.35	1.35	1.36	1.32
Diesel locomotive	t/(106 tkm)	2.58	2.57	2.59	2.54	2.50	2.46	2.43	2.46	2.49	2.52
	tce/(106 tkm)	3.76	3.74	3.77	3.70	3.64	3.58	3.54	3.58	3.63	3.67

Note: One-ton standard coal of diesel locomotive/million-ton kilometers is calculated by diesel/standard coal coefficient: 1 kg diesel = 1.4571 kgce. One-ton standard coal of electric locomotive/million-ton kilometers is calculated by electric equivalent/standard coal coefficient: 1 kw·h = 0.1229 kgce.

Data Source: Energy Statistics of National Bureau of Statistics, China's Energy Statistics Yearbook 2009 (China Statistical Publishing House, Beijing, 2010); China communications and transportation association, *China Transportation Yearbook 2001–2009* (China Transportation Yearbook Clubs, Beijing, 2001–2009).

(2) Highways: During the 11th FYP period, China's highway mileage is increased to 3,860,800 km in 2009 from 3,345,200 km in 2005, and highway technology and pavement behavior are further enhanced. National road density continues to increase to $40.22\,km/10^2km^2$ in 2009, an increase of $1.36\,km/10^2\,km^2$ comparing with that in the previous year (Ministry of Transport, People's Republic of China, 2010). Since 2000, the average fuel consumption intensity of cars in China is gradually increased. Though still lagged behind by developed countries such as Europe and Japan, with the introduction of "Vehicle fuel consumption limits standards", the overall fuel consumption of passenger cars in China has been greatly improved. Though rapid development of transport sector and rapid adoption of private cars lead to an increase in overall energy consumption, the promulgation of the policy for fuel consumption limits bring about a significant decrease in the overall average fuel consumption intensity during mid-11th FYP period comparing with that of late 10th FYP period. Hundred-ton km fuel consumption of gasoline freight vehicles is increased from 8 L in 2005 to 8.3 L in 2007; besides, the intensity of diesel trucks during the late 10th FYP was maintained; the consumption per transport-unit of petrol passenger cars was dropped to 13.1 L in 2007 from 13.2 L in 2005; the consumption per transport-unit of diesel passenger cars was dropped from 11.6 L in 2005 down to 11.2 L in 2007; the energy consumption intensity of gasoline and diesel passenger cars was respectively dropped by 0.8% and 3.4% during the 11th FYP period (Table 10.3).

(3) Waterway: Waterway transport occupies an important place in transportation. Carbon consumption in waterway transport mainly lies in energy consumption of freight transportation; consumed fuels are mainly diesel and fuel oil. The waterway turnover volume of traffic in 2008 is dropped greatly for various factors such as the global financial crisis; however, waterway transport in 2007 rose to record amount since 2000. It is shown by intensity data of waterway carbon consumption that average thousand-ton km fuel consumption in 2007 was raised by 240% comparing with the year of 2006 (see Table 10.4).

In addition to the impact of changes in statistical coverage of data, expert opinion believes that another main reason for the above phenomenon is the increase in freight rates brought by good momentum of water transportation in 2007, which in turn brings a low fuel costs relative to income. This led to the increasing horsepower in shipping industry, improving the speed of the transportation and production. The ship fuel consumption is

Table 10.3. Consumption per transport-unit of gasoline passenger cars and trucks versus that of diesel passenger cars and trucks.

Item	Measurement	2000	2001	2002	2003	2004	2005	2006	2007
Gasoline passenger cars	$L/(10^2\,tkm)$	11	12	10	12	12	13.2	12.8	13.1
	$tce/(10^6\,tkm)$	118.2	128.9	107.4	128.9	128.9	141.8	137.5	140.7
Diesel passenger cars	$L/(10^2\,tkm)$	9	8	10	11	11	11.6	11.2	11.2
	$tce/(10^6\,tkm)$	111.5	99.1	123.9	136.2	136.2	143.7	138.7	138.7
Gasoline freight vehicles	$L/(10^2\,tkm)$	7	8	8	8	8	8	7.9	8.3
	$tce/(10^6\,tkm)$	75.2	85.9	85.9	85.9	85.9	85.9	84.9	89.2
Diesel trucks	$L/(10^2\,tkm)$	5	6	6	6	6	6.3	6.5	6.3
	$tce/(10^6\,tkm)$	61.9	74.3	74.3	74.3	74.3	78.0	80.5	78.0

Note: 1 L gasoline equals to 0.73 kg; gasoline/standard coal coefficient: 1 kg gasoline = 1.4714 kgce; 1 L diesel equal to 0.85 kg; diesel/standard coal coefficient: 1 kg diesel = 1.4571 kgce; in 2008, big changes occurred in highway statistics caliber and related data is not suitable for historical comparison.
Data Source: China communications and transportation association, *China Transportation Yearbook 2001–2009* (China Transportation Yearbook Clubs, Beijing, 2001–2009).

Table 10.4. Unit waterway consumption.

Item	Measurement	2000	2001	2002	2003	2004	2005	2006	2007
Unit waterway consumption	$Kg/(10^3\,tkm)$	9	2	6	6	6	7	5	12
	$tce/(10^6\,tkm)$	12.9	2.9	8.6	8.6	8.6	10.0	7.1	17.1

Note: Fuel oil/standard coal coefficient: 1 kg fuel oil = 1.42866 kgce; in 2008, big changed occurred in waterway statistics caliber and related data is not suitable for historical comparison.
Data Source: China communications and transportation association, *China Transportation Yearbook 2001–2009* (China Transportation Yearbook Clubs, Beijing, 2001–2009).

positively correlated with the speed cube, leading to an increase in fuel consumption per transport-unit of waterway transport. Furthermore, although domestic shipping companies have worked consistently toward energy saving under the influence of government's advocacy, the effort to reduce energy consumption through technological progress is inefficient, lacking input of new technology and equipment as well as independent innovation incentive mechanisms.

In summary, waterway transport shows great potential in energy saving, which can be realized through technological improvements and innovation and optimization of shipping management means. Box 10.1 introduces the successful experience of Qingdao Ocean Shipping Company. The Company has long put energy conservation as an important indicator for company

Box 10.1 Embarking on the long road of energy saving.

Since 1980s, to improve host combustion conditions, thermal efficiency and mechanical efficiency, Qingdao Ocean Shipping Company (hereinafter referred to as "the Company"), who was faced with problems such as degenerating conditions of mechanic of old ships and thermal efficiency reducing, strengthened contact and cooperation with scientific research sectors, aimed at latest development of the international shipping technologies, combined with the actual situation of each ship, and adopted a series of energy saving measures, including transformation of host supercharger energy-saving technology, adjusting of host compression ratio, propeller edge cutting, installation of stern pod and energy-saving devices, leading to good energy saving effect. Research and analysis show that the contribution rate of technological progress to energy saving is 40%–60%.

Staff of the Company also stated frankly that there is still a long way to achieve the goal; there is still a wide gap in energy efficiency and energy consumption per product unit when comparing with that of advanced enterprises at home and aboard, and the Company's potential to save energy is still huge. Energy saving actions of shipping companies can only be seen as an initial awareness of good sense; there still leaves great potential to energy conservation; with people's in-depth insight into the content of energy saving, the use of new technology and new equipment will bring a bright future to energy efficiency of shipping companies.

Faced with a future of increasing energy scarcity, energy-saving awareness, technological innovation and science potential tapping will become an inexhaustible energy treasure. For energy saving, there rises an issue of whether awareness comes first or technology comes first. We urgently appeal that awareness and technology must go hand in hand, because energy saving of Chinese enterprises still has a long way to go.

(*Data Source*: Teng Chunyan, Exploring a road to green shipping — Energy saving of Qingdao Ocean Shipping Company, China Water Transport, No.8, (2007),) operation. Under combined effects of various measures, Qingdao Ocean Shipping Company made good progress in energy saving.

(4) Civil aviation: With people's living standards improving and the demand for convenience in long journeys further expanding, air transport become more and more people's traveling choice. Therefore, rapid development arose in the civil aviation industry; its business mileage grew from 1,998,500 km in 2005 to 2,461,800 km in 2008. Growing air passenger traffic and voyage gained the aviation industry with more importance in the government's policy of energy conservation. Under guidance of energy saving and low-carbon economy policies, the rapid-developing air transport industry has made considerable progress in energy conservation. As can be seen from Table 10.5, since 2000, fuel consumption per ton km of air transportation has been declined steadily year by year.

Until 2008, consumption per ton km of civil aviation has been reduced by 7.1% comparing with the year of 2005 and by 21.2% comparing with the year of 2000, bringing a saving of 3,000 tce, 9,000 tce, and 9,000 tce respectively in 2006, 2007, and 2008, accumulating to 21,000 tce (Fig. 10.6).

Energy efficiency improvement in air transport in China mainly comes from optimization of air transportation routes and improvement of operational efficiency and operational management levels. In 2009, various measures were taken in China's civil aviation to actively implement the "Airline Industry Energy Conservation and Emission Reduction Plan", achieving remarkable results. Then a series of measures were taken in China's air traffic control system, such as optimizing the main route and opening up "temporary nonstop flight", shortening the annual total flight distance of 15,800,067 km, saving fuel of 88,000 tons, saving fuel costs of 440 million yuan ($66.66 million), reducing CO_2 emissions of 275,000 tons (China Aviation News, 2009). While developing refined management, airline companies promoted energy conservation and emission reduction through design process and technological innovation, etc.

10.1.3 *Quick growth in transportation demand and rising proportion of traffic modes with high energy consumption*

Currently, carbon emission from global transportation accounts for 20% to 25% of total emissions. According to a report from Proceedings of the National Academy of Sciences (PNAS), global carbon emissions from transport will account for 30%–50% of the total global carbon emissions in 2050

Table 10.5. Consumption per transport-unit of civil aviation.

Items	Measurement	2000	2001	2002	2003	2004	2005	2006	2007	2008
Consumption	Kg/(tkm)	0.396	0.381	0.364	0.354	0.341	0.336	0.327	0.309	0.312
	tce/(10^6 tkm)	575.7	553.9	529.2	514.7	495.8	488.5	475.4	449.3	453.7

Note: Jet fuel/standard coal coefficient: 1 kg jet fuel = 1.4539 kgce.
Data Source: China communications and transportation association, *China Transportation Yearbook 2001–2009* (China Transportation Yearbook Clubs, Beijing, 2001–2009).

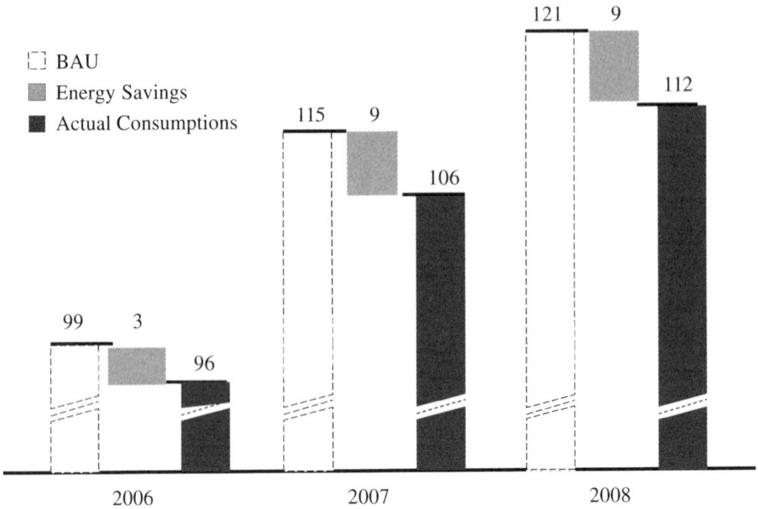

Fig. 10.6. Energy savings of civil aviation.

Note: 1 people·km = 0.072 tkm; the consumption per transport-unit in 2005 is used as the energy consumption benchmark, actual energy consumption is calculated using the actual consumption per transport-unit of each year, and the energy savings are obtained by their difference.

Data Source: China communications and transportation association, *China Transportation Yearbook 2007–2009* (China Transportation Yearbook Clubs, Beijing, 2007–2009).

(Fuglestved *et al.*, 2008; Nakicenovic *et al.*, 2000). Traffic emission reduction has become a global problem to be solved for energy saving. For China, the world's most populous country, rising domestic demand for passenger and freight transport and the development of energy-intensive modes of transportation in traffic structure have an adverse effect on traffic emission reduction during the 11th FYP period; the traffic emission reduction in China still has a long way to go in the future.

(1) *Growing domestic passenger and freight turnover volume of traffic in China*

Increasing passenger and freight turnover volume will bring a quick growth of energy consumption in traffic in the next decade. Table 10.6 shows that China's overall turnover volume of domestic passenger in 2008 was raised by 32% comparing with the year of 2005 and by 88.1% comparing with the year of 2000. Passenger turnover volume of water transport was decreased by 12.7% comparing with the year of 2005 and by 41.1% comparing with

Table 10.6. Comparison of passenger traffic of all transportation modes in 2007 comparing with the years of 2000 and 2005.

Transportation modes	Overall of passenger volume in 2008 (hundred million people·km)	Growth rate comparing with the year of 2005 (%)	Growth rate comparing with the year of 2000 (%)
Railway	7,779	28.3	71.6
Highway	12,476	34.3	87.4
Waterway	59	−12.7	−41.1
Civil aviation	2,306	44.8	212.5
Total	22,619	32.9	88.1

Data Source: China communications and transportation association, *China Transportation Yearbook 2001–2009* (China Transportation Yearbook Clubs, Beijing, 2001–2009).

the year of 2000[5]; air passenger turnover volume rose by 44.8% comparing with the year of 2005 and by 212.5% comparing with the year of 2000. Highway passenger turnover volume was also increased over the same period by 34.3% and 87.4%; railway was increased by 28.3% and 71.6%.

The turnover volume of passenger transportation in China has been growing year by year since 2000; with social life being enriched and tourism rising, people's traveling frequency and trip distance are increased, therefore the turnover volume of passenger transportation will continue to rise rapidly.

The current market demand for freight traffic continues to grow. The rapid development of engineering, agriculture, and logistics industry is driving demand of domestic transportation to grow. Substantial growth in volume is shown in industrial raw materials transportation, product distribution, and agricultural production and marketing locations. For transportation of freight, as shown in Table 10.7, until 2008, freight turnover volume rose by 85.7% comparing with the year of 2005 and by 184.3% comparing with the year of 2000. Freight turnover volume of various modes of transport is also greatly increased. Among them, the air freight turnover volume rose by 23.5% comparing with the year of 2005 and by about 100.0% comparing with the year of 2000; turnover volume of highway traffic was also increased over the same period by 278.1% and 436.2%; waterway turnover

[5]Water transportation is mainly transportation of bulk freight; the water traffic proportion becomes smaller except for traveling and touring.

Table 10.7. Comparison of turnover of freight traffic of all transportation modes in 2007 comparing with the year of 2000 and 2005.

Transportation modes	Total turnover volume in 2008 (hundred million tkm)	Growth rate comparing with the year of 2005 (%)	Growth rate comparing with the year of 2000 (%)
Railway	25,106	21.1	82.3
Highway	32,868	278.1	436.2
Waterway	17,412	56.6	161.4
Civil aviation	42	23.5	100.0
Tunnel	1,944	78.7	205.7
Total	77,373	85.7	184.3

Data Source: China communications and transportation association, China Transportation Yearbook 2001–2009. (China Transportation Yearbook Clubs, Beijing, 2001–2009).

volume was increased by 56.6% and 161.4% over the same period, and railway turnover volume by 21.1% and 82.3%, respectively. Freight turnover volume in China's transportation has been increasing since 2000 and the growth rate during the 11th FYP period becomes faster.

(2) *A trend toward transportation modes of high consumption per transport-unit*

During the 11th FYP period, significant changes occurred in China's traffic modes. High demand for timeliness and convenience of transportation leads to a gradual change in traffic modes. It is shown in Fig. 10.7 that the changes are mainly reflected in shrinkage of the total turnover volume of traffic means of lower consumption, such as railway and waterway, and the gradual increase in that of traffic means of higher consumption, such as aviation and highways. Highway passenger turnover volume was increased from 54.61% in 2005 to 55.16% in 2008; the largest increase is in air turnover volume, from 9.36% in 2005 to 10.36% in 2007. Shipping turnover volume, influenced by the financial crisis, shrank to 10.19% in 2008. With the increase in flight times and reduction in fares, the number of people choosing to fly continues to increase.

Changes in freight traffic modes are more obvious comparing with that in passenger traffic modes (Fig. 10.8). From 2005 to 2008, the proportion of railway freight traffic dropped from 49.75% to 32.45%; water freight traffic dropped from 26.69% to 22. 50% (taking into account the influence of the financial crisis in 2008); at the same time, highway freight traffic rose from

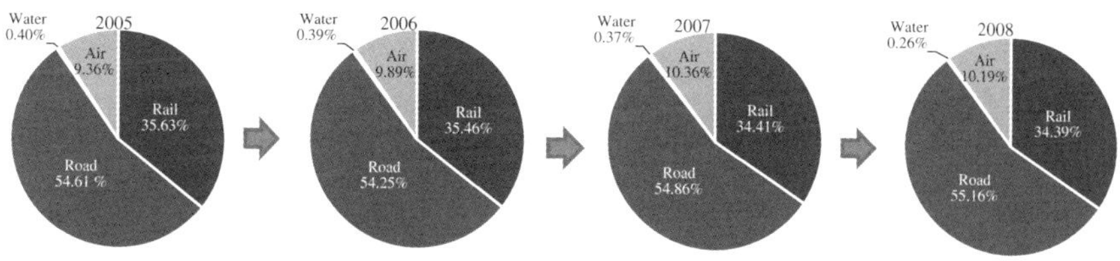

Fig. 10.7. Changes in passenger turnover volume of traffic means from 2005 to 2008 in China.
Data Source: China communications and transportation association, China Transportation Yearbook 2006–2009. (China Transportation Yearbook Clubs, Beijing, 2006–2009).

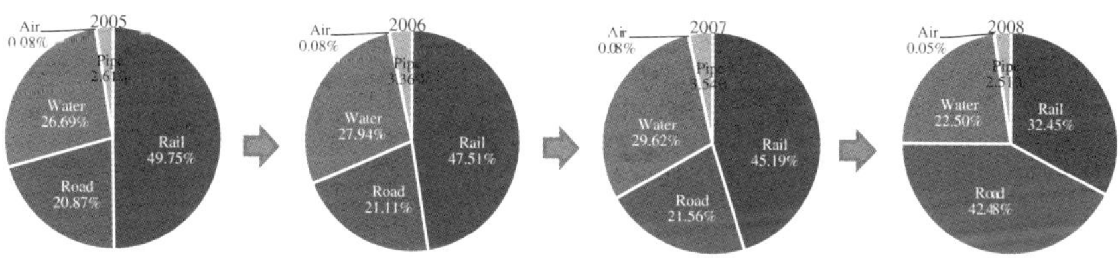

Fig. 10.8. Changes in freight turnover volume of traffic means from 2005 to 2008 in China.
Data Source: China communications and transportation association, *China Transportation Yearbook 2006–2009* (China Transportation Yearbook Clubs, Beijing, 2006–2009).

20.87% to 42.48%; there are no obvious changes in air freight traffic and the freight turnover volume in 2009 was decreased slightly under the influence of the financial crisis.

In general, the rise in highway and air passenger turnover volume and the increase in highway freight turnover drive China's transportation to develop toward the traffic structure of high energy consumption, putting a challenge to energy saving during the 11th FYP period.

10.2 Analysis on Major Policies

10.2.1 *Classification and key points of energy saving policies*

Factors influencing carbon consumption and carbon emission in transportation sector are classified into four kinds based on the analysis of IEA: Transportation activity (A — Activity); transportation structure (S — structure); energy consumption intensity (I — Intensity of energy use);

fuel structure (F — Fuel structure). China's policies are classified into these four kinds using ASIF model.

Current traffic policies in China make effort from the above four aspects. Effective policies are made to reduce traffic activity (A), which can radically reduce carbon consumption and carbon emissions. Since China is in the middle of industrialization and urbanization, the increase in transportation activity will sustain a long time. The key point of these policies is to reduce unnecessary traffic. For example, people's demand to traffic can be reduced through urban layout optimization (fewer traveling times and shorter traveling distance), and freight turnover volume can be reduced through optimizing logistics and industry layout.

The Transportation Structure (S) in China now enters into an important strategic period. From the current overall situation, railway traffic has been lagged behind by highway traffic in transport length, network coverage and turnover volume of passenger transportation; at the same time, the expansion of public transport is faced with great challenges posed by rapid growth of private transport. China adjusted "long-term railway network planning" during the 11th FYP period, adopted the ideas of appropriate advance planning, improved objectives of railway mileage planning, and at the same time broadened investment in and financing of rail transit construction, which should be the road to take in the future. Meanwhile, the urban public transport priority, as the important concept of a low-carbon

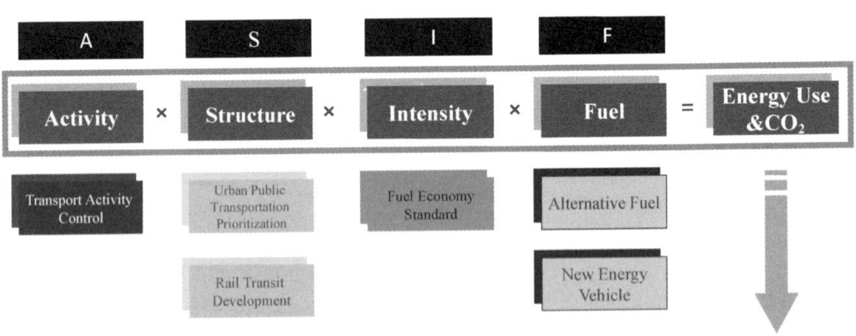

Fig. 10.9. Classification methods and examples of traffic policies.

Note: For key policy points during the 11th FYP period, please refer to attachment of this Chapter.

transport and low-carbon city, was adopted by a growing number of cities and was put into practice.

China's policies on transport energy consumption intensity (I) were focused on fuel economy of vehicles. During the 11th FYP period, the policy of fuel consumption limits has achieved great results, and will continue to serve as the main policy of energy saving in the future. Currently, relevant departments are working on the third-phase limit standard of passenger car fuel economy standards.

Improvements in fuel structure (F) can also reduce carbon emission. For example, the use of clean energy or renewable energy achieved significant results to reduce carbon consumption and carbon emission. New energy, as a transportation fuel, is still under development and is the focus of future energy conservation policies (Box 10.2).

Box 10.2. Policies and effects of new-energy vehicles.

New energy vehicles have won widespread attention in China, and the Chinese government has promulgated a series of policies and interim measures on trial run of new energy vehicles. During the last decade, a total of 42 standards have been developed for energy-saving and new-energy vehicles, and a number of 1,600 invention patents were applied and granted. "Automobile Industry Development Policy" encourages automobile companies to develop and produce new fuel vehicles. "New Energy Vehicles Production and Admittance Management Rules" state that new energy vehicles include hybrid electric vehicles, battery electric vehicles (BEV, including solar cars), fuel cell electric vehicle (FCEV), hydrogen engine vehicles and vehicles using other new energy sources, such as efficient storage devices and dimethyl ether.

Vehicles using natural gas enjoyed a rapid development and the amount of these vehicles reached 57 thousand in 2007, gas emission meeting Euro ? standard. Currently, natural gas vehicles that can meet Euro IV standards and are expected to be on the market soon are being developed in China. Development of bio-liquid fuel vehicles also received great attention; using nonfood bio-liquid fuel is the main development direction, and the main technology is still under development. China's coal-based synthetic fuel development has also made progress, developing a more mature coal-to-methanol and dimethyl ether production technology; demonstration of

(*Continued*)

Box 10.2. (*Continued*)

vehicles using methanol is carried out in some places, and demonstration operation of dimethyl ether car has been started.

(*Data Source*: International New Energy Net, Automobile alternative fuel development status and prospects (2010), *http://www. in-en.com/enweneergy/html/newenergy-1406140698613076.html.* (accessed on 10 October 2010)).

For electromobiles, the "Pilot interim measures for the administration of financial subsidy funds of private buy new energy vehicles" has been implemented and its practicability has yet to be seen.

With further process of global decarbonization, electric vehicles become a hot spot for energy saving and emission reduction in traffic. In order to better promote the industrialization of electric vehicles, starting from 2009, the Ministry of Finance, National Development and Reform Commission, Ministry of Industry and the Information and the Ministry of Science and Technology jointly launched the demonstration project of "1,000 new power vehicles in 10 City". This project includes the demonstration operation and charging station building and has gained support from various departments and local government. Now, 98 charging stations and 325 charging piles have been built, including two discharging bus stations.

(*Data Source*: Press center-China Internet, (2010), "1,000 new power vehicles in 10 City" demonstration project goes well.)

Seen from function modes, China's transport policies mainly focus on improving traffic share and energy consumption intensity. These two types of policies exert the greatest impact on current carbon consumption and carbon emission from transport in China, and are the major policy achievements in energy saving and emission reduction during the 11th FYP period. With the rapid popularization of vehicles, the proportion of carbon emission and energy consumption of road transport will continue to rise; policies on traveling modes and energy consumption intensity provide an effective instrument to solve this problem. The implementation effect is most obvious during the 11th FYP period; most efforts are made to implement policies on urban public transport priority and fuel economy standards.

10.2.2 *Urban public transport priority*

(1) *General*

Urban public transport is one of the patterns used in cities and plays an important role in the whole transportation system. Based on data estimates from "China Statistical Yearbook" since 1996, annual passenger turnover volume of urban transport is equivalent to nearly 20% of an intercity passenger turnover volume. For the recent 30 years, since the continuous progress of urbanization, the amount of cars owned in cities is rapidly increased; however, the city's resources for road construction are very limited. Thus, the growth rate of urban road never keeps up with the growth rate of motor vehicles, resulting in growing urban traffic congestion. In this context, the Ministry of Construction jointly proposed the "Policy on urban public transport development priority," (Jiancheng, 2004, No. 38), and then subsequently introduced a number of policies to promote public transport priority (Table 10.8). Driven by these policies, the development of urban public transport is accelerated.

Table 10.8. Public transport policies.

Policies	Issuing unit	Key points	ASIF features
Notice on urban public transport development priority (Guobanfa, 2005, No. 46), Policy on urban public transport development priority, (Jiancheng, 2004, No. 38)	State Development and Reform Commission; Ministry of Science and Technology; Ministry of Public Security; the Ministry of Finance; Ministry of Land and Resources; Ministry of Construction	• The priority of urban public transport development. • Tasks and goals of urban public transport development priority (speed, accuracy, coverage, average traveling time, share rate). • Strengthening the instruction of urban planning. • Constructing public traffic road system and BRT.	SA

(*Continued*)

Table 10.8. (*Continued*)

Policies	Issuing unit	Key points	ASIF features
Opinions on economic policies on urban public traffic development priority (Jiancheng, 2006, No. 288)	Ministry of Construction; National Development and Reform Commission; Ministry of Finance; Ministry of Labor and Social Security;	• Formulating related economic policies on urban public transport development priority. • Promoting reformation of urban public traffic industry. • Raising industry technology and service quality. • Building a standard system for sound urban public traffic laws. • Increasing input in urban public transport. • Building subsidies mechanism for low bus fare. • Implementing bus fuel allowance and other subsidies. • Standardizing economic compensations for bus. • Ensuring bus industry stability.	SI

Policies on urban public transport priority positively contribute to reduction of carbon emissions. Carbon emission of urban public transport is positively correlated to carbon consumption of transport modes, while energy consumption and carbon impact depend on the total amount of urban traveling (related with people's frequency of activities in the city), the share rate of urban public transport, the production efficiency of urban public transport (passenger turnover volume completed in unit energy consumption) and the use of the energy structure.

(1) Total urban traveling amount

Total urban traveling amount is the total sum of various traveling amount produced by urban population for production and living, including traveling of various modes (such as walking, bicycling, motorcycling, and traveling by cars, public electric car, subway, etc.) and it is the base number for public traffic traveling. Total urban traveling amount depends on average traveling per capita. The traveling level per capita is one of the measurements for urban development; it mainly includes two aspects, namely, the number of traveling per capita and the average traveling distance. In general, the indicators increase with urbanization and raise of income per capita. In 2008, China's urban amount of inter-city traveling per capita was 47 times/year, while intra-city public transport travelling is 116 times, increased respectively by 41.6% and 40.7% comparing with the year of 1996. From 1996 to 2008, the level of urbanization in China reached from 29.37% to 45.68%; the income per capita of urban residents increased from 4844.78 yuan ($734.058) to 17,067.78 yuan ($2586.027). In addition to the city's development level, factors influencing traveling include urban layout structure and land use patterns. The impact of these factors on motorized intra city traveling is particularly significant. Studies show that high density, intensive, transit-oriented development (TOD) and mixed-use land development model will significantly reduce the level of motorized traveling. China's traditional small-scale neighborhood design and fine grid street design also help to reduce vehicle traveling mileage and increase the proportion of low-carbon traveling mileage, while large-blocks urban form not only leads to high car traveling mileage, but also increases travel mileage of other forms (See Fig. 10.10).

Many factors can affect the average traveling distance through various means, such as city size, development mode, road network density, public transport network structure and distribution. The simultaneous increase of traveling rate and average traveling distance will directly lead to growth in carbon emission.

(2) Share rate of public transport

The total traveling amount in a city can be calculated based on the average traveling amount and total population, and the proportion of traveling

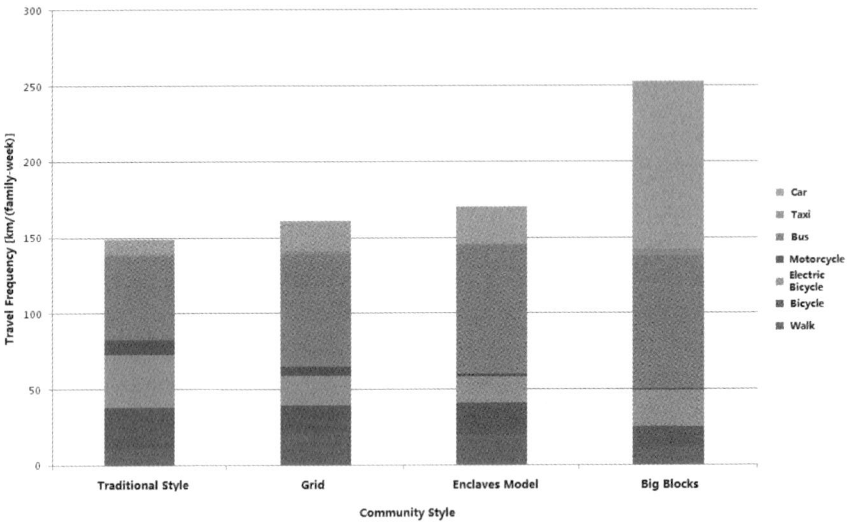

Fig. 10.10. Relations between residents' traveling characteristics and block forms. *Data Source*: Y. Jiang, *Does Energy Follow Form* (Massachusetts Institute of Technology, Cambridge, 2010).

amount by urban public transport to the overall traveling amount is called the share rate of public transport. Share rate is calculated by traveling times generally, and yet sometimes by traveling distance (people·km), i.e., the proportion of people km by public transport to people km of all traveling modes. The latter is valuable for calculating carbon emission; however, there is no statistics of traveling distance by public transport in public data, thus this method is only adopted in few studies.

The energy consumption and carbon emission of public transport are much smaller comparing with that of cars. According to a study by the Ministry of Transport, the energy consumption per capita of bus in every hundred km is 8.4% of that of cars; electric cars 3.4% to 4% and subway 5%. Under circumstances that green traveling modes such as bicycling and walking do not change, the increase in share rate of public transport means the reduction of carbon emission.

(3) Unit energy consumption of public transport

For unit vehicle kilometer, the energy consumption of public transport vehicles is larger comparing with that of cars, and the carbon emission is higher. However, single-time passenger carrying amount transported by public traffic is much larger comparing with that by cars; for each person kilometer, the carbon emission of public transport vehicles is much smaller. Therefore, the amount of passengers transported by public transport becomes one of the important factors to reveal the advantages of public transport in carbon emission. This indicator is measured by load ratio in public transport. Besides, the advantage in carbon emission also depends on the speed of public transport and capital productivity of vehicle utilization of public transport (passenger amount of each public vehicle in unit time).

(4) Fuel structure of public transport

Different from other transport modes, the use of different fuels generates different carbon influence for public transport. Currently, fuels used by public transport are mainly diesel, gasoline, electricity, compressed natural gas (CNG) and liquefied petroleum gas (LPG).

Electricity is mainly used in city subway, light rail systems and trolley buses. Without regard to electricity production, Carbon emission of electrical energy is zero. Diesel and petrol have different density and calorific value; calorific value of gasoline per volume is approximately 90% of diesel of the same volume (Feng and Sauer, 2004), while its carbon emission is about 15% less comparing with that of diesel. On the other hand, the energy efficiency of diesel engine is 20% to 40% higher comparing with that of gasoline engine. Therefore, unit carbon emission produced by diesel transportation is at least 10% less than that of gasoline (This includes carbon emission during the refining process) (OECD/ITF, 2008).

The European Union has analyzed the energy conversion efficiency and carbon emissions during the whole process of technical routes (from well to wheel). The results showed that advanced diesel vehicle and hybrid vehicles, compared with conventional gasoline vehicles, can improve energy efficiency and reduce carbon emissions; besides, the energy efficiency of advanced diesel vehicles and hybrid vehicles is higher than that of most alternative fuel vehicles; the carbon consumption and carbon emissions of LPG and CNG vehicles are lower than that of conventional gasoline vehicles; a report by Swiss Federal Testing and Research Laboratory states that carbon emission of natural gas vehicles can be reduced by 20% to 30%.

(2) *Effects of public transport priority policies*

(1) Steady improvement of traveling level and rapid increase in traveling amount of public transport:

From a global point of view, the process of urbanization is often accompanied by growth in people's demand for traveling. In China, it is estimated

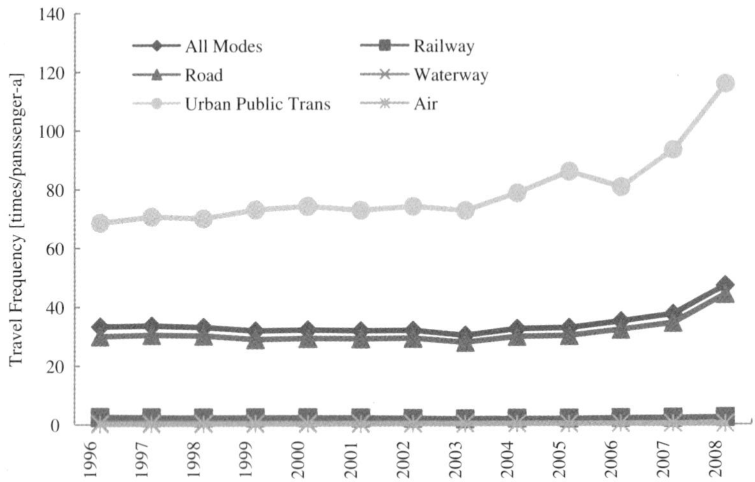

Fig. 10.11. Average annual traveling amount of urban residents in China.
Note: Traveling amount = total annual traveling amount of all modes (ten thousand person-time)/current-year urban population (ten thousand people); the traveling modes include railway, highway, waterway, and civil aviation, all referring to long-distance traveling; urban public transport refers to traveling within the city through public transport.
Data Source: National Bureau of Statistics, China Statistics Yearbook 1996–2008. (China Statistical Publishing House, Beijing, 1997–2009).

based on available statistics that, from 1996 to 2006, the average long-distance traveling amount of residents in China has lingered between 30 to 35 times a year, and rose quickly to 47 times per year since 2006. During the same period, traveling amount of public transport maintained a steady increase from 68 times per year in 1996 to 115 times a year in 2008; in particular, the growth rate began to accelerate since 2004 (see Fig. 10.11).

It can also be seen from Fig. 10.11 that the long-distance traveling amount by railway, water transport, and civil aviation has been relatively small, with the annual railway traveling amount per capita being about 2 times while that of water transport and air transport less than 0.5 times. In comparison, the traveling amount of road traffic is much higher than that of the above modes, and the traveling amount of public transport is much higher than that of highway traveling, the former being about twice the latter. Moreover, in recent years, the growth rates of highway traveling amount and public transport traveling amount have been accelerated, and the growth rate of public transport is faster than that of highway traveling, widening the gap in traveling amount.

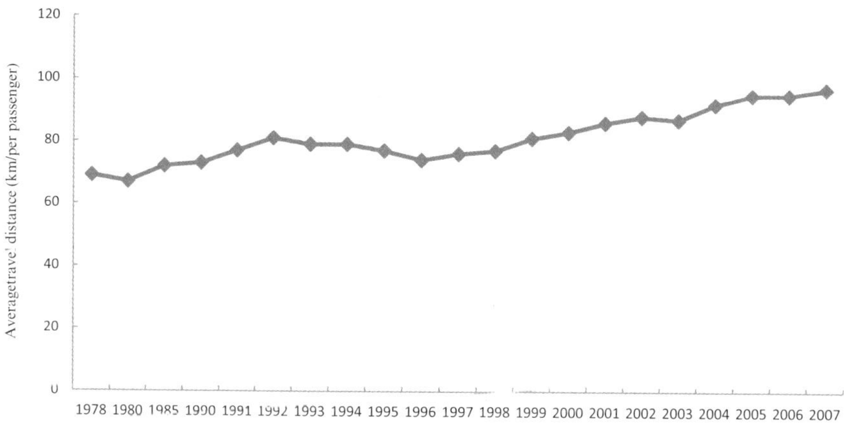

Fig. 10.12. Average long-distance traveling distance of Chinese residents. *Data Source*: National Bureau of Statistics, *China Statistical Yearbook 1996–2008* (China Statistical Publishing House, Beijing, 1997–2009).

Historical data from developed countries show that during urbanization process, people's long distance traveling will increase. From the reform and opening up in 1978 to 2007, Chinese residents' long-distance traveling is largely in line with this law. The trend of slow growth in traveling distance can be seen from the Fig. 10.12.

It is generally believed that, the traveling distance of public transport will increase with city expansion. However, the traveling distance of urban public transport is influenced, to a large extent, by city planning, spatial distribution and equilibrium level of housing and employment, coverage of public transport services and composition of groups enjoying public transport. Taking Beijing for example, because of irregular city expansion and inadequate public transport networks, bus traveling distance per capita has been rising slowly and reached to 11.4 km in the year 2000. Then it gradually decreased, particularly after the implementation of low fare policy at the end of 2006, when large number of cycling, walking and other short-distance traveling were attracted, making the average traveling distance dropping from 9.5 km in 2005 to 7.27 km in 2009 (Meng Fei, 2010). However, this is a rather special case. At present, except for Beijing, no other cities cut fares by large-scale to greatly increase the share rate of public transportation. Therefore, there exists the trend that average traveling distance increases with city expansion.

Table 10.9. Average traveling distance of public transport in China. (measurement: km).

City	Beijing	Shanghai	Shenzhen	Jinan	Kunming	Chongqing
Average public transport travelling distance	7.27	1.30	7.70	1.11	5.45	6.33

Data Source: Meng fei, Influence of urban traveling modes to energy and environment, Master's thesis, Department of Environmental Science and Technology, Tsinghua University, Beijing (2010).

As for average traveling distance of public transport, there is a lack of statistical standards in various cities across China. This is mainly because that it is technically more time-consuming, often requiring large-scale household survey or traveling by bus, resulting in higher cost. The average traveling distance listed in Table 10.9 is calculated based on research projects in various cities. It is generally believed that the average traveling distance by public transport in China is between 5 km and 10 km.

Figure 10.13 shows the annual traveling amount (person-time) of urban public transport in China since 1998. It is shown that the total amount rises continuously in two obvious speeding phases, i.e., from 2004 to 2005 and from 2006 to 2009. This may be caused by a series of policies on public transport priority implemented during these two periods.

It should be noted that comparing with the year of 2005, the total traveling amount in 2006 is decreased, which is caused by weakening promotion of public transport priority policies on the one hand and rising amount of private cars since 2004, on the other hand.

The traveling indicator combining average traveling distance and average traveling times is the turnover volume amount. Figure 10.14 shows the trend curve of passenger turnover volume calculated using total traveling person-time and traveling distance per capita from the Statistical Yearbook. In the absence of statistics on average traveling distance of public transport, the value of 7 km/person-time is adopted in reference to Table 10.9. This value will affect the absolute turnover volume value, but the turnover volume trend is less affected. It can be seen from the turnover volume trend curve that until the year 2003, turnover volume of public transport grew slowly along with turnover volume of long-distance traveling; after 2003, the growth rates of the two were accelerated, turnover volume of long-distance

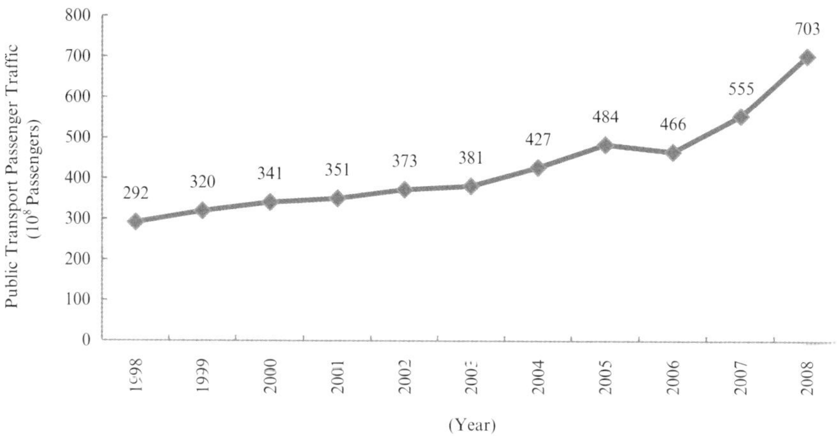

Fig. 10.13. Total public transport traveling amount.
Data Source: National Bureau of Statistics, *China Statistical Yearbook 1998–2008* (China Statistical Publishing House, Beijing, 1999–2009).

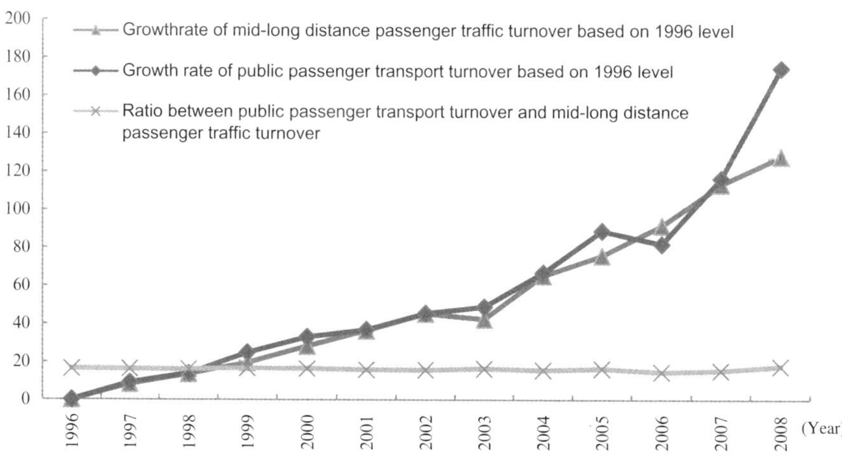

Fig. 10.14. Turnover volume curve of public passenger transport (measurement: %).
Note. passenger transport turnover volume growth = (current-year passenger turnover volume − passenger turnover volume in 1996)/passenger turnover volume in 1996; bus passenger turnover volume to long-distance passenger turnover volume = bus passenger turnover volume/long-distance passenger turnover volume; the average public transport traveling distance is supposed to be 7 km/person-time.
Data Source: National Bureau of Statistics, *China Statistical Yearbook 1996–2008* (China Statistical Publishing House, Beijing, 1997–2009).

Table 10.10. Annual traveling turnover volume growth rate (measurement: %).

Year	1997	1998	1999	2000	2001	2002
Long-distance traveling	8.13	4.84	5.23	7.10	6.10	6.22
Public transport traveling	8.93	4.50	9.65	6.63	2.83	6.31
year	2003	2004	2005	2006	2007	2008
Long distance traveling	−1.88	15.16	6.00	8.30	10.67	6.30
Public transport traveling	2.29	12.05	13.20	−3.67	19.04	26.75

Note: Annual traveling turnover volume growth rate = (current-year passenger turnover volume − passenger turnover volume of previous year)/passenger turnover volume of previous year.
Data Source: National Bureau of Statistics, *China Statistical Yearbook 1997–2002* (China Statistical Publishing House, Beijing, 1998–2003).

passenger grew relatively gently while that of public transport enjoying two rapid increase respectively from 2003 to 2005 and from 2006 to 2008; the growth rates of the two periods were faster than the growth rate of long-distance passenger, widening the gap between the two.

Table 10.10 lists the growth rate of long-distance and public transport traveling turnover volume comparing with that of last year. As can be seen, residents' average traveling turnover volume witnessed a growth in each year except for long distance traveling in 2003 and public transport traveling in 2006. The growth rate of public transport traveling was increased by more than 19% in 2007 and 26% in 2008, showing an upward trend. Yet since 2000, large fluctuations arise in growth of public transport traveling. Statistics show that in turnover volume of long-distance traveling, turnover volume of highway traveling has been accounting for more than 80% of the total (including highways, railway, aviation, waterways) since 1978. It is believed that it has a strong positive correlation with amount of privately-owned cars. It is shown in Table 10.11 that the amount of privately-owned cars in 2006 reached the highest since 2000. The growth of privately-owned cars often indirectly reflects the decline in attractiveness of public transport traveling and reduction in its share rate. It can be seen from Table 10.10 that public transport traveling volume in 2006 was decreased by 3.67% comparing with the year of 2005. Another reason may be that in 2004 and 2005, excessive growth arose in volume of public transport, and conditions to maintain the stable growth (for example, the growth rate of bus was slowed down in 2005 and 2006; or growth of passenger flow was slowed when the service level was dropped while the load factor was improved) were not kept in pace, resulting in turnover volume decreasing.

Table 10.11. Growth rate of private cars (measurement: %).

Year	2001	2002	2003	2004	2005	2006	2007	2008
Growth comparing with previous year	23.3	25.7	25.8	21.5	24.7	26.3	23.3	21.7
Growth comparing with the year 2000	23.3	55.0	95.0	136.9	195.5	273.1	360.0	459.9

Data Source: National Bureau of Statistics, *China Statistical Yearbook 2001–2008* (China Statistical Publishing House, Beijing, 2002–2009).

(2) The share rate of public transport is raised in some cities:

The share rate of public transport is an important measurement for its development. As stated above, the share rate of public transport can be measured by the proportion of public transport volume to all transportation volume as well as the proportion of public transport traveling times to all transportation traveling times. The data of each method is hard to obtain and sample survey is needed. The share rate of the whole country should be the mean value of traveling amount weighing after conducting sample survey in each city. There is a lack of statistical standards because it is technically more time-consuming. However, data, though unsystematic and not comprehensive, can be found in annual traffic reports and special studies. For example, changes in share rate of public transport can be seen in share rate of travelling modes in Fig. 10.15. Except for individual cities, the share rate of public transport is slightly increased or maintained the same. According to Ministry of Housing and Urban–Rural Development, the share rate of current public transport in China ranges from 10% to 20%

It can be seen from Table 10.12 that since 2005, the share rate of public transport in Beijing and Kunming have been rising greatly. Both the two cities are leaders in promoting public transport priority. Kunming first adopted bus lanes and Beijing first built BRT. At the same time, the low bus fare strategy in Beijing and subway and light rail building in recent years all lead to the rise in share rate of public transport.

It should be noted that during rapid motorization, cars entered into common people's life quickly, the traveling times per capita and traveling distance are increased continuously, yet the share rate of public transport did not rise but remained steady, which is positive for emission deduction.

(3) Supply of public transport is raised steadily and transit capital productivity is improved continuously:

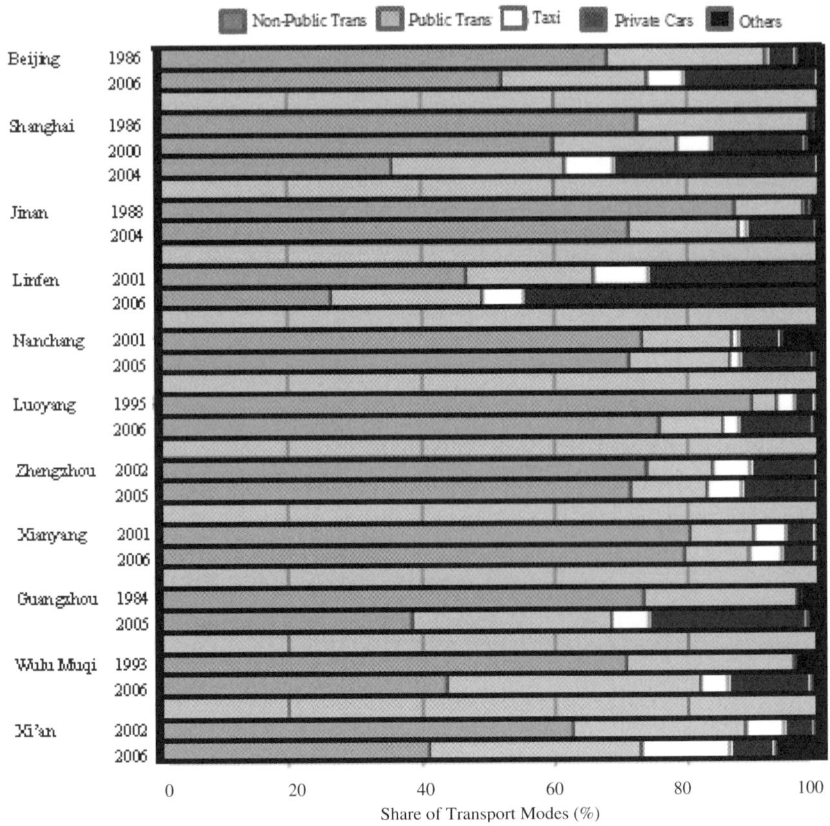

Fig. 10.15. Changes in passenger public transport volume (share rate changes according to per capita traveling times).
Data Source: World Bank, Urban traffic and CO_2 emission: Some characteristics of Chinese cities, World Bank, Washington (2001).

Public transport supply determines whether demand can be met. If supply is inadequate, the service level may be decreased, thus suppressing demand. The prerequisite of supply is the quantity and quality of mobile production facilities (mainly public transportation vehicles) and fixed production facilities (including bus lanes, bus stations, hub, etc.) owned by bus companies. Shown in Fig. 10.16 are changes in number of standard buses owned by every ten thousand people since 2000. A steady rise was shown, especially from 2003 to 2004 and from 2006 to 2008. This shows that the amount of buses, one of the most important means of production for public transport, is consistent with the trend of demand for transit traveling.

Table 10.12. Changes in share rate of public transport (measurement: %).

Year	2005	2006	2007	2008	2009
Beijing (subway only)	29.8 (5.7)	30.2 (5.8)	34.5 (7)	—	39.3 (9)
Shanghai	26	24	23.5	23.2	—
Kunming	14.7	18.7	21	23.7	—
Chongqing	—	—	35.1	34.3	33

Data Source: Annual traffic development reports and transit studies of these cities.

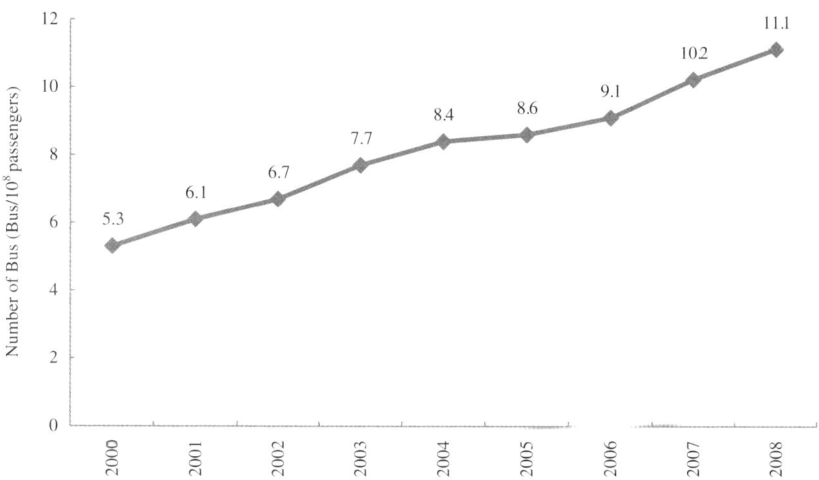

Fig. 10.16. Variation tendency of bus amount.

It can also be seen from Fig. 10.17 that the growth rate of bus is relatively slow in comparison with the growth rate of private cars. From 2000 to 2008, the number of private cars owned by per ten thousand people was increased by 460% with an average annual increase of 24%; at the same time, the number of bus per ten thousand people was increased by only 110% with an average annual increase of 9.8%. This means that if you want to have the passenger turnover volume of public transport grow in step with or exceed middle- and long-distance passenger transport (as mentioned above, 80% of which is completed by private cars), the capital productivity of public transport (i.e., number of passengers transported in unit time by each bus) has to be constantly improved (Darido, 2009). The trend showed in Fig. 10.13 confirmed this point to some extent.

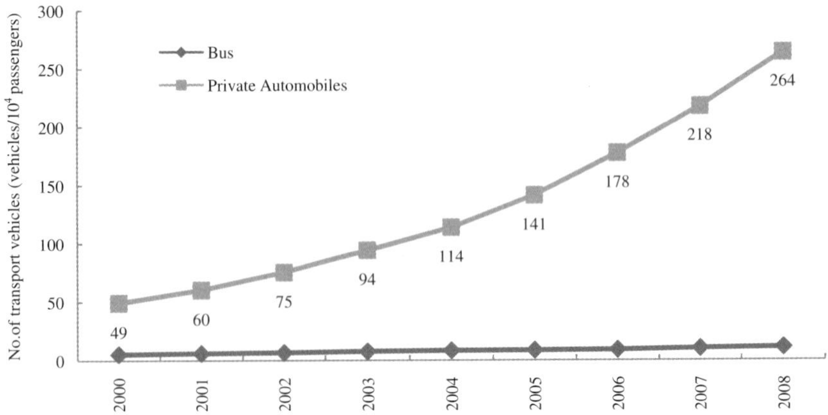

Fig. 10.17. Comparison of changes in bus amount and private vehicle amount.

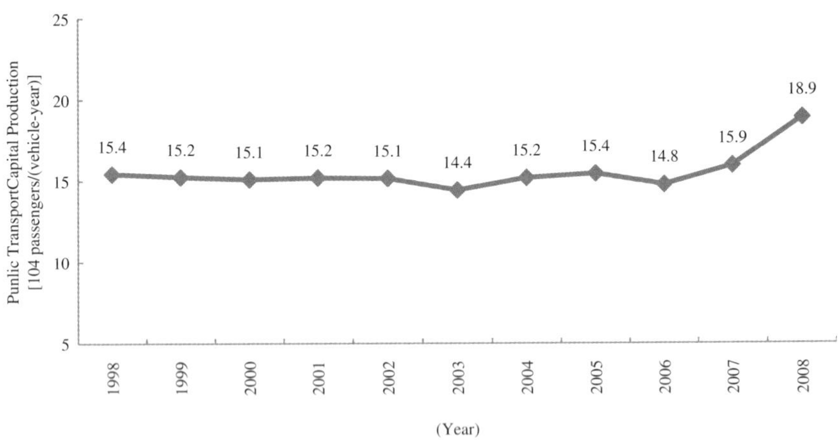

Fig. 10.18. Transit capital productivity.

Note: Transit capital productivity = public transport volume/year-end public transport operational vehicle amount.

Data Source: National Bureau of Statistics, *China Statistics Yearbook 1998–2008* (China Statistical Publishing House, Beijing, 1999–2009).

It can be seen from Fig. 10.18 that there exists a slight downward trend in transit capital productivity from 1998 to 2002, with the lowest point in 2003, and a slow rise in 2004 and 2005. After the fall of 2006 (as mentioned above, one of the reasons may be the rise in private vehicles in 2005 and

Table 10.13. BRT achievements.

City	BRT Length (km)	Passenger flow volume (ten thousand per day)	BRT special vehicles ( )
Beijing	56	23	87
Guangzhou	22.5	80	—
Xiamen	64	—	208
Hangzhou	51	55	104
Jinan	53	25	165
Kunming	50	20	25
Changzhou	46	17	60
Zhengzhou	30	7	170
Dalian	14	—	64
Chongqing	13	25	10
Hefei	30	4	75
Suzhou	8	—	—
Taiyuan	10	—	—
Yancheng	25	—	20
Zaozhuang	27	2	—

Data Source: Organized according to data from Energy Foundation Sustainable Urban Projects.

2006), there showed a rapid rise. This is consistent with the structure of urban traveling and the implementation timing of public policy.

The improvement in supply of urban public transport is also reflected in hundreds of kilometers of Bus Rapid Transit (BRT) lines built in many cities since 2005. It is proved at home and abroad that, BRT can significantly improve service levels and transport efficiency of public transport. Not only in the corridor itself, BRT can optimize the structure of the whole city's public transport network and help to maintain or enhance the share rate of the city's public transport. This is one of the major achievements since the implementation of bus priority policy in China. Table 10.13 lists the status of some cities' BRT system.

In addition to BRT, more than 3,000 km of public transport lanes has been designed all over the country (Zuomin, 2008). With rapid growth in urban vehicle population and growing road congestion, these bus lanes play an important role in increasing operational speed, improving public services, ensuring the share rate of public transport.

(4) The exploration to clean fuel in public transport structure:

Traditionally, diesel and gasoline vehicles are seen as the mainstream of public transport vehicles. However, accurate statistical data on their respective

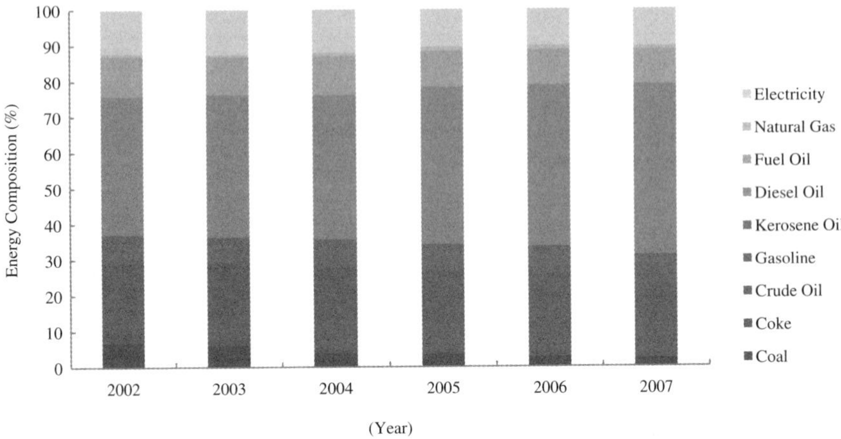

Fig. 10.19. Fuel structure of China's posts and telecommunications and transportations sector.
Data Source: National Bureau of Statistics, China Statistical Yearbook 2002–2007 (China Statistical Publishing House, Beijing, 2003–2008).

proportions are lacking. This section tries to understand the fuel structure of public transport through analyzing the fuel structure of China's posts and telecommunications and transportation sector. Since diesel or gasoline is used by above 90% of public transport vehicles, the ratio of diesel to gasoline should be about 7:3 based on data in Fig. 10.19. Fuels used in public transport vehicles should be similar to this ratio, which is used as one of the basic assumptions to calculate carbon emission.

Although CNG, LPG and hybrid vehicles have been tried on in public transport and witnessed rapid growth during a certain period, their share is still small in general terms, about 5% to 3% of the total. The amount of CNG vehicles has been slowly rising since 2003, while the amount of LPG vehicles has been in decline after 2005 (see Fig. 10.20). This may be because that the environmental friendliness and economical efficiency of this kind of vehicle has been questioned. For example, according to some bus operators in Guangzhou, fuel cost of LPG vehicles is higher comparing with that of diesel vehicles, and cost of their spare parts and maintenance is higher too. In addition, although the carbon emissions of LPG vehicles are lower comparing with that of diesel and petrol vehicles, emissions of other pollutants such as hydrocarbons and nitrogen oxides are high.

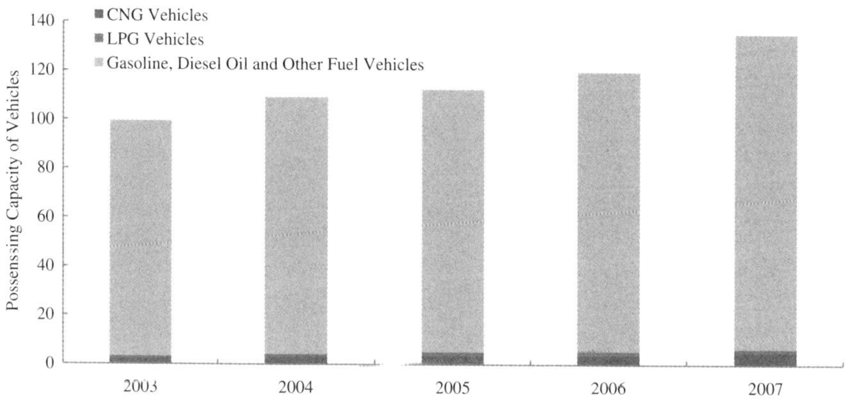

Fig. 10.20. Amount of CNG and LPG vehicles.
Data Source: L. Zuomin, Green transport for growing cities, (2008), http://ltaacademy.lta.gov.sg/doc/BjGT01%20Li-Green%20Transport%20Policies%20in%20China%20(Chinese).pdf (accessed on 8 October 2010).

(5) Scenario analysis:

The emission of public transport (except subway and light rail) from 2000 to 2008 is calculated in this section based on the above analysis and assumption. This calculation only accounts for the utilization phase rather than the influence of life cycle. As shown in Table 10.14, some of the assumptions are based on current studies with relatively higher accuracy and some values need sensitivity analysis.

Emissions based on the above assumptions are shown in Table 10.15. The results show that, the total carbon emissions decreases with transport load factor increasing; otherwise, emission rises; there exists an inverse relationship between the two. If the load factor is certain, the increase of average traveling distance will lead to increase of emissions; there is a positive correlation between the two (Fig. 10.21). Under such assumptions, it is calculated that from 2000 to 2008, carbon emissions from public transport vehicles range from a few million tons to thousands of millions of tons. For example, if the average traveling distance is 717 and the average load factor is 60, the carbon emissions are about 20.16 million tonnes (Table 10.15).

If public transport traveling is partly completed by cars from 2000 to 2008 or fuel structure is changed, then carbon emission will be greatly influenced. In this section, extra assumptions and nine different assumptions are raised to calculate the upper limit and lower limit of carbon emission.

Table 10.14. Assumptions about carbon emissions of public transport.

Parameters	Values	Sources
Average traveling disance	5–10 km	Based on small-sample urban transportation annual report
Average load factor	40–100 people/vehicle	Based on the actual load capacity of standard vehicle and the surveyed load factor
Portortion of gasoline vehicles	30%	Based on Fig. 10.19
Portortion of diesel vehicles	66%	Based on Fig. 10.19
Ratio of CNG vehicles to LPG vehicles	4%	Mid-value based on Fig. 10.20
Standard machine fuel consumption of gasoline vehicles	25 L/100 km	Based on the weight of standard vehicle and in reference with fuel economy standard
Carbon emission factors of gasoline vehicles	1,961 g/L	Mid-value between 2002 and 2008 in reference with Zhang Qingyu's (2010) study of historical emission data in southern China
Unit vehicle km carbon emission difference between diesel vehicles and gasoline vehicles	−15%	Mid-value based on the above analysis
Unit vehicle km carbon emission difference between CNG and LPG vehicles	−25%	Mid-value based on the above analysis

In these assumptions, fuel structure is adjusted[6] and the average share rate of public transport traveling from 2000 to 2008 is assumed to be 20%. Considering the share rate of cars, based on related studies, the average

[6]Since the proportion of CNG and LPG is quite small with little influence, following calculations are based on the assumption that public transport only consist of gasoline vehicles and diesel vehicles.

Table 10.15. Carbon emissions of public transport between 2000 and 2008 (measurement: ten thousand t).

Average traveling distance / Average load factor (people/vehicle)	5	6	7	8	9	10
40	2,160	2,592	3,023	3,455	3,887	4,319
60	1,440	1,728	2,016	2,304	2,592	2,880
80	1,080	1,296	1,512	1,728	1,944	2,160
100	864	1,037	1,209	1,382	1,555	1,728

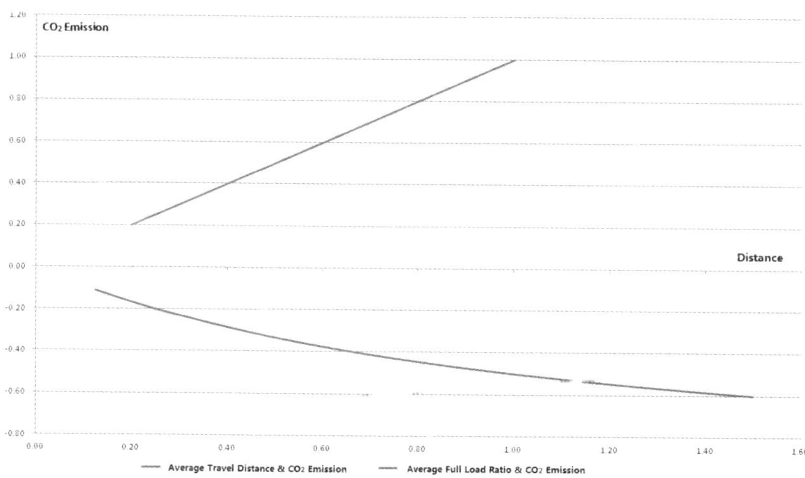

Fig. 10.21. Relationship between emissions and average traveling distance and average load factor.

load factor of cars is assumed to be 1.2 people/vehicle; emission factor of cars is assumed to be 2110 g/L; and the fuel consumption is assumed to be 8.6 L/100 km, the mid-value of statistic from 2002 to 2008.

Besides, public transport does not provide door to-door transportation; people have to walk or bicycle about 1 km to catch buses, and this distance is called traveling distance increment here. This distance is zero emission for public transport but produces corresponding carbon emission for cars. The average traveling distance of public transport is assumed to be 5 to 10 km and the traveling distance increment is assumed to be 10% of traveling distance, i.e., about 500 to 100 km. Of course, for the same starting place and destination, farther or nearer routes may be chosen than that by public transport. In this section, the in-vehicle distance of cars and public transport is assumed to be the same.

Based on the assumption that the share rate of public transport is 20%, every 5% changes of public transport traveling amount will bring 1% to share rate of public transport traveling.[1] The following calculations assume that changes in share rate of public transport are caused by cars and other traveling modes (walking, bicycling, and motorcycling) remain unchanged.

It can be seen in Table 10.16 that every 1% increase in the share rate of public transport means tens of thousands tons of emission reduction.

It is shown in Table 10.17 that when the share rate of public transport is raised by 5%, if the average traveling distance of public transport is 7 km, carbon emission deduction will be hundreds of millions of tons. Different from Table 10.15, Table 10.16, and Table 10.17 reflect carbon emission deduction, i.e., amount of reduced car carbon emission by public transport. Therefore, the higher the load factor of public transport and the longer the average traveling distance, the bigger the potential of carbon emission deduction.

Table 10.16. Carbon emission reduction when the share rate of public transport is increased by 1% from 2000 to 2008 (measurement: ten thousand t).

Average traveling distance / Average load factor (people/vehicle)	5	6	7	8	9	10
40	1 533	1 840	2 147	2 453	2 760	3 067
60	1 570	1 883	2 197	2 511	2 825	3 139
80	1 588	1 905	2 223	2 540	2 858	3 175
100	1 598	1 918	2 238	2 558	2 877	3 197

Table 10.17. Carbon emission deduction from 2000 to 2008 if the share rate of public transport is raised by 5% (measurement: ten thousand t).

Average traveling distance / Average load factor (people/vehicle)	5	6	7	8	9	10
40	7 667	9 201	10 734	12 267	13 801	15 334
60	7 848	9 417	10 987	12 557	14 126	15 696
80	7 938	9 526	11 114	12 701	14 289	15 876
100	7 992	9 591	11 189	12 788	14 386	15 985

Table 10.18. Carbon emission deduction caused by different fuel structure and share rate of public transport.

Scenario	Fuel structure	Increased share rate (%)	Carbon emission difference (ten thousand t)	Mid-value of carbon emission difference (ten thousand t)
Scenario 1	Public transport: gasoline vehicles 40%, diesel vehicles 60%; cars: gasoline 100%	1	1 497–3 168	2 333
Scenario 2		5	7 631–15 956	11 794
Scenario 3		10	15 298–31 940	23 619
Scenario 4	Public transport: gasoline vehicles 30%, diesel vehicles 70%; cars: gasoline 100%	1	1 533–3 197	2 365
Scenario 5		5	7 667–15 985	11 826
Scenario 6		10	15 334–31 970	23 652
Scenario 7	Public transport: gasoline vehicles 20%, diesel vehicles 80%; cars: gasoline 100%	1	1 570–3 226	2 398
Scenario 8		5	7 703–16 014	11 859
Scenario 9		10	15 370–31 999	23 685

Table 10.18 shows the carbon emission deduction that may be caused by different fuel structure and different share rates of public transport. The assumptions in these calculations are different from the above, i.e., the load factor of public transport is between 40 and 100, the average traveling distance ranges from 5 km to 10 km and the traveling distance increment is 10%. It is found that the share rate of public transport has a big impact on carbon emission. Every 5% changes of share rate leads to huge difference in carbon emission. Taking scenario 1, 2, and 3 for example, with the same fuel structure, if the share rate of public transport is increased respectively by 1%, 5%, and 10%, carbon emissions (mid-value) will be respectively 2,333 ten thousand tons, 11794 ten thousand tons and 23619 ten thousand tons. In contrast, fuel structure does not have such a big impact; 10% changes in

fuel structure may lead to hundreds of thousands of tons changes in carbon emission.

The above scenario analysis discloses the impact on carbon emission from different fuel structure and share rate. Though there is no data to confirm the increase in share rate of China's public transport since the implementation of public transport priority policies, it is a fact that deteriorated motorization and urban slow system lead to decreased share rate of walking and bicycling. Judging from this sense, it is a recognized fact that the increase in traveling amount of public transport is partly from the increase in the share rate of public transport. Since the implementation of public transport priority policies, the abnormal increase in public transport amount (surpassing historical growth rate and urban traveling growth rate of the same period) may mainly be from the increase in share rate.

It can be seen from Table 10.19 that before the implementation of public transport priority policies (1997–2003), the average annual growth rate of the traveling amount of public transport is 5.9%; since 2004, with the implementation of public transport priority policies, the average growth rate reaches to 13.5%. Assuming that without public transport priority policies and growing with the average rate of 5.9%, the passenger transport amount of public transport is about 36.4 billion people-time less (defined as hypernormal increment hereafter). The contribution of this hypernormal increment lies in mode shifting structure, i.e., whether it is shifted from cars or from other modes. Based on the above fuel structure parameters, assuming that the average load factor of public transport is 70, the average traveling distance is 7 km and other parameters remain unchanged, the carbon emission deduction from public transport priority policies from 2004 to 2008 is about 40.72 million tons. Of course, the share rate of walking and bicycling is declining and part of the hypernormal increment may come from walking and bicycling and the actual carbon emission deduction may be less than the above number.

(3) *Policy and system analysis*

In China, a series of public transport priority policies is introduced against the backdrop of deteriorating traveling environment, the rapid growth in amount of private cars and increasing congestion in cities. Recalling data from Table 10.11, the growth rate of amount of private cars in 2003, 2005, and 2006 are much higher comparing with that in other years. Meanwhile, the growth in traveling times and the total volume of public transport in

Table 10.19. Changes in traveling amount of public transport before and after the implementation of public transport priority policies.

Year	Passenger transport amount (hundred million people-time)	Growth comparing with that of the last year (%)	Average annual growth (%)	Normal passenger transport amount (hundred million people-time)	Hypernormal increment (hundred million people-time)
1996	256	—	—	—	—
1997	279	8.9%		—	—
1998	292	4.5%		—	—
1999	320	9.7%		—	—
2000	341	6.6%	5.9%	—	—
2001	351	2.8%		—	—
2002	373	6.3%		—	—
2003	381	2.3%		—	—
2004	427	12.0%		404	24
2005	484	13.2%		427	56
2006	466	−3.7%	13.5%	453	13
2007	555	19.0%		479	75
2008	703	26.7%		507	196

2003 was declined. In the first quarter of 2004, the strategic decision to launch public transport priority, though to some extent passively, is timely. The policy in 2006, implemented when the driving force of previous policies is declining, is also very timely.

These policies have the following characteristics:

(1) The strategic positioning of public transport development priority:

Based on experience of some developed countries, in the process of rapid motorization and with people's living standards increasing, the proportion of car traveling will increase and urban public transport is faced with the danger of being marginalized. China is no exception, which can be seen from the more than 20% of growth in amount of private cars in China for several years. Document Jiancheng (2004) No. 38 states: "to establish the dominant position of public transport in urban traffic for five years." Public transport development priority will help to establish the status of public transport in urban development process, ensure most people's traveling choice in most cases and protect public transport from being marginalized.

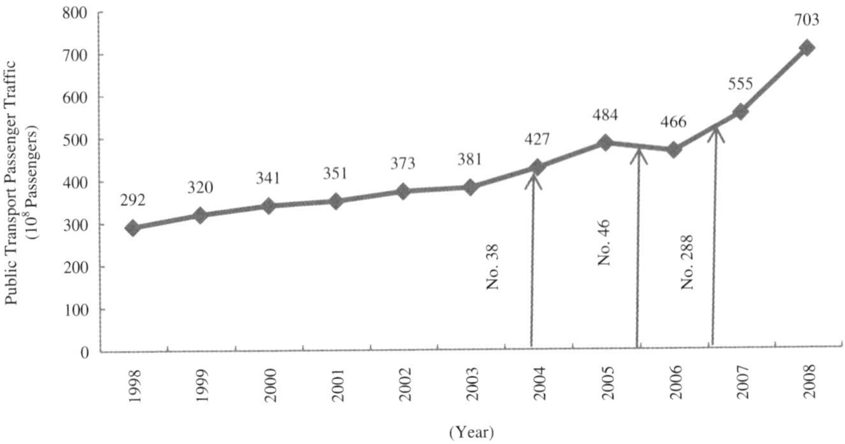

Fig. 10.22. Changes in public transport amount and implementation timing of these policies.

From the perspective of reducing carbon emissions, this positioning is the foundation of all other measures.

(2) Clarifying requirements of public transport supply and helping to raise the share rate of public transport:

These documents provided quantitative requirements to the target of public transport development priority; for example, an requirement was specified that within five years "300 m-radius public electric vehicles stops coverage will reach more than 50% in built-up area and more than 70% downtown. In large cities, large volume public transport should be the backbone of traveling, conventional public electric cars should be the main part and taxi and other public transport should be used as supplement". For infrastructure, requirements of improving station facilities of urban public transport and constructing private road system for public transport are raised. Policy effects analysis shows that during the rapid development of urbanization, the growth in traveling demand is beyond doubt. How to increase supply to meet those needs determines the impact on carbon emission. To increase the coverage of public transport and infrastructure for public transport is the prerequisite to attract growing traveling demand to public transportation, reduce traffic energy consumption and reduce carbon emissions. Although the policy did not make request on amount of public transport vehicles, lanes, BRT corridors, the construction of bus stations and the rapid growth in demand will naturally encourage carriers to expand production and increase investment in vehicles. The curve of public transport ownership above also illustrates this point.

(3) Clarifying the target of share rate by public transport:

The Document proposes to raise the share rate of public transport in large cities to above 30% in five years and above 20% in medium cities. Although this goal has not been realized, it can be seen from the above analysis on carbon emission estimations that the share rate of public transport has a significant impact on carbon emissions. 1% increase in the share rate of public transport leads to hundreds or even thousands of tons of reductions in carbon emission. So, even though the target has not been fully realized, this policy contains the downward trend in the share rate of public transport, and the resulted carbon reduction effect should not be underestimated either. Therefore, the demands on share rate are necessary.

(4) Requirements on sector reform and public transport operation quality help to improve the production efficiency of public transport and reduce fuel consumption:

The policy of deepening reform in state-owned enterprise, introducing competition mechanism and improving the technological level of industries is essentially making requirements on the operation management level and production efficiency of public transport. In addition, it is requested that the average operating speed of public transport should reach 20 km/h in 5 years, punctuality rate reach 90%, the transport time between any two points of large cities no more than 50 minutes, and medium cities less than 30 minutes. These objectives directly or indirectly make requests on the efficiency of public transport. These requirements can help to increase direct traveling and attractiveness of public transport, reduce mileage and carbon emissions, and save fuel.

These Policies, on the whole, play a very positive role in promoting supply of public transport, increasing its attractiveness to improve or maintain its share rate and improving its productivity to reduce energy consumption. However, these policies also have some shortcomings as to carbon emissions.

First of all, the original intention to develop these policies is to ease urban congestion, improve residents' traveling environment and achieve sustainable development of cities through developing public transport. Therefore requests on increasing traffic supply, improving public transport quality and increasing the share rate of public transport are made clear but there is no specific consideration of energy-saving and emission reduction. For example, the low fare strategy of public transport's public service orientation greatly improved the share rate of public transport; in Beijing since the implementation of low fare policy in early 2007, the share rate jumped by 4%. However, according to Beijing Traffic Development Research Center, this part of increase is mainly from walking or bicycling groups. In this sense, the increase in share rate of public transport not only failed to reduce carbon emissions but may increase carbon emissions. In another example, the fuel subsidies are essentially conducive to the development of public transport, but the implementation all over the country is different, with some in accordance with volume, some by amount of standard vehicle ownership and some by fuel consumption. Measures of these subsidies could lead to very different effects on carbon emissions. Subsidy by standard vehicle amount might encourage operators to purchase vehicles, but cannot guarantee the improvement of productivity, while productivity is the controlling factors of reduction. Subsidy by fuel consumption may actually encourage enterprises to consume more fuel and produce more emissions.

Secondly, standards of performance evaluation, especially evaluation methods related to carbon emission reduction, are lacking for public transport enterprises. Public transport companies are generally large and medium state-owned enterprises; its own production activities are complicated; coupled with its public service positioning, these companies do not belong to general commercial business, making the examination of public transport enterprises very difficult. For example, the cost structure of public transport is complicated, its production activities are throughout the city, making it difficult for outsiders to understand and grasp its true volume and structure. Plus its subsidy policies are varied widely, making it difficult to decide whether this policy is positive or negative from the perspective of carbon emission. For example, increased direct public transportation can improve direct bus traveling, help reduce traveling mileage and reduce carbon emissions; however, if the transfer is not free, operators will not be committed to this reduction. Again, if the subsidy is calculated according to vehicle miles, the operator may not have the power to compress vehicle mileage, reduce departure intervals and improve the load factor, etc. Current policies are in lack of a scientific and effective means to prevent the occurrence of these phenomena.

Again, assessment data collection methods, assessment continuity and incentive measures are lacking. For example, although objectives of share rate for the public transport traveling are required, there exists no explanation about how to adopt a consistent and comparable way to this target;

and judging from existing data, there is no continuity requirements, though the value of certain annual share rate is spontaneously measured by individual cities and individual studies. This makes it difficult to understand the implementation effect and uncontrollable to measure whether cities can meet requirements. Moreover, no incentives are provided in current policies.

Finally, specific operational methods of some policies need support from detailed rules. For example, the policy requires cities to "increase the input in public transport," but does not provide quantity requirements. A city may double its input which was quite small and the total investment comparing with other cities can still be very small. Another example is that for the requirement of "improving public transport station facilities", there exists no quantitative demand of public transport station facilities relative to the size of the cities and construction scale. In addition, there are no explicit requirements on construction scale and speed of bus lanes or BRT. This requires the development of local rules based on the national policy to quantify the implementation of public transport development priority. Listed in Table 10.20 are several policies on public transport development priority in Beijing this year. These policies are very specific and workable, but most cities have no such policies with quantitative indicators.

(4) *Conclusions and suggestions*

(1) Conclusions:

It can be concluded from policy effect analysis that development of China's public transport enjoys the following characteristics:

1) The traveling amount of China's public transport grows slowly before 2003, keeping pace with the growth of long-distance traveling amount. This reflects the common rule that traveling amount develops along with economy.
2) From 2003 to 2005 and between 2006 and 2008, the capital productivity and traveling amount of public transport witnessed a faster growth.
3) In 2006, the total traveling amount and capital productivity of public transport were slightly declined. The reasons are complicated but may be related to the following: After a fast development of two years, the growth bottleneck may arise, for example, the growth rate of traveling capacity was declined in 2005 after the quick growth in 2003 and 2004; besides, the growth rate of amount of privately-owned cars broke historical record.

Table 10.20. Policies of public transport development priority in Beijing.

Policies	Key points	ASIF features
The guideline of giving priority to public transport development	• Public transport's public benefit positioning • Speeding up railway construction (competing 561 km of rail transit by the year 2015 and raising its share rate from 6% to 23%) • Fully upgrading ground public transport system and implementing "four priorities": Land for facilities, investment arrangement, right distribution and fiscal support (in 2010, Beijing will realize 450 km of public transport lanes and continue to increase the investment on public transport, with the total investing reaching 71.5 billion yuan ($10.74 billion), accounting for 45% of total investment on traffic and infrastructure facilities)	ASIF
	The preferential policy of 60% discount for ground public transport (January 1, 2007)	S
	The policy of "2 yuan for the whole trip" for subways (October 1, 2007)	S

4) Seen from the amount of vehicles owned by ten thousand people, the growth rate of public transport is much lower comparing with that of private cars.
5) Since 2005, the share rate of public transport has been greatly raised in some cities.
6) Since 2005, BRT and bus lanes system has been greatly developed.
7) Since 2003, LPG and CNG buses have gained attention and been developed.
8) Since 2000, the stability and fast development of public transport lead to considerable carbon emission deduction from about millions of tons to hundreds of millions of tons.

(2) Suggestions:

From the above analysis, it can be seen that the development of public transport after 2003 is not accidental; during this period, a series of public transport priority policies is issued. This indicates that against the background of motorization, especially with the rapid growth of the amount of private cars, the policies of public transport priority did maintain the position of public transport and even promoted the development of public transport. Still, from the perspective of emission deduction, these policies have potentials in the following areas:

1) Different cities should develop practical and detailed rules based on their own actual conditions. National policies, as against the prevalent situation of the country, cannot provide realistic quantitative requirements for some of the key indicators. Local cities should develop practical and detailed rules in accordance with principles established in national policies and based on their own actual conditions such as public transport development and government financial resources. Detailed rules should be based on current situation and provide quantitative indicators for development goals of public transport; for example, the quantity of financial and material resources should be specified to ensure supply of public transport grow with demand during the process of urban development in the future and to ensure that scale of infrastructure for public transport (such as BRT, bus lanes, stop stations, the number of standard vehicle owned by ten thousand people, line length, coverage, etc.) grow with or exceed the growth of infrastructure (road area, parking spaces) for private cars.

2) To establish quantifiable performance evaluation system for the above policies, especially systems linked with fuel conservation. Specifically, an index system to measure the development of public transport should be established. This system should provide indicators for carbon emissions control and ensure its assessment and incentives not only to help to improve supply of public transport and improve the attractiveness of public transportation, but also reduce emissions. For example, in evaluating the share rate of public transport, structure of the urban traveling patterns should be considered to make urban traveling shift toward a greener mode. Second, integrated quantitative assessment standards for public transport enterprises linked with the companies' financial subsidies should be established to promote efficiency optimization of the public transport services and improvement of the image as well as the attractiveness of public transport. Finally, the development level of public transport should be used as one of the measurements for urban development and be linked with local political achievements. The information reflected in index system should be collected according to national standardized approach and calculation methods should be provided by national departments in charge. The results

should be made public each year, the best in annual reports on regional development, etc.

3) Attach importance to the guidance of city planning policies. The denominator to calculate public transport share rate is the total amount of traveling in various modes. While enlarging the numerator, the denominator should be reduced. The approach is to change existing urban planning concepts, so that the unnecessary motorized traveling can be reduced in the future. Seen from the world's new theories on and practice of urban planning, people-first and pedestrians-, bicycles- and public transport-oriented urban planning and development model can significantly reduce the total traveling amount and at the same time, significantly improve the traveling ratio of public transport. These advanced concepts and practice should be reflected in public transport priority policies in the future.

10.2.3 *Fuel economy standards*

1. *General*

(1) Fuel economy policies in China:

Automobile fuel economy standards are indicators for measuring the amount of fuel consumed by vehicle traveling a certain distance. China has introduced fuel consumption limits and related policies for light passenger cars, light commercial vehicles, operating passenger/freight vehicles and mopeds, motorcycles, tricycles, and slow speed trucks that are newly produced or are under production (See Fig. 10.23).

Main policies are as follows:

1) "Vehicle fuel consumption limits standards": The "Vehicle fuel consumption limits standards", China's first state-level mandatory standard, was formally approved and promulgated in 2004 by General Administration of Quality Supervision, Inspection and Quarantine and the State Standards Committee, and implemented on July 1, 2005. Since then, developing and using fuel consumption limits become another new means of energy saving, effectively coordinated with the State Council's initiative to build a conservation-oriented society. Passenger car manufacturers actively improve existing technologies and the energy conservation effect is remarkable

China's vehicle fuel economy standards are to some extent different from fuel economy standards in other countries. The fuel economy standards are applicable to petrol and diesel passenger cars with less than 3,500 kg in

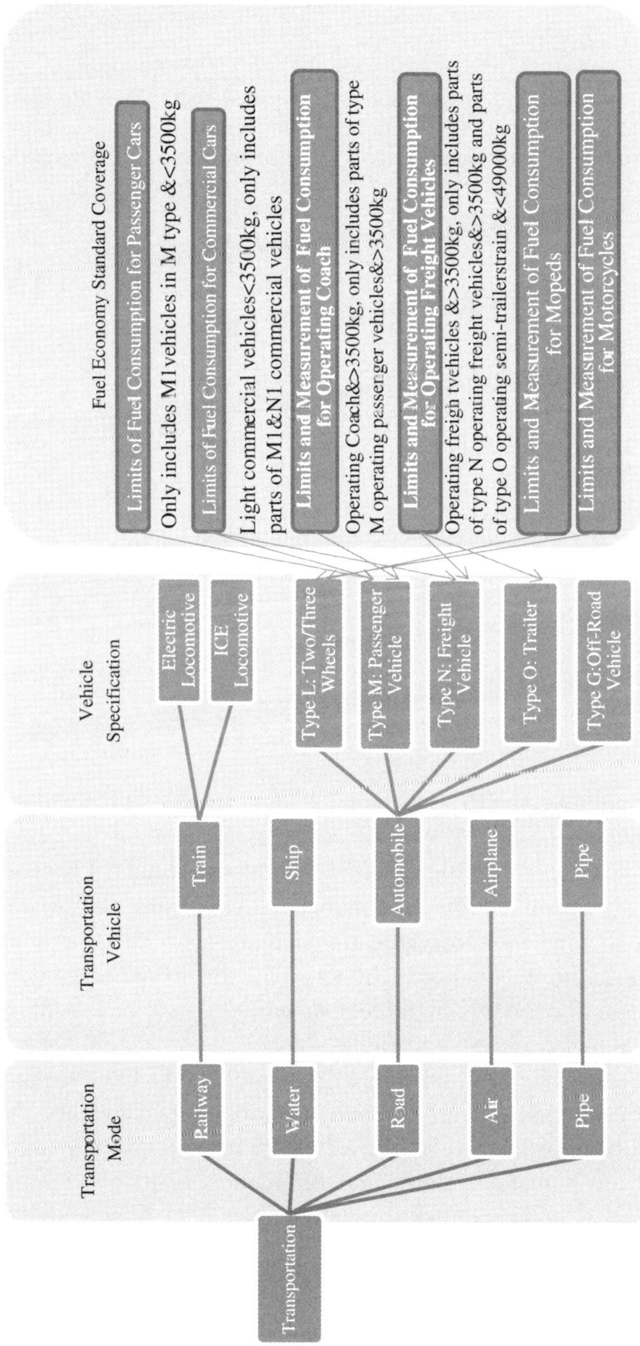

Fig. 10.23. Coverage of fuel economy policies in China.

Table 10.21. First-phase and second-phase vehicle fuel consumption limits standards (measurement: L/100 km).

Curb weight (kg)	Phase one Manual transmission	Phase one Automatic transmission/SUV	Phase two Manual transmission	Phase two Automatic transmission/SUV
750	7.2	7.6	6.2	6.6
865	7.2	7.6	6.5	6.9
980	7.7	8.2	7	7.4
1,090	8.3	8.8	7.5	8
1,205	8.9	9.4	8.1	8.6
1,320	9.5	10.1	8.6	9.1
1,430	10.1	10.7	9.2	9.8
1,540	10.7	11.5	9.7	10.3
1,660	11.3	12	10.2	10.8
1,770	11.9	12.6	10.7	11.3
1,880	12.4	13.1	11.1	11.8
2,000	12.8	13.6	11.5	12.2
2,110	13.2	14	11.9	12.6
2,280	13.7	14.5	12.3	13
2,510	14.6	15.5	13.1	13.9
3,500	15.5	16.4	13.9	14.7

Data Source: Vehicle fuel consumption limits standards (GB19578 — 2004).

curb weight, equipped with nine seats, and a maximum design speed of not less than 50 km/h in MI class[7]; the measurement for fuel consumption limits is liters per hundred kilometers and the standard is stricter as vehicle's weight with full equipment increases. At present, the gross weight of vehicles is divided into 16 segments; each segment with corresponding fuel consumption limits standards. In addition, within each segment, a 6% of grace is given to relatively high fuel consumption cars such as automatic transmission vehicles and SUVs (Table 10.21).

China's vehicle fuel consumption limits standards have been implemented in two phases respectively: for newly developed models, the first phase was fully implemented on July 1, 2005 and the second phase was implemented on January 1, 2008; and for models under production, the implementation is postponed for a year, i.e., the first phase starting from

[7]M class is passenger vehicles and N class is freight vehicles.

July 1, 2006 and the second phase January 1, 2009. Currently, relevant departments are preparing for the introduction of third-phase limits standards.

2) Fuel consumption limits standards for light commercial vehicles. Inspired by good effects from "Light Passenger Vehicle fuel consumption limits standards", the State Administration of Quality Supervision and Standardization Committee again issued a "Fuel consumption limits standards for light commercial vehicles" on July 19, 2007 and implemented it on February 1, 2008. The objectives of this document are light commercial vehicles with a total mass below 3.5 t and usually 9 to 12 seats, mainly including light trucks, light passenger buses and so on. To reduce the fuel consumption of light commercial vehicles is of great significance to fuel conservation throughout the automotive industry. The document requires that since Febuary 1, 2008, new certificated vehicles and their transformed vehicles should be consistent with requirements of the second phase; from January 1, 2009, vehicles under production and their transformed vehicles should be consistent with requirements of the first phase; from January 1, 2011 onwards, all vehicles should meet the requirements of the second phase.

3) "Limits and measurement methods of fuel consumption for commercial vehicle for passenger transportation" and "Limits and measurement methods of fuel consumption for commercial vehicle for freight transportation". In 2008, the Ministry of Transport successively promulgated the "Limits and measurement methods of fuel consumption for commercial vehicle for passenger transportation" and "Limits and measurement methods of fuel consumption for commercial vehicle for freight transportation" and implemented them on September 1, 2008. These two documents are respectively applicable to commercial vehicles for passenger transportation and commercial vehicles for freight transportation using diesel or gasoline as fuel and with a maximum total mass of over 3,500 kg. Commercial vehicles for freight transportation include trucks and dump trucks weighted 3,500 kg to 31,000 kg and tractor semi-trailers with a maximum total mass of 49,000 kg. Dump trucks and trucks are graded every 2,000 kg; tractor semi-trailers are divided into four levels according to their total mass. In 2009, "Methods on supervision and administration of operating vehicle fuel consumption" was promulgated and requires that vehicles defined in the above two documents should conduct business activities in compliance with the fuel consumption standards.

1975	**The American Government passed fuel economy standards**
1976	**Canada launched the program of fuel economy standards**
1978	**Australia formulated voluntary agreements of fuel economy standards for new model passenger vehicles**
1998	**Europe started the voluntary agreements of carbon emission limits for passenger vehicles**
1999	**Japan prepared relaxed fuel economy standards**
2000	**Canada formulated voluntary agreements of gas emission reduction for passenger vehicles**
2001	**China started the program to study policies and regulations on vehicle fuel economy standards**
2004	**China adopted the fuel economy standards for passenger vehicles**
2004	**Republic of Korea adopted fuel economy standards**
2006	**Japan raised its fuel economy standards for passenger vehicles and added fuel economy standards for commercial vehicles**
2007	**China implemented limits and measurement measures of fuel emissions for commercial vehicles**
2009	**Europe implemented compulsive carbon emission limits for passenger vehicles**

Fig. 10.24. Countries implementing fuel economy standards.

(2) Vehicle fuel economy standards of other countries:

In the past 30 years, vehicle fuel economy standards are adopted by many countries and directly or indirectly reduced the emission of greenhouse gases. Currently, fuel economy standards have become one of the most effective means to control vehicle fuel consumption. Besides China, America, Japan, EU, Australia, Canada, and the Republic of Korea all have their own fuel economy standards (Fig. 10.24), providing political basis for about 70% of the traffic emission all over the world.

In view of the historical, cultural, and political differences of different countries, the adopted national fuel economy standards are also different in limit objectives, the unit of measurement, limit parameters and control objectives (as shown in Table 10.22). In some countries, the implementation of fuel economy standards starts from voluntary agreements; however, except for Australia and Canada, to ensure the effectiveness and efforts of implementation, these policies are made mandatory. Other major greenhouse gas emitting countries such as Mexico and India have begun to consider similar policies (Climate works, 2010).

The fuel limit standards and reference objective are different for different countries. Currently, the strictest limit standard is that of Europe and Japan; China's limit standard, though adopted later, is stricter comparing with that of America, Canada, and Australia. Fuel economy is converted

Table 10.22. Vehicle fuel economy standards of different countries.

Country or region	Limit objective	Measurement	Limit parameter	Control objective	Measurement cycle
China	fuel	L/100 km	curb weight	Single vehicle	NEDC
EU	CO_2	g/km	curb weight	ACEA	NEDC
Japan	fuel	Km/L	Unloaded vehicle	the same kind of vehicle produced by the same manufacturer	JC08
America	fuel	mpg	Footprint area	Single vehicle manufacturer	U.S.CAFE
Canada	GHG	5.3 mt emission reduction	GHG emission	Vehicle fleet	U.S.CAFE
Austrilia	fuel	L/100 km	Unit standard	Vehicle fleet	NEDC
The Republic of Korea	fuel	Km/L	Engine size	Automakers fedaration	U.S.EPA cities

Note: Measurement cycle refers to the measurement release of fuel economy, including US CAFÉ (comprehensive average fuel economy), NEDC (new European driving cycle) and JC08 and vertical projected area of footprint area.

Data Source: J. Yuefu and W. Zhao, Study and formulation of vehicle fuel consumption standards of the next phase. China's sustainable energy project team, Beijing (2010); A. Feng and Sauer, 2004.

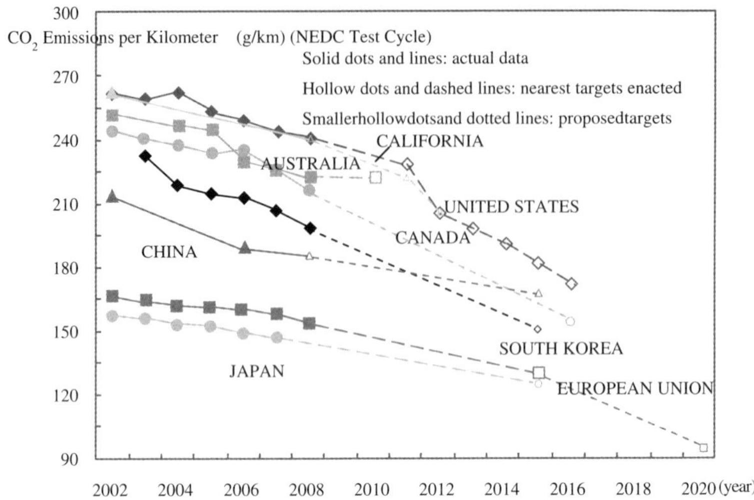

Fig. 10.25. Average carbon emissions per thousand kilometers and standards of all regions.
Data Source: International Council of Clean Transportation, Global passenger vehicle fuel economy and GHG emissions standards, (2010), http://www.theicct.org/2010/04/ghg-fe-standards-update/ (accessed on 12 October 2010).

into carbon emission standards per thousand kilometers in Fig. 10.25 and it can be seen that Japan and Europe have the lowest carbon emission standards per thousand kilometers, that of China is placed in the middle and that of America and Australia are the highest.

2. *Effects of fuel economy policies*

In China's many policies on fuel economy standards, the "Vehicle fuel consumption limits standards" was first implemented, and the effect is most significant. Since its promulgation and implementation in 2004, the average fuel consumption of new vehicles was reduced by 11.5%, to 8.06 L/100 km in 2006 from 9.11 L/100 km in 2002, resulting in an average decline of 3.02%. At the end of 2006, the cumulative fuel savings was about 1.18 million tons and the direct economic benefits was up to 7 billion yuan ($1.06 billion) (National Development and Reform Commission, 2007).

The annual report in 2009 from Innovation Center for Energy and Transport, based on vehicle fuel economy standards, calculated that the average fleet fuel consumption was 7.79 L/100 km in 2008 (A. Feng, 2010).

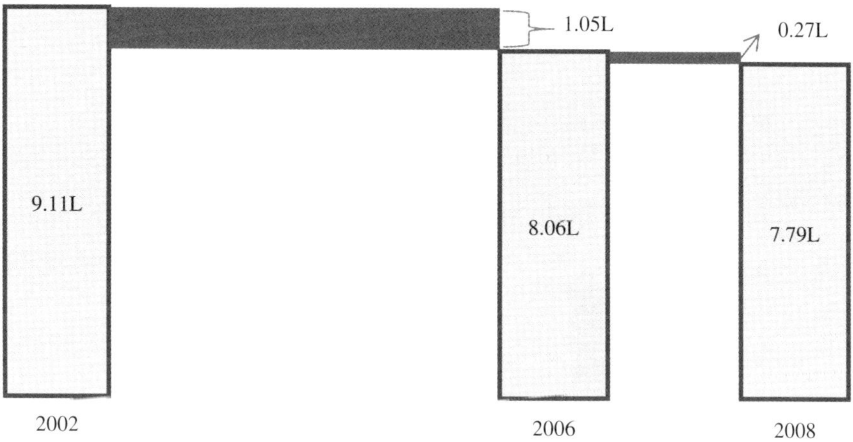

Fig. 10.26. Changes of vehicle fuel consumption per hundred kilometers. *Data Source*: J. Yuefu and W. Zhao, Study and formulation of vehicle fuel consumption standards of the next phase, China's sustainable energy project team, Beijing (2010); A. Feng, Annual Report 2009, (2010), http://www.icet.org.cn/news3.asp?Cataid=A00040001 (accessed on 15 October 2010).

Based on the above data, after new vehicle fuel consumption was significantly decreased in the first phase, the dropping of average fuel consumption was relatively slow, by 3.3% in the two-year period with an average annual reduction of 1.69 % (Fig. 10.26)

Fuel economy standards are very effective for most countries, particularly China. Although the engine technology in China is left behind by other developed countries, with the historical changes of China entering the auto society, institutional changes and technological improvements will lead to greater net returns. As shown in Fig. 10.27, using fuel consumption and emission of vehicles confined by limit standards during 11th FYP calculated with the fuel economy in 2005 as the reference value and using the fuel consumption and emission calculated with actual fuel economy as the actual value, the standard coal saved in 2006, 2007, and 2008 are respectively 220 thousand tons, 420 thousand tons and 600 thousand tons; carbon emission reduction are respectively 500 thousand tons, 910 thousand tons and 1,300 thousand tons. In the first few years of 11th FYP, the accumulated energy saving is 1.24 Mtce and carbon emission reduction is 2.71 million tons.

The Chinese government is currently preparing for the introduction of third-phase fuel economy standards, which target at promoting innovation in vehicle technology, further improving fuel economy and narrowing the

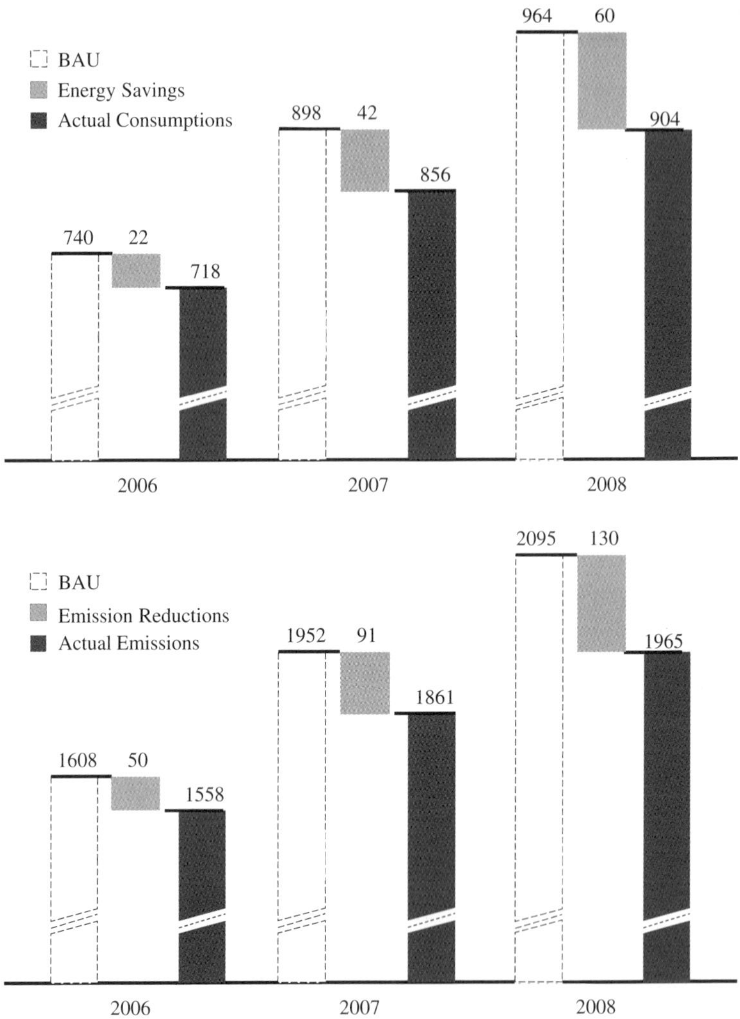

Fig. 10.27. Emission reduction achievements from light passenger cars during 11th FYP. *Note*: The fuel economy in 2005 and 2007 is calculated based on the average deduction rate and is converted into carbon emissions based on ICCT fuel index. If the average traveling distance of passenger vehicles is 20,000 km and all vehicles are gasoline vehicles (the proportion of diesel vehicles is quite small and this neglecting will not affect the results), 80% of annual sales is in compliance with the definition of passenger cars in "Vehicle fuel consumption limits standards" that curb weight is smaller than 3.5 t and the vehicle belongs to MI.

gap between foreign advanced countries, and will make efforts to promote the application of advanced vehicle energy-saving technologies and advanced engine and promote the upgrading of vehicle and engine products (China Quality News Network, 2009). In 2009, China's Ministry of Industry issued the draft standard of "Third-phase fuel consumption evaluation method and target for passenger cars". The fuel consumption limit is planned to be reduced by 20%, to 7 L/100 km, i.e., 167 g CO_2/km, in 2015 for new passenger vehicles (State Council Development Research Center of Information Network, 2010). As shown in Fig. 10.28, Innovation Center for Energy and Transportation, based on this document, estimated that the average fuel consumption of new passenger vehicles in 2015 is about 6.67 L/100 km (Feng, 2010). China continues to lower fuel consumption

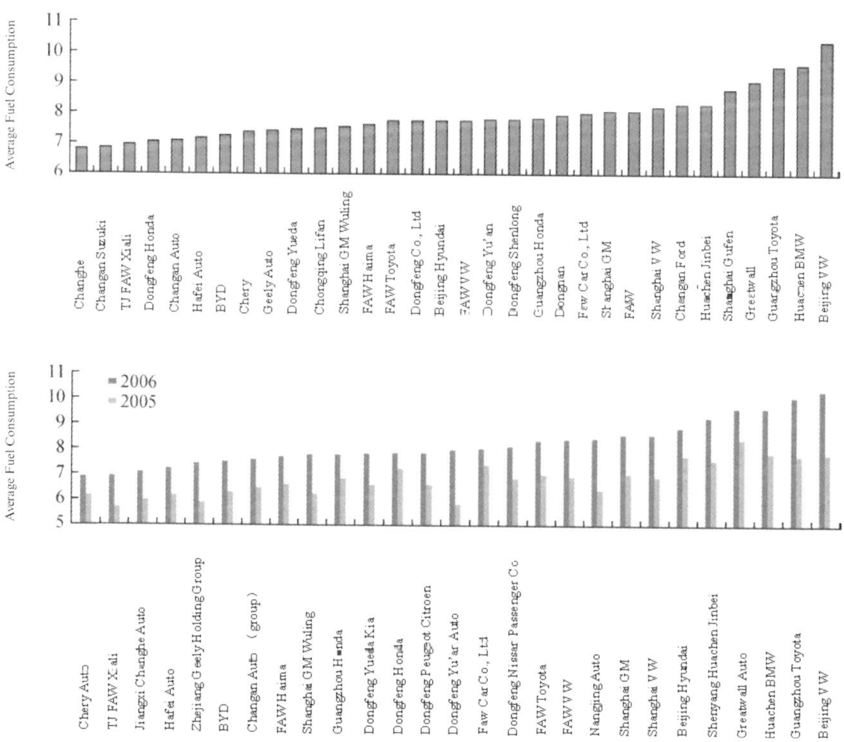

Fig. 10.28. Average fuel consumption in China in 2008 and predictions in 2015. *Data Source*: A. Feng, Annual report 2009. Innovation Center for Energy and Transportation, (2010), http://www.icet.org.cn/news3.asp?Cataid=A00040001 (accessed on 15 October 2010).

limits, indicating the intensity and determination of Chinese government to reduce greenhouse gas emissions in the transport sector.

Based on past experience and experts' predictions, Climate Works determines that if China, the United States and the European Union can raise their fuel economy annually by 3% to 4%,[8] carbon emission all over the world in transport will be reduced by 20% in 2030. If China can reduce vehicle fuel economy standards to 4.70 L/100 km in 2025, carbon emissions can be reduced by 280 million tons in 2030 (Climate Works, 2010).

It can be seen that fuel economy standards have great potential. However, in the past three years, the trend of large passenger cars balanced out most of the progress in fuel efficiency brought by fuel economy standards. The average curb weight of passenger vehicles in 2002 was only 1,230 kg, and it increased to 1,356 kg in 2006, an increase of about 10%. Starting with vehicles with a curb weight of 1,230 kg, the corresponding fuel consumption is 9.11 L/100 km, and when curb weight is increased to 1,360 kg, the relative fuel consumption should be increased to 9.73 L/100 km. In other words, without considering other factors, vehicle fuel consumption increased by weight increase from 2002 to 2006 is 6.8% (Yuefu, 2010). The engine displacement, power and vehicle curb weight have witnessed different degrees of growth and sales of imported cars have also gone up. Seen from imported models, in 2007, the increase of imported cars is faster compared to the domestic cars, in which the proportion of large displacement cars reached the highest level in history and the proportion of 2.5 L displacement models is close to 70% (Yang, 2010). Policy-makers should continue to improve fuel economy policies against the trend of large, heavy and large-displacement vehicles. Box 10.3, taking the United States for example, pointed out that large vehicles have adverse impact on fuel economy, providing reference for China.

3. *Fuel economy policy analyses*

During the 11th FYP period, energy saving and emission reduction become the priority for all industries and all levels of government. For the automotive industry, energy conservation is seen as a great challenge and a more important issue to be solved. The policies and regulations related to fuel economy standards introduced in China have their own characteristics. In this chapter, the characteristics of "Passenger car fuel consumption

---

[8]The annual increase rate of European Union's economy is 3.5%.

Box 10.3. The increase in vehicle mass runs counter to the raise of fuel economy.

In US, fuel economy standards are stalled. The reasons can be easily found: from 1985 to 2008, the actual energy-saving technology has been moved forward, but car manufacturers are making use of new energy technology to develop heavier high-horsepower engine vehicles. This completely balanced out the effect of technological progress in energy conservation, making meaningless more stringent fuel economy standards. During this period, the average mass of vehicles in the US is increased by about 42% and the engine power was increased by about 26% over the same period.

Data Source: Climate Works Foundation and the International Council on Clean Transportation. 2010. Policy Efficacy Project-Transportation.

limits", China's mandatory national standards, is analyzed to understand the unique characteristics of phase, policy priorities, and policy objectives and operation efforts.

(1) Phased implementation plan

Different from policies on fuel economy standards in other countries, China is the only country using phased strategy to raise vehicle fuel efficiency. This strategy is closely connected to China's current situation, based on current technology development and is beneficial to China.

"Vehicle fuel consumption limits standards" was successfully implemented in two phases: the limit standard in phase one is equal to the average vehicle fuel consumption amount in 2002; the limit standard of fuel efficiency in phase two was raised by 10%. Currently, related departments are preparing for third-phase fuel economy standards to gradually restrict the fuel economy in China's auto industry.

"Fuel consumption limits for light commercial vehicles" is implemented on Febuary 1, 2008, and phase-two limits will be applied since January 1, 2011. Phase-one limits in "limits and measurement methods of fuel consumption for commercial vehicle for passenger transportation" are applied since its implementation and phase-two limits are to be applied 19th month after. "Limits and measurement methods of fuel consumption for commercial vehicle for freight transportation" also provides limits in two phases.

The limits in phase one makes 10% to 20% of high fuel consumption vehicles not suitable for transportation; limits in phase two will be implemented one year and a half later; fuel economy in phase two will be raised by 10%.

The implementation of China's fuel economy standards has the characteristics of announcing in advance and implementing by phases. Compared with other countries, this method has three major advantages. First, the phased fuel economy standards provide automobile manufacturing industry with sufficient preparation time, preventing car manufacturing industry from losing benefit while gradually increasing vehicle fuel efficiency. The base limit point of the average vehicle fuel consumption amount in 2002 also fully takes into account the current level of technology, greatly improving their viability. Second, the phased fuel economy standards provided reference to technical indicators for enterprises to eliminate outdated products and achieve technical progress targets for energy saving. The automobile manufacturers can develop energy-saving technology and innovation based on fuel consumption limits of the next phase. Finally, the phased fuel economy standards effectively guaranteed the sustainability of automotive industry to improve energy-saving technologies. The retreat from fuel efficiency caused by America's stalling fuel economy standards after 1985 and separation from standards made policy-makers recognize the importance of policy sustainability.

(2) Focusing on major issues and highlights the key point

The increase in fuel economy standards is based on vehicle curb weight, and increase in fuel efficiency limits go with increase in vehicle mass. The reason is that higher mass brings higher fuel consumption. To encourage miniaturized and light vehicles and inhibit the scale of large-mass vehicles with high fuel consumption, fuel consumption limit for low-mass vehicles is looser, and limits for high-mass vehicles are stricter. As shown in Fig. 10.29, from the perspective of limits standards for passenger vehicles, the fuel consumption limit for the lowest mass in the first phase as well as mass in a higher phase is 7.2 L/100 km, thus enjoying relaxed restrictions. For vehicle curb weight from 750 kg to 3,500 kg, the fuel consumption limits of the highest mass section in phase one and phase two are respectively 115.3% and 124.2% higher comparing with that of the lowest mass section. However, the limit for the highest mass segment is 366.7% higher than that of the minimum mass section. It can be seen from Fig. 10.29 that the curb weight of the last two sections ranges larger than that of other sections but the limit growth difference is not so much.

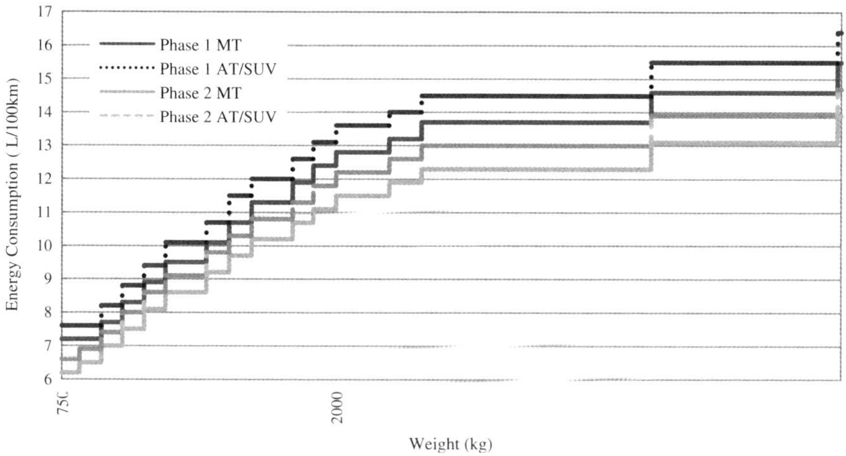

Fig. 10.29. Limit standards for passenger vehicles with different curb weights. *Data Source*: Vehicle fuel consumption limits standards (GB19578 — 2004).

Seen from current situation, the strategy of focusing on major issues is effective, with fuel consumption deduction in different curb weight sections ranging from 10% to 20% and fuel consumption deduction of large curb weight quite obvious (Feng *et al.*, 2008).

(3) Single vehicle management and advancement with times

Different from other countries with vehicle type, vehicle manufacturers or automobile manufacturers association as objective of fuel economy standards, China's implementation objective of fuel economy standards is fuel economy amount of single vehicle. This kind of objective setting helps to promote vehicles' fuel economy and reduce management costs. According to statistics, there were more than 130 vehicle manufacturers as early as in 2008. Due to the independence of each manufacturer, using manufacturer as the control objective at that time would greatly increase the management costs. From the effect assessment of the first two phases, the advantages of using single vehicle as the control objective are in that phased requirements of fuel economy standards can be timely completed, unqualified vehicles abolished, outmoded products eliminated and energy saving technology developed and progress promoted. However, the flexibility of this strategy cannot be compared to strategies using vehicle model or car dealers / association as control objective. This is because China's auto industry is still in its initial stage of development, and to limit fuel consumption efficiency is not conducive to the diversification of automotive products. If car dealer is used as control objective, then the formula to calculate fuel economy should be:

$$\text{Average enterprise fuel economy} = (F_aN_a + F_bN_b + F_cN_c + F_dN_d)/(N_a + N_b + N_c + N_d), \quad (10.1)$$

wherein a, b, c, and d refer to different modes of the same manufacturer; N refers to the production quantity of one mode and F refers to its fuel economy.

China's current fuel economy standards require that single vehicle of F_a, F_b, F_c, and F_d meet the standard for their curb weight section; Using car manufacturers as the limit objective allows car dealers to have more freedom, capable of planning to produce different vehicles to meet different needs of consumers while ensuring the overall fuel economy standards.

Although too rigid, using single vehicle as control objective is necessary in the initial stages of fuel economy standards in China. With gradual development of car manufacturing industry and gradual unification of vehicle fuel consumption standards, while guaranteeing the goal of energy efficiency, relevant departments are considering using larger units as control objectives in fuel economy standards in the future.

(4) Flexible standards and encouraging creativity

The passenger vehicle fuel economy standards not only changes with the mass of vehicles, but also gives about 6% of grace to vehicles using advanced technologies of higher fuel consumption, such as automatic transmission cars, to encourage technological innovation. In addition, drawing lessons from the US, SUVs with high fuel consumption are required to meet standards of automatic transmission vehicles to prevent sharp increase of vehicles with large mass and high fuel consumption.

In addition, China's fuel economy standards only provide provisions for gasoline and diesel vehicles, excluding gas and alcohol fuel vehicles. Although the vast majority of China's automobile uses gasoline and diesel, this policy has the intention to encourage innovation and development of vehicles using new energy-saving fuel.

(5) Positive effects and easy operation

Vehicle fuel economy standards in China are made entirely in accordance with technical characteristics and the actual situation of the Chinese automotive products, targeting at "eliminating backward products and promoting technological progress," playing a crucial role in promoting development and applications of automotive energy-saving technologies, improving vehicle fuel economy standards, controlling and addressing energy and environmental issues caused by vehicle fuel consumption. The policy not only produces positive results but also operates easily. The policy of vehicle fuel economy standards is easier to enforce than other policies; from the perspective of controlling scope, this policy only addresses a relatively small number of car manufacturers instead of tens of thousands of consumers, so the operation is relatively simple.

4. *Fuel economy institutional analyses*

The successful implementation and satisfactory results of China's vehicle fuel economy standards not only benefited from the policy formulation itself, but also benefited from the continuous improvement of system environment. This section will analyze the strength and weakness of current institutional environment for China's vehicle fuel economy standards.

(1) Strength of institutional environment

The average fuel consumption of passenger cars is dropped by 11.5% in the first phase of China's passenger vehicle fuel consumption limits. In the second phase, fuel efficiency is raised by of 10%, and the fuel efficiency in the next phase is expected to witness a substantial increase, reaching the target of 2015 7 L/100 km in 2015. In contrast, the United States implemented fuel economy standards from 1975 to 1985 under the impact of Middle East oil embargo, when fuel consumption efficiency of motor vehicles and light trucks was nearly doubled; however during 1985 and 2008, the US federal government took no new initiatives and the fuel economy standards remained 8.39 L per hundred kilometers (28 miles per gallon). Until recently, the US federal government proposed new regulations requiring that the average fuel economy should achieve 35.5 miles per gallon by 2016 instead of by the original 2020 (The White House, 2009).

China made great progress in fuel economy in a very short time due to the following main reasons in terms of its institutional environment: China's entering into the automotive society, complementarities with other policies and active cooperation of policy stakeholders.

1) China started to enter automobile society: Compared with large industrial countries such as Europe and America, the development of vehicles in

China started late. Not until the early 21st century has China's automobile industry grown rapidly, when cars really started to enter ordinary people's family. 110 years have passed from the first Auto Show in the United States in 1900. The vehicle industry in America has a long history and is relatively mature, and the automotive industry associations have considerable influence. In relation to policies and regulations on fuel economy and new energy, the auto industry and oil industry brought considerable resistance to the policy-making process of the United States, making old fuel economy standards not updated for a long time. When China started to formulate fuel consumption limits for passenger cars, the automotive industry was still in a mess and fuel economy standards are varied. The automotive industry was not familiar with the concept of "fuel economy standards", so not much impediment was met when China started to implement fuel economy standards.

Considering China's large population, China's automotive industry is blessed with unlimited potential in its future development. China's total auto sales in 2009 surpassed that of the United States, but in 2007 when every thousand people in the US owned 821 cars and every thousand people all over the world owned 145, every thousand people in China only owned 33 cars (Japan Automobile Manufacturers Association, 2009). Although the total amount of owned vehicles was increased, the amount of cars owned per capita was very low, leaving huge room for development. The rapid development of vehicle sales and ownership also shows the extraordinary significance of fuel economy standards in energy saving.

As shown in Fig. 10.30, the amount of vehicles owned in China reached 199 million, in which 85 million vehicles are cars, according to the latest data from Traffic Management Bureau of Ministry of Public Security in October, 2010 (Economic Observer Online, 2010). That means a 12% increase than that of the year of 2009. In the past decade, the amount of vehicle ownership in China has been growing rapidly and steadily. Forecasts at home and abroad are that there will be over 120 million vehicles in 2020; the Equipment Industry Bureau of Ministry of Industry and Information Technology forecasts that there will be over 200 million vehicles in China in 2020 (Xinhua News Agency, 2010).

The same growth momentum can be seen from production and sales of China's auto industry. According to statistical data processing in this report, growth in China's car production and sales in 2007 over the previous year were 22.0% and 21.7%. In 2008, influenced by the global financial crisis, the growth than that of the year of 2007 was only 4.7% and 6.7%; however,

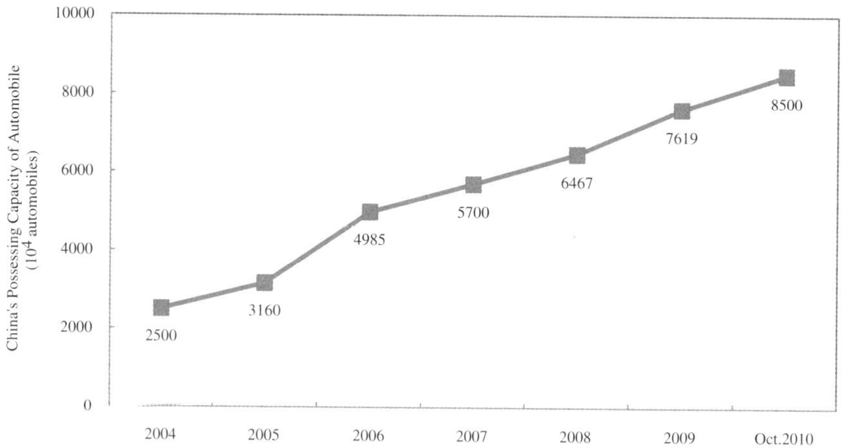

Fig. 10.30. Amount of vehicles owned in China.

Note: The amount in 2010 is an estimated number in October according to news report.

Data Source: China Automotive Technology and Research Center, please refer to http://www.autoinfo.gov.cn/autoinfo_cn/cszh/qcbyl/A012101index_1.htm.

another sales peak ushered in 2009, when the automobile production and sales amounts were respectively 13.791 million and 13.644800 million, a record-high growth of 48.30% and 46.15%. China Automobile Association forecasts that the growth rate in 2010 will be around 10% with 15 million vehicles by the end of 2010 (China Automobile Industry Association, 2010). The sales in 2010 are expected to be 15.125 million with an increase of 15.2% (Sinotrust, 2010). The strong momentum in 2009 drove the share of China's sales in the global market to 18.5% in 2009 from 13.5% in 2008. The share is forecasted to be 20.9% in 2013 (Xinmin Auto Net, 2010).

As shown in Fig. 10.31, in production and sales increase in 2009, the production and sales of passenger vehicles are respectively 10.3838 million and 10.3313 million, an increase of 54.11% and 52.93%; those of commercial vehicles are 3,407,200 and 3,313,500, an increase of 33.02% and 28.39% (China Automobile Industry Association, 2010). Sales of passenger vehicles in 2010 are expected to reach 11,328,100, an increase of 15.2%; sales of commercial vehicles are expected to reach 3,796,700, an increase of 15.3% (Sinotrust, 2010). The great proportion of passenger vehicles in China determines the great contribution of fuel consumption limits on passenger vehicles to energy saving in China. The later released fuel consumption limits for commercial vehicles will also vigorously promote the development and progress of energy efficiency of commercial vehicles.

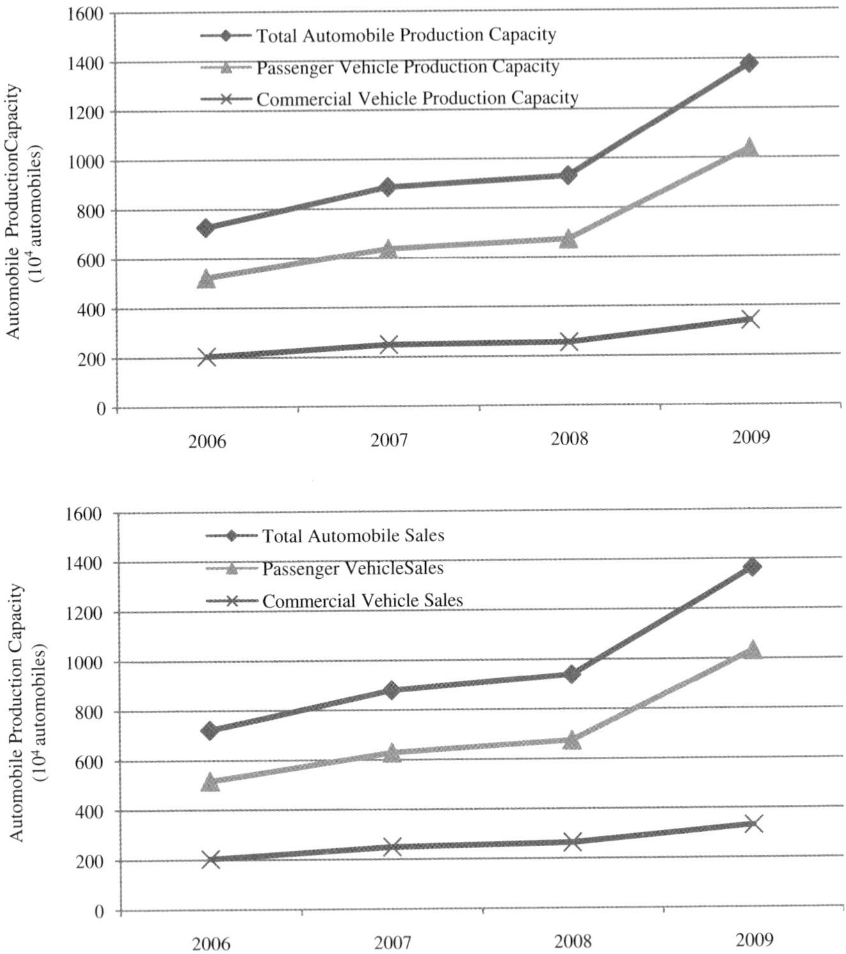

Fig. 10.31. Chart of passenger vehicles and commercial vehicles in China's vehicle productions and sales.
Data Source: CAAM, please refer to http://www.caam.org.cn/newslist/a40-1.html; Sinotrust, please refer to http://bjyouth.ynet.com/article.jsp?oid=61980922; China's vehicle production and sales data platform, http://auto.sohu.com/xcsj.

2) Complementarities with other policies: China's first national compulsory standard "Vehicle fuel consumption limits standards", officially released by the State General Administration of Quality Supervision, Inspection and Quarantine and the State Standards Committee in 2004, opened the door to energy saving through fuel consumption limits.

The following "Light commercial vehicle fuel consumption limit standards", "limits and measurement methods of fuel consumption for commercial vehicle for passenger transportation" and "limits and measurement methods of fuel consumption for commercial vehicle for freight transportation" brought a new round of changes in energy saving technology for automobile manufacturers.

Before the release of "Vehicle fuel consumption limits standards", the Chinese General Administration of Quality Supervision, Inspection and Quarantine issued in 2003 "Light vehicle fuel consumption test method" and implemented it at the end of the same year, laying a good foundation for China's release of fuel consumption limits standards. In 2009, the release of "Light vehicle fuel consumption labeling regulations," and "Methods on supervision and administration of operating vehicle fuel consumption" and other supporting policies improved the management measures for vehicle supply and played a positive role in successful implementation of China's fuel economy standards.

Effective fiscal policy is an indispensable part to establish and improve vehicle fuel economy standards and will effectively promote the overall fuel efficiency of vehicles. Policy preference is shown in changes in vehicle excise tax in recent years. The latest adjustment on vehicle excise tax by the Ministry of Finance and State Administration of Taxation in 2008 shifted the excise rate of small vehicles with the displacement below 1.0 L from 3% to 1%, vehicles with the displacement between 3.0 L and 4.0 L from 15% to 25%, and vehicles with the displacement above 4.0 L from 20% to 40%. After this adjustment, sales of passenger vehicles with displacement

Table 10.23. Changes of automobile excise tax in China.

Displacement	Tax rate before April 1, 2006 (%)	Tax rate before August 31, 2008 (%)	Tax rate after September 1, 2008 (%)
<1.0	3	3	1
1.0–1.5	5	3	3
1.5–2.0	5	5	5
2.0–2.5	8	9	9
2.5–3.0	8	12	12
3.0–4.0	8	15	25
>4.0	8	20	40

Data Source: "Notice on adjusting passenger vehicle excise tax policies" (Caishui [2008], No. 105).

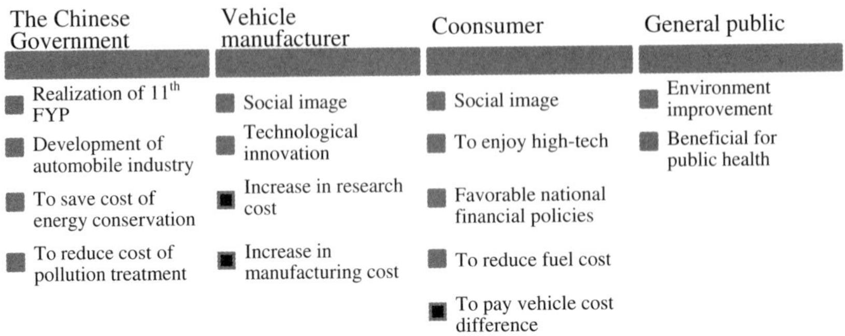

Fig. 10.32. Analysis chart of policy stakeholders.

below 1.6 L reached 7,196,000 in 2009, an increase of 71.3%, accounting for 69.7% of total passenger vehicle sales, an increase of 8% in its market share (Economic Daily, 2010), which is significant for vehicle structural optimization in China.

3) Positive cooperation of stakeholders: The effective implementation of China's fuel economy policy is inseparable from the positive cooperation of all stakeholders. These stakeholders include the Chinese government, auto dealers, consumers and the masses (Fig. 10.32). First of all, from the perspective of the government, i.e., the policy makers, the raise of vehicle fuel economy efficiency is beneficial to its emissions-reduction targets during the 11th FYP period. In addition, the auto industry can take this opportunity to optimize and upgrade and to eliminate outdated companies. Finally, in the absence of this policy, fuel consumption cost would be greatly increased, and treatment of greenhouse gases and carbon emissions would become a considerable expenditure for the government. Therefore, the Chinese government is not only the policy maker, but also the beneficiary of fuel economy standards. The Chinese government created an excellent institutional environment for the implementation of fuel economy standards and advocated a positive response from all parties.

Automobile manufacturers are direct audience of fuel economy standards and vehicle consumers are indirect audience. For these two parties, higher fuel economy standards will undoubtedly set a positive social image. Vehicle dealers, whether to show its social responsibility, public dedication or to update its technology and attract consumers, are willing to turn its brand into a green brand. However, vehicle fuel efficiency improvements also required manufacturers to increase costs on research and development as well as manufacturing and shifting.

For consumers, vehicle is a continuation of their social status. Thus, energy-saving vehicles can earn them a higher social status. Consumers can not only save fuel costs and enjoy advanced automotive technology while driving but also demonstrate their tastes. The cost of environment-protecting vehicles is higher than ordinary vehicles, based on their economic status, some consumers may not want to accept this cost increases because of their social image. Yet there are consumers purchasing low-consumption and new-energy vehicles because of financial and tax policy preference.

Finally, for the masses, the implementation of fuel economy standards means the optimization of environmental quality, which is beneficial to public health. This will get the support from most people. Therefore, the stable improvement of China's vehicle fuel efficiency is greatly related to the cooperation of policy stakeholders.

(2) Institutional environment weakness

Despite considerable improvement in China's vehicle fuel efficiency, the average fuel consumption of automotive products is still pretty high compared to that of Europe and Japan, countries with the world's highest levels of fuel economy. From the perspective of institutional environment, factors threatening China's continued rapid development of fuel efficiency are not few. Among them, Chinese people's preference to large-displacement vehicles and limitations of energy-saving technology become the two biggest problems.

1) Consumers' preference for large vehicles. The greater the mass of vehicles, the greater the energy consumption required for driving the same distance. Thus, large-scale vehicles will no doubt hinder the rapid increase of China's overall fuel efficiency. In 2002, the average curb weight of China's passenger vehicles is only 1,230 kg, which was increased to 1,356 kg in 2006, an increase of 10% (Yang, 2010). So consumers still prefer high-consumption, largedisplacement and multifunctional vehicles such as SUV. Although influenced by policy preferences in the last two years, the sales of small vehicles have increased significantly, the consumption market of high-consumption, large-displacement and multifunctional vehicles is also increased year by year, which is very unfavorable for the optimization of China's vehicle structure. Since 2002, the growth rate of large vehicles has been higher comparing with that of small vehicles for 7 years, and the growth rate this year is three times more comparing with that of small

vehicles, market share exceeding 10% for the first time (National Passengers' Association, Sinotrust, 2010).

2) Emission reduction "bottleneck." First, China's auto industry started relatively late, and energy saving technology is lagged behind by developed countries; although independent research has been carried out, there still exists a big gap. According to China's Sustainable Energy Program "Study and formulation of vehicle fuel consumption standards of the next phase", the elimination of backward high-consumption products and technologies and the application of variable valve, overhead camshafts and multi-gear transmissions help to improve the wide application of new technologies of vehicle fuel economy. Yet this document also pointed out that there still exists a big gap between developed countries, especially in the application of high-end technologies. Therefore, the research and development of China's passenger vehicle energy saving technology still has a long way to go.

Second, China's fuel quality significantly affects vehicle efficiency; China's fuel quality remains to be further improved to raise vehicle fuel economy through promoting energy conservation and applications of new technologies. If not raised with development of energy saving technologies, the fuel quality will undoubtedly restrict and hinder further improvement of fuel economy in China's entire auto industry.

Finally, urban traffic structure and road conditions are another important factor affecting the efficiency of fuel consumption in China. With rapid economic growth and growing urbanization in China, the urban population is rapidly expanding; people's living standards continue to improve, people's economic and cultural life becomes richer, and purchasing vehicle is incorporated within the scope of consumption. Large number of vehicles inevitably causes traffic congestion and vehicles will frequently accelerate, decelerate and stop to avoid accidents. This traveling condition and driving behavior exert negative effects on fuel consumption efficiency, greatly increasing fuel consumption and carbon emissions. How to ensure smooth traveling becomes an issue requiring unremitting efforts. In recent years, investment in China's urban transport infrastructure has been expanded, there are more and more viaducts and overpasses, and transportation network becomes more and more complicated. These improvements increased road area, temporarily alleviating traffic congestion in some cities. However, from the rapid increase in road traffic, high-density urban land use patterns easily limit the utilization of road space resource in the future. More rational urban transport planning is of utmost importance in addressing traffic expansion and traffic congestion.

5. *Suggestions*

From the above analysis it can be seen that successful implementation of China's fuel economy standards is benefited from good policy formulation and institutional environment. However, both the policy itself and the institutional environment have their weakness. The third-phase vehicle fuel economy standards are about to be implemented, and suggestions are given based on policy implementation and national conditions.

(1) Policy suggestions

1) From "focusing on major issues" to "giving consideration to more things". Seen from current situation, guided by the policy of "focusing on major issues" during the first two phases, the fuel consumption of high-mass passenger vehicles in China was reduced much greatly comparing with that of low-mass vehicles. The inadequate fuel consumption reduction of low-mass small-displacement vehicles reflected the inadequate technique advancement of such vehicles. Compared with developed countries with high fuel economy, China' fuel consumption economy is lagged behind, especially that of mini-vehicles with curb weight under 1,00 kg. Manufacturers' research and development of fuel efficiency of small vehicles is inadequate, lack of achievements and in need of the guiding of relevant policies. Therefore, when setting standards for the next phase, large vehicles will still be the focus of policy norms, and attention should also be given to fuel efficiency management of light vehicles.

2) Policy control objective should shift from single vehicle to vehicle fleet in order to make the Chinese auto market more flexible and freer. To ensure improvement of all vehicles' fuel economy and reduction of management cost, the control objective of the first- and second-phase passenger vehicle fuel economy standards is the fuel economy of single vehicle. However, this policy is less flexible than policies using vehicle model or Dealer Association as control objectives. Single vehicle control policies not only inhibit the diversification of automotive products, but also hinder the improvement of fuel efficiency of individual models at the lowest cost by the market automatically.

With the progressive development of China's automobile manufacturing industry and the gradual unification of vehicle fuel consumption standards, under the premise of achieving overall energy efficiency, using larger units as the control objectives of fuel economy standards in the next phase will be more beneficial to the overall goals.

3) To use carbon emission to replace fuel consumption per transport-unit. Using fuel efficiency as measurement provided car manufacturers with reference standard to improve vehicles' energy-saving technology, but it is not linked to carbon emissions. There are two advantages to measure unit energy efficiency by carbon directly: First, diesel-fueled vehicles have higher carbon emission than gasoline-fueled vehicles, thus avoiding confusion and speculation of different carbon intensity in diesel and gasoline, more conducive to emission reduction goals; Second, the use of carbon as limit measurement can build more profound reduction awareness for vehicle manufacturers and even the whole society. Using carbon emission to replace fuel consumption per transport-unit can encourage vehicle manufacturers to develop more effective emission-reduction technologies by limiting vehicles' carbon emissions and indirectly increase consumers' emission-reduction awareness on vehicles and all aspects of life.

4) To strengthen the implementation of China's fuel economy standards on imported vehicles. Because of the management functions and other reasons, enforcement of China's current fuel economy standards on imported vehicles is very weak. Imported vehicles generally have higher quality and emissions, so inadequate implementation of fuel-efficient standards on imported vehicles will greatly affect China's current energy efficiency achievement. Vehicles of the United States are known for their large quality and displacement, and their average quality is 40% higher comparing with that of vehicles produced in China (Climate Works, 2010). So in the next phase, when implementing fuel economy standards, the Chinese import and export sectors, such as Customs and General Administration of Quality Supervision, Inspection and Quarantine, should strengthen management force, deepen management functions and provide testing and formulate reasonable reward and punishment system to better achieve the objectives of the next phase.

(2) Institutional suggestions

1) Establish and improve economic incentives and regulatory mechanisms for supporting fuel economy standards. The supervision and management mechanism for China's current fuel economy standards is not fully completed; there are no corresponding economic incentives and punishment

measures in particular. Vehicle manufacturers that can meet fuel standards and continuously improve fuel efficiency should be given appropriate incentives, and those who do not meet fuel standards should be appropriately punished. The regulation and reward and punishment system of fuel economy standards in France has achieved a very good effect, and is expected to play a greater role in the future. United States also punish vehicle manufacturers that do not meet fuel consumption limit in the form of "gas guzzler tax". China should make up for deficiencies in supervision, use economic incentives and punishment mechanism to complete policies on fuel economy standards and automobile consumption tax in favor of small vehicles, and encourage and promote the improvement of vehicles' overall fuel efficiency.

2) Further establish and improve fuel economy standards in other areas of the transportation sector. Recalling Fig. 10.23, China did not make mandatory fuel consumption standards on other areas of the transportation sector. For example, there are no national mandatory fuel efficiency requirements for heavy-duty passenger vehicles and in the areas of aviation and water transport. At present, apart from Japan, no other country has provided requirements for medium to heavy vehicles. Exhaust emissions of medium to heavy-duty vehicles in China accounted for 85% of the entire transport emissions in 2005; and it is expected that the heavy-duty vehicles will account for 12% of the total vehicles in 2030, producing 60% of all emissions (Climate Works, 2010). Mandatory fuel economy standards in all areas of the transportation sector can improve the overall transport fuel efficiency in China, accelerate the emissions-reduction targets, and enable people from different fields to compete, learn from each other and exchange experiences. In addition, there are social vehicles without enough information and not counted in statistics of energy consumption of China's road transport. Large number of social vehicles complicated China's fuel economy situation. Therefore, strict control of vehicles and vehicle information management cannot be ignored.

3) Enhance independent innovation capacity and develop reasonable and effective transport planning. The limitations of energy-saving technologies and road transport planning of China's auto industry have become the "bottleneck" to further improve fuel economy. Reports indicate that, despite that the energy-saving fuel efficiency technologies become more complicated, there is no substantial improvement in costs (National Research Council, 2010). The ability of independent innovation of China's auto industry should be improved, and the long-term production, learning and

research cooperation mechanism is still very weak. The establishment of National Engineering Laboratory for Automotive Energy Saving and Conservation can help to establish advanced test facilities for energy saving technology research and development and form a research and development platform for industry-leading vehicle energy-saving technologies, new materials and new energy (China Economic Net, 2010). As for road transport planning, scientific and rational planning for urban road traffic should be provided to optimize urban land layout, improve urban operational efficiency and ensure an efficient, economic and environment-friendly traffic environment, thus avoiding fuel efficiency reduction and unnecessary traffic congestion caused by vehicles and pedestrians.

10.2.4 *Rail transit*

(1) *The replacement of civil aviation by rail passenger transport has been reflected in local areas*

To strengthen construction of rail transit is one of the characteristics of low-carbon development in the transportation sector. The development of high-capacity rail transit can effectively meet the growing traveling demand in China, and reduce carbon-intensive traveling, such as reducing the proportion of vehicle-miles-traveled (VMT) and air-miles-traveled, thus to some extent slowing down the rapid growth in carbon emissions and energy consumption. China has formed a railway network of a considerable size connecting the East to the West and North to the South, but this network is faced with challenges from high energy-consumption modes of transport. During the 11th FYP period, China adopted a series of policies to promote construction of rapid passenger rail network, exhibiting unprecedented policy support. Despite that in a short term the contribution of these policies to China's low-carbon development of the transport sector is difficult to measure, these policies have put a certain pressure on civil aviation, and their actual influence has been initially revealed. In 2009, the General Office of Civil Aviation Administration of China organized the research of "The impact of high-speed railway on civil aviation development and policy suggestions" to analyze the impact of "high speed" rail transport, marked by the six great speedups and construction of large-scale passenger line, on air traffic. Seen from analysis results, the rise of high-speed rail transport may bring major structural changes in aviation in the future market, change the way consumers

Table 10.24. Comparison of data on Beijing–Taiyuan flight segment from China Eastern Airlines before and after the operation of high speed rail.

Time	Flight quantity	Number of available seats	Total booked tickets	Passenger load factor (%)	Flight income (yuan)	Average ticket price (yuan)
March, 2009	420	60 842	47 522	78.11	24 817 230	522
April, 2009	279	40 603	29 259	72.06	13 475 336	461
Reduction	34%	33%	38%	60%	46%	12%

Data Source: Discussion group of Civil Aviation Administration of China. 2009b. The impact of high-speed railway on civil aviation development and policy suggestions (continued). *Civil Aviation*, 105:41–44.

travel, change airline's business model, and at the same time change the air route network structure. Realistic competition impact is directly reflected in three aspects: ① the loss of time advantage of civil aviation in short-distance routes within 800 km; ② reduction of airlines transport capacity; ③ decrease of flight load factors and profitability levels. These effects have been actually reflected in local areas. For example, China Eastern Airlines accounted for 54% of the Beijing–Taiyuan air-traveling market, which declined by 34%, losing nearly 38% of customer source since April 1, 2009, when the Shijiazhuang–Taiyuan line was put into use (Table 10.24).

(2) *Political analysis of trail transport development*

At present, the length, network coverage and volume of passenger transportation of rail transit lines are lagged behind by highway traffic, and the growth rate of rail transit is also faced with challenges from aviation (Table 10.25). The average annual growth rate of highway mileage from 1978 to 2008 is 4.7%, building freeway of 60,300 km; and the average annual growth rate of civil aviation mileage (repeated distance excluded) is 9.5%, while the average annual increase of railway network is only 1.4%, from 51,700 km to 79,700 km.

During 11th FYP period, China has adopted "moderately advanced" planning ideas to improve the planning objectives of railway mileage and broaden investment and financing channels of rail transit construction. In 2009, the national railway mileage reached 86 thousand km, raising the length of mileage to the second in the world. In that year, 5557.3 km new

Table 10.25. Coverage of China's transportation network.

Passenger transport network	Policy documents	coverage
High-speed rail passenger transport	Long-term rail network planning (2008 adjustment)	Planning: To establish fast passenger passage between capital cities and medium and large cities, plan to build "four vertical and four horizontal" passenger lines and inter-city passenger transport system in the economically developed and densely populated regions.
National road transport	Layout and planning of national highway transportation junction (2007)	Planning: A total of 179 transportation hubs, including 12 combination hubs and containing a total of 196 cities.
Freeway	National highway network planning	Current status: Covering capital cities and big cities with more than 500 thousand population, connecting 60% of medium-sized cities with an urban population more than 200 thousand. Planning: Connecting all medium-sized cities with an urban population more than 200 thousand, forming a highly efficient transport network.
Cruise way	National inland waterway and port layout planning (2007)	Planning: In 2020, high-grade inland waterways and major ports cover 20 provinces (autonomous regions and municipalities), connecting 56 cities with a population more than 500 thousand.
Civil aviation	National civil airport layout plan (2007)	Current status: As of the end of 2006, using 100 km of ground transportation or 1.5 hours of traveling as radius of the airport service, existing airport can provide air services for 52% of the county administrative units.

(Continued)

Table 10.25. (*Continued*)

Passenger transport network	Policy documents	coverage
		Planning: By 2010, 75% of the country administrative units can enjoy the air services within 100 km of ground transportation or 1.5 hours of travelling distance; by 2020, over 80% of the county administrative units can enjoy the air services within 100 km of ground transportation or 1.5 hours of traveling distance.

lines, 4128.8 km double-tracks and 8448.3 km electrified railway were put into operation, respectively increasing by 2.2 times, 1.1 times, and 3.3 times, record high in history. Main policies are listed below.

1) "Long-term railway network planning" is adjusted to improve the planning objectives of railway mileage. In 2004, the State Council examined and approved the "long-term railway network planning". Since 2005, the Ministry of Railways successively introduced a number of policies to encourage investment of foreign and private capital in railways, but achieved little. In order to accelerate development of railway network, in 2008, China adjusted the "long-term railway network planning" and adjusted the objective of 100 thousand km length of railroad lines in service in 2020 to 12 thousand km, passenger line adjusting from 12 thousand km to 16 thousand km, and electrification rate from 50% to 60%, adding 4,00 km length of passenger line while maintaining the original "four vertical and four horizontal" framework.

Seen from networking, service target and functions, China's rail transportation network can be divided into three levels, namely national trunk rail transit network, regional rail transport network and urban rail transport network (Pu Aihua, 2009). In 2004, "Regional Intercity Rail Transit Planning" is included in the "long-term railway network planning" compiled by Ministry of Railways, planning a total of 15 routes, a total road network length of 1,659 km, and an investment of about 137.3 billion yuan ($20.616 billion). In 2008, 18 intercity rail lines were added to the adjusted

Table 10.26. Rail transport planning of major city groups in China.

City groups	Planning peroid	Intercity rail	Planning mileage
Beijing–Tianjin–Hebei region	2005–2020	Beijing–Tianjin–Tanggu	160
		Beijing–Shijiazhuang	263
		Beijing–Tangshan–Qinhuangdao	287
The Yangtze River Delta	2005–2020	Nanjing–Zhenjiang–Wuxi–Suzhou–Shanghai	295
		Shanghai–Hangzhou	160
		Hangzhou–Ningbo	158
		Changzhou–Jiangyin–Changshu–Suzhou	124
		Suzhou–Jiaxing	78
Pearl River Delta	2005–2020	Guangzhou–Dongguan–Shenzhen	131
		Guangzhou–Zhuhai	115
		Guangzhou–Foshan	37
		Xiaolan–Humen	70
		Jiangmen–Xiaolan	30
		Guangzhou–Zhaoqing	116
		Dongguan–Huizhou	89

planning, increasing 3,887 km mileage and an investment of 235.72 billion yuan ($35.39 billion) (Dong Yan, 2008). At present, China has formed an intercity rail transit construction structure based on city groups of Beijing–Tianjin–Hebei region, the Yangtze River Delta and Pearl River Delta (Table 10.26). These regions are not only regions with most developed economy, high opening-up degree, high personnel volume and active demand on passenger transport, but also regions with highest concentration of civil aviation routes, largest volume, most gold routes and fieriest competition between airlines (Research Group of CAAC, 2009a). If the rail network can play its role as expected, intercity rail will occupy an important position in intercity passenger transport.

2) Increased investment in railway construction. The 11th FYP period is a period with the largest investment in railway in China's history; especially in 2008 and 2009, the growth rate of fixed asset investment is more than 60% and investment in capital construction was increased by 80% over the previous year (Fig. 10.33). In 2009, the investment in fixed assets of national railway (including capital construction, renovation and purchase of rolling

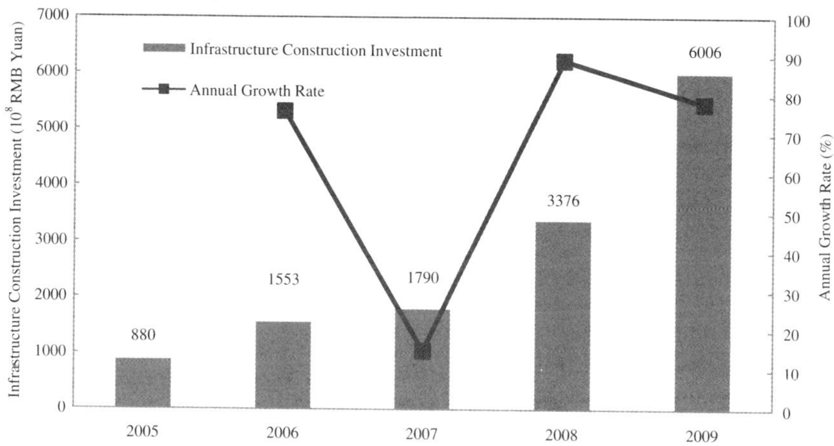

Fig. 10.33. Investment in rail infrastructure during the 11th FYP period. *Data Source*: MOR Statistics Center of People's Republic of China, 2006–2010. 2005–2009 Railway Statistical Bulletin, http://www.china.mor.gov.cn/ (accessed on 20 November 2010).

stock) reached 701.321 billion yuan ($105.30 billion), 80.3% of the total investment during the 9th FYP period, and the 10th FYP period, hitting a record high.

3) Development of the railway energy-saving technology was enhanced. With living standards improving, people's demands on comfortableness and speed are gradually increased, which to some extent raised the energy-saving pressure on technology from the railway industry. The railway sector on the one hand speeded up the program of "substituting electricity for oil" to improve the use of electric locomotive, emphasizing the goal of more than 45% of electrification rate in 2010 and 60% in 2020, and on the other hand, improved and enhanced the technical equipment of diesel and electric locomotives to encourage the use of energy-saving technologies. At the same time, special research on energy saving and emission reduction of high-speed railway was launched. According to data from Comprehensive Department, Transportation Bureau of MOR, CRH2-300, and CRH3 train set with power car, having adopted lightweight technology and streamlined body, weight 30% lighter than common passenger trains, exhibiting the effect of reducing energy consumption. CRH3 train set with power car has a consumption of 15 KW per hour per capita; the Beijing–Tianjin inter-city train set with power car, with a maximum operating speed of 350 km, consumes only 7.5 kilowatt hour per capita from South Beijing Railway Station to Tianjin

Railway Station (China Communications and Transportation Association, 2009).

(3) *Institutional analysis of the development of rail transit*

Investment and financing system has been a major problem in China's rail construction. Compared with highway construction, construction of the railway network in China is lagged behind and the participation of external capital in the rail industry is faced with many obstacles and barriers.

To promote the reform of investment and financing of railway industry, the Ministry of Railways issued "Railway investment and financing system reform program during the Eleventh Five-Year period" in August 2006, clarifying the guiding ideology and overall objectives of railway investment, establishing key projects such as expanding sizes of roads constructed by joint capital, actively promoting railway companies holding reform and listing, expanding railway construction bond issuance, establishing railway industry investment fund, expanding the use of foreign capital and using rationally bank loans and providing necessary policy measures. In May 2010, the State Council issued "Several Opinions on Encouraging and Guiding the Healthy Development of Private Investment", which clearly pointed out the need to "encourage the participation of private capital in the construction of main lines and by-lines of railways, railway car ferries and railway stations, yards and other facilities, explore and establish the investment funds for the railway industry, actively support railway enterprises in accelerating their shareholding structure reform and going public" and so on. So far, the Ministry of Railways increased cooperation with some provinces, autonomous regions and municipalities in railway construction and expanded the scale of roads constructed by joint capital. In 2008, a total capital investment of 337.554 billion yuan ($50.684 billion) was completed for national railways, and an investment of 334.97 billion yuan ($50.296 billion) was completed for national railways and joint-venture railways, in which investment of local governments and enterprises is 91.176 billion yuan ($13.690 billion), accounting for 27%, the most over the years (Fig. 10.34).

The construction of main trunk lines of passenger transport and intercity rail lines of passenger transport comes into step with regulation and reform of the investment system in China's railway industry. The rail industry is

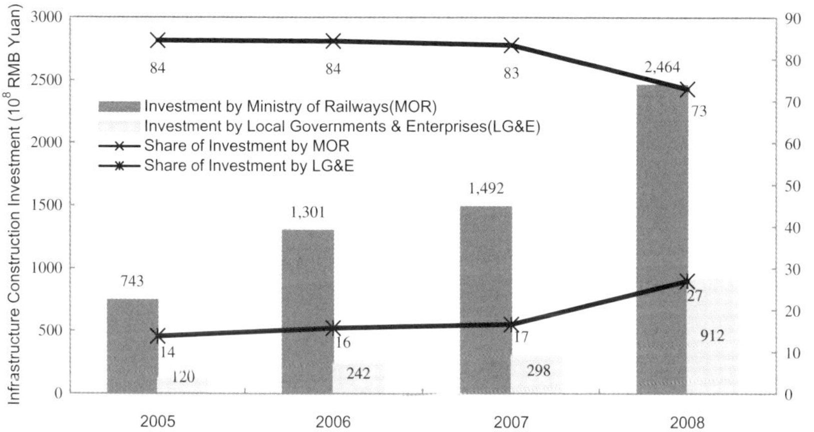

Fig. 10.34. Primary Structure of investment in rail infrastructure during the 11th FYP period.
Data Source: Ministry of Railways, PRC, 2005–2008. Railway General 2005–2008, http://www.china-mor.gov.cn/ (accessed on 20 November 2010).

actively seeking for domestic and foreign investors, using stock to accumulate social capital and exploring a new way to railway construction and operational management, which embodies China's commitment to financing reform of rail construction and at the same time exhibits incentive compatibility.

10.3 Conclusion

With passenger and freight turnover volume of China's transportation growing rapidly, transportation structure shifting to high consumption modes and arriving of automotive society, the overall fuel consumption and carbon emissions in the transportation industry are increasing year by year. This situation will not be fundamentally improved in the near future, making the task of energy-saving and emission reduction in the transport sector in China more difficult.

Although the overall fuel consumption and carbon emissions in the transport sector are on the rise, fuel consumption per transport-unit and carbon intensity of most modes of transport has been steadily declining during the 11th FYP period. A series of conferences and documents on energy saving in China request various government departments and organizations attach great importance to energy conservation of the transportation sector. Under the joint efforts of all parties, the traffic departments made significant achievements during the 11th FYP period, contributing a lot to the emission reduction targets of the 11th FYP period. Energy saved by railway and air traffic in the first three years of the 11th FYP period was respectively 5.83 Mtce and 21 thousand tce. At present, the transportation industry in China is shifting from extensive mode to intensive and environment-friendly mode, so as to realize the harmonious economic development, resource utilization and environmental protection while meeting huge traffic demand caused by increasing volume of passenger and freight traffic.

The Chinese government attaches great importance to energy-saving and emission reduction of the transport sector in recent years, and promulgated a series of policies and regulations, which can be generally divided into four different types: transportation activity (A), share rate structure (S), energy consumption intensity (I), and fuel structure (F). Among these four factors, share rate structure (S) and energy consumption intensity (I) exerted greater impact on China's energy consumption and carbon emissions in transportation during the 11th FYP period. With rapid popularization of cars, the share of carbon emissions and fuel consumption of cars in total fuel consumption and carbon emissions is rising, so policies on traveling modes and energy consumption intensity will provide effective means to fuel consumption increase caused by increase in amount of vehicles owned.

It can be concluded from analyzing policies on share rate modes (urban public transport priority and developing rail transit) and policies on energy consumption intensity (fuel economy standards) that under the context of vehicle rapid development, the policy of public transport priority maintained the position of public transport and promoted the development of public transport. Maintaining or improving the share rate of public transport is of particular significance. Preliminary estimation indicates that over the past few years, carbon reduction brought by this policy may be a few thousand tons, even millions of tons. The effect of fuel economy standards on China's energy saving is also very prominent. Although there is a large gap between China's energy-saving engine technology and those of developed countries, China's passenger car fuel economy is greatly improved, narrowing the gap with foreign advanced countries. Fuel consumption per hundred kilometers of light passenger vehicles has been falling, saving 1.24 Mtce of energy and reducing 2.71 million tons of CO_2 in the first three years of the 11th FYP period. With the arrival of automobile society, the historic change in China, institutional reform and technological improvements will lead to greater net returns. China is now working on developing phase-three standards for passenger vehicle fuel economy, and the average fuel consumption of China's new passenger vehicle will be 6.67 L/100 km in 2015. China's "public transport priority" policies and fuel consumption limits standards reflected the determination of the Chinese Government to reduce greenhouse gas emissions in the transport sector. Generally speaking, during the 11th FYP period, policies on fuel economy and urban public transport priority made great contributions in terms of energy saving. Policies on railway transport and support of new energy automotive industry witnessed an initial development during the 11th FYP period, and will determine the layout of low-carbon transport modes in the future.

Low-carbon economy requires China's transport sector to use energy effectively and rationally, allocate all modes of transport optimally and reduce energy consumption. China is now in the transition from 11th FYP to 12th FYP; Gao Hongfeng, Vice Minister of Ministry of Transportation, hosted "Preliminary results briefing of building a low carbon transport system," and asked to incorporate low-carbon transport development into the key work of "Twelfth Five-Year" to effectively highlight the importance, difficulty and urgency of energy saving. The decarbonization in transportation sector has a long way to go; there is a large gap between China' technology and internationally advanced technologies, therefore, China's low-carbon transport relies on policy guidance and technological innovation combined with energy saving experience from abroad, so as to enhance the capacity of independent innovation and ensure the effectiveness of energy conservation.

Appendix: Main Policies

This appendix summarized main national policies on energy conservation and emission reduction and policies related to fuel economy during the 11th FYP, and pointed out their direct influencing factors (Table 10.27).

Table 10.27. List of main policies.

Name	Issuing department	Key points	ASIF features
Exposure Draft of Urban Public Transport Protocol (issued by Legal Affairs Office in October 2010)	State Council	• Explicitly propose to increase capital investment, ensure development priority and leading position of public transport in urban transportation.	S
Pilot interim measures for the administration of financial subsidy funds of Private buy new energy vehicles (2010–2012)	National Development and Reform Committee, Ministry of Finance, Ministry of Science and Technology, Ministry of Industry and Information Technology	• Granting subsidies of 3,000 yuan/kw·h to qualified new energy vehicles. Providing a maximum grant of 5 million yuan for plug-in hybrid vehicles and a maximum grant of 6 million for each pure electric passenger vehicles. • Financial assistance adopted the mechanism that when an enterprise has received subsidies for 50 thousand vehicles within the pilot period (2010–2012), the central government will appropriately reduce the subsidy standards. • The central government grants subsidy according to the number of private purchase in pilot cities and required standards. For battery-leasing enterprises, the number of subsidy is determined according to the number of new energy vehicles receiving their service.	F

(Continued)

Table 10.27. (*Continued*)

Name	Issuing department	Key points	ASIF features
Circular of the State Council of P.R.C. on Further Strengthening the Efforts to Meet the Purpose of "Eleventh Five-Year" Energy Saving and Emission Reduction (Guofa [2010] No.12)	State Council	• Strengthening vehicle fuel scale assessment, strictly implementing standards for vehicle fuel consumption limits, prohibiting lines with actual passenger load rate less than 70% from putting new capacity. Promoting highway drop and pull transport, speeding up railway electrification construction and transport equipment upgrading, optimizing air routes. • Comprehensively promote energy-efficient vehicles.	AI
Automobile Industry Development Policy(revised in 2009)	National Development and Reform Committee, Ministry of Industry and Information Technology	• Guide and encourage the development of energy-saving environment-friendly small vehicles. • Support the research and development of new vehicle fuels such as alcohol fuel, natural gas, mixed fuel, hydrogen fuel and others, encourage manufacturers to develop new-fuel vehicles. • Automotive industry and related industries should attach importance to the development and application of new technology to improve vehicle fuel economy. The average fuel consumption of new passenger vehicles should be decreased by 15% before 2010 comparing with the year of 2003.	FI

(*Continued*)

Table 10.27. (*Continued*)

Name	Issuing department	Key points	ASIF features
Light Vehicle Fuel Consumption Logo (GB 22757 — 2008)	General Administration of Quality Supervision, Inspection and Quarantine, Standardization Administration of China	• It is required that passenger vehicles and light commercial vehicles sold from January 1, 2010 in China should paste vehicle fuel consumption logos to provide consumers with accurate information on fuel consumption.	I
Methods on supervision and administration of operating vehicle fuel consumption	Ministry of Transport	• Implemented from November 1, 2009. • Vehicles that do not meet the standard for fuel consumption limits shall not be used for operations. • Urban bus, taxi and passenger and freight vehicles with a total mass not exceeding 3,500 kg are not within the application scope of this document. • Provincial road transport authority is responsible for the supervision and management of the fuel consumption of road transport vehicles within its administrative region. • Road transport authorities above the county level should issue "road transport permits", vehicles that are not listed in the "vehicle model list" or not consistent with the "vehicle model list" should not be issued with "road transport permit".	I

(*Continued*)

Table 10.27. (*Continued*)

Name	Issuing department	Key points	ASIF features
Limits and measurement methods of fuel consumption for commercial vehicle for passenger transportation (JT711 — 2008)	Ministry of Transport	• Effect from September 1. • This standard specifies the limit and measurement method of fuel consumption for passenger vehicles, applicable to passenger vehicles using diesel or gasoline as fuel with the maximum gross mass over 3,500 kg. • Limits of the first phase are implemented since its implementation and limits of the second phase are to be implemented 19 months after its implementation.	I
Limits and measurement methods of fuel consumption for commercial vehicle for freight transportation (JT719 — 2008)	Ministry of Transport	• Effect from September 1. • This standard covers vehicles and dump trucks weighted 3,500 kg to 31,000 kg and tractor semi-trailer with a maximum total mass of 49,000 kg. Vehicles, and dump trucks are graded every 2,000 kg; articulated train is divided into four levels according to the total mass; • Fuel consumption limits in two phases are provided: in the first phase, 10% to 20% of high fuel-consumption vehicles are made not qualified for operation; limit standard of the second phase is implemented one year after and the fuel economy of operating vehicles is raised by 10% based on that of the first phase	I

(*Continued*)

Table 10.27. (*Continued*)

Name	Issuing department	Key points	ASIF features
Limits and measurement methods of fuel consumption for low-speed commercial vehicle for freight transportation. (GB 21378 — 2008)	General Administration of Quality Supervision, Inspection and Quarantine, Standardization of Administration of China Standardization Administration of China	• Issued on January 21, 2008 and implemented on June 1, 2008. • This standard specified the evaluation of fuel consumption, fuel consumption limit, measurement conditions, measurement methods, measurement repeatability test and measurement data processing norms for low-speed trucks in Phase one.	I
Limits and measurement methods of fuel consumption for three-wheeled automobile (GB21377 — 2008)	General Administration of Quality Supervision, Inspection and Quarantine, Standardization of Administration of China	• Issued on January 21, 2008 and implemented on June 1, 2008. • This standard specified the evaluation of fuel consumption, fuel consumption limit, measurement conditions, measurement methods, measurement repeatability test and measurement data processing norms for tricycle in Phase one.	1I

(*Continued*)

Table 10.27. (*Continued*)

Name	Issuing department	Key points	ASIF features
Fuel consumption of Light Motorcycle (GB16486 — 2008)	General Administration of Quality Supervision, Inspection and Quarantine, Standardization of Administration of China	• Implemented on July 1, 2009 for newly certified models, and implemented on January 1, 2010 for models under production. • Specified the limit of 2.0 L fuel consumption per hundred kilometers for two-wheeled light motorcycles with engine displacement less than or equal to 50 ml, and the limit of 2.3 L fuel consumption per hundred kilometers for three-wheeled light motorcycles with engine displacement less than or equal to 50 ml.	II
limits and measurement methods of fuel consumption for motorcycles (GB15744 — 2008)	General Administration of Quality Supervision, Inspection and Quarantine, Standardization of Administration of China	• Implemented on July 1, 2009 for newly certified models, and implemented on January 1, 2010 for models under production. • Provided limits by displacement segments for two- and three-wheeled motorcycle with displacement more than 50 ml.	II

(*Continued*)

Table 10.27. (*Continued*)

Name	Issuing department	Key points	ASIF features
Notice on Adjusting the Consumption Tax Policies on Passenger Vehicles (Caishui [2008] No. 105) Announcement on Adjustment of Import Consumption Tax on Some Passenger Cars (Caiguanshui [2008] No. 73)	Ministry of Finance, State Administration of Taxation	• The tax rate of passenger cars with cylinder capacity (displacement) below 1.0 L (including 1.0 L) is adjusted from 3% to 1%. • The tax rate of passenger cars with cylinder capacity (displacement) between 3.0 L to 4.0 L is adjusted from 15% to 25%. • The tax rate of passenger cars with cylinder capacity (displacement) above 4.0 L is adjusted from 20% to 40%.	II
Measures for Implementing the Energy Conservation Law of the People's Republic of China by Highway and Waterway Transport (issued in 2008)	Ministry of Transport	• Implemented on September 1, 2008 to improve statistics on transport energy consumption. • Provided energy efficiency standards and the requirement that traffic administrative departments should adopt measures to strictly enforce the implementation of fuel consumption limit standards. • Ministry of Transport should formulate (revise) energy consumption limits of unit product for transport equipment with installed power of more than 300 KW and the limits should be implemented by competent department of transportation at all levels.	II

(*Continued*)

Table 10.27. (*Continued*)

Name	Issuing department	Key points	ASIF features
Civil Aviation Industry Energy Conservation and Emission Reduction Plan (issued in 2008)	Civil Aviation Administration of China, National Development and Reform Committee	• The planning base period is the year 2005 and the planning period is from 2005 to 2015. The Planning is divided into five parts: • current situation of energy use and emissions of the aviation industry, • challenges faced by energy conservation in civil aviation industry, • guiding ideology, • principles and standards of the aviation industry for energy conservation, • the main task of civil aviation industry for energy saving, security measures for energy saving of civil aviation industry.	II
Fuel Consumption Limits for Light Commercial Vehicles (GB2099 — 2007)	General Administration of Quality Supervision, Inspection and Quarantine, Standardization of Administration of China	• Scheduled for implementation on February 1, 2008, the executive objectives are light commercial vehicles with a total mass less than 3.5 t and usually 9 to 12 seats, including the light truck and light passenger vehicles.	II

(*Continued*)

Table 10.27. (*Continued*)

Name	Issuing department	Key points	ASIF features
Rules on the (National Production Admission Administration of New Energy Automobiles (Announcement of State Development and Reform Committee, [2007] No. 72, issued on October 17, 2007)	National Development and Reform Committee	• Classification of new energy vehicles • Provides access conditions for new energy automobile companies	FF
Opinions on Building Energy-saving Transportation (Jiaoguihuafa, 2006, No.140)	Ministry of Transport	• The overall goal of 20% reduction of energy consumption per t-km for operating vehicles and vessels	I
Guiding Opinions of Comprehensively Implementing Notice of the State Council on Promoting Energy Conservation by the Transport Sector (Jiaotifafa, 2006, No.592)	Ministry of Transport	• Specify the energy saving target of the transport sector, identify the building of energy-efficient transportation infrastructure. • Strengthen the focus on transportation management for energy conservation. • Vigorously promote energy-saving technological progress and further establish and improve the regulatory system of energy saving. • Improve the working mechanism of energy-saving security of the transport sector.	SS

(*Continued*)

Table 10.27. (*Continued*)

Name	Issuing department	Key points	ASIF features
Opinions on Policy of urban public transport development priority (Jiancheng, 2006, No.288)	Ministry of Construction, National Development and Reform Committee, Ministry of Finance, Ministry of Labor and Social Security	• Increase investment in urban public transport. • Establish bus fare subsidy mechanism and put bus fuel subsidies and other subsidies into practice. • Specify special bus economic compensation. • Ensure the stability of public transport industry.	SSI
vehicle fuel consumption limits standards (GB 19578 — 2004)	General Administration of Quality Supervision, Inspection and Quarantine, Standardization of Administration of China	• China's fuel consumption limits standards for passenger vehicles are mainly applicable to passenger vehicles (M1) with maximum design speed not less than 50 km/h, maximum design mass not exceeding 3,500 kg and with nine seats. • The standard is implemented in two phases: the first phase is for newly developed models and is implemented on July 1, 2005 and the second phase January 1, 2008; the implementation of the newly produced vehicles is one year later comparing with that of newly developed models, the first phase on July 1, 2006 and the second phase on January 1, 2009.	II

(*Continued*)

Table 10.27. (*Continued*)

Name	Issuing department	Key points	ASIF features
Circular of the State Council Concerning Opinions of Ministry of Construction on Giving Priority to Public Transport Development (Guo, 2005, No. 46) (Jiancheng, 2004, No.38)	State Council, National Development and Reform Committee, Ministry of Science and Technology, Ministry of Public Security, Ministry of Finance, Ministry of Land and Resources, Ministry of Construction	• Priority to the development of urban public transport, main tasks and goals (speed, punctuality rate, coverage, average travel time, share rate) of urban public transport development priority. • Strengthening the guiding role of urban planning. • Construction of public transport road system and BRT. • Formulating relevant economic policies on urban public transport development priority. • Promoting the reform of urban public transport sector. • Improving the technological level and service quality. • Establishing a sound system of regulations and standards for urban public transport regulations and standards.	SSA
Measurement methods of fuel consumption for light-duty vehicles (GB/T9233 — 2003)	General Administration of Quality Supervision, Inspection and Quarantine, Standardization of Administration of China	• This standard was issued in 2003 to regulate automobile fuel consumption testing.	I

CHAPTER 11

Agriculture and Forestry

11.1 Performance of Greenhouse Gas Emission

Comparing with other sectors, agriculture and forestry sector[1] has its particularities that it is both the source of greenhouse gas emission and important sink for CO_2. This sector emits fossil energy-related CO_2, and other greenhouse gases —- CH_4, N_2O while the agriculture and forestry ecosystem absorbs CO_2. Chemical fertilizer, pesticides and agricultural machinery productions lead to embedded carbon emission of upstream production industries. The analysis in this chapter includes direct and indirect carbon emission, the absorbing of CO_2 and the emission of other greenhouse gases (Fig. 11.1).

(1) Direct CO_2 emission: Emission of CO_2 from agriculture and forestry sector's energy consuming.[2]
(2) Emission of other greenhouse gases[3]:
Greenhouse gas emission from farm production: CH_4 emission from paddy fields, N_2O emission from soil as a result of using chemical fertilizer[4];

[1]China's agriculture sector includes agriculture, forestry, livestock farming, fishery and related service industry. Forestry is considered to be a part of this sector in the statistic data. Considering the great carbon sink importance, this report makes forestry a separate sector from agricultural sector, calling it "agriculture and forestry sector".
[2]Energy consumption of agriculture and forestry sector does not include energy from biomass (straw and firewood), methane. This energy is mainly used in rural areas rather than for agricultural production. So it is not included in this chapter.
[3]*Tokyo Protocol* lists six kinds of greenhouse gases emitted by human including CO_2, CH_4, N_2O, HFCs, PFCs, and SF6. This report calculates only the emission of CO_2, CH_4, and N_2O. According to Global Warming Potential, greenhouse effect caused by CH_4 is 21 times higher than that by CO_2, N_2O 310 times higher than CO_2 (IPCC, 2006).
[4]The N_2O emission from farmland soil refers to the only emissions caused by using pesticides fertilizer.

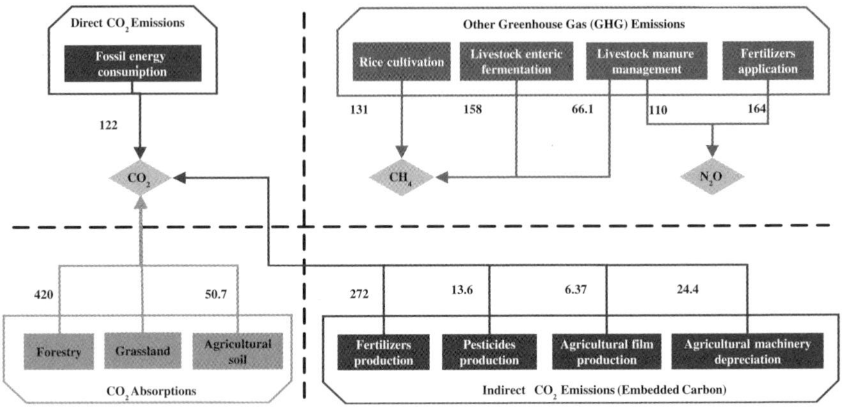

Fig. 11.1. Main process of carbon emission and absorbing in agriculture and forestry sector.

Note: The numbers in the figure are emission and absorbing of greenhouse gas sources and sink, and the unit is $MtCO_2$-eq. The calculation method will be explained in detail and grassland carbon sink is not calculated in this chapter.

Greenhouse gas emission from livestock farming: CH_4 emission from ruminant's enteric fermentation, CH_4 and N_2O emission from animal wastes.

(3) Indirect CO_2 emission (embedded carbon): CO_2 emission in the production process of chemical fertilizer, pesticides, agricultural film and agricultural machinery.

(4) Absorbing of CO_2[5]: Annual increase of carbon sink in the soil of farmland and forest.

11.1.1 *Significant decrease in energy consumption and the direct energy-related CO_2 emission*

(1) *Decrease in energy consumption and related CO_2 emission*

Energy consumption in agriculture and forestry sector is relatively low. In 2008, the total energy consumption was 35.1 Mtce, accounting only 1.7%

[5]Carbon sink of agriculture and forestry sector does not involve that from crop cultivation and wood soil. China's crops mainly are Panicum miliaceum and the CO_2 emission and absorbing between the crops and human strike a balance within the whole ecosystem, so carbon sinks from crop cultivation is not evaluated in this chapter.

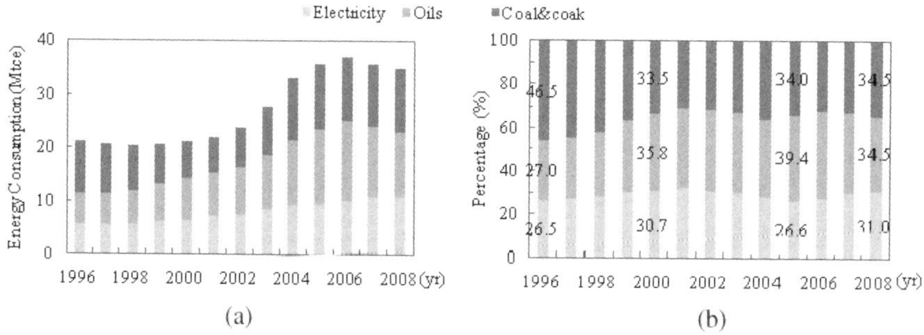

Fig. 11.2. Amount of energy consumption and its structure in agriculture and forestry sector (a) energy consumption; (b) energy structure
Source of the data: *China Energy Statistical Yearbook* (2009).
Note: The amount of coal = coal + coke; the carrier vehicles in agricultural machinery (the four wheeler and three wheeler) consume a lot of diesel, patrol, so they are assigned to the transport sector. The diesel consumed in agriculture and forestry sector is mainly for agricultural machinery and 25% of this part of diesel was used for transporting in 2007 (Tan, 2008). So oils consumed in agriculture and forestry sector in this book = oils (end consumption in agriculture, forestry, livestock farming and fishery) — petrol — 25%*diesel. The heating power consumption in agriculture and forestry sector is omitted since it takes up less than 0.1% of the total energy consumption in this sector.

of that year's total energy consumption and a 2.2% decrease compared with that of 2005. There are three phases of annual trend for energy consumption: from 1996 to 2000, the energy consumption stayed at about 21 Mtce; from 2001 to 2006, energy consumption had an annual increase of 13.6%, increasing from 22.1 Mtce to 37.1 Mtce; from 2007 to the present, energy consumption of agriculture and forestry sector decreased: a 3.2% decrease from 2006 to 2007, a 2.1% decrease from 2007 to 2008. The average annual decrease rate reached 2.6% (Fig. 11.2a).

Viewed from the structure of energy consumption, from 1996 to 2008, the ratio of coal consumption to energy consumption decreased from 46.5% to 34.5%; the ratio of oils increased from 27.0% in 1996 to 39.4% in 2005. In 2006, this ratio reached a highest level of 40.6%, and then decreased annually to 34.5% in 2008. (Fig. 11.2b).

As to energy-related carbon emission, CO_2 emissions of agriculture and forestry sector in 2008 were 122 $MtCO_2$, accounting 1.9% of that year's total energy-related CO_2 emissions and was about the same as that in 2005. From 1996 to 2001, the CO_2 emissions related with energy was relative stable, staying at about 80 $MtCO_2$; from 2002, the amount began to increase rapidly and in 2006 it reached a peak of 128 $MtCO_2$. But in 2007 and 2008,

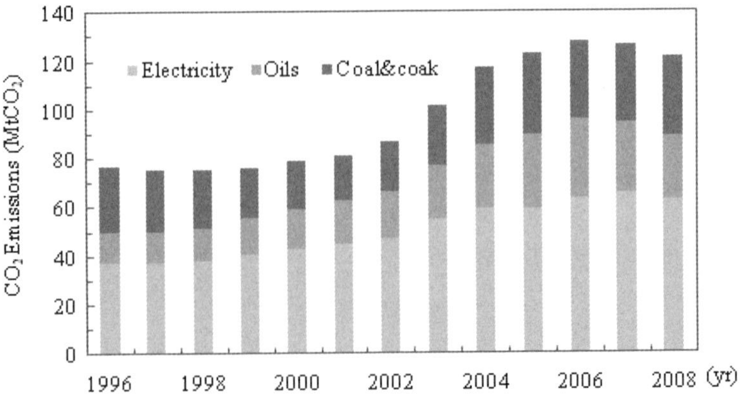

Fig. 11.3. CO_2 emission from agriculture and forestry sector's energy consumption. *Note*: CO_2 emission factor from electrical power is calculated according to the year's coal consumption and energy structure.

a continual decrease in CO_2 emissions was observed and the emission in 2008 decreased 4.7% compared with that in 2006 (Fig. 11.3).

(2) *Significant decrease in energy consumption intensity and CO_2 emission intensity*

From 1996 to 2006, energy consumption intensity of added value per ten thousand yuan showed fluctuation (Fig. 11.4a). There was first an increase and then a decrease during that period, but it stayed at about 120 kgce per ten thousand yuan[6]; from 2003, this energy consumption intensity increased obviously and in 2005 it reached 160 kgce per ten thousand yuan. Since the beginning of the 11th five-year plan, energy consumption intensity has decreased significantly especially in 2007 and 2008, which had a decrease of 8.1% and 15.0% compared with that in 2005 respectively.

CO_2 emission intensity changed in accordance with energy consumption intensity (Fig. 11.4b). In 2004, the highest CO_2 emission intensity of 552 kg CO_2 per ten thousand yuan[7] was observed and then a continual decrease year by year. In 2008, the CO_2 emission intensity was about 473 kg CO_2 per ten thousand yuan which decreased 14.3% compared with that in 2005.

[6]Energy consumption was calculated by electro thermal equivalent method.

[7]This was calculated according to the emission factor of various types of energy. Emission factor of CO_2 from electrical power was calculated according to the coal consumed and power supply in that very year.

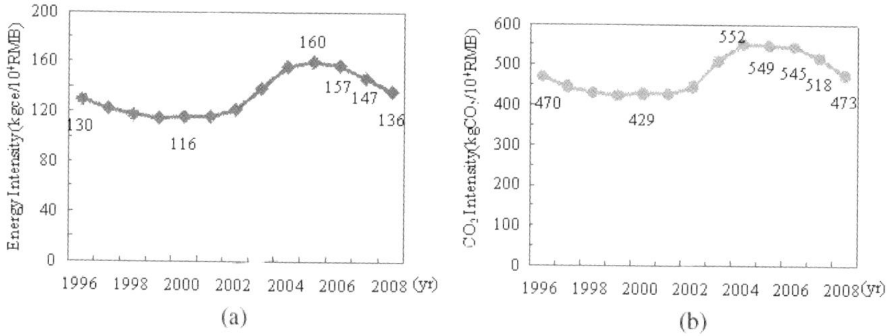

Fig. 11.4. Energy consumption intensity for added value (ten thousand yuan) (a) energy consumption intensity; (b) CO_2 emission intensity.
Note: The added value was converted into the constant price in 2005.
Sources of the data: *China Statistical Yearbook 2009*, *China Energy Statistics Yearbook 2009*.

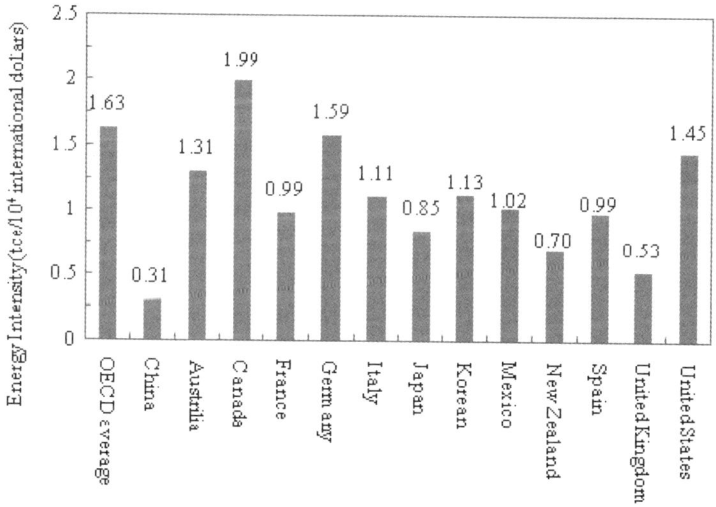

Fig. 11.5. Energy consumption intensity of agriculture added value in some countries. *Note*: Energy consumption intensity = energy consumption volume of agriculture and forestry sector/[total output value by purchasing power parity × first industry proportion (agriculture)]. *Data Source*: IEA (2009), *International Statistical Yearbook 2007*.

International comparison data (Fig. 11.5) shows that China's energy consumption intensity per unit agriculture's added value is 0.31 tce per ten thousand international dollars (purchasing power parity), which is 19.0% of that in OCED member countries. It is 24.4%, 22.1%, and 36.5% of that in Australia, America, and Japan respectively.

China's agricultural modernization is lagging behind compared with developed countries, especially in agriculture machinery. This is the reason that energy consumption intensity of unit agriculture's added value is relatively low. America achieved mechanization for grain crop in 1960s and for cash crop in 1970s; the mechanization for the main crop sugarcane reached 90% in Australia; in 1996, 85% of Japan's machine-harvested rice was by combine harvester (China Machinery Network, 2007). Developed countries' processing rate for agricultural products was about 90% while China's was only about 45% (*Development Plan for Farming Industry during the 11th five-year Plan*). With the modernization of China's agriculture, energy consumption for agricultural production will increase. So in the meantime realizing the mechanization of agriculture, saving the energy consumed by farming machines might be a major future development direction of agriculture energy saving.

11.1.2 CH_4 and N_2O emission kept stable for a long time

There are two phases for CH_4 and N_2O emission: From 1988 to 1996, the total volume of CH_4 and N_2O emissions increased continually with a total increase of 1/3 within the 8 years; from 1997 to 2008, CH_4 and N_2O emissions kept stable (Fig. 11.6). In 2008, the total emissions of CH_4 and N_2O reached 629 $MtCO_2$-eq, which was 5.2 times of the CO_2 emissions from energy consumption in the same year.

The People's Republic of China Initial National Communication on Climate Change (1994) showed that in 1994 CH_4 and N_2O emissions from agriculture sector were 605 $MtCO_2$-eq. USEPA (2006) estimated China's CH_4 and N_2O emissions from paddy field, chemical fertilizer, livestock enteric fermentation, and animal wastes were respectively 694 $MtCO_2$-eq and 704 $MtCO_2$-eq. The estimate result in this report was close to the above given data.

In 2008, CH_4 emissions were 356 $MtCO_2$-eq, decreasing 3.26% compared with that in 2007. The sources of CH_4 emission include paddy fields, animal wastes, and livestock enteric fermentation. In 2008, CH_4 emissions from paddy fields accounted 1/3 of the total CH_4 emissions, 1/6 from manure management (animal wastes). These two sources kept relatively stable. Since 1997, CH_4 emission from enteric fermentation has decreased, contributing most to the total decrease of CH_4 emission. In Fig. 11.7, we can see that the total number of swine and poultry increased while the number of sheep kept stable.

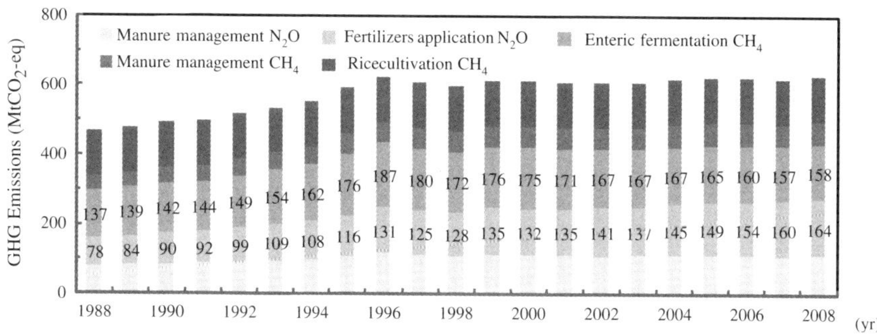

Fig. 11.6. CH_4 and N_2O emission from agriculture and forestry sector.

Note: ① N_2O and CH_4 emission from various sources (except paddy fields) is calculated with IPCC greenhouse gas emission inventories tier 1 (IPCC, 2006); ② N_2O emission from manure management includes mainly direct N_2O emission from manure pile and indirect emission from nitrogen volatilizing and leaching. N_2O emission from manure management $= \sum$ number of animals raised $\times$ average weight $\times$ N excretion rates $\times$ (N_2O direct emission factor + nitrogen volatilizing loss rate $\times$ indirect emission ratio of nitrogen volatilizing + nitrogen leaching loss rate $\times$ indirect emission ratio of nitrogen leaching); ③ N_2O emission from chemical fertilizer = total amount of fertilizer used $\times$ N_2O direct emission factor of nitrogen input; ④ CH_4 emission from livestock enteric fermentation, $s = \sum$ average annual number of animals raised $\times$ CH_4 emission ratio of enteric fermentation. CH_4 emission from livestock enteric fermentation involves swine, cows, sheep, camels, horses, donkeys, and mules; CH_4 emission from manure management $= \sum$ average annual number of animals raised $\times$ CH_4 emission ratio of manure management. CH_4 emission from manure management involves the animals mentioned above as well as poultry; ⑤ Since 1998, China's areas of paddy fields has kept stable, so the amount of CH_4 emission from paddy fields adopts constant value. From 2000 to 2005, CH_4 emission from paddy fields was 6.25 tons, which equals to 131 $MtCO_2$-eq (Wang *et al.*, 2009).

Source of the data: Number of domestic animals and amount of chemical fertilizers are from *China Agriculture Yearbook* (1989–2009).

However, the number of cows has shown decline since 1997. The CH_4 emission ratio of cow's enteric fermentation is 11 times of a sheep and 55 times of a swine. So the reduction in cows raised is the main reason for the decrease of CH_4 emission.

N_2O emission is mainly from manure management and the use of chemical fertilizer (Fig. 11.7). Since 1997, the amount of N_2O emissions from animal wastes have stayed at about 110 $MtCO_2$-eq. But the amount of N_2O emissions from chemical fertilizer increased significantly from 1988 to 2008 and the emission reached 164 $MtCO_2$-eq in 2008. The main reason is the increasing use of chemical fertilizer (mainly nitrogenous fertilizer).

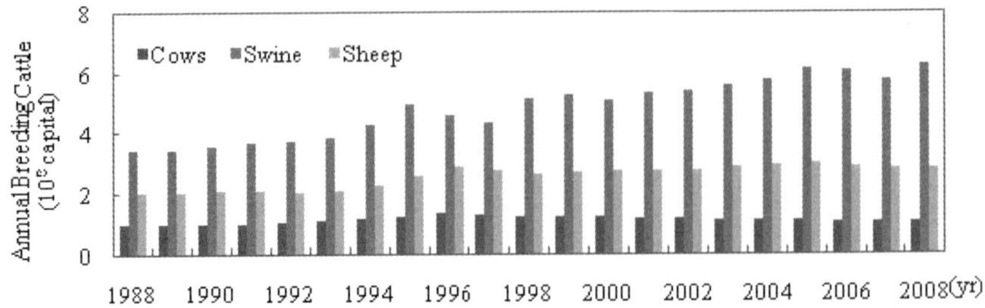

(a)

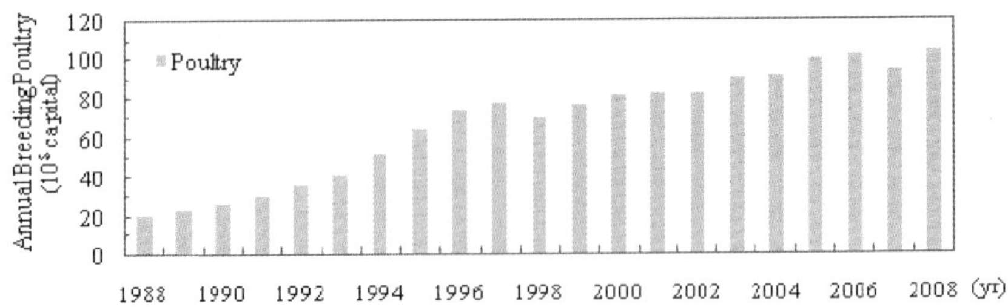

(b)

Fig. 11.7. Average annual number of livestock and poultry.

Note: Average annual number of livestock = number of livestock raised every year × breeding days/365 (IPCC, 2006); *China Agriculture Yearbook* counts the number of livestock and poultry according to the number of in stock and slaughtered ones. If counting only the in stock, then the emission from domestic animals whose sale rate (365/breeding days) are more than one (swine and poultry) will be overlooked; if counting only the number of slaughtered domestic animals, then the emission whose sale rate (365/breeding days) are less than one (cows, horses, sheep, mules, and donkeys) will be overlooked; if counting both, then there will be overlaps. So suppose that the domestic animals raised are divided into three parts: Number of animals slaughtered, newly added number and the number of animals in stock for the whole year. The weight of the former two is 0.5 and the last one is 1. The average annual number of domestic animals = the number slaughtered/2 + (in stock number at the end of year + the number slaughtered — in stock number at end of last year)/2 + max (in stock number at the end of last year — number slaughtered, 0).

Source of the data: Number of slaughtered domestic animals and number of in stock animals are from *China Agriculture Yearbook* (1989–2009).

11.1.3 *Continual increase of embedded carbon emission from agricultural means of production*

The production of any products directly or indirectly uses energy. The energy consumed in the whole production chain to get the end products is called "embedded energy". Likewise, the CO_2 emission in the whole production chain is called "embedded carbon". The embedded carbon emission in agriculture and forestry sector refers to the total carbon emission from the production and using of chemical fertilizer, pesticides, agricultural film, and agricultural machinery. Although embedded carbon emission in agriculture and forestry sector is not counted into this sector in statistics scope, data of this emission is of vital importance to have an in-depth understanding of this sector.

With the usage of agricultural means of the production increasing, embedded carbon emission from agriculture and forestry sector shows a trend of increase (Fig. 11.8). In 2008, the total amount of embedded carbon emissions were 317 $MtCO_2$-eq, increasing 10.1% compared with that in

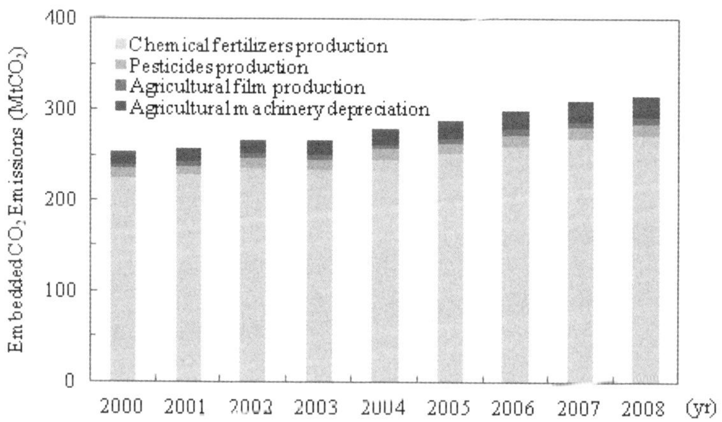

Fig. 11.8. Embedded carbon emission in agriculture and forestry sector.

Note: ① 3 tce coals will be used to produce 1 t pesticides and 8.13 t CO_2 will be emitted (Zhang, 2010).

② The CO_2 emission during the production of agricultural film is calculated in accordance with the polythene production process. The material for polythene production is ethane. 1.009 t ethane and 521 kWh electrical power will be used to produce 1 t polythene (China Agricultural Film Network, 2010). According to *China Energy Statistical Yearbook 2009*, the energy consumed to produce ethane in 2008 was 1,003 kgce/t, then it can be estimated that the emissions will be 3.17 t CO_2 in producing one ton agricultural film.

③ 90% of the material for agricultural machinery is iron and steel. The carbon emission of agricultural machinery depreciation is calculated in accordance with the carbon emission ratio of iron and steel industry. The amount of iron and steel is calculated according to the agricultural machinery's service life (usually 10 years) and the number and weight of the agricultural machinery. And since there are a variety of farm machines, we can only get a rough calculation. According to *China Energy Statistical Yearbook 2009*, the amount of emissions are 1.97 t CO_2 to produce one ton steel calculated with the standard of 727 kgce/t in 2008.

④ CO_2 Emission from the production of chemical fertilizer is calculated by nitrogenous fertilizer, phosphate fertilizer and potash fertilizer, respectively. CO_2 emission differs

2005. In the same year, the amount of embedded carbon emission was 2.6 times of the energy-related CO_2 emission. Embedded carbon emission from chemical fertilizer was the most in various agricultural means of production. In 2008, embedded carbon emission from chemical fertilizer accounted 85.8% of the total embedded carbon emission.

11.1.4 *Chemical fertilizer is the main contributor to the increase of greenhouse gas emission in agriculture and forestry sector*

(1) *Chemical fertilizer causes the increase of greenhouse gas emission in agriculture and forestry sector*

Greenhouse gas emission from use of chemical fertilizer includes embedded carbon emission in the production of chemical fertilizer and N_2O emission during the use of chemical fertilizer. From 2000 to 2008, greenhouse gas emission from the use of chemical fertilizer increased continually. In 2008, the emissions were 436 $MtCO_2$-eq, accounting 40.8% of the total emission from agriculture and forestry sector. It was an increase of 8.7% compared with that in 2005 and was 3.6, 1.3, 3.3 times of that from energy consumption, domestic animal raising and paddy fields respectively. It is the largest emission source in agriculture and forestry sector (Fig. 11.9).

Fig. 11.8. (*Continued*). for various nitrogenous fertilizers. According to *China Chemical Fertilizer Industry Yearbook 2005–2006*, China's major nitrogenous fertilizer includes urea and ammonium bicarbonate, accounting 60% and 25% of the total fertilizer respectively. The other 15% comprises mainly ammonium nitrate. Considering different production technology and energy consuming structure, the CO_2 emissions of producing one ton nitrogen of urea and ammonium bicarbonate are 6.75 t and 6.42 t respectively. CO_2 emission from ammonium nitrate uses the average emission amount of urea and ammonium bicarbonate and with the extra 4.12 tCO_2-eq N_2O emission in producing ammonium nitrate. The total emissions of ammonium nitrate will be 10.8 tCO_2-eq. The production of phosphate and potash fertilizer uses mainly electrical power. About 9 GJ and 8 GJ energy will be consumed to produce 1 t P_2O_5 and 1 t K_2O (IFA, 2009), which are equal to 2,512 kW.h and 2225 kW.h electricity power consumption. Then the CO_2 emissions will be 2.06 t and 1.82 t respectively. The consumption structure of nitrogenous fertilizer, phosphate fertilizer and potash fertilizer refer to Zhang *et al.* (2008). The degree of self sufficiency of the three kinds of fertilizer is 100%, 95%, and 29% respectively, so the embedded carbon emission from these fertilizers are not totally from China. But to have the whole picture of the agricultural means of production development, this part is all included into the total carbon emission.
Sources of the data: The data of chemical fertilizer, pesticides, agricultural film and agricultural machinery are from *China Agriculture Yearbook 1989–2009*.

Table 11.1. Energy consumption of various nitrogenous fertilizers in China.

Type of nitrogenous fertilizer	Nitrogen content	Energy consumption type	Energy consumption per ton of output (t or m^3)	Electricity consumption per ton of output (kWh)	
				Large enterprise	Medium and small enterprise
Urea	46%	coal	1.56	60	1,032.57
		Natural gas	1,000	60	900
		oil	0.8	600	
Ammonium bicarbonate	17%	coal	0.5	400	

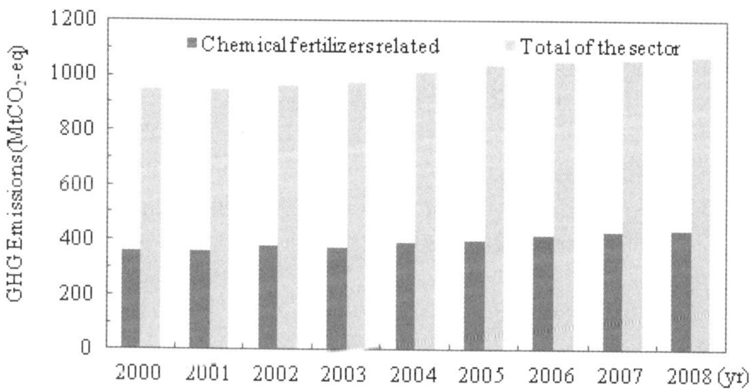

Fig. 11.9. Greenhouse gas emission from the use of chemical fertilizer.
Note: Greenhouse gas emission from agriculture and forestry sector (CO_2-eq) = CO_2 emission from energy consumption + CH_4 emission + N_2O emission + embedded carbon (CO_2) emission.

(2) *Greenhouse gas emission intensity per unit chemical fertilizer is decreasing*

Greenhouse gas emission intensity related with chemical fertilizer (the amount of greenhouse gas emission per unit chemical fertilizer) decreased from 9.85 t CO_2-eq/t in 1980 to 8.32 t CO_2-eq/t in 2008, which was a 15.5% decrease (Fig. 11.10a). Embedded carbon emission intensity from chemical fertilizer production decreased 14.9% and N_2O emission intensity from the use of chemical fertilizer decreased 15.7%. To show the performance of agriculture and forestry sector, this report uses the embedded carbon emission

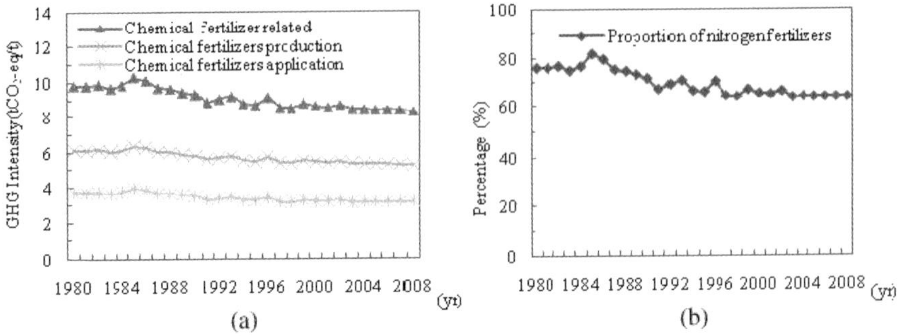

Fig. 11.10. Greenhouse gas emission intensity related with chemical fertilizer and the proportion of nitrogenous fertilizer (a) greenhouse gas emission intensity related with chemical fertilizer; (b) the proportion of nitrogenous fertilizer.

intensity per unit fertilizer in 2005. Decrease of embedded carbon emission from chemical fertilizer was caused by change of chemical using structure, mainly the decline in nitrogenous fertilizer using. The proportion of nitrogenous fertilizer declined from 75.9% in 1980 to 64.1% in 2008 (Fig. 11.10b). Since the 11th five-year plan, proportion of nitrogenous fertilizer had no significant change and greenhouse gas emission from chemical fertilizer had very little decline. Emission in 2008 decreased only 0.95% compared with that in 2005.

(3) *Chemical fertilizer-related greenhouse gas emission intensity per unit crop output is increasing*

There are two phases for this emission intensity (Fig. 11.11): Before 1995, the grain crop and cash crop continued to increase, with an annual increase rate of 3.5% and 4.0% respectively; from 1996, greenhouse gas emission intensity kept stable while chemical fertilizer used in per unit area increased gradually. Emission intensity was 549 kgCO_2-eq/t in 2008, which were lower than that in 2007, the same with that in 2006. The reason for higher emission intensity in 2007 was because chemical fertilizer used in per unit area was more than that in 2008 while the crop output per unit area was lower than that in 2008. After 1996, sown area of cash crop was in decline and greenhouse gas emission intensity from per unit output cash crop was 1,255 kgCO_2-eq/t in 2008, which was 6.4% lower than that in 2005.

Chemical fertilizer is of vital importance to the output of crops. Chemical fertilizer used per unit area every year is increasing and the variation trend can be divided into three phases (Fig. 11.12a): From 1988 to 1995,

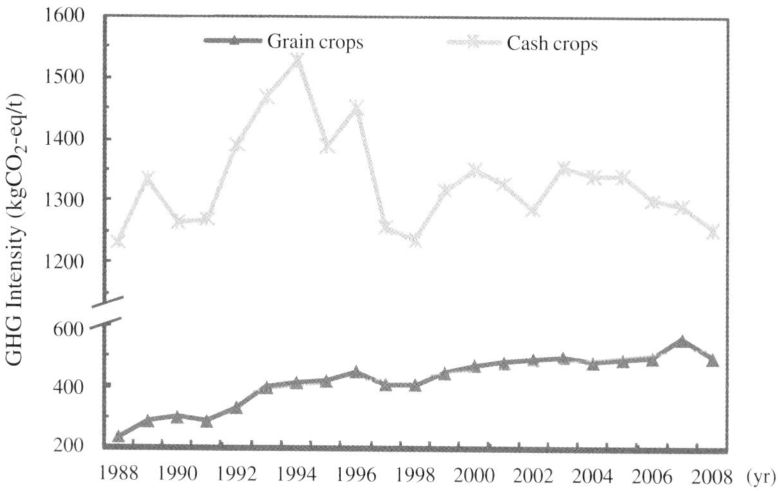

Fig. 11.11. Greenhouse gas emission intensity related with chemical fertilizer used per unit output.

Note: ① Grain crop and cash crop are the main crops in China. Grain crop include cereal, beans, potato and cash crop in a narrow sense include oil plants, cotton, hemp, sugar crop, tobacco leaf and medicinal herbs and in a broad sense also include fruit and vegetable. Cash crop in this report refers to the broad sense. It is reported that the ratio of chemical fertilizer using for grain crop and cash crop is 1:2. The total amount of chemical fertilizer used for the two kinds of crops should be calculated according to their sown areas. ② The output of grain crop and cash crop are very different, especially the output of sugar plant, fruit and vegetables which contain a lot of water. So the unit output of crop in this report is calculated by dry matter.

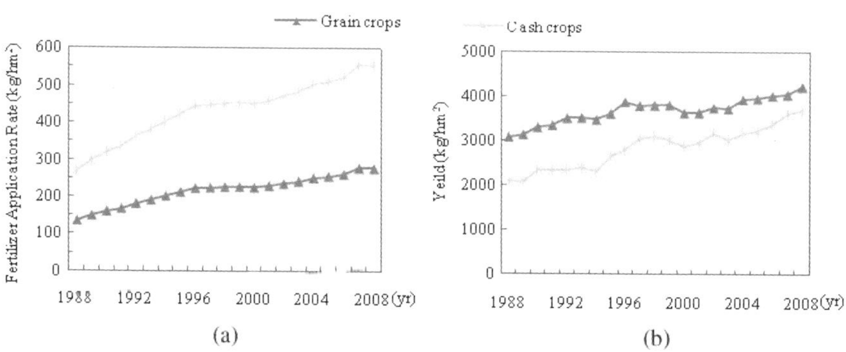

Fig. 11.12. Chemical fertilizer used per unit area and the grain yield (a) fertilizers application rate; (b) yield.

the increase was very significant with an annual increase rate of 8.3%; from 1996 to 2001, the annual increase rate slowed down and stayed at about 1.4%; from 2002 to 2008, the rate rose a bit about 3%. The changing trend of grain yield per unit area shows fluctuation. This is mainly caused by change of natural conditions such as flood, drought, plant diseases, and pests and so on. Grain yield per unit area is in accordance with the amount of chemical fertilizer used (Fig. 11.2b). From 1988 to 1995, output of grain crop and cash crop increased at an annual rate of 2.4% and 3.8% respectively, which were far slower than the increase rate of chemical fertilizer using; from 1996 to 2001, grain yield decreased because the increase rate of chemical fertilizer using slowed down; from 2002 to 2008, grain yield increased again with an annual rate of 2.1% and 2.7% respectively which were close to the increase rate of chemical fertilizer using. It is worthwhile to note that from 2008, chemical fertilizer used per unit area has begun to decrease.

Although chemical fertilizer is of vital importance to grain yield, much waste exists in chemical fertilizer using (Column 11.1).

Column 11.1. Severe waste in chemical fertilizer using and its harm.

China feeds 22% of the world's population with only 7% of the world's farmland, consuming 35% of the world's chemical fertilizer. The nitrogenous fertilizer using in China is three times of that in America, 1.5 times of France and 1.6 times of Germany. In North China, 588 kg nitrogenous fertilizers are used per acre area annually, and 227 kg excess nitrogen is released into the environment. The average utilization ratio of China's nitrogenous fertilizer is only 30% to 35%, and in some high-yield region this ratio is less than 30%. The loss ratio is as high as 52% to 60%. More than 10 million tons of nitrogen is wasted annually because of improper fertilizing. With consideration of farmers' labor cost, the direct economic loss as a result of excess nitrogenous fertilizer using in 2008 was about 15.55 billion yuan, which was 10% of the primary industry's added value.

China is the largest producer and consumer of chemical fertilizer. Nitrogenous fertilizer takes up 80% of the total chemical fertilizer output in China, and about 1/3 of the world's total chemical fertilizer output. As early as 1997, China's nitrogenous fertilizer output has begun to exceed its usage amount, and nitrogenous fertilizer output is still increasing at a rate of 6.7%. Nitrogenous fertilizer industry is an energy intensive industry

(*Continued*)

Column 11.1. (*Continued*)

and 70% of China's nitrogenous fertilizer production relies on coal, consuming about 100 million tce. This amount is still increasing at a speed of 10 million tce annually. In 2008, nitrogenous fertilizer consumed 3.85% of the nation's primary energy output, which was 3.51% of the total energy consumption. According to some reports, the fossil energy consumed by agricultural production and agricultural chemicals industry takes up 15% of total energy consumption.

There are multiple negative externalities in excess nitrogenous fertilizer using: Long term excess chemical fertilizer using leads to impoverishing and hardening of lands and soil erosion, and the backwards fertilization often leads to ineffectiveness of the fertilizer; nitrogenous fertilizer using also causes eutrophication of lakes, rivers and shallow waters along the coast, and making fish die from lack of oxygen results great loss to fishery; the large amount of greenhouse gas emission from the production process of nitrogenous fertilizer will also be aggravating climate change which will have negative effects on agriculture; furthermore, nitrogenous fertilizer will pollute soil with heavy metal, leading to the enrichment of nitrite which may do harm to human health through food chain.

Additionally, subsidy from the government for agricultural production and nitrogenous fertilizer may also promote excess use of nitrogenous fertilizer.

Source of the data: Cheng Cunwang, Shi Yan and Wen Tiejun (2010). Greenpeace — real cost of nitrogenous fertilizer, Greenpeace — real cost of nitrogenous fertilizer II.

11.1.5 *Terrestrial carbon sink increases significantly*

(1) *Forestry carbon sink stock increases rapidly*

"Carbon sink" is a concept opposite to "carbon source", referring to carbon carriers in the nature. Forest, grass land, and agricultural soil can all serve as carbon carrier, playing an important part in carbon sink. Carbon sink stock reflects CO_2 absorbing and fixing ability of the carbon carrier.

Forest carbon sink refers to the process, activity or mechanism of absorbing and fixing of CO_2 in the atmosphere by forest through tree planting, strengthening the management of forestry, reducing deforesting, forest conservation, and combining with carbon sink trade in accordance with

Table 11.2. Outline of all the forest census.

Forest census	Period	Forest area (ten thousand hm^2)	Forest coverage (%)	Forest growing stock (100 million m^3)	Artificial forest area (ten thousand hm^2)	Artificial forest growing stock (100 million m^3)
1st	1973–1976	12,186	12.7	86.6	1,139	1.6
2nd	1977–1981	11,528	12.0	90.3	1,273	2.7
3rd	1984–1988	12,465	13.0	91.4	1,874	5.3
4th	1989–1993	13,370	13.9	101.4	2,137	7.1
5th	1994–1998	15,894	16.6	112.7	2,914	10.1
6th	1999–2003	17,491	18.2	124.6	3,229	15.0
7th	2004–2009	19,545	20.4	133.6	6,169	19.6

Source of the data: The data of 1st to the 6th forest census are from http://www.cfsdc.org/index.htm; the data of the 7th forest census is from Chinese Government Net (2010b).

related rules. *Tokyo Protocol* promotes two ways of emission reduction: One is direct CO_2 and other greenhouse gas emission reduction by industries through production reduction or technology updating; the other is indirect emission reduction through increasing forest carbon sink.

The forest census conducted every five years (Table 11.2) shows that from 1973 to 1984, the area of forest shrinked and after 1984 China's area of forest increased from 13.9% in 1984 to 20.4% in 2009. This increase is mainly from artificial forest. China has the largest area of artificial forest in the world. From 1990 to 2005, China's artificial forest took up 73% of the world's total (Chinese Government Net, 2010a). Although the world's forest area is decreasing, China's area of artificial forest increases continually, contributing a lot to CO_2 emission reduction.

According to the forest census, Chinese researchers estimated carbon stock of forest and artificial forest with the aid of field survey. Although there are differences in the results (Fig. 11.13, Fig. 11.14), they all show a trend of increasing in carbon sink of China's forest. This report estimates carbon stock of forest and artificial forest in the 7th forest census on the basis of method adopted by Li *et al.* (2009).

From 2004 to 2009, carbon sinks of China's forest were about 22.2 $GtCO_2$, increasing 10.4% comparing with that from 1999 to 2003, with an annual increase of 420 $MtCO_2/a$. From 2004 to 2009, carbon sinks of

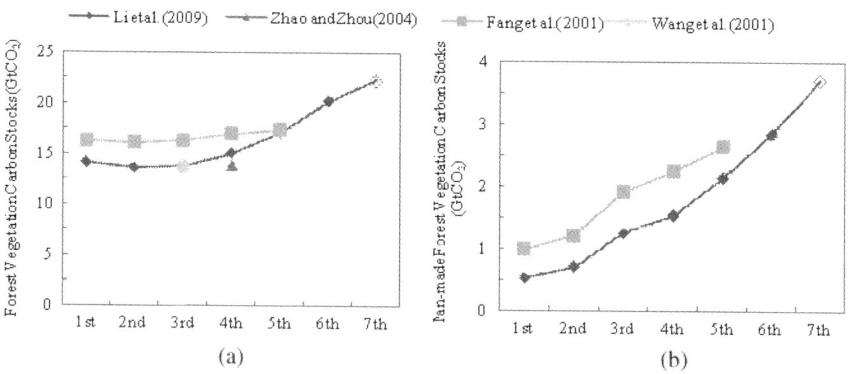

Fig. 11.13 Different studies' estimating results of carbon stock of the forest in the forest census.
Note: The hollow data maker represents the predictions of this report on the basis of the former censuses estimated results by establishing carbon stock-growing stock fitting equations.

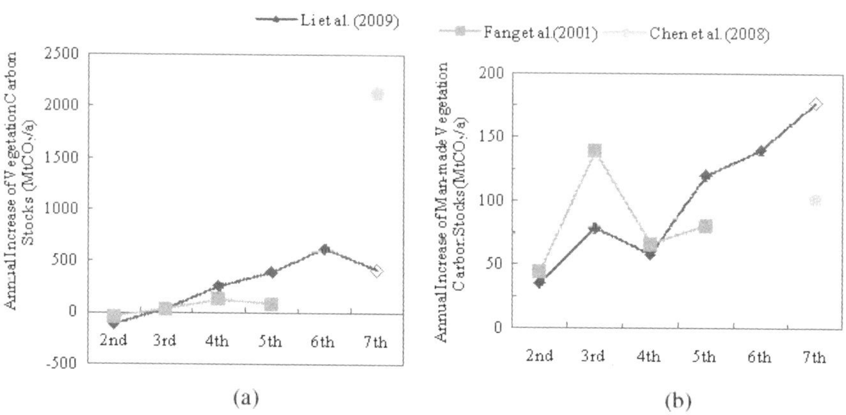

Fig. 11.14. Different studies' carbon stock changes of the forest in the forest census.
Note: Estimating result of Chen *et al.* (2008) is based on the survey data in 2004. The hollow data marker represents the predict result of this report.

China's artificial forest were about 3.72 $GtCO_2$, increasing 91.0% comparing with that from 1999 to 2003, with an annual increase of 177 $MtCO_2/a$.

(2) *Carbon sink of agricultural soil increases*

Carbon sequestration of agricultural soil means to absorb and convert CO_2 in the atmosphere into soil organic matter through plants. And CO_2 in the atmosphere is transferred into long-term carbon pool to form carbon sink.

Long-term plowing makes organic matter in soil (organic carbon) decrease and agricultural soil becomes sources of CO_2 emission. Paustian *et al.* (1997) pointed out that decrease of organic matter in the soil was a main reason for the increase of CO_2 in the atmosphere. CO_2 density now increases to 360 ppm from 280 ppm[8] before industrial revolution, increasing 165 GtC. The CO_2 emission from agricultural soil is about 30%; Li (2000) estimated that the loss of carbon in agricultural soil was about 73.8 MtC every year in China and pointed out that this was mainly because of improper use of agricultural machinery. In 1950s, organic matter in the soil of Northeast China was about 5%–8% (Qu, 1999). Owing to improper chemical fertilizer using and management, China's agricultural soil fertility is declining. In 1980s, the 2nd soil census showed that 38.3% of the soil contained 1%–2% organic matter and 26.0% contained less than 1%. (Huang, 2005).

On the other hand, agricultural soil has great potential of absorbing carbon. Many studies showed that organic matter in soil could be increased through proper agriculture management, regain and even improve the organic carbon content. Paustian *et al.* (1998) estimated that global agricultural soil's carbon absorbing potential was 0.4 to 0.9 GtC/a. Lal (2004) estimated this potential to be 0.4 to 0.6 GtC/a. No-till, organic fertilizer and straw return management can improve soil's carbon absorbing ability. Refer to the following table for carbon absorbing potential of various agricultural measures.

The soil's carbon absorbing potentiality under measures of No-till and straw return management is roughly estimated according to area of No-till and straw return management and the results in Table 11.3 (Fig. 11.15). In 2008, agricultural soil's carbon absorbing potentiality was (16.5 ± 4.8) $MtCO_2/a$ under measures of No-till, which was one time more than that in 2005. Increasing rate between 2000 and 2005 was relatively low, increasing 14.7% annually; from 2005 to 2008, the increasing rate speeded up with an annual increase of 31.8%; in 2008, agricultural soil's carbon absorbing potentiality under measure of straw return management was $(34.2\% \pm 15.9\%)$ $MtCO_2/a$, increasing 17.7% comparing with that in 2005. From 2000 to 2003, the potentiality kept stable and in 2004, it increased again. From 2005 to 2008, the annual increase rate was 5.8%.

[8] $1\,\text{ppm} = 10^{-6}$.

Table 11.3. The carbon absorbing potential of various agricultural measures.

Agricultural measures	International studies: Carbon absorbing rate of the soil [tC/(hm^2·a)]	International studies: Sources	Domestic studies: Carbon absorbing rate of the soil [tC/(hm^2·a)]	Domestic studies: Sources
No-till/ conservation tillage	Humid temperate zone 0.5–1.0 semiarid temperate zone 0.1–0.5 humid tropical zone 0.2–0.5 semiarid tropical zone 0.1–0.2	Lal, 2004	0.40	Li *et al.*, 2009
	Europe 0.2–0.4	Smith *et al.*, 2000a	0.28	Chen *et al.*, 2008
	US 0.38	Jarecki *et al.*, 2005	0.51	Jin *et al.*, 2008
	Canada 0.14	Smith *et al.*, 2000b	—	—
	the world 0.57±0.14	West and Post, 2002	—	—
Straw return management	Europe 0.1–0.3	Smith *et al.*, 2000a	0.31	Li *et al.*, 2009
	Europe 0.15	Vleeshouwers and Verhagen, 2002	0.22	Chen *et al.*, 2008
	—	—	0.60	Jin *et al.*, 2008
Organic fertilizer using	US 0.39 Europe 0.15	Jarecki *et al.*, 2005 Vleeshouwers and Verhagen, 2002	0 32	Chen *et al.*, 2008
			0.55	Jin *et al.*, 2008

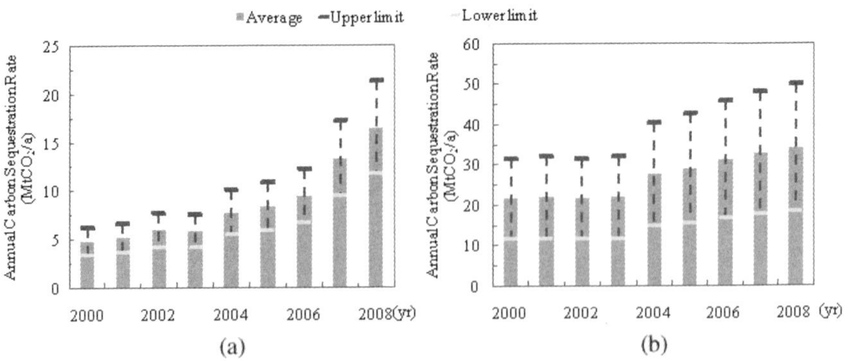

Fig. 11.15. The soil's average annual carbon absorbing potentiality under the agricultural management measures in China (a) no-till; (b) straw return management.
Data Sources: The areas of No-till and straw return management comes from *China Agricultural Machinery Yearbook 2001–2004* and *China Agricultural Machinery Industry Yearbook 2005–2009.*

(3) *Grassland's carbon sink still needs to be explored*

According to some materials, forest takes up 39%–40%, grassland 33%–34% and farmland 20%–22% of carbon stocks within the world's ecosystem. So carbon sinks of grassland and forest are of the same importance.

China has the 2nd largest area of grassland, which takes up 40% of the total territory and natural grassland covers 20% of the land area. So increasing carbon sink of grassland ecosystem through improving grassland management is a low-cost way to reduce greenhouse gas emission. Grassland deteriorating is a severe problem in utilizing natural resources for China. According to Chen *et al.* (2008), 37.41 million hm^2 of China's typical grasslands, 20.41 million hm^2 desert grasslands and 29.19 million hm^2 alpine grasslands deteriorate. If these three parts of deteriorated grasslands can be recovered, then the carbon absorbing potentiality can reach 3162 MtC–6691 MtC.

11.2 Policy Evaluation of Greenhouse Gas Emission

In recent years, China has paid much attention to sustainable development of natural resources and environmental conservation in the process of economic development. Energy saving and greenhouse gas emission reduction have become a vital topic. Items concerning agricultural production in several major policy documents related with energy saving and emission

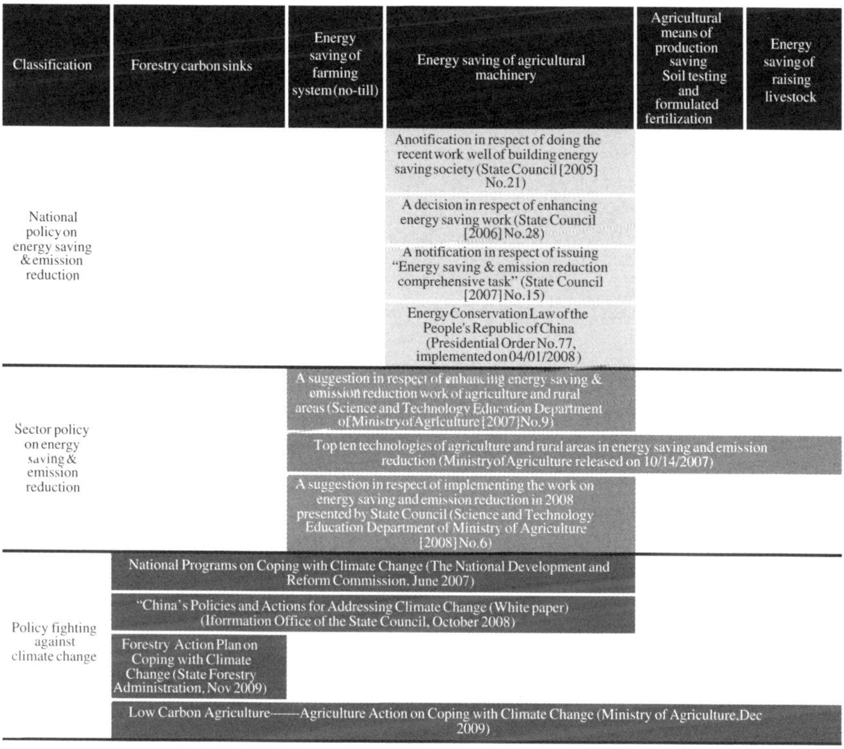

Fig. 11.16. Policy documents of energy saving and emission reduction and coping with climate change.

Note: As to energy saving in rural life, methane, coal-saving stove, hydroenergy, wind energy source, solar energy and other renewable resources are promoted. This chapter focuses on production in agriculture and forestry sector rather than rural life.

reduction focus basically on energy saving of agricultural machinery. These items aim to achieve the goal of energy saving and emission reduction which supported by technology of producing agricultural machinery (Fig. 11.16). By contrast, these documents pay more attention to industry, building and transportation. In response to the energy saving and emission reduction policy, Ministry of Agriculture put forward three major policy documents, which not only make clear the energy saving of agricultural machinery, but also of livestock raising (ecological raising mode), farming system (No-till) and saving of agricultural means of production.

In *National Programs on Coping with Climate Change* drew up by National Development and Reform Commission in June 2007 and *China's*

Policies and Actions for Addressing Climate Change issued by Information Office of the State Council in October 2008, polices concerning energy saving and emission reduction in agriculture and forestry focus on the following four aspects:

(1) Energy saving of agricultural machinery. Specific measures include abandoning backward machines, using oil saving agricultural machinery, promoting minimal tillage/No-till and machines capable of combined operation. Encourage the use of electric motor, wind and solar energy source and other renewable energy sources for agricultural machinery.
(2) Construction of intensified ecological agriculture. This includes using chemical fertilizer and pesticides scientifically, promoting organic fertilizer, soil testing and formulated fertilization to improve the soil quality and reduce N_2O emission.
(3) Promoting technological development and its utilization. This includes selecting rice varieties with high yield and low emission, promoting semiarid cultivation techniques and scientific irrigation techniques to reduce CH_4 emission in paddy fields, developing technologies to breed ruminants with desirable qualities and technologies of large-scale farming to reduce CH_4 emission, developing technologies of environment-friendly fertilizer to reduce N_2O emission and promoting straw return and minimal tillage (No-till) to improve the agricultural soil's carbon stocks.
(4) Increasing forestry carbon sinks. Further the development of forestry through laws and rules, industrial policies and construction of major projects to increase forest resources and forestry carbon sink.

State Forestry Administration issued *Forestry Action Plan on Coping with Climate Change* in November 2009 and put forward the development goals of 2010, 2020, and 2050 (in 2010, 4 million hm^2 of forest will be planted, national forest coverage rate will reach 20% and forest growing stock will reach 13.2 billion m^3). 22 major actions will be taken, including 15 actions to mitigate climate change such as promoting voluntary tree planting and afforestation action of major forest projects and seven forestry actions to adapt to climate change which include improving the adaptability of artificial ecosystem and building natural reserves to protect certain forest species.

Edited and issued by Ministry of Agriculture in December 2009, *Low Carbon Agriculture — Agriculture Action on Coping with Climate Change* pointed out many agricultural measures to cope with the climate change. Measures to mitigate greenhouse gas emission include promoting intermittent irrigation techniques, enhancing utilization efficiency of ruminants' feed, fertilizer saving technology (Soil testing and formulated fertilization) and other technologies for cleaner production, carrying forward energy saving of agricultural machinery, using organic fertilizer and promoting straw return, conservation tillage, grassland conservation (forbidding grazing or using rotation grazing, returning grazing land to grassland) and afforestation. Measures to adapt to climate change include strengthening the construction of agricultural infrastructure, developing water-saving agriculture, and adjusting planting structure and planting system and so on.

Combining the performance of greenhouse gas emission in agriculture and forestry sector and the policies promoted, this chapter analyses six major forest projects, conservation tillage/straw return, agricultural mechanization and soil testing and formulated fertilization.

11.2.1 *Six major forest projects*

The six major forest projects include Natural Forest Protection Project (NFPP), Conversion of Farmland to Forests Project (CFFP), Wind and Sand Storm Source Control Project (WSSSCP), Fast Growth, High Yield Forests Project (FHFP), Three Northern and Yangtze River Basin Shelter Forest Systems Project (TYRBSFSP), and Wildlife and Nature Reserve Establishment Project (WNREP). The Program Outline of National Economy and Social Development for the 10th Five-year Plan passed by the 4th meeting of the 9th National People's Congress in 2001 brought "Comprehensive Treatment Projects for the Building of Ecological Environment in Key Regions" which includes six major forest projects into the 11th five-year plan.

(1) *Goals of the six major forest projects*

(1) Natural Forest Protection Project (NFPP)

NFPP was formally launched in October 2000, aiming to protect natural forest resources along the upper reach of Yangtze River, in Northeast China and Inner Mongolia through forbidding lumbering in natural forest and reducing the yield of commercial lumber.

NFPP include two parts: One is along the upper reach of the Yangtze River and the middle and upper reaches of Yellow River and the other is state owned forest region in Northeast China and Inner Mongolia. The aim of the former project is to protect the 61.18 million hm^2 forests along the upper reach of Yangtze River and the middle and upper reaches of Yellow River and to reduce the annual forest consumption by 61.08 million m^3 by 2010. Also, from 2000 to 2010, 12.73 million hm^2 forests shall be planted, including 3.67 million hm^2 forests by closing hillsides to facilitate afforestation, 7.13 million hm^2 forests by aerial sowing and 1.93 million hm^2 forests by artificial afforestation. The aim of the latter is to reduce 7.52 million m^3 lumber yield of state owned forests in Northeast China and Inner Mongolia by 2003. And from 2000 to 2010, 96.2 billion yuan would be invested to NFPP, of which 80% — 78.4 billion yuan would be from the central government's subsidies and the remaining 17.8 billion yuan would be from local governments.

(2) Conversion of Farmland to Forests Project (CFFP)

CFFP began its trial implementation in 1999 and in March 2001, it was listed in *Program Outline of National Economy and Social Development for the 10th Five-year Plan*. In 2002, it was launched all over China. CFFP, mainly addressing soil erosion is related with many policies and procedures and involves many areas and people.

The aim of CFFP is to return 13.88 million hm^2 farmlands to forests, planting 16.80 hm^2 forests in barren mountains and land suitable for afforestation. The subsidies for afforestation during 2001 and 2010 is 23 billion yuan, living subsidies is 25.8 billion yuan and food subsidies is 161 million tons.

(3) Wind and Sand Storm Source Control Project (WSSSCP)

WSSSCP is the key project in the ecosystem construction around Tianjin and Beijing, addressing sandstorm problems around the capital. Green Olympic and capital image are the unique international and political backgrounds for this project which lasts 10 years from 2001 to 2010.

WSSSCP is to plant 4.94 million hm^2 forests (including not the area by CFFP). The total investment is 57.7 billion yuan, including 27.6 billion yuan financial investment from the central government and 30.1 billion yuan (including central subsidies and local supporting fund) investment in infrastructure construction.

(4) Fast Growth, High Yield Forests Project (FHFP)

FHFP is the backbone projects for China's agriculture and forestry system and is also the "Hope Project" for enhancing forestry development, addressing mainly issues concerning lumber and forest products. This project lasts from 2001 to 2015.

The aim of FHFP is to build 9.20 million hm^2 of fast-growing and high-yield forests. The total investment is 71.8 billion yuan and investment from 2001 to 2010 is 50.2 billion yuan.

(5) Three Northern and Yangtze River Basin Shelter Forest Systems Project (TNYRBSFSP)

TNYRBSFSP involves the widest area and is the most complicated shelterbelt project. This project is to prevent and control desertification in the three-North area and address the ecological issues in other areas. This project lasts from 1978 to 2050 and is divided into three periods and eight phases. The fourth phase of the project lasts for ten years from 2001 to 2010.

TNYRBSFSP includes three-North shelter forest, shelter forest along Yangtze River, shelter forest along the coast, shelter forest along Perl River, afforestation around Taihang Mountain and afforestation on the plain. The fourth phase of the project ends in 2010 and will plant 9.50 million hm^2 forests on the basis of good preservation of forest resources within the project area. The 9.50 million hm^2 forests include 6.30 million hm^2 artificial forests, 1.94 million hm^2 forest by closing hillsides to facilitate afforestation, 1.26 million hm^2 forest by aerial planting. The total investment of the fourth phase is 35.4 billion yuan ($5.36 billion), of which 25.2 billion yuan ($3.81 billion) is invested by the central government. 16.1 billion yuan ($2.44 billion) of the central government's investment is investment of infrastructure construction, which is 64% of the central government's investment and 9.1 billion yuan ($1.38 billion) is special fiscal investment, which takes a proportion of 36%. Of the investment of infrastructure construction, 13.8 billion yuan ($2.09billion) is for afforestation, 1.87 billion yuan ($0.28 billion) is for seeding infrastructure construction and 470 million yuan ($71.2 million) for technological support.

(6) Wildlife and Nature Reserve Establishment Project (WNREP) WNREP was launched in June 2001 with approval of former State Development Planning Commission. This project is an ecological protection project with strategic significance and is also a "diplomatic project" for improving China's image and enhancing cooperation with the whole world, addressing mainly issues such as gene preservation, biodiversity protection, natural conservation, and wetland preservation. The project is divided into three phases: 2001 to 2010, 2011 to 2030, and 2031 to 2050. Protection of wild animals and plants and their habitats is the project's main aim, but aim of artificial afforestation is not clarified in it.

Table 11.4. Construction of five major forest projects (unit: ten thousand hm^2).

Year	Total	NFPP	CFFP	WSSSCP	FHFP	TNYRBS FSP
2000		42.6				
2001	316	94.8	87.1	21.7	8.89	104
2002	678	85.6	442	67.6	4.57	77.6
2003	826	68.8	620	82.4	2.04	53.4
2004	480	64.1	322	47.3	2.23	44.8
2005	311	42.5	190	40.8	0.95	36.8
2006	281	77.5	105	41.0	0.91	56.7
2007	268	73.3	106	31.5	0.34	57.4
2008	344	101	119	46.9	0.40	76.6
Total	3504	650	1990	379	20.3	507

Source of the data: China Rural Statistical Yearbook 2009.

(2) *Construction of five forestry projects*

Please refer to Table 11.4 for the five major forest projects of NFPP, CFFP, WSSSCP, FHFP, and TNYRBSFSP. By 2008, 6.5 million hm^2 forests had been planted through NFPP, 19.90 million hm^2 forests through CFFP, 3.79 million hm^2 forests through WSSSCP, 2030 thousand hm^2 through construction of FHFP and 5.07 million hm^2 through TNYRBSFSP.

(3) *Carbon sequestration effect of the five major forest projects*

Chen *et al.* (2008) pointed out on the basis of former studies that carbon absorbility of China's artificial forest was between 0.89 $tC/(hm^2{\cdot}a)$ and 1.40 $tC/(hm^2{\cdot}a)$. According to this, increases of carbon sink since the beginning of the five major forest projects are shown in the following figure (Fig. 11.17). Since carbon sequestration of forest land is not obvious at the beginning of the projects, the estimated results may be higher than actual ones.

By 2008, carbon sink increase of NFPP in the nine years since its initiation was (131 ± 29.2) $MtCO_2$, which was 17.4% of the total carbon sink increase of the five projects, and the annual increase was (14.6 ± 3.25) $MtCO_2$; the carbon sink increase of CFFP in the eight years since its initiation was (442 ± 98.4) $MtCO_2$, which was 58.5% of the total increase and the annual increase was (55.2 ± 10.9) $MtCO_2$; carbon sink increase of WSSSCP in the eight years was (74.5 ± 14.6) $MtCO_2$, which was 9.9% of the total increase and the annual increase was (9.31 ± 1.84) $MtCO_2$; carbon sink

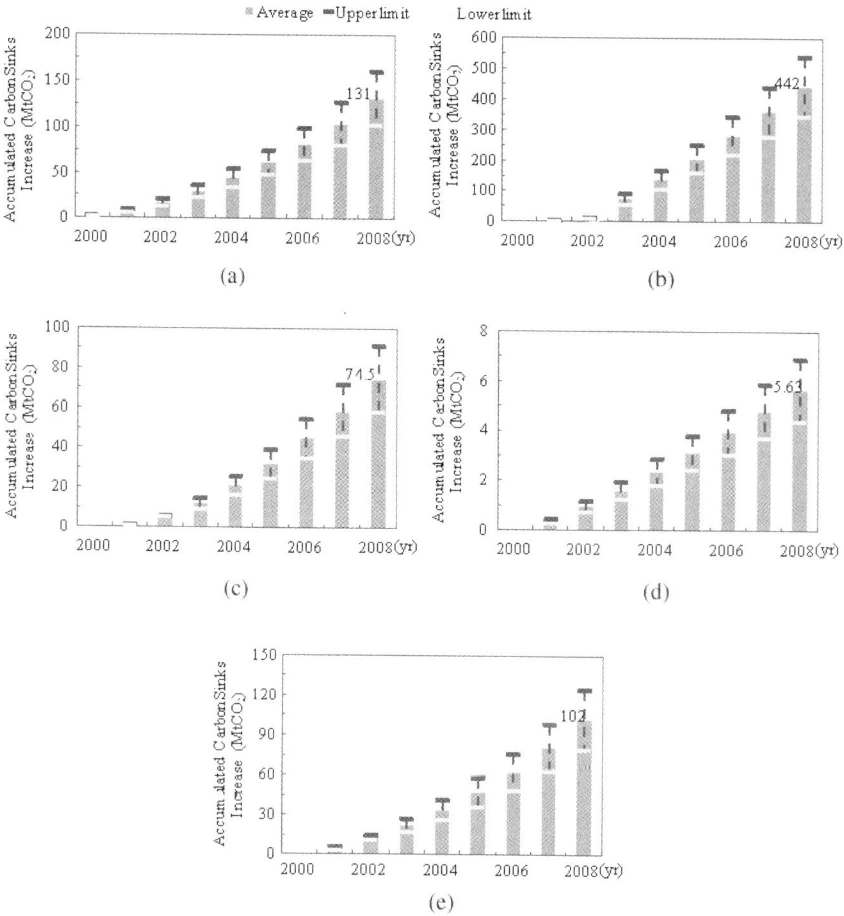

Fig. 11.17. Carbon sink increases of the five major forest projects since their initiation (a) NFPP; (b) CFFP; (c) WSSSCP; (d) FHFP; (e) TNYRBSFSP.

increase of FHFP in the eight years was (5.63 ± 1.25) $MtCO_2$, which was 0.8% of the total increase, and the annual increase was (0.70 ± 0.14) $MtCO_2$; carbon sink increase of TNYRBSFSP in the eight years was (102 ± 22.7) $MtCO_2$, which was 13.5% of the total increase, and the annual increase was (12.7 ± 2.52) $MtCO_2$.

From the above results, carbon sink increase of the project of CFFP contributes the most; carbon sink increase of NFPP was more or less the same with the project of TNYRBSFSP; and carbon sink increase FHFP contributes least.

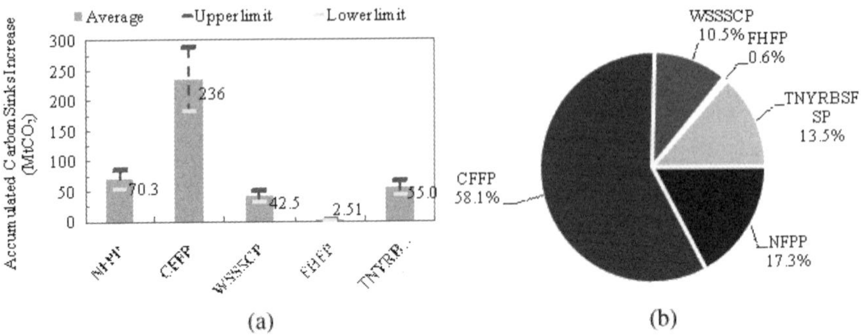

Fig. 11.18. Carbon sink increases and their contribution rates of the five projects from 2006 to 2008 (a) Total carbon sink increase; (b) carbon sinks contribution of each project.

The total carbon sink increase of the five projects from 2006 to 2008 was (407 ± 90.6) $MtCO_2$, with an annual increase of (136 ± 30.2) $MtCO_2$. The contribution rates of the five projects were 58.1%, 17.3%, 13.5%, 10.5%, and 0.6%. Please refer to Fig. 11.18 for the results.

11.2.2 *Conservation tillage/straw return management*

Conservation tillage is a major technology of sustainable development in technology both at home and abroad. Through No-till/minimal tillage and straw return management, soil erosion and sand dust pollution will be controlled, soil fertility and drought resistance ability of the field will be improved and meanwhile, energy and cost will be saved. Conservation tillage is originated from the shocking "American Black Storm". Between 1920s and 1930s, because of repeated tillage by giant machines and the arid climate, desertification became more and more severe. And after the severe soil erosion and sand storms, conservation tillage was developed.

Straw return management is to smash the straw and scatter it onto the field. Ploughed into the soil, the straw will then rot and can fertilize the field and thus the production will increase. According to statistics, 600 million tons of straw is produced every year in China. The straw contains 3 million tons of nitrogen, 0.7 million tons of phosphorus, and 7 million tons of potassium, which is more than a quarter of China's current chemical fertilizer usage. Besides, 90 million tons of organic matter is also contained (Zhao, 2010). Straw return management is important for fertilization of the field and adding organic matter into the field. It is one of major means to use straw.

China's conversation tillage is closely related with straw return management. Conservation tillage involves mainly no-till/minimal tillage, stubble mulch and returning straw to fields.

(1) *Policies concerning conservation tillage/straw return management*

(1) Administrative laws and regulations in China
Since 2005, conservation tillage has been in the No. 1 document of the central government. In 2005, No. 1 document of the central government put forward that "traditional tillage should be reformed and conservation tillage be developed"; in 2006, demonstration project for conservation tillage was started; in 2007, subsidies for No-till promotion was speculated in No. 1 document of the central government; 2008 to 2010 was a period of fast development for conservation tillage (Table 11.5).

In 2007, No. 1 document by the central government encouraged farmers to return straw to the field for conservation and cultivation of the soil's fertilization; in 2009, pilot program for offering subsidies and prizes to farmers returning straw to fields was launched in China; in 2010, in order to further encourage returning straw to fields, No. 1 document by the central government stipulated the measures for using farm machines to support it (Table 11.5).

There are many ways for comprehensive utilization of straw. The use of straw in No. 1 documents by central government between 2004 and 2006 mainly concentrated on straw gasification (methane) and straw used for raising domestic animals; in 2007, besides returning straw to fields, the document also stressed the development of material industry and put forward that research of producing biomass fuels, fertilizer and feed by straw should be accelerated and pilot projects for producing fuels through straw gasification and solidification should be launched. No. 1 documents by central government between 2008 and 2010 emphasized the importance of returning straw to fields and energetically advocated it.

Besides No. 1 document by the central government, *Notice by the State Council to Transpond 'Development Outline for National Program of Raising Livestock with Straw and Using Animal Wastes by Ministry of Agriculture'* (No. 43 [1996] issued by General Office) in 1996 had already mentioned that raising livestock on straw and returning the straw to the fields in the form of manure would increase beef and mutton production, enrich "vegetable basket" market, increase organic matter in the soil and

Table 11.5. Policies concerning conservation tillage/straw return management.

Name of the policy	Reference number/issuing time	Conservation tillage	Straw return management
Advice on policies about further strengthening rural work and improving agricultural production capacity by CCCP and the state council	[2005] No. 1 issued by CCCP	Reform traditional tillage and develop conservation tillage	—
Advice on furthering the construction of new socialist countryside by CCCP and the state council	[2006] No. 1 issued by CCCP	Carry out demonstration project for conservation tillage and pilot project for subsidies used to improve organic matter in the soil	—
Advice on enhancing modern agricultural development and carrying out construction of new socialist countryside	[2007] No. 1 issued by CCCP	Reform farming system and planting method and carry out pilot project for No-till subsidies	Encourage the farmers to use green manure, farmyard manure and promote returning straw to fields
Advice on promoting agricultural development increasing farmers' income by strengthening infrastructure construction	[2008] No.1 issued by CCCP	Carry out the project of conservation tillage	Encourage farmers to use organic fertilizer and green manure and return straw to the field

(*Continued*)

Table 11.5. (*Continued*)

Name of the policy	Reference number/issuing time	Conservation tillage	Straw return management
Advice on promoting stable agricultural development and increasing farmers' income by CCCP and state council	[2009] No. 1 issued by CCCP	Further the project of conservation tillage and quicken the construction of demonstration projects for dry farming	Carry out pilot projects of offering subsidies to farmers using organic fertilizer, green manure and returning straw to fields
Advice on balancing urban and rural development and strengthening the foundation for rural development	[2010] No. 1 issued by CCCP	Promote conservation tillage, carry out demonstration projects for dry farming and offer subsidies for using dry farming.	Encourage returning straw to fields and the use of farm machine in paddy fields
The 11th five-year plan for national development of rural and agricultural economy (2006–2010)	Issued in June, 2006 by Ministry of Agriculture	Reform the traditional tillage and develop conservation tillage and dry farming; promote good seeds and farming methods, especially deep ploughing, conservation tillage and precision/semi-precision seeding; build demonstrative zones in the three Norths and around Beijing and Tianjin and promote technologies such as No-till, minimal tillage and returning straw to fields; carry out demonstration projects.	Strengthen the promotion of good seeds and farming methods and returning straw to fields (raising livestock with straw)

the soil's fertilization, reduce cost for agricultural production and promote sustainable agricultural development. The aim of this policy is to strengthen the development of animal husbandry and adjust agricultural structure. 12 years later, in 2008, *Advice on Promoting the Comprehensive Utilization of Straw by General Office of the State Council* (No. 105 [2008] issued by General Office) mentioned that comprehensive utilization of straw in animal husbandry and planting should be promoted and returning straw to fields through rapid decomposition, animals' digestion and machines should be promoted, too. It also mentioned that farm owners and feed industries should use straw to produce high quality feed. Biomass energy made from straw and processing industry with straw as its material should be developed systematically in order to resolve problems of resource waste and environmental pollution originating from straw abandoning and illegal burning of straw.

(2) Department regulations
In 2002, Ministry of Agriculture issued *Report on Accelerating the Demonstration and Promotion of Conservation Tillage* (No. 130 [2002] issued by Ministry of Agriculture). Since 2005, a series of regulations on conservation tillage (Table 11.6) has been made, including the choice of supporting agricultural machinery, supervision of the effect, encouragement and incentive and other aspects. From the perspective of regulations, *Regulations on Projects of Conservation Tillage (on trial)* and *Regulations on the Supervision of Conservation Tillage Effect (on trial)* were drawn to supervise the implementation and effect of conservation tillage; from the perspective of technology, *Main Points of Technology for Conservation Tillage (on trial)* was drawn to guide conservation tillage activities; from the perspective of promotion, *Textbook for Conservation Tillage Training, Picture Album for the Promotion of Conservation Tillage, Reference of Farm Machines for Conservation Tillage and Propaganda Film for Conservation Tillage were* compiled to guide conservation tillage activities.

In 2000 and 2001, two notices on returning straw to fields by machines were issued by Ministry of Agriculture (Table 11.7), and farm machines were used to return straw to the field. Department regulations concerning returning straw to fields are reflected in the comprehensive utilization of straw. Returning straw to fields causes more and more attention. At the beginning, it had to compete with other means such as straw gasification, raising livestock with straw, using straw as industrial material and straw biomass and then in 2009, a principle was put forward which

Table 11.6. Department regulations on conservation tillage.

Policies and documents	Reference number/ issuing time
Report on Accelerating the Demonstration and Promotion of Conservation Tillage	No. 130 [2002] issued by Ministry of Agriculture
Notice on Printing and Distributing "Key Points of Implementation of Conservation Tillage Technologies" and "Regulations on Implementation of Projects and Supervising the Effects" by General Office of Ministry of Agriculture	No. 3 [2003] issued by General Office of Ministry of Agriculture
Notice on Announcing the Selection Results of Corn Planter for Conservation Tillage	No. 8 [2005] issued by General Office of Ministry of Agriculture
Notice on Technological Innovation Projects for Conservation Tillage in 2006 by Agricultural Mechanization Department of Ministry of Agriculture	No. 7 [2006] issued by Branch of Farm Machine
Notice on Implementation of Conservation Tillage Projects in 2006 by Agricultural Mechanization Department of Ministry of Agriculture	No. 7 [2006] issued by Branch of Farm Machine
Advice on development of Conservation Tillage by Ministry of Agriculture	No. 10 [2007] issued by Branch of Farm Machine
Notice on Recognition of Advanced Unit and Individual for Promoting Conservation Tillage by the General Office of Ministry of Agriculture	May 16th, 2007
Notice on the Printing of *Evaluation Plan of Conservation Tillage Projects* by Agricultural Mechanization Department of Ministry of Agriculture	No. 3 [2009] issued by Branch of Farm Machine
Notice on the Printing of *Construction Planning of Conservation Tillage Projects* by Ministry of Agriculture and NDRC	No. 7 [2009] issued by Planning Committee of Ministry of Agriculture
Notice on Checking and Accepting of Conservation Tillage Projects in 2010 by Agricultural Mechanization Department of Ministry of Agriculture	No. 56 [2010] issued by Branch of Farm Machine

Table 11.7. Department regulations on returning straw to fields.

Policies and documents	Reference number/issuing time	Items
Notice on Issuing Management Methods of Straw Burning and Comprehensive Utilization by State Environmental Protection Administration, Ministry of Agriculture, Ministry of Finance, Ministry of Communications and General Administration of Civil Aviation of China	No. 98 [1999] issued by SEPA	Various regions should promote returning straw to fields, develop straw feed, straw gasification, straw composition with microorganism and high temperature and use straw as industrial material and other ways of comprehensive utilization
Notice on Returning Straw to Fields with Machines and Forbidding Burning Straw in 2000 by Ministry of Agriculture	No. 4 [2000] Issued by Branch of Farm Machine	Please refer to the title
Notice on Returning Straw to Fields with Machines in 2001 by Ministry of Agriculture	No. 3 [2001] issued by Branch of Farm Machine	Please refer to the title
Notice on Furthering Comprehensive Utilization of Straw by Ministry of Agriculture	No. 4 [2003] issued by Branch of Farm Machine	Returning straw by smashing it with machines and conversation tillage and other direct returning technology should be utilized to handle the straw in large quantity in the suburban areas of large and medium-sized cities and plains since the mechanization foundation there is solid; returning straw to fields by stacking and composition should be promoted in hilly areas and less developed areas; straw ensiling, ammonization, straw rubbing and other demonstration technologies should be promoted in areas where herbivore are abundant and enhance the adjustment

(Continued)

Table 11.7. (*Continued*)

Policies and documents	Reference number/issuing time	Items
		of agricultural structure and increase farmers' income through raising livestock with straw and returning it to fields in the form of manure; straw resources should be fully used in developed areas and biomass energy should be developed to promote cultural and ideological progress in rural areas.
Notice on Strengthening Bans on Straw Burning and Comprehensive Utilization of Straw by the State Environmental Protection Administration	April 29th, 2003	Guide the farmers to carry out and promote mechanization of returning straw to fields, returning straw to fields through ensiling, ammonization, stacking and composition and processing, conservation tillage, energy transformation and other technologies.
Emergency Notice on Strengthening Bans on Straw Burning and Comprehensive Utilization of Straw by Ministry of Agriculture	No. 20 [2007] issued by Branch of Farm Machine	In recent years agricultural and farm machine systems have explored actively ways of utilizing straw and promoted nationally technologies such as "direct returning to fields", "conservation tillage", "raising livestock with straw", "process straw into chunk or grain", "biological composition", "straw gasification", "cultivating mushroom with straw", and "using straw as industrial material".
Notice on Printing Suggestions on Compiling Planning of Comprehensive Utilization of Straw by State Development and Reform Committee and Ministry of Agriculture	No. 378 [2009]	Stick to the principle that returning straw to fields should be combined with industrialization, encourage large-scale and industrialized production and guide farmers to use straw comprehensively.

said returning straw to field should be combined with industrialization. *Notice on Strengthening Comprehensive Utilization of Straw by Ministry of Agriculture* (No. 4 [2003] issued by Branch of Farm Machine) offered guidelines for returning straw to fields through three means in different areas: Smashing straw with machines, returning straw to fields through stacking and composition and returning straw to fields in the form of animal wastes.

(2) *Current situation of conservation tillage/straw return management*

In 2008, no-till fields were 11.4 million hm^2, increasing nearly one time compared with that in 2005. Between 2000 and 2005, no-till fields were increasing relatively slowly, with an annual increase of 14.7%. But from 2005 to 2008, the increase was very fast, with an annual increase was 31.8%. The proportion of no-till fields increased from 2.6% in 2000 to 4.8% in 2005, and then to 9.4% in 2008 (Fig. 11.19a).

Compared with other countries, the ratio of fields using conservation tillage (no-till) is relatively low and still has room for promotion. In 2002, 60% of America's fields used various conservation tillage, equaling almost all the suitable areas for conservation tillage; Canada's fields using conservation tillage was 30%; Brazil 60%, Australia 73%. Conservation tillage in Europe starts late but develops fast. 16%–28% of the fields had used conservation tillage in 12 countries (Chang *et al.*, 2008).

In 2008, fields using straw return management in China was 22.77 million hm^2, increasing 17.7% compared with that in 2005. From 2000 to 2003, this area had basically no change, then in 2004, it leaped to 18.36 million hm^2. Since 2004, this area had increased rapidly. From 2005 to 2008, the average annual increasing rate was 5.19%. The ratio of straw return management was higher than that of conservation tillage, increasing from 11.2% in 2000 to 15.8% in 2005 and then to 18.7% in 2008 (Fig. 11.19b).

(3) *Carbon sequestration of conservation tillage/straw return management*

Setting year 2000 as the benchmark, agricultural soil carbon sinks increased (77.5 ± 22.5) $MtCO_2$ in 2008 because of the promotion of no-till and the annual carbon sequestration was (8.6 ± 2.5) $MtCO_2$. Increase of agricultural soil carbon sinks owing to straw return management was (242 ± 112) $MtCO_2$ with an annual increase of (26.9 ± 12.4) $MtCO_2$. The total soil

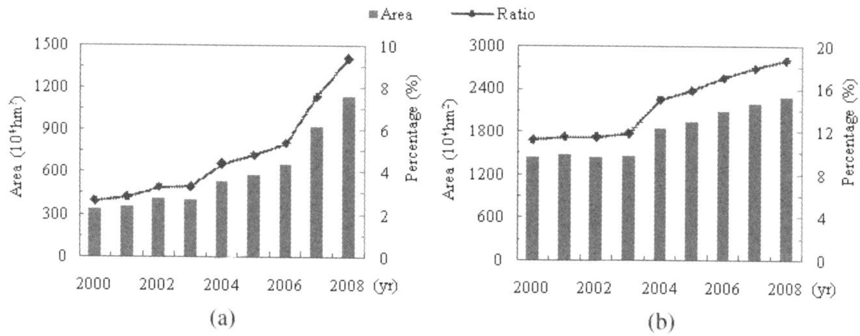

Fig. 11.19. Conservation tillage/returning straw to fields from 2000 to 2008. *Note*: Ratio refers to that of areas using conservation tillage/returning straw to fields to farmland area (a) no-till; (b) straw return management.

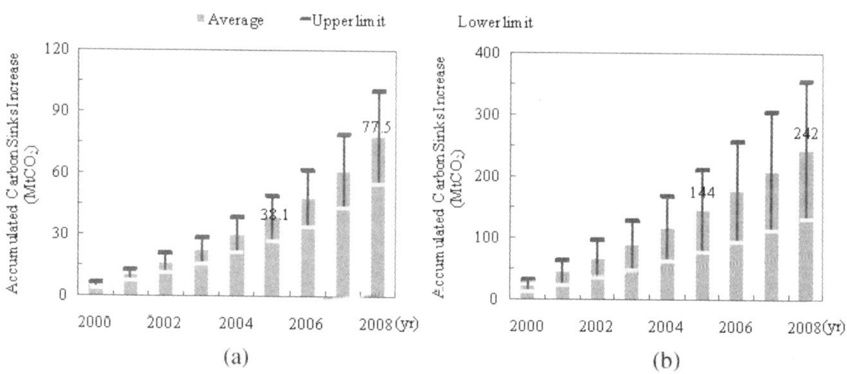

Fig. 11.20. Soil carbon sink increase by using zero tillage and returning straw to fields (a) no-till; (b) straw return management.

carbon sink increase owing to no-till and straw return management from 2006 to 2008 were (39.3 ± 16.9) $MtCO_2$ and (98.3 ± 47.7) $MtCO_2$ (Fig. 11.20).

11.2.3 *Agricultural mechanization*

The influence of agricultural mechanization on greenhouse gas emission is of double blades:

Promoting greenhouse gas emission: the use of farm machines increases energy consumption

Reducing greenhouse gas emission: energy-saving farm machines can improve energy utilization rate; greenhouse gas emission because of livestock enteric fermentation and animal waste management may reduce since farm machines replace draft animals; embedded carbon may also decrease since machines that can save water, seeds, fertilizer and pesticides are used; carbon sink increases since farm machines related with no-till and straw return management are used.

(1) *Policies concerning agricultural mechanization*

(1) National administrative regulations

Agricultural Mechanization Promotion Act of PRC (presidential order No. 16, implemented on November 1st, 2004) (hereinafter referred to as *Promotion Act*) implemented on November 1st, 2004 which set foundation for agricultural mechanization. The *Promotion Act* encouraged and supported farmers and agricultural production organizations to use advanced and suitable farm machines and made clear that the so-called "advanced and suitable" referred to "farm machines that were energy saving, of high quality, multifunctional, produced with advanced technologies and material at a lower cost and sold at a proper price." Besides, the *Promotion Act* also mentioned that "demonstration sites for agricultural mechanization should be established" and that accrediting agencies for farm machines should announce the accrediting results of farm machines' applicability, safety and reliability; in order to improve farm machine's energy utilization rate, "the state encouraged farm machines' trans-regional service", and "encouraged cooperation among farmers in the use of farm machines"; it also mentioned "tax preferential policies for farm machine research and development", "subsidies for buying farm machines", "service tax preferential policies for production operation with farm machines" and "financial subsidies for fuel cost by farm machines".

Both the *11th Five-year Planning Outline for Economic and Social Development of PRC* (2006–2010) and the *11th Five-year Plan for National Development of Agriculture and Rural Economy* (2006–2010) offer specific aims and guidance for the development of agricultural machinery: By 2010, the total power for agricultural machinery should reach 800 million kW, mechanization for ploughing, sowing and harvesting should reach 45% and energy-saving farm machines and other processing equipment for agricultural products should be promoted.

Since 2004, No. 1 document by the central government had continually emphasized agricultural mechanization which was a reflection of its importance. No. 1 document by the central government in 2004 first mentioned subsidies for buying farm machines; in 2005, the document mentioned that local government should contribute to the subsidies for buying farm machines; in 2006, the document emphasized again fast development of agricultural mechanization. From 2007 to 2009, a series of policies and measures were put forward to quicken the development of agricultural mechanization. In 2010, the document emphasized the main aspects of agricultural mechanization such as increasing subsidies for buying farm machines, enriching the types of subsidies, promoting conservation tillage and reform of agricultural technologies and supporting straw return management (Fig. 11.8).

(2) Department Regulations

The *11th Five-year Plan for National Development of Agricultural Mechanization* (2006–2010) (No. 3 [2006] issued by Branch of Farm Machine) issued by Ministry of Agriculture in August 2006 further made clear the aim and measures for agricultural mechanization and put forward five aims including "stable increase in agricultural machinery equipments", "improvement of agricultural mechanization", "rapid development in industrialization of agricultural service", "improvement in technology and innovation ability for agricultural machinery", and "obvious improvement in supervision in farm machines' safety"; from the perspective of measures, the Plan put forward five measures including "improvement in subsidies for buying farm machines", "financial support for technological innovation for farm machines", "tax preferential policies for research and development of farm machines and production of new farm machines", "tax preferential policies for income from farm machine service" and "improvement in subsidy policies concerning fuel cost for farm machines".

Together with other departments, Ministry of Agriculture put forward some policies concerning agricultural mechanization, involving executive order, technology promotion and economic incentive. Various polices have different effects for greenhouse gas emission (Fig. 11.21). From the perspective of policy orientation, policies during the 11th five-year plan emphasized improvement of agricultural equipment, i.e., the increase of total power for agricultural machinery; the main type of policy was incentive policy of which subsidies for buying farm machines caused most attention, and with continual improvement, policies tended to focus on increase of energy utilization rate, subsidies for water, pesticides and fertilizer saving farm machines. To promote agricultural mechanization in hilly areas, policies

Table 11.8. Items in No. 1 document by the central government concerning agricultural mechanization.

Name of the policy	Reference number	Items related with promotion of agricultural mechanization
Advice on policies for increasing farmers' income by the State Council	No. 1 [2004] issued by the central government	Improve agricultural mechanization and offer subsidies for farmers, workers on the farm, farm machine specialists and agencies involving direct buying of farm machines.
Advice on policies for strengthening rural work and improving agricultural production capacity by the State Council	No .1 [2005] issued by the central government	The central government will increase subsidies for fine seeds and the buying of farm machines and local governments should also allot some subsidies for fine seeds and the buying of farm machines.
Advice on promoting construction of new socialist countryside by the State Council	No. 1 [2006] issued by the central government	Promote agricultural mechanization, especially mechanization of main crops, key link in the production chain and major grain producing areas. Encourage producing and using electricity and fuel saving farm machines and processing machines for agricultural products. Increase subsidies for buying farm machines.
Advice on developing modern agriculture and constructing new socialist countryside by the State Council	No. 1 [2007] issued by the central government	Equip agriculture with modern equipments; promote agricultural irrigation, mechanization and informatization; enrich the types and scopes of subsidies for buying farm machines; develop new agro-industry and multifunctional farm machines; strengthen technological innovation and structural adjustment in farm machine industry, especially large and medium-sized tractors and multifunctional combine harvesters and other farm machines; improve the structure and level of farm equipment and make

(*Continued*)

Table 11.8. (*Continued*)

Name of the policy	Reference number	Items related with promotion of agricultural mechanization
		it suitable for the current national situations; accelerate mechanization in grain production, enlarge the service areas of farm machines and promote trans-regional ploughing, sowing and harvesting in busy seasons; build demonstration sites for agricultural mechanization; encourage cooperation in farm machine usage and the establishment of agricultural service agencies and promote the industrialization and marketization of farm machine service; strengthen the supervision and management of farm machine safety.
Advice on strengthening agricultural foundation, promoting agricultural development and increasing farmers' income	No.1 [2008] issued by the central government	Enrich the types of subsidies for buying farm machines and increase the amount of subsidies, making the subsidies cover all the agricultural machines and tools; accelerate agricultural mechanization. Agricultural mechanization is an urgent need in transforming agricultural production modes and offers opportunities for promoting the secondary industry. Promote mechanization in grain production and accelerate mechanization in cash crop and husbandry; strengthen the research and development of suitable farm machines, especially in main grain production areas, hilly areas in the south and schistosome-infected areas; improve the tax

(*Continued*)

Table 11.8. (*Continued*)

Name of the policy	Reference number	Items related with promotion of agricultural mechanization
		preferential policies concerning farm machines, reduce or remit the tax for farm machine service such as road toll for tractors and farm machines doing trans-regional service; carry on the conservation tillage projects; develop agricultural cooperative, service agencies for farm machines; strengthen the supervision and management of farm machine safety.
Advice on promoting stable agricultural development and increasing farmers' income by the State Council	No.1 [2009] issued by the central government	Enlarge the range of subsidies, covering all the farms in the country and offering subsidies for energy saving and reliable farm machines with high safety, suitable service and of advanced technologies, and thus promote the popularization and development of farm machines; accelerate agricultural mechanization; launch projects for farm machine promotion, promote construction of demonstration sites, roads for farm machines, improve popularization and supervision of farm machine; popularize mechanization in sowing and harvesting of main grain and oil crop, accelerate research and development of farm machines that are suitable for hilly areas and large-scale operation; support technological reform of farm machines, improving the durability, applicability

(*Continued*)

Table 11.8. (*Continued*)

Name of the policy	Reference number	Items related with promotion of agricultural mechanization
		and after service of farm machines; establish pilot projects of subsidies for farm machine use in key links of the production chain; offer credit aid to major grain and farm machine producers and farm machine service agencies; improve the supplying system of fuel used for farm machines and set up compensation system for upgrading and dumping of energy-intensive farm machines.
Advice on further balancing urban and rural development and reinforcing the foundation for agricultural development	No. 1 [2010] issued by the central government	Further increase subsidies for buying farm machines, enrich types of subsidies and offer subsidies to drought resisting and water saving machines and machines concerning animal husbandry and forestry; promote and improve the dynamic adjusting systems of the subsidy for agricultural means of production; offer larger ratio of the newly added subsidies to large grain producers and special cooperatives for farmers; improve subsidy policies suitable for pasturing areas, forest regions and reclamation areas; accelerate agricultural mechanization and promote mechanization in soil loosening, straw return management and rice seeding and transplanting; establish modern agricultural demonstration zones.

<table>
<tr><td></td><td>Increase greenhouse gas emissions</td><td colspan="3">Mitigate greenhouse gas emissions</td></tr>
<tr><td>Policy Category</td><td>Promote the growth of total agricultural machinery</td><td>Improve the energy utilization of agricultural machinery</td><td>Develop agricultural machinery to save water, pesticides and fertilizers</td><td>Machine to replace Cattle</td></tr>
<tr><td>Executive Order</td><td></td><td colspan="2">On Further Improving the Energy Saving of Agricultural Mechanization No. 34 (2010) issued by Ministry of Agriculture</td><td></td></tr>
<tr><td>Technique Promotion</td><td></td><td>On Promoting Energy Saving Technology of Agricultural Machinery Repair. No. 2 (2009) issued by Ministry of Agriculture</td><td></td><td></td></tr>
<tr><td rowspan="2">Economic Incentives</td><td colspan="4">Subsidies for the Purchase of Agricultural Machinery
➢ Temporal Method for Management of Agricultural Machinery Purchase Subsidies. No. 11 (2005) issued by Ministry of Agriculture
➢ Special Implementation of Agricultural Machinery Purchase Subsidy Program in 2005. No. 11 (2005) issued by Ministry of Agriculture
➢ Emergent Notice on Further Strengthening the Management of Subsidies for the Purchase of Agricultural Machinery. No. 8 (2006) issued by Ministry of Agriculture
➢ Notification on Further Improving the Management of Agricultural Machinery Purchase Subsidy. No. 24 (2007) issued by Ministry of Agriculture
➢ Emergent Notification of Declaration Preparation on Applying Agricultural Machinery Subsidies in 2008. No. 6 (2008) issued by Ministry of Agriculture.
➢ Implement Programs of Agricultural Machinery Purchase Subsidies in 2009. No. 90 (2008) issued by Ministry of Agriculture
➢ Advice on Implementation of Agricultural Machinery Purchase Subsidies Policy. No. 28 (2010) issued by Ministry of Agriculture</td></tr>
<tr><td colspan="4">Fuel Subsidies of Agricultural Machinery(Without Any Special Subsidies)
➢ Notification of Adding Comprehensive Direct Subsidies on Diesel Fuel and Chemical Fertilizers for the Farmers. No. 33 (2006) issued by the cable of Ministry of Finance
➢ Notification on Perfect the Work of Farmer's Comprehensive Direct Subsidies in 2007. No. 1 (2007) issued by the cable of Ministry of Finance
➢ Notification on Perfect the Work of Farmer's Comprehensive Direct Subsidies in 2008. No. 1 (2008) issued by the cable of Ministry of Finance</td></tr>
</table>

Fig. 11.21. Policy of relevant departments concerning greenhouse gas emissions of agricultural mechanization.

for those areas were relatively loose. National subsidy policies for farm machines' fuel cost were not issued, but since 2006, direct subsidies[9] had been offered to grain producers; in 2009 and 2010, two policies concerning energy saving of agricultural mechanization were issued. They were optional policies, focusing mainly on education and technology promotion; national tax preferential policies concerning research and development of farm machines are not yet issued.

This book mainly analyzes series of policies concerning agricultural machinery purchase subsidy. The purpose of those polices is to encourage and support the use of advanced and practical agricultural machineries, accelerate the process of agricultural mechanization, improve the comprehensive productivity in farming, increase agricultural efficiency, save peasant's cost and increase incomings.

Central Document No. 1 of 2004 first proposed subsidy policy for the purchase of agricultural machinery, and in 2005, the government promulgated *Measures for the Administration of Subsidies used for the Purchase of Agricultural Machineries (For trial implementation)* (No. 11 (2005) issued by Ministry of Agriculture), meanwhile, every year from then on, the government would issue a special plan for the management of agricultural machinery purchase subsidy. With the development of this policy, the amount of subsidies for agricultural machinery was gradually increasing, the scope of subsidies expanding, and the types of subsidies were continuously improved (Table 11.9).

(1) The Amount of Subsidies. Central and local finance bureau increased the purchase of agricultural machinery year by year, the amount of investment surged up from 2004's 0.47 billion yuan to 2008's 5.7 billion yuan. Relatively, the increase of local government was stable, which increased from 2004's 0.4 billion yuan to 2008's 1.7 billion yuan. The financial input of central government increased rapidly, which rose from 2004's 0.07 billion yuan to 2008's 4 billion yuan. And the amount in 2009 and 2010 was as much as 13 billion yuan and 15.5 billion yuan.
(2) Coverage Scope of Subsidies. In 2004, the coverage scope of agricultural machinery purchase subsidy was only 16 provinces and 66 counties, when it came to 2005, the provinces coverage rate arrived at 100%,

[9]It refers to comprehensive direct subsidy policy concerning additional cost for diesel, chemical fertilizer, pesticides and other agricultural means of productions on the basis of current direct subsidies for grain producers.

Table 11.9. Change of Agricultural Machinery Purchase Subsidy Policy.

Year	Amount of subsidy (billion) Central	Local	Coverage scope of subsidies	Subsidies categories	Total amount of subsidies for agricultural machinery(set)
2004	0.07	0.4	16 provinces 66 counties		100,000
2005	0.3	0.8	34 provinces 500 counties	6 big categories 18 small ones	200,000
2006	0.6	1.1	1,126 counties	6 big categories 19 small ones	300,000
2007	2.8	1.3	1,716 counties	7 big categories 24 small ones	600,000
2008	4.0	1.7	2,653 counties	9 big categories 18 small ones	1,200,000
2009	13	—	—	12 big categories 38 small ones	3,000,000
2010	15.5	—	—	12 big categories 45 small ones	—

500 county-level administrative regions benefited from that policy, and another 2,653 county farms were brought into that benefit circle in 2008.

(3) Benefit Group of Subsidies. In 2004, subsidies were aimed at the "individual farmers and service organizations which directly engaged in agricultural production, agricultural machinery", then in 2005, people who purchased "suits of agricultural machinery"[10] was absorbed into the group, which aimed at improving agricultural efficiency and realizing the multifunction of machine. As to the additional term "technology demonstration households listed in Agriculture Department of Engineering Science and Technology", it severed to encourage farmers or agricultural service organizations to adopt advanced technology machinery and equipment, such as the agricultural machinery which could save fuel, electricity, water, pesticides, fertilizers and seeds. The newly term of 2010, which brought person or organization who declared to update their agricultural machinery into the benefit circle, aimed at eliminating backward machines and improving energy efficiency.

[10]Suit of agricultural machinery refers to the host machine and matching equipments, such as tractors and harrow, plow, etc.

(4) Categories of Subsidies. Machinery purchase subsidies in 2004 mainly applied on the plant of wheat, rice, corn and soybean. The subsidies of 2005, however, mainly focused on the joint operating equipments which were used on controlling seeding system for precise seeder, no-till, and mulching, etc., and that change aimed at attaching more attention on improving agricultural machinery efficiency and saving agricultural means of production. Since 2007, a subsidy category named "walking tractors and paddy field farming machines" was proposed, so as to implement the farming model "machine to replace cattle", to promote agricultural mechanization in hilly area, to reduce the human and animal operations, reduce the spread of cattle-borne diseases, to cut off the transmission of schistosomiasis, and to reduce enteric fermentation of farming animals and greenhouse gas emission.

(5) The Subsidy Standards. Subsidy standards in 2004 only formulated that central government subsidy could not be more than 30% of machine price, and then more specific standards were added on the aspects of average county subsidies, single machine subsidies, large equipment subsidies, etc. In 2005, the amount of average county subsidy scope was 300,000 yuan to 700,000 yuan, in 2007, the scope was from 300,000 yuan to 1,300,000 yuan, the highest subsidy in some big agricultural county whose cultivated land was over 1.5 million mu (247,105 acres) or the infected area of schistosomiasis was up to 1.5 million yuan. The subsidy for the purchase of a single machine raised from 2005's 30,000 yuan to 2009's 50,000 yuan, what's more, the subsidy standard of 100 plus horsepower tractors, high-performance forage harvester, large no-till planter and milking machines arrived as much as 120,000 yuan. When it came to 2010, the subsidy of a single machine like large cotton picking machine, sugar cane harvesting machine and tractors of 200 plus horsepower arrived to 200,000 yuan. It was clear that subsidy policy was leading the agricultural machinery to a multi-functional, large-scale direction.

(2) *The effectiveness of agricultural mechanization*

(1) Improvement of equipment and farming mechanization lead to the rapid growth of energy consumption:

The National Agricultural machinery power of 2008 was up to 822 million kW, which was increased by 6.9% comparing with that of 2007, and contributed to the advanced achieved the goal of 800 million kW target.

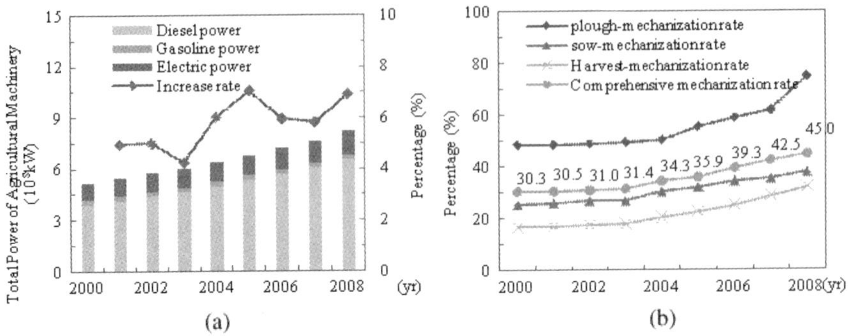

Fig. 11.22. Level of agricultural machinery and mechanization.
Source: "China Agricultural Machinery Yearbook" (2001–2004); "China Agricultural Machinery Industry Yearbook" (2005–2009). (a) Total Power of Agricultural Machinery and Its Growth Rate; (b) Level of Agricultural Mechanization.

The subsidy for the purchase of agricultural machinery was applied since 2004, and the average annual growth rate of its total power was about 6.4% between 2004 and 2008, which was significantly higher than 4.7% from 2001 and 2003 (Fig. 11.22a). In 2008, the national farming mechanization was 45%, which increased by 2.5% compared with that in 2007. The average annual growth rate of farming mechanization between 2004 and 2008 was 2.81%, among which the rate of 2006 and 2007 was over 3%. On the contrary, before the implement of subsidy policy, the annual growth rate between 2001 and 2003 was just 0.54% (Fig. 11.22b).

The diesel-powered agricultural machinery ratio of 2008 was about 80%. Before 2006, the consumption of diesel fuel increased with the use of diesel-powered agricultural machinery, the increase rate of the former even surpassed the latter (Fig. 11.27), which indicated that the increasing usage of agricultural equipment was the main reason for the rapid growth of energy consumption.

(2) Development of large-scale and multi-functional agricultural machinery and improvement of energy utilization:

With the implementation of subsidies for the purchase of agricultural machinery, agricultural machinery headed for the high-powered and multi-functional direction. The average power of tractors and combine harvesters was significantly increased since 2004 (Fig. 11.23a). The quantitative ratio of tractors implements to tractors also presented a growth trend, which raised from 2000's 1.4:1 to 2007's 1.7:1, in 2008, however, the ratio decreased to 1.6:1. There was still a wide gap comparing with developed

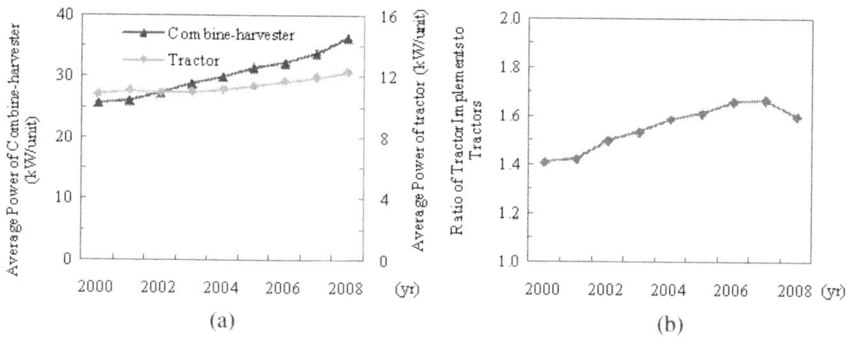

Fig. 11.23. Average power of agricultural machinery and quantitative ratio of tractor implements to tractors.
Source: "China Agricultural Machinery Yearbook" (2001–2004); "China Agricultural Machinery Industry Yearbook" (2005–2009).

countries whose ratio was 3:1 (Fig. 11.23b). To some extent, the upsizing of agricultural machinery could improve the fuel efficiency, and the increase of quantitative proportion proved that agricultural machinery was heading for a multi-functional direction.

Energy consumption showed that in the initial implementation period (2004–2005) of agricultural machinery purchase subsidy policy, the benefit group mainly focused on the level of agricultural machinery and the rapid increase of farming mechanization, while put litter emphasis on promoting energy efficiency of agricultural machinery, which caused the diesel consumption to take an increasing trend, and arrived at its summit 26.2 kgce/kW in 2006. Since then, the subsidies for large machines gradually increased, hence, diesel consumption of per kilowatt diesel-powered agricultural machinery began to decline in 2007, and it dropped to 18.4 kgce/kW (Fig. 11.24) in 2008. Besides, with the attention attached on energy saving and emission reduction, the elimination of backward agricultural machinery also pushed forward the decrease of diesel fuel consumption.

(3) "Machine to replace Cattle" reduce CH_4 and N_2O emission:

Statistics showed that the number of China's draught animals increased from 1988's 76.06 million to 1996's 91.82 million, and then declined to 50.53 million in 2008. On one hand, "machine to replace cattle" would increase CO_2 emissions; on the other hand, it would also reduce the emission of CH_4 and N_2O which were generated in the process of draught animals' enteric fermentation and manure management.

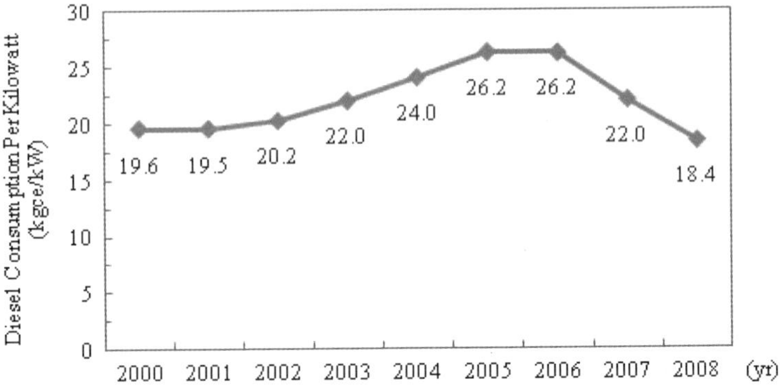

Fig. 11.24. Diesel consumption of per kilowatt diesel-powered agricultural machinery.

From 1996 to 2008, with the agricultural mechanization, the total power output of agricultural machinery increased by about 120%, while the breeding of draft animals declined by 40%. Take the situation in 1996 as the basis, in 2008, the newly added emission of CO_2 released by using agricultural machinery was up to 39.2 Mt CO_2, correspondently, the greenhouse gas emission generated in the process of draft animals' gastrointestinal fermentation and manure management fell by 51.6 Mt CO_2-eq (Fig. 11.25).

(3) *Energy-saving and emission-reduction of agricultural machinery*

Judging from the data of energy consumption, the amount of diesel consumption of agricultural machinery in 2007 and 2008 respectively dropped 0.98 Mtce and 2.73 Mtce compared with that in 2005, which proved that the energy saving of agricultural machinery had won initial success. The diesel consumption of agricultural machinery in 2007 and 2008 separately reduced by 2.6 Mtce and 5.2 Mtce, and the emission of CO_2 reduced by 5.6 MtCO_2, and 11.2 MtCO_2 (Fig. 11.26).

11.2.4 *Soil testing and formulated fertilization*

(1) *Soil testing and formulated fertilization policies*

Based on soil testing and fertilizer field experiment, and according to crop's nutrition requirement rules, the capability of soil fertility and fertilizer effect, and with the reasonable application of organic fertilizer, the government proposed some policies which were related to the application

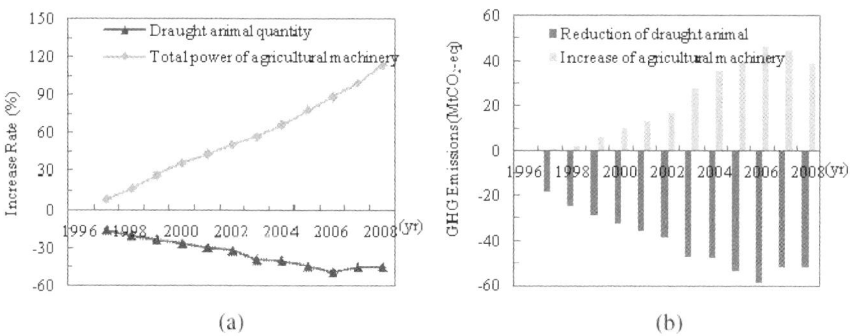

Fig. 11.25. Development trends of "machine to replace cattle" and its impact on greenhouse gas emissions (a) Increase rate change of agricultural machinery and draught animals; (b) Greenhouse gas emissions under the influence of "machine to replace cattle" (take the data of 1996 as the basis).

Note: Draught animals include cattle, buffalo, horses, donkeys, mules, and camels, and the number of draft animals also includes the number of horses, donkeys, mules, and camels recorded in the statistical yearbook. In that figure, assuming the energy consumption of agricultural machinery is diesel and all the electric power, but in fact, a large part of electricity consumption comes from other services business concerning farming, forestry, husbandry, and fishing. Therefore, the estimated greenhouse gas emissions increased in the mechanization of agriculture is larger than the actual number.

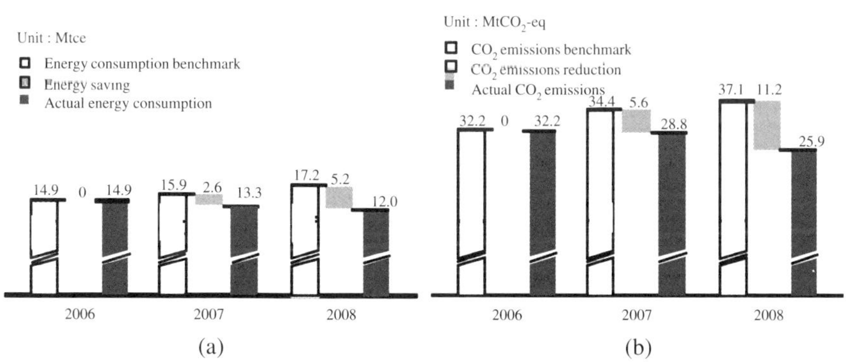

Fig. 11.26. Energy-saving and emission-reduction by using diesel-powered agricultural machines (2006–2008) (a) Energy savings, (b) Greenhouse gas emissions reduction.

Note: The calculation basis of agricultural machinery energy savings: benchmark means the assumption that each year's (from 2006 to 2008) energy consumption and CO_2 emissions of per diesel-powered agricultural machines remains at the level of 2005.

number, application time and application methods of nitrogen, phosphorus, potassium, as well as trace element fertilizers. The core of soil testing and formulated fertilization technology is to resolve the contradiction between requirement of fertilizer and capability of soil fertility, so as to balance the supply of nutrients and meet needs of crops, and thus to archive the purposes of improving efficiency, reducing the amount of fertilizer, increasing crop yield, reducing the cost and improving the efficiency.

(1) National Administrative Regulations

For six continuous years since 2005, the promotion of soil testing and formulated fertilization was discussed in Central Document No. 1 (Table 11.10). In 2005, Central Document No. 1 for the first time proposed the "Promotion of Soil Testing and Formulated Fertilization", and then in 2006, the "Subsidy Policy of Soil Testing and Formulated Fertilization" came out, and the scope and scale of subsidy was continuously expanded between 2007 and 2010. Meanwhile, soil testing and formulated fertilization was listed in two important documents of "National Economic and Social Development Eleventh Five-Year Plan" and "Development of Agriculture and Rural Economy Eleventh Five-Year Plan".

(2) Guided by Central Document No. 1 of 2005, the Ministry of Agriculture vigorously promoted soil testing and formulated fertilization policy, and unveiled about 36 regulations concerning soil testing and formulated fertilization, which included four main categories — implementation, technology and equipment support, efficiency checking, scientific management (Table 11.11). The number of regulation on implementation covered the most, which was as many as 14. As to the aspect of technology and equipment support, it not only published the document "Revised Technical Regulation of Soil Testing and Formulated Fertilization (for trial implementation)", but also conducted training for agricultural technology personnel. In 2007, Ministry of Agriculture started to pay attention to the effect of soil testing and formulated fertilization, and unveiled series of regulations on reporting, spot checking and rechecking. Since the subsidy of soil testing and formulated fertilization in most counties were relaying on the central and local finance, Ministry of Agriculture issued the relevant subsidy regulations; and in order to improve the accuracy of soil testing, it also proposed data management measures.

Capital subsidy was the most important means to promote soil testing and formulated fertilization. Currently, the national level subsidy aimed at the organizations which were responsible for the techniques promotion of soil testing and formulated fertilization and enterprises which produced

Table 11.10. Provisions of soil testing and formulated fertilization in central document No. 1.

Policy name	Document no/Publish time	Soil testing and formulated fertilization
CPC Central Committee and State Council on Further Strengthening Rural Work and Improving Comprehensive Agricultural Production Capacity	No. 1 (2005) issued by National Development and Reform Commission	Promote the policy of soil testing and formulated fertilization, implement organic fertilizer utilization and safe disposal method, guide farmers to use farmyard manure and increase soil organic matter
CPC Central Committee and State Council on Promoting the Construction of the New Socialist Countryside	No. 1 (2006) issued by National Development and Reform Commission	Increase subsidy of soil testing and formulated fertilization
CPC Central Committee and State Council on Developing Modern Agriculture and Pushing forward the Construction of New Socialist Countryside	No. 1 (2007) issued by National Development and Reform Commission	Expand the scope and scale of soil testing and formulated fertilization subsidy, further promote the advanced technology of fertilization diagnosis, precision fertilization
CPC Central Committee and State Council on Further Strengthening the Promotion of Agricultural Infrastructure, Further Developing Agriculture and Increasing Farmer's Income	No. 1 (2008) issued by National Development and Reform Commission	Speed up the pace of implementing "fertile soil project", expand the scale of soil testing and formulated fertilization

(*Continued*)

Table 11.10. (*Continued*)

Policy name	Document no/Publish time	Soil testing and formulated fertilization
CPC Central Committee and State Council on Promoting Steady Development of Agriculture and Continuously Increasing Farmer's Income	No. 1 (2009) issued by National Development and Reform Commission	Continue to promote the "fertile soil project" and expand the scope of soil testing and formulated fertilizer
CPC Central Committee and State Council on Strengthening the Coordination Development of Urban and Rural and Further Consolidating Agricultural and Rural Development	No. 1 (2010) issued by National Development and Reform Commission	Expand the scale and scope of subsidy on soil testing and formulated fertilization and soil organic matter enhancement
National Economic and Social Development : The Eleventh Five-Year Plan (2006–2010)	Approved on the fourth meeting of the Tenth National People's Congress on March 14, 2006	Put forward the applicable techniques on Soil testing and formulated fertilization, balanced fertilization, slow the release of nitrogen fertilizer, biological pest control
National Agricultural and Rural Economic Development: the Eleventh Five-Year Plan (2006–2010)	Issued by Ministry of Agriculture in June 2006	Strengthen the work of soil testing and formulated fertilization, promote fertilizer saving techniques; implement soil testing and formulated fertilization subsidy; carry out the project as soil testing and formulated fertilization, and push forward scientific fertilization

Table 11.11. Regulations concerning soil testing and formulated fertilization I issued by ministry of agriculture.

Policy category	Policy name	Document No/publish time
Implementation	*Implementation of Soil Testing and Formulated Fertilization for Ten million Households.*	March 25, 2005, issued by General Office of Ministry of Agriculture
	Ministry of Agriculture on Emergent Notification of Carrying Out Soil Testing and Formulated Fertilization in Spring.	No. 8 (2005) issued by Ministry of Agriculture
	Ministry of Agriculture on the Issuance of "Autumn Soil Testing and Formulated Fertilization Action".	No. 16 (2005) issued by Ministry of Agriculture
	Ministry of Agriculture on Further Strengthening the Work of Soil Testing and Formulated Fertilization.	No. 10 (2006) issued by Ministry of Agriculture
	Ministry of Agriculture on the Issuance of "National Soil Testing and Formulated Fertilization Program in 2007".	No. 4 (2007) issued by Ministry of Agriculture
	General Office of Ministry of Agriculture, and Ministry of Finance on Approving 2008's Subsidy Plan of Soil Testing and Formulated Fertilization.	No. 54 (2008) issued by Ministry of Agriculture
	General Office of Ministry of Agriculture on Strengthening the Propaganda of Soil Testing and Formulated Fertilization.	No. 60 (2008) issued by Ministry of Agriculture
	General Office of Ministry of Agriculture on Emergent Notification of Carrying Out Soil Testing and Formulated Fertilization in Autumn and Winter.	October 16, 2008

(*Continued*)

Table 11.11. (*Continued*)

Policy category	Policy name	Document No/publish time
	General Office of Ministry of Agriculture on Supporting the Implementation of Soil Testing and Formulated Fertilization in Tibet Autonomous Region.	No. 8 (2009) issued by Ministry of Agriculture
	General Office of Ministry of Agriculture on the Implementation Plan of Soil Testing and Formulated Fertilization in 2009.	No. 26 (2009) issued by Ministry of Agriculture
	General Office of Ministry of Agriculture on the Issuance of National Spring Program of Applying Fertilization Rationally and Promoting Production.	No. 33 (2009) issued by Ministry of Agriculture
	General Office of Ministry of Agriculture, the Ministry of Finance on the Issuance of Implementing Soil Testing and Formulated Fertilization Subsidy in 2009.	No. 32 (2009) issued by Ministry of Agriculture
	General Office of Ministry of Agriculture on the Issuance of National Popularization of Soil Testing and Formulated Fertilization in 2010.	No. 18 (2010) issued by Ministry of Agriculture
	General Office of Ministry of Agriculture, the Ministry of Finance on the Issuance of Guidance for National Implementation of Soil Testing and Formulated Fertilization Subsidy in 2010.	No. 47 (2010) issued by Ministry of Agriculture
Technology and Equipment Support	*Ministry of Agriculture on the Issuance of Revised Technical Regulation of Soil Testing and Formulated Fertilization (for trial implementation).*	No. 5 (2006) issued by Ministry of Agriculture

(*Continued*)

Table 11.11. (*Continued*)

Policy category	Policy name	Document No/publish time
	General Office of Ministry of Agriculture on Further Enhancing the Application Training of Soil Testing and Formulated Fertilization in Winter.	No. 164 (2008) issued by Ministry of Agriculture
	General Office of Ministry of Agriculture on Guidance for Scientific Fertilization in Spring, 2009	No. 23 (2009) issued by Ministry of Agriculture
	General Office of Ministry of Agriculture on Strengthening the Guidance Service of Soil Testing and Formulated Fertilization in Autumn and Winter.	No. 095 (2009) issued by Ministry of Agriculture
	General Office of Ministry of Agriculture on Adjusting and Organizing Its Technical Experts of Soil Testing and Formulated Fertilization.	No. 45 (2010) issued by Ministry of Agriculture
	General Office of Ministry of Agriculture on the Issuance of Guidance on the Building of Counties (Farms) Laboratory for Soil Testing and Formulated Fertilization.	No. 82 (2009) issued by Ministry of Agriculture
	General Office of Ministry of Agriculture on the Issuance of Guidance on Scientific Fertilization of Main Crops in 2010	No. 36 (2010) issued by Ministry of Agriculture
	Ministry of Agriculture on the Issuance of "The Building of Technology Demonstration Counties (Farms) on Soil Testing and Formulated Fertilization".	No. 3 (2010) issued by Ministry of Agriculture

(*Continued*)

Table 11.11. (*Continued*)

Policy category	Policy name	Document No/publish time
Efficiency Checking	*General Office of Ministry of Agriculture on Carrying Spot Checks and Rechecks of Soil Testing and Formulated Fertilization Subsidies in the Experimental Counties in 2005.*	No. 44 (2007) issued by Ministry of Agriculture
	Department of Crop Farming Administration, Ministry of Agriculture on Emergent Notification of Scheduling Work Progress Concerning Soil Testing and Formulated Fertilization in the First Half of 2007.	No. 10 (2007) issued by Ministry of Agriculture
	Department of Crop Farming Administration, Ministry of Agriculture on Carrying Out Investigation of Rice Planting's Soil Testing and Formulated Fertilization.	No. 62 (2007) issued by Ministry of Agriculture
	General office of Ministry of Agriculture on Investigating the Application Effect of Rice and Corn Planting's Soil Testing and Formulated Fertilization.	No. 114 (2008) issued by Ministry of Agriculture
	General Office of Ministry of Agriculture on Carrying Out Effect Evaluation of Soil Testing and Formulated Fertilization.	No. 78 (2009) issued by Ministry of Agriculture
	General Office of Ministry of Agriculture on Summarizing the Work of Soil Testing and Formulated Fertilization in the Last Five Years.	No. 81 (2010) issued by Ministry of Agriculture
	General Office of Ministry of Agriculture on Carrying Out the Spot Check of Popularizing the Soil Testing and Formulated Fertilization in Demonstrational Counties.	No. 91 (2010) issued by Ministry of Agriculture

(*Continued*)

Table 11.11. (*Continued*)

Policy category	Policy name	Document No/publish time
	General Office of Ministry of Agriculture on the Issuance of Pilot Implementation of Evaluation Program Concerning Provincial Soil Testing and Formulated Fertilizction.	No. 95 (2010) issued by Ministry of Agriculture
Scientific Management	*Ministry of Finance, Ministry of Agriculture on the Issuance of Interim Measures of Subsidy Management Concerning Soil Testing and Formulated Fertilization.*	No. 101 (2005) issued by Ministry of Agriculture
	Ministry of Agriculture on the Issuance of Interim Measures of Subsidy Management Concerning Soil Testing and Formulated Fertilization.	No. 8 (2006) issued by Ministry of Agriculture
	Department of Crop Farming Administration, Ministry of Agriculture on Enabling the Statistics Management System of Soil Testing and Formulated Fertilization.	(2006) issued by Ministry of Agriculture
	General Office of Ministry of Agriculture on Further Standardize Data Management of Soil Testing and Fertilizer.	No. 94 (2008) issued by Ministry of Agriculture
	Ministry of Agriculture on Further Strengthening the Management of Soil Testing and Formulated Fertilization.	No. 6 (2009) issued by Ministry of Agriculture
	General Office of Ministry of Agriculture on Holding a National Meeting Discussing Soil Testing and Formulated Fertilization and the Management of Cultivated Land Quality.	No. 103 (2010) issued by Ministry of Agriculture

fertilizer by applying the formula, instead, the farmers could not get the subsidy directly.

The targets of promoting soil testing and formulated fertilization can be found from each year's program: In 2005, the initial target of covering area was 50 million mu (8,236,846 acres) and the number of household was 10 million, when it came to 2009, the covering area was 900 million mu (148,263,228 acres), which took up about 50% of the total of arable land in China, and the number of household increased to 120 million. The crops also extended from grain crops to the cash crops such as cotton, oil crops, etc. In 2009 and 2010, it even extended to vegetables and fruits. What should be noteworthy was that 2010's implementation program not only mentioned the promotion area of soil testing and formulated fertilization, but also required that the area[11] should be more than 400 million mu (658,947,683 acres). (Table 11.12)

(2) *The promotion effect of soil testing and formulated fertilization*

In 2005, the central government invested 200 million yuan on soil testing and formulated fertilization and made 200 counties benefit form that project, and the promotion area reached to 5.33 million hm^2, then in 2006, the number of counties increased to 600, and the central financial input was up to 500 million yuan. Financial input and the promotion area were continuously increased in 2007 and 2008(Table 11.13). The promotion project of soil testing and formulated fertilization in the demonstration counties saved cost but raised efficiency, saved fertilizer but raised production (Column 11.2).

According to national statistics, the total consumption and application rate per unit area of China's chemical fertilizer continuously increased. Since the promotion of soil testing and formulated fertilization policy in 2005, however, the application rate of chemical fertilizer immediately declined (Fig. 11.27). Without the promotion of soil testing and formulated

[11]The soil testing and formulated fertilization policy proposed by Ministry of Agriculture is free of charge to test the soil nutrients for farmer, allocate the usage quantity of nitrogen, phosphate, and potash. In the initial period of the policy, there is no obligation to its actual implementation; hence, it depends on the farmers' willing. The acceptance of that policy, however, is affected by many factors, in addition to the educational level of them; the price of formula fertilizer also matters a lot. If the price is too high, then the farmers would chose to add the quantity of common fertilizer.

Table 11.12. Each year's promotion target of soil testing and formulated fertilization project.

Year	Promotion area	Implementation area	The service number of household	The growth of fertilizer efficiency (%)	Cost saving (yuan/mu)	The crops applied soil testing and formulated fertilization
2005	50 million mu (8,236,846 acres)	—	10 million	3–5	>20, the total saving is more than 5 billon yuan	Rice, wheat, corn, soybean
2007	640 million mu (105,431,629 acres)	—	>100 million	>3	25	Mainly wheat and rice
2008	900 million (148,263,223 acres)	—	120 million	—	—	grain, cotton, and oil crops
2009	1 billion mu (164,736,920 acres)	—	150 million	—	—	Grain, cotton, oil, vegetable, fruit
2010	1.1 billion mu (181,210,613 acres)	>400 million mu (65,894,768 arces)	160 million	—	30	Grain, cotton, oil, vegetable, fruit

Table 11.13. Promotion and implementation effect of soil testing and formulated fertilization.

Year	Promotion scope	Financial input	Implementation area (million hm^3)	Reduced unnecessary usage of fertilizer (million t)	Cost saving yuan/hm^2
2005	200 counties	200 million yuan	5.33	2.4	375
2006	600 counties	500 million yuan	17.33	0.5	>375
2007	1,200 counties	900 million yuan	42.67	1.1	450
2008	1,861 counties	1,150 million yuan	60	1.2	>450

Source: "China Agriculture Yearbook" (2005–2009).

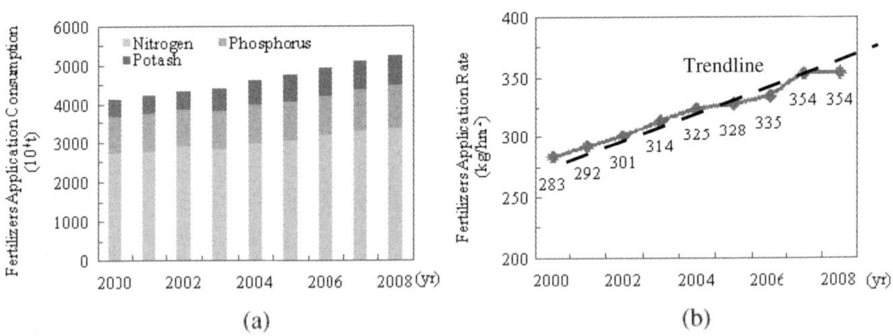

Fig. 11.27. Application rates of chemical fertilizer and that of unit area (a) Application consumption of chemical fertilizer; (b) The application rate of chemical fertilizer per unit area.

fertilization, the use of chemical fertilization would have been increased as the growth speed between 2000 and 2004. Each year from 2005 to 2008, the use of chemical fertilizer separately reduced by 975,000 t, 1,550,000 t, 290,000 t, and 1,82,000 t. This number here was slightly higher than that in the Fig. 11.13. The reason contributed to the decrease of chemical fertilizer in 2007 was that the reduction of cultivation area, besides, the chemical fertilizer usage was higher than that of 2008, hence, the chemical fertilizer application rate per unit area was higher, and comparatively, the reduction amount was less.

On one aspect, the policy of soil testing and formulated fertilization could adjust the structure of fertilizer application. The statistics showed that the application structure of chemical fertilizer did not change between 2005 and 2008, the proportion of nitrogen fertilizer, phosphate fertilizer and potash fertilizer remained to be 1:0.34:0.22. The adjustment of fertilizer structure was a long process, and it was very hard to see the obvious result in the short term. On the other aspect, the result of the above figure was obtained based on market consumption (the consumption of chemical fertilizer = output + import − export), therefore, some certain inaccuracy was possible. However, in some soil testing and formulated fertilization area like Jiangsu province, the proportion of nitrogen fertilizer, phosphate fertilizer and potash fertilizer was 1:0.30:0.32, while actually, the proportion 1: 0.28:0.22[12] was commonly applied.

Column 11.2. The obvious effect of implementing soil testing and formulated fertilization policy.

Yitong County of Jilin province started the soil testing and formulated fertilization policy since 2005, then in 2006, the agricultural technology station of Yitong County did the sample collection in 11 villages and towns, tested 8,500 soil samples, analyzed samples for 52,000 times, delivered 9,200 guiding cards, saved 6,500 t fertilize and 700,000 yuan, increased the crop production 19.5 million kg, added 23.4 million yuan output value, and converted the traditional concept of fertilizing.

Source: http://www.natesc.gov.cn/Html/2007_02_16/52871_53741_2007_02_16_74337.html.

Yishui County of Shandong Province implemented the subsidy policy of soil testing and formulated fertilization for two consecutive years since 2007. During those two years, that county collected 5,530 samples, delivered 100,000 advising cards, trained 12,256 technicians and demonstration household, extended the area of formulated fertilization to 120,000 mu (19,768 acres). The production of peanut increased by 5.3%, saved 3 kg fertilizer per mu, and reduced cost 77.14 yuan per mu. Meanwhile, the production of corn increased by 5.4%, saved 3.5 kg fertilizer per mu, and reduced cost 67.88 yuan per mu. The production of wheat increased

(*Continued*)

[12]Source: "China Agriculture Yearbook 2009".

Column 11.2. (*Continued*)

by 5.4%, and saved 4 kg fertilizer per mu, reduced cost 71.25 yuan per mu. And the total addition amount of crops and oil was up to 7691.6 t, the total cost saving was up to 30,225,200 yuan, and the total reduction of unnecessary fertilizer was 1,470 t.

Source: http://www.natesc.gov.cn/sites/sfb/, http://nc.mofcom.gov.cn/news/7758790.html.

Dafeng City of Jiangsu Province was listed as one of the first demonstration counties on the implementation of soil testing and formulated fertilization policy in September, 2005. From 2006 to 2007, it established the soil nutrient database and indicator system for the soil testing and formulated fertilization, and divided different areas, planted the crops based on the division. In those two years, the cost saving was up to 38.46 million yuan, and the saving of fertilizer was 2,148 t.

Source: http://www.natesc.gov.cn/Html/2007_11_30/52871_53741_2007_11_30_76888.html.

Yaozhou District of Tongchuan City, Shaanxi Province was one of the subsidy places by the Ministry of Agriculture. Since 2008, the soil testing and formulated fertilization policy was promoted in that area, which mainly applied on the plant of wheat, corn, rapeseed and fruit trees. In 2010, the application area was 612,000 mu (100,818acres), the total saving amount of chemical fertilizer was 1,600 t. The promotion area of soil testing and formulated fertilization policy between 2008 and 2010 was 1.828 million mu (301,139 acres), which saved 80 million yuan for farmers.

Source: http://www. people.com.cn/GB/186331/11840186.html.

In 2006, Fuliang County of Jiangxi Province started soil testing and formulated fertilization technology and got promoted in the whole county in 2009, and the total promotion area was 766,900 mu (126,336 acres), which increased production 23,000 t, saved 114.89 t fertilizer, and saved the cost 40,046,900 yuan. In order to comprehensively promote soil testing and formulated fertilization technology, the county organized a leading group named "the soil testing and formulated fertilization subsidy project in Fuliang County", which was composed by technical experts from "Soil and fertilizer Station", "Agricultural Technology Station", "Plant Protection Station". By the full use of television, newspapers, broadcasting van, wall advertisement and posters, the soil testing and

(*Continued*)

Column 11.2. (*Continued*)

formulated fertilization was widely propagated. The departments concerning agriculture edited "The Textbook of Soil Testing and Formulated Fertilization Technology Training" and "High-yield Cultivation Technology of Clover", and hold several training courses, carried out the technical guidance and advisory activities so as to help farmers to choose fertilizer and help them to confirm the fertilizer amount and chose favorable fertilizing time.

Source: http://www.natesc.gov.cn/Html/2009_03_19/52871_53741_2009_03_19 _78048.html.

(3) *Greenhouse gas emission reduction since the implementation of soil testing and formulated fertilization policy*

According to statistics published on "China Agriculture Yearbook", the reduction amount of greenhouse gas emission was approximately calculated (Fig. 11.28). The reduction amount of 2006, 2007, and 2008 was 4.3 MtCO_2-eq, 9.5 MtCO_2-eq, and 10.3 MtCO_2-eq, which respectively took up 1.0%, 2.2%, and 2.4% of that year's chemical fertilizer related greenhouse gas.

Although the optimal adjustment structure of fertilizer application was not obvious from 2006 to 2008, in the long term, the optimal adjustment structure of fertilizer application could effectively reduce the emission of greenhouse gas. While China was increasing the total usage amount of fertilizer in 1980, the structure of fertilizer was also changing, the use of nitrogenous fertilizer decreasing, and the increased use of phosphate fertilizer and potash fertilizer contributed to the reduction of embedded carbon emission intensity.

The proportion of nitrogenous fertilizer, phosphate fertilizer and potash fertilizer increased from 1980's 1:0.25:0.22 to 1990's 1:0.3:0.10, and then to 2000's 1:0.35:0.18, 2008's 1:0.34:0.22. Assuming that the total use of chemical fertilizer remained at the level in 2008, and application structure remained that of 1990 and 2000, or remained at the most proper proportion of 1:0.5:0.4 (http://www.zgny.com.cn/, 2006), the greenhouse gas emission of chemical fertilization related could be approximately calculated (Fig. 11.29). If the structure of chemical fertilizer remained at the level of 1990, the reduction of greenhouse gas emission was 41 MtCO_2-eq. If remained at the level of 2000, the reduction was 8 MtCO_2-eq. Therefore, it was clear that the soil testing and formulated fertilization policy and its

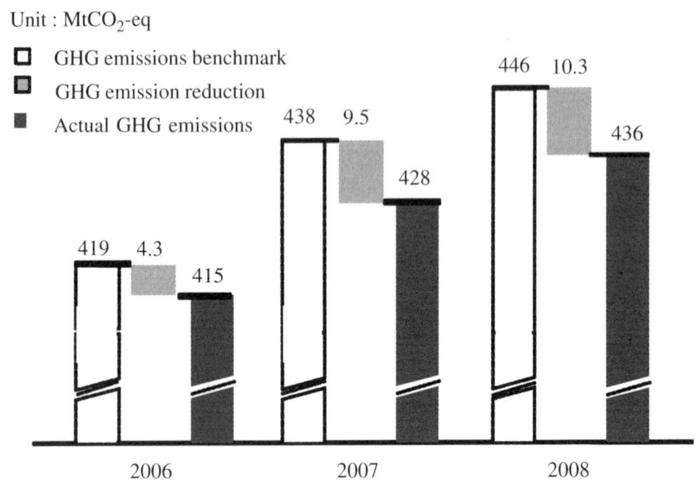

Fig. 11.28. Greenhouse gas emission reduction since the implementation of soil testing and formulated fertilization policy.
Note: The structure of chemical fertilizer (NPK application rate proportion) is calculated based on its consumption proportion.

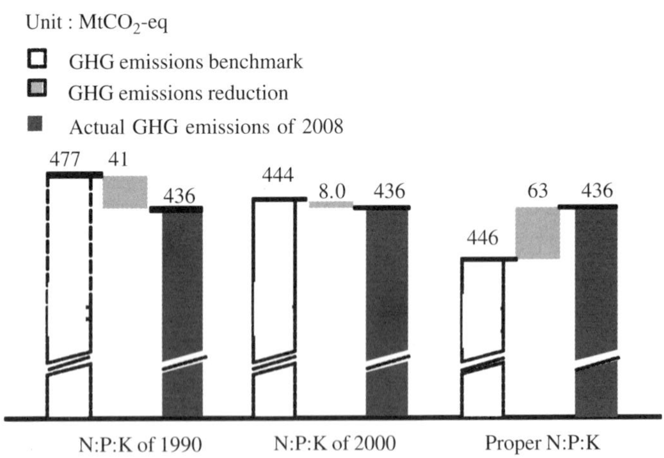

Fig. 11.29. Situation of 2008's greenhouse gas emission reduction under different fertilizer application structure.

optimal adjustment on the fertilizer structure could effectively reduce the greenhouse gas emission.

11.3 Institutional Analysis on Low Carbon Development in Agriculture and Forestry

11.3.1 *Institutional background*

(1) *Acceleration of agricultural modernization*

Agricultural modernization aiming at mechanization, fertilization, and irrigation developed quickly in the past several decades. The total power of agricultural machinery was increased nearly by 680 times from 1.2 million kW in 1957 to 822 million kW in 2008. The number of medium and large tractors rose from 14,674 to 2,995,214 units, increased by more than 200 times. The effective irrigated areas doubled, reaching 58.47 million hm^2. The amount of fertilizer applied increased nearly by 140 times from 0.373 million tons to 52.39 million tons. Despite those achievements, China still has a long way to fully modernizing its agriculture. As the chemical fertilizers were widely used, China has become the biggest fertilizer consumption country in the world. However, the heavy use of fertilizers caused a series of environmental problems. Thus, how to reduce fertilizer amount and increase its efficiency became necessary in the process of sustainable development of agriculture. *The 11th Five-Year Plan for National Economy and Social Development of China* issued in 2006 pointed out, during the "Eleventh Five-Year" period, measures taken for agricultural modernization included reforming traditional farming methods, promoting agricultural standardization, developing energy-efficiency agriculture, as well as adopting scientific technologies to fertilization, pesticides and agricultural film, promoting fertilization based on soil test, balanced fertilization, slow-releasing nitrogenous fertilizer and technologies to biological prevention against disease and insects. It also mentioned widely using advanced and appropriate agricultural machines and appliances, and fertilizing scientifically. According to the *11th Five-Year Plan for the Development of National Agricultural Mechanization* issued in 2006, the comprehensive mechanization rate (ploughing, sowing and harvesting) in 2005 was 36% and should reach 45% during the "Eleventh Five-Year" period. In 2010, agricultural modernization was identified as an important part in the *Twelfth Five-Year Plan*, which accelerated its development.

(2) *Urbanization accelerated and farmers' income mainly from nonagricultural activities*

With the agricultural modernization, urbanization also developed quickly. From 1978 to 2008, the percentage of rural population was reduced from 82.1% to 54.3%. At the same time, a huge number of farmers went to cities or towns as migrant workers. According to statistics, the number of migrant workers in China reached 225.42 million by the end of 2008 and their income was mainly from nonagricultural activities.

Before the implementation of household contract responsibility system, farmers' income was determined by the production team they were in and allocated according to the work points they earned. Under such a system, farmers had no other methods to increase their income. After that, they were greatly motivated. During the period between 1978 and 1993, farmers' income was closely connected with grain production that played a crucial role in their income increase. It first happened in 1993 that farmers went to cities or towns as migrant workers, and the number became larger and larger in the following years. Their major income source shifted from grain production to wages gained by working in cities or towns. From 1999 to 2008, the percentage of wage income in their total net income was increased gradually, from 28.5% to 38.5%. Comparatively, the percentage of their income from farming decreased from 65.5% to 51.1% (Fig. 11.30).

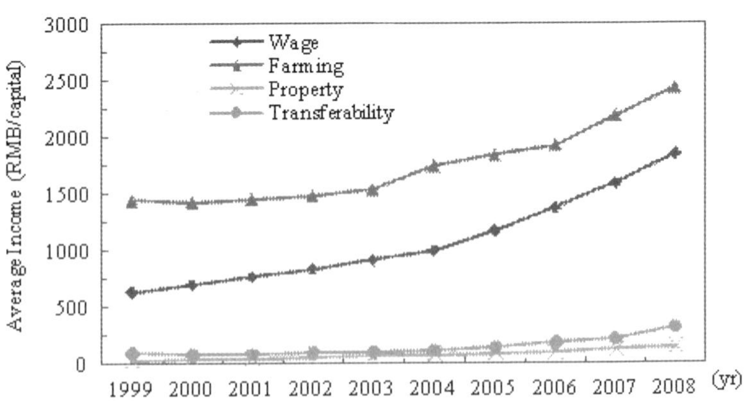

Fig. 11.30. Changes of farmers' income sources.
Source: *China Statistical Yearbook 2009*.

(3) *From "Agriculture Supporting Industry" to "Industry Nurturing Agriculture"*

In order to achieve rapid industrialization, China adopted a policy in favor of industry development for a long time. This enlarged the gap between agriculture and industry, and prevented the sustainable development of agriculture and the increase of farmers' income. In 2005, the *No. 1 Document* issued by the Communist Party of China Central Committee claimed adjustment of income distribution structure, enhancement of agricultural fundamental position, deep reform of rural areas and more support for agriculture, rural areas and farmers, according to the requirements of "Industry Nurturing Agriculture and Cities for Rural Areas". It urged government of all levels to change policies at different development stages and ensure the development of agriculture and rural economy in the whole process of national economic development. This was the first governmental document clearly requiring industry to nurture agriculture. On October 12, 2008, the Communist Party of China (CPC) Central Committee approved *Decisions on Major Issues Concerning Rural Reform and Development* at the third plenary session conference, further defining building a new socialist countryside as a strategic task on the road towards agricultural modernization with Chinese characteristics. In order to realize rapid unification of urban–rural economic and social development, *The Decision* encouraged industrial support for agriculture and cities for rural areas so as to strengthen the fundamental position of agriculture, increase farmers' income, protect farmers' rights and create a harmonious countryside through flexible policies and institutional innovation. This motivated farmers and boosted rapid and healthy development of rural economy. These documents set a benchmark for future policies of "Industry Nurturing Agriculture".

In 2006, China's legislature approved a motion on abolishing regulations on agricultural tax (*The Decision of the Standing Committee of the National People's Congress on Abolishing the Regulation of the People's Republic of China on Agriculture Tax, President Order No. 46*), which meant China's 2,600-year-old agricultural tax would no longer exist. In addition, the government strengthened support for agriculture, and the national finance used on agriculture was 245 billion, 317.3 billion, 431.8 billion, and 595.5 billion RMB from 2005 to 2008, respectively covering 7.2%, 7.9%, 8.7%, and 9.5% of total fiscal expenditure of that year.

(4) *Food security and increase of farmers' income still a main task*

(1) Food Security: It is necessary that China makes agricultural development a primary task because of the fact that China uses only 7% of world's land to feed 22% of the global population. *The Agriculture Law of the People's Republic of China* (*President Order No. 81*) amended in 2002 particularly added clauses on food security, which showed the country's great concern. In order to avoid the hidden dangers of food security arising from the quick reduction of arable lands, *The 11th Five-Year Plan for National Economy and Social Development of China* officially listed *Keep the Current Cultivated Land Amount* as a binding target in 2005. The central government in 2008 put forward a strictest cultivated land protection system so as to ensure at least 1.8 billion mu (*120 million* hm^2) of arable land. The reduction of arable land made food security a serious problem. Moreover, climate change may reduce the food production, which worsened the problem of food security. Anyhow, food security will always be a primary goal of China's agricultural policy.

(2) Increase of Farmers' Income: The household contract responsibility system implemented at the beginning of China's reform and opening up motivated farmers and increased their income while increasing grain production. From 1978 to 1985, the per capita net income of rural residents rose from 134 yuan up to 398 yuan, the urban–rural income ratio dropped from 2.6 to 1.9 — lowest since the establishment of the New China. However, the urban–rural income gap was enlarged because of the rapid development of industrialization since 1986, reaching 2.5 — international warning line in 1992, 3.0 in 2002 and 3.3 in 2008, which affected social stability and China's long-term development. The amended *Agriculture Law* in 2002, compared to the 1993 version, further stressed the importance of increasing farmers' income, of which the first, seventh, thirty-eighth and seventy-ninth clauses all mentioned the issue of farmers' income. In 2004, the State Council, aiming at the problem that farmers' net income grew slowly, issued *Suggestions to Increasing Farmers' Income*, saying that the slow growth of farmers' income affected the improvement of their living standard, the grain production and supply of agricultural products. The document also mentioned that this situation may hinder the development of rural and national economy, affect rural progress and the construction of a better-off society. It is an important economic issue and an important political issue. Besides, an annual 5% increase of farmers' net income was stipulated in the *11th Five-Year Plan for National Economy and Social*

Development of China. The central government put this problem into *No. 1 Document* for the second time in 2009. All these documents implied that the increase of farmers' income was definitely one of the major issues in China's agricultural policies.

(5) *Resources and environment protection as a major goal in the development of agriculture and forestry:*

In order to meet the growing demands of agricultural products, there appeared reclamation land from lake, forest, grassland, and sloping land because of the policies "taking grain as the key link" policy, resulting in the overuse of agricultural resources and ecological and environmental destruction. Meanwhile, pollution became more and more severe due to the heavy use of chemical fertilizers, pesticides and agricultural film. Therefore, "Protection of Agricultural Resources and Environment" was put in Chapter One in *Agriculture Law* adopted in 1993, followed by five detailed clauses. The amended version in 2002 added five additional clauses about the protection. In 2006, the *11th Five-Year Plan (2006–2010) for National Agriculture and Rural Economic Development* defined the harmonious development of rural areas as the third goal, following the goals of supplying sufficient agricultural products and increasing farmers' income. The protection of agricultural resources and environment was an important part of harmonious development in rural areas and its goals in "Eleventh Five-Year" included 5% increase of utilization rate of chemical fertilizer and pesticides, 27% households using methane, 50% emission reduction from agricultural pollutants, and 50% reduction of polluted areas by comprehensive controls.

Regarding the low-carbon development in agriculture and forestry, a binding target that forest coverage should increase from 18.2% in 2005 to 20% in 2010 was set in the *11th Five-Year Plan for National Economy and Social Development* issued in 2005. The Ministry of Agriculture began fertilization program based on soil test in order to improve fertilizer efficiency the same year and later issued 36 related policies. In 2007, the Ministry of Agriculture issued *Suggestions to Energy Savings and Emission Reduction in Agriculture and Rural Areas*, listed detailed requirements. The State Council issued *Suggestions to Rapid and Healthy Development of Agricultural Mechanization and Agricultural Machinery Industry* in 2010, requiring the popularity of environment-friendly mechanical technologies that helped improve production and efficiency and save resources while removing old and high energy consuming machines. Meanwhile, the Ministry of Agriculture distributed *Notice on Energy Savings and Emission Reduction of*

Agricultural Machinery, further pointing out the direction and measures of energy saving. In 2008, the State Council issued *Notice on Promoting Comprehensive Utilization of Crops Straws*, regarding the comprehensive utilization of straws as an important way to save energy and protect environment. The next year, the Ministry of Agriculture and the National Development and Reform Commission jointly issued Conservation Tillage Construction Plan 2009–2015, pointing out the development direction of conservation tillage. As the work to save energy and reduce emission was carried out nation-widely, resources and environment protection in agriculture has become a major development task.

11.3.2 *Institutional analysis*

Forestry plays a crucial role when China fights against climate change. Chinese government has already embarked on reforestation projects and large scale of tree-planting in order to prevent the ecological and environmental deterioration since 1998. This improved ecology and environment, as well as created numerous carbon sequestrations. The government regards carbon sequestration as an important factor in forestry development in recent years. Guangxi Reforestation Project was the first successful project to achieve forest carbon sequestration under the Clean Development Mechanism (CDM). The Chinese government spent a huge amount of money in forestry, the cumulative number of which in returning the reclaimed land to forest reached 233.2 billion from 1999 to 2009. The forestry development improved ecological environment and also benefited millions of farmers. However, it must be coordinated with arable land use, considering China's limited land. Under such conditions, its development pace would slow down and develop steadily in the following years. For the next several decades, forest management would play a more important role in China's forestry development because of the reduction of suitable reforestation land and more difficulties and expenditure of tree-planting.

The low-carbon development in agriculture has a positive impact on food safety, farmers' income increase and protection of agricultural resources and environment. The heavy use of pesticides and chemical fertilizers hindered the increase of farmers' income, polluted environment, degraded land and affected grain production. Therefore, sustainable agricultural development became a strategic necessity and farmers' appeal. In operation, the rapid development of China's agricultural mechanization is helpful in saving water, reducing fertilizers, conservation tillage and establishing advanced farming system that can strengthen the ability against climate change. Agriculture is greatly influenced by climate and its low-carbon development can reduce carbon emissions and promote carbon sequestration, which is a significant step in fighting climate change. A modernized agricultural system can strengthen the ability of agricultural departments against climate change.

China must put high agricultural production as a primary task due to the fact that China has much population but less arable land. In order to achieve high production, too much was spent on China's agriculture, but with little output. Food safety and farmer's income increase are two major tasks in China's agricultural policies. Therefore, China's agricultural low-carbon development should move forward stably, provided that it does not affect food safety and the increase of farmers' income. During the "Eleventh Five-Year" period, Chinese government issued a series of policies on agricultural low-carbon development, but, in general, China's agricultural low-carbon development was still at its initial stage. In this period, the energy conservation and emission reduction in agriculture did not draw much attention because of its comparatively low energy consumption and high degree of dispersion of energy use. Within agricultural sector, the most important was to improve grain production and increase farmers' income, which did not require much energy and resource consumption. As a result, the low-carbon development would not affect food safety and the increase of farmers' income.

Agricultural low-carbon development relies on government support and hundreds of millions of farmers' cooperation and required farmers to change the traditional farming methods. It is a long process because of farmers' dispersion, the long-term traditional farming system and the limitations of farmers' quality. Also, agricultural low-carbon development and the sustainable development in rural areas require long-term government investment considering the limitations of support from industry.

The book discussed several measures taken in low-carbon development of agriculture and forestry.

Column 11.3. Guangxi Reforestation Project.

Guangxi Reforestation Project was the first CDM reforestation carbon sequestration project in the world. It was part of Guangxi Integrated Forestry Development and Conservation Project (GIFDCP), using loans

(*Continued*)

Column 11.3. (*Continued*)

from the World Bank. The Project was started in 2004 and approved by CDM committee in 2006 after survey, feasibility research, quality evaluation and project design.

Project components to: (1) establish 4,000 hm^2 multifunctional forest (Cangwu County and Huanjiang County); (2) gather reforestation and carbon sequestration experience; (3) monitor the overall performance of the project activity; (4) explore the carbon finance mechanism of restoration of degraded land; (5) strengthen local construction ability through training.

Operating modes includes reforestation: (1) by individual household; (2) by a group of farmers; (3) by shareholding agreements between local farmers/communities and the forest company.

Project Implementation. The construction period was from 2006 to 2009, including land acquisition, breeding, planting, fertilizing, weeding, and tending. 1,660 hm^2 plantation was established in 2006, and 2,340 hm^2 in 2007, followed by a three-year tending. A 30-year forest management included prevention and treatment of forest insects and disease, prevention of forest-fire, forest protection, forest harvesting, reforestation, management and surveillance of emission reduction.

The total investment was US$22.7 millions, US$1.13 millions of which were long term loans from Guangxi local commercial bank, US$4.7 million from local farmers and the forest company (sales revenue of present and future timbers, turpentine and other forest products) and US$15.74 millions short term loans from local banks.

The total cumulative amount of emission reduction of CO_2 reached 191, 000 tons of CO_2 from 2006–2010, and 773,000 tons from 2006–2035.

Benefits of the Project. (1) About 5,000 households benefited from the project. The total income was estimated at US$21.1 million, including US$15.0 million from employment; US$3.5 million from sales of wood and nonwood products and US$2.0 million from sales of certified emission reductions (CERs). The annual per capita income was increased by US$34 compared to the one with that in 2004. The project created 40 long-term job positions, sustainable fuel-wood supply and provided technical training and demonstration opportunities. (2) The environmental benefits included enhancing biodiversity and ecosystem integrity, controlling soil erosion, regulating hydrological flows that in turn alleviated droughts risk and reduced flooding risks.

Source: Li (2007).

(1) *Returning the reclaimed lands to forests*

It is an important step to return the reclaimed lands to forests in ecological and environmental protection. The land reclamation in ecologically fragile areas resulted in serious soil erosion. On August, 1997, Jiang Zemin made an important instruction in *Survey Report on Soil Erosion Control in North Part of Shanxi Province and Building Ecological Agriculture*. He said it was possible to recreate a beautiful landscape in Northwest Territories through the persistence of generations of people. Chinese government planned to return the reclaimed lands to forests after the 1998 Yangtze River Flood in order to prevent the deterioration of China's ecology and environment.

China spent a huge amount of money in this ecological project. In 1999, the country spent 336 million yuan RMB, and the number was increased for the following years, reaching 1.47 billion, 2.49 billion, 10.6 billion, 19.3 billion and 19.3 billion respectively in the year of 2000 to 2004. The cumulative investment by the central government was 233.2 billion from 1999 to 2009, and over 200 billion more in the next decade from 2010 to 2021. Besides, the government will subsidize farmers for a long time to ensure the implementation of this project.

On September 2000, the State Council issued *Suggestions about Returning the Reclaimed Land to Grasslands or Forests*, clearly defined the subsidy standards for farmers. The subsidy included: (1) annual 150 kg grain per mu of arable land for farmers living in upper reaches of Yangtze River and 100 kg for those in upper and middle reaches of Yellow River; (2) annual 20 yuan RMB per mu; (3) 50 yuan RMB planting subsidy per mu of derelict hills. The subsidy duration period was two years for grasslands, five years for economic forests, and eight years for ecological forests. *The 2007 State Council Notice on Improving Policies on Returning the Reclaimed Lands to Forests* further defined subsidy standards: Annual 105 yuan RMB per mu in Yangtze River and Southern regions and 70 yuan RMB in Yellow River and Northern regions while keeping the annual 20 yuan RMB subsidy.

For one thing, the subsidy for returning the reclaimed lands to forests made up for farmers' economic losses due to the low productivity of the reclaimed lands. For the other, the project of returning the reclaimed lands to forests liberated farmers from barren lands and transferred labor forces. The project was welcomed by farmers since its implementation. The planted forest area was quickly expanded from 1999–2003, from 448,000 hm^2 to 6,196,000 hm^2 per year. But, the central government adjusted the strategy and reduced the annual amount of planted forests in 2004, considering the food safety problem arising from decreasing arable lands. Therefore, the planted forest area was 3,220,000 hm^2 and 1,900,000 hm^2 in 2004 and 2005, and would keep an annual 1,000,000 ha increase in the following years.

Followed by reforestation, more efforts should be made regarding forest management to make sure the forests will not be reclaimed again. The 2002 *Regulations on Returning the Reclaimed Lands to Forests* claimed that violators who reclaimed the lands without authorization, planted crops in the forests, or made deforestation in project-implemented areas would be severely punished economically, or even criminally. In addition, the government should give more incentives to farmers and help them increase income by various ways while subsidizing them, because their working environment and income sources would affect the effect of reforestation.

(2) *Conservation tillage*

The conservation tillage originated from the US in 1950s where nontillage technology was widely used. Later, nontillage technology was introduced to Europe and Canada in 1960s, and Australia and Latin America in 1970s. At present, conservation tillage has become a dominant technology in those countries and regions. It proves that conservation tillage has such benefits as improving soil quality and maintaining sustainable agricultural development.

In 1991, China launched a pilot project of conservation tillage in Shanxi and Hebei provinces. In 2002, the Ministry of Agriculture carried out demonstration projects of conservation tillage in 38 counties in 12 provinces. The central government has been requiring the development of conservation tillage in *No. 1 Document* for six consecutive years since 2005. In 2009, the Ministry of Agriculture and the National Development and Reform Commission jointly issued *Construction Plan for Conservation Tillage (2009–2015)*. It required conservation tillage should be applied to 600 fields effectively and efficiently, with a total area of 20 million mu, covering 3.1% of total lands in those demonstration counties. Then with reference to the successful experience of pilot project, conservation tillage will be applied to other fields and the newly added area will reach 170 million mu. This marks that China's conservation tillage enters widely promotion stage.

Now the promotion of China's conservation tillage was achieved mainly through: ① pilot project to research and resolve potential technical difficulties and make farmers familiar with conservation tillage. The central authorities founded a special fund in 2002 and annually invested 30 million yuan to demonstrate conservation tillage by technical training, publicity, consultancy, subsidy and purchase of prototype machines. A total number of 170 million yuan had been invested till the end of 2007. ② preferential subsidies to conservation tillage machines. The 2007 *Suggestions to Developing Conservation Tillage* required purchase subsidies to machines related to conservation tillage. ③ direct subsidy to farmers. This policy was carried out only in several provinces. In 2009, the Liaoning provincial government subsidized farmers working in conservation tillage fields 20 yuan RMB per mu. Generally, China's conservation tillage was still at the beginning stage.

Conservation tillage means innovation of farming techniques, as well as changes of crop cultivation system, farm management, traditional farming practices and management models. It takes time for farmers to deeply understand and accept it, because a majority of farmers are used to traditional methods and more concerned about grain productions, economic benefits and risks arising from innovation. Besides, conservation tillage needs lots of agricultural machines, which cannot be fully achieved due to the low stage of China's mechanization. All these factors determine that the application of conservation tillage is a long process.

(3) *Agricultural mechanization*

It is a necessity to achieve agricultural mechanization in the process of agricultural modernization. Agricultural mechanization can greatly improve agricultural efficiency and productivity, liberate farmers from the back-breaking toil, boost urban development and increase farmers' income. Moreover, the advanced machinery like drainage and irrigation appliances and nontillage equipment is helpful for the transformation of modes of production, the sustainable agricultural development and environmental protection. In fact, agricultural mechanization has been a key part in agricultural development since the establishment of the New China, but lagged behind because of financial limitations.

The wide usage of agricultural machinery definitely consumes more energy and this is unavoidable because agricultural mechanization plays an important role in agricultural modernization and helps achieve balanced social development. The agricultural low-carbon development does not mean the absolute quantitative reduction of energy consumption, but consuming the least energy to achieve development goal. Also, the low-carbon development is closely dependent on agricultural machinery.

The slow development of agricultural mechanization in the past several decades is partly because of the insufficient government investment.

Farmers cannot afford the high price of machines, therefore, the Chinese government began to give discounts to farmers when they purchased agricultural machines and appliances, and directly subsidized those manufacturing companies since the year of 2004.

Considering farmers' financial difficulties, the government allows them to purchase agricultural machines at a discounted price, i.e., at a price detracting subsidies. Moreover, farmers can provide machines to be purchased as guarantee to get loans from banks or pay by installments. This greatly motivates farmers to purchase agricultural machines.

Through the purchase subsidy of agricultural machines, the government develops low-carbon agriculture through: ① removing old high-energy-consumption machines; ② promoting energy-saving machines; ③ increasing efficiency by using large-scale and multifunctional machineries; ④ developing machineries that saved water, reduced fertilizers and pesticides, as well as straw manure and conservation tillage machineries. Besides, the Chinese government issued documents to provide ten energy-saving and emission-reducing technologies for machinery maintenance. All these guarantee agricultural development towards saving energy.

(4) *Soil test fertilization*

The heavy use of chemical fertilizers brought about serious environmental pollution and farmers' income was affected by fertilizer price due to the high dependency on fertilizer. Meanwhile, the long-term incorrect usage of fertilizers eroded soil and affected grain production. Therefore, the promotion of soil test fertilization can protect environment and help increase farmers' income and a necessity in agricultural production.

Soil test fertilization included four procedures: ① measurement of soil properties; ② determination of chemical fertilizer proportions according to soil properties and crops planted; ③ production of proportioned chemical fertilizers; ④ application of fertilizers. It involves governmental policy-making, agricultural technology institutes' testing soil and putting forward proportions plans, fertilizer companies' production and farmers' using of fertilizers.

In this process, the central government subsidizes agricultural technology institutes to encourage them to better serve the farmers. According to the *2010 Suggestions on Soil Test Fertilization* issued by Ministry of Agriculture and Ministry of Finance, the subsidy standards were: 160,000–500,000 yuan RMB for demonstration counties (fields) started in

2005, 2006, and 2007; 300,000–650,000 yuan RMB for those started in 2008 and 2009.

Some provinces also issued subsidy policies for farmers using proportioned chemical fertilizers. For instance, Beijing municipal government subsidized farmers 400 yuan RMB per ton in 2008. Shanghai municipal government subsidized agricultural households and institutes planting rice and vegetables 200 yuan RMB per ton. The Qingpu District of Shanghai City additionally subsidized 370 yuan per ton. In 2010, Lanxi City in Zhejiang Province subsidized Farmers Association, farmers having large agricultural land and agricultural companies 400 yuan per ton of proportioned chemical fertilizers. But, up to now, few places gave direct subsidy to farmers fertilizing based on soil test.

It was obvious that the central government policy mainly focused on agricultural technology institutes and paid less attention to fertilizer manufacturing companies and farmers. This resulted in the rapid expansion of counties using proportioned chemical fertilizer, but not a harmonious situation with the quantity of fertilizer. Therefore, the country should take measures to change the present situation and improve the production of proportioned chemical fertilizer and motivate farmers to use such fertilizer.

References

F. An, Annual Report 2009, The Innovation Center for Energy and Transportation (2010), http://www.Icet.org.cn/news3.asp?Cataid=A00040001 (accessed on 15 October 2010).

F. An and A. Sauer, Comparison of passenger vehicle fuel economy and greenhouse gas emission standards around the world, Energy and Transportation Technologies LLC, World Resources Institute, Washington (2004).

SERC Report, Different Electricity Price Effectively Restraining High Energy Consuming Industries, http://news.bjx.com.cn/html/20080912/145261.shtml.

The Oriental Environmental Institute, Energy Research Institute, National Development and Reform Commission, Tracking evaluation on renewable energy law and related environmental policy research, Energy Foundation, Beijing (2009).

The Oriental Environmental Institute, Energy Research Institute, National Development and Reform Commission, Implementation of renewable energy law and industry development research review and evaluation of renewable energy law, Energy Foundation, Beijing (2009).

The Oriental Environmental Institute. Energy Research Institute, National Development and Reform Commission, Tracking evaluation on renewable energy law and related environmental policy research, Energy Foundation, Beijing (2010).

Beijing UnBank Consulting Center, Research Report on China's Wind Energy Development 2009 (Beijing, 2010).

C. L. Chang, L. P. Liu, L. F. Zhang *et al.*, Development Situations and Comments on Conservation Tillage, *Chinese Agricultural Science Bulletin* **24**(2) (2008) 167–172.

G. H. Chen, Study on the Opportunities and Challenges of China's Ethylene Development (2007), http://wenkn.baidu.com/view/eqzclf5f-804d2b160b4ec000.html (accessed on 4 December 2010).

P. Q. Chen, X. K. Wang and L. M. Wang, Carbon Budget and Sink Promotion in China's Terrestrial Ecosystem (Science Press, Beijing, 2008).

Y. Chen and C. Chen, The Application and Importance of Contract Energy Management to Public Buildings. *Environmental Protection* **4** (2009) 53–56.

Y. Dong, Study on China's Inter-city Railway and Existing Problems. Comprehensive Transportation, **5** (2008) 27–30.

X. Z. Feng, J. Zou and G. Q. Xu, Economic Research on China's Fuel Economy Standard. *Environmental Protection* **392**(6) (2008) 23–26.

J. F. Gao, The Development and Reform Commission Reducing Thermal On-Grid Electricity Price in 8 Provinces/Districts/Cities (2007), http://stock.Stockstar.com/GA2007121000716724.shtml (accessed on 28 October 2010).

A. X. Guo, Report on Energy Reduction by Vice-governor Zhang He on Behalf of Provincial Government (2008), http://www.ynet.com/hbqnb/article.jsp?oid=27228609 (accessed on 4 December 2010).

Center for Industrial Energy Efficiency (CIEE), Report on China's Industrial Energy Reduction 2008 (Chemical Industry Press, Beijing, 2009).

Newsenergy, Development and Prospect of Alternative Fuel for Vehicles (2010), http://www.in-en.com/newsenergy/html/newsenergy1406140-698613076.html.

State Electricity Regulatory Commission, Regulatory Report on Electricity Price 2008, State Electricity Regulatory Commission (SERC), Beijing (2009).

State Electricity Regulatory Commission, Regulatory Report on Electricity Price 2009, SERC, Beijing (2010).

Wind Energy Research Team, SERC, Survey Report on Wind Energy Development, Wind Energy Research Team, SERC, Beijing (2009).

National Development and Reform Commission, The Annual Report of Policies and Actions on Climate Change 2010, National Development and Reform Commission (NDRC) Beijing (2010).

Energy Bureau, NDRC, Energy Usage Communiqué of Thousand Companies 2007, Energy Bureau, NDRC, Beijing (2007).

Energy Bureau, NDRC, Energy Usage Communiqué of Thousand Companies 2008, Energy Bureau, NDRC, Beijing (2008).

Energy Bureau, NDRC, Energy Usage Communiqué of Thousand Companies 2009, Energy Bureau, NDRC, Beijing (2009).

Energy Bureau, NDRC, Energy Usage Communiqué of Thousand Companies 2010, Energy Bureau, NDRC, Beijing (2010).

Energy Research Institute, NDRC, Financing Guide: Energy Efficiency and Renewable Energy Project (China Environment and Science Press, Beijing, 2010).

National Development and Planning Commission, Achievements in Rural Power Grid Construction and Transformation (2002), http://www.sdpc.gov.cn/xwfh/t2005070828030.htm (accesseed on 6 December 2010).

National Bureau of Statistics of China, *China Statistical Yearbook* (1996–2009), http://www.stats.gove.cn/tjsj/mdsj/ (accessed on 20 October 2010).

National Bureau of Statistics of China, *China Statistical Yearbook 2006* (China Statistic Press, Beijing, 2006).

National Bureau of Statistics of China, *China's Energy Statistical Yearbook 2009* (China Statistic Press, Beijing, 2009).

Energy Statistics of National Bureau of Statistics of China, *China's Energy Statistical Yearbook 2009* (China Statistic Press, Beijing, 2010).

Energy Statistics of National Bureau of Statistics of China, *China Statistical Yearbook 2010* (China Statistic Press, Beijing, 2010).

Institute of Guo Tai Ju'an Securities, Alternative Energy Strategy Biweekly. Stock Research Biweekly 5:2 (2009).

Energy Institute, State Grid Corporation of China, Analysis Report on China's Energy and Electricity Reduction 2010 (China Electricity Press, Beijing, 2010).

State Council, Notice on Contract Energy Management, http://www.gov.cn/zwgk/2010-04/06/content_1573706.htm (accessed on 2 December 2010).

Development Research Center, State Council, *Popularization of New Energy Vehicles*, NdDaily (2010).

H. T. He, Carbon dioxide emission in cement production and the quantitive research, *Cement Engg*, (1) (2009) 61–65.

H. X. Huang, China's soil situation, problems and suggestions, *Soil and Fertilizer*, (1) (2005) 3–6.

Research on China's Solar Energy Photovoltaic Development (2010), http://www.jdol.com.cn/jdnews/detailInfo.aspx?id=247921 (accessed on 8 December 2010).

Reform Conference on Heat Metering, *Construction Science Technology*, **22** (2009) 8–9.

L. Jin, Y. E. Li, Q. Z. Gao *et al.*, Estimate of carbon sequestration under cropland management in China, *Scientia Agricultura Sinica*, **41**(3) (2008) 734–743.

Y. F. Jin and Z. Wang, Research on Vehicle Fuel Consumption Standard (2010), http://www.chinanews.com.cn/auto/news/2010/03-16/2171193.shtml (accessed on 7 December 2010).

China's Motor Vehicles over 200 Million 85 million Automobiles, http://www.eeo.com.cn/industry/real_estate/2010/10/09/182241.shtml (accessed on 28 October 2010).
Kunming Transport Institute, *Kunming Urban Transport Development Yearbook 2008*, Transport Institute, Kunming (2009).
C. S. Li, Loss of soil carbon threatens Chinese agriculture: A comparison on agroecosystem carbon pool in China and the U.S., *Quaternary Science*, **20**(4) (2000) 345–349.
J. F. Li, Y. Liu, S. C. Wang *et al.*, Development Report on Cross-Straits Photovoltaic Industry (World Wildlife Fund, Beijing, 2010a).
J. F. Li, P. F. Shi and H. Gao, China's Wind Energy Development Report 2010 (Hainan Press, Haikou, 2010b).
J. F. Li, S. C. Wang, M. J. Zhang *et al.*, China's Photovoltaic Development Report 2007 (China Environment and Science Technology Press, Beijing, 2007).
K. R. Li, M. Huang, B. Dai *et al.*, Research on China's Terrestrial Ecosystem Process (China Meteorological Press, Beijing, 2009).
N. Y. Li, China's Forestry Carbon Sequestration (China Forestry Press, Beijing, 2007).
Z. J. Li and Y. Jiang, Pondering over the situation of domestic generalized building energy consumption, *Architectural Journal*, **7** (2006) 30–33.
Z. M. Li, China Green Transport Policy (2009), http://Itaacademy.ita.gov.sg/doc/BjGT01%20Li-Green%20Transport%20Policies%20in%20China%20 (Chinese).pdf.
United Nations Development Programme, Renmin University of China, China Human Development Report: Low-carbon Economy and Sustainable Development, (China Translation & Publishing Corporation, Beijing, 2010).
Z. Lin and B. Hao, Thinking on contract energy management of building energy conservation (2010), *Construction Science and Technology*,16:26.
B. S. Liu, Five Companies Punished for Not Achieving Energy Saving and Emission Reduction Goals in Shandong Province (2007), http://news.xinhuanet.com/newscenter/2007-10/20content_6914869.htm (accessed on 4 December 2010).
C. Liu, X. L. Kong and T. M. Gao, Empirical analysis of changes in China's industrial sector energy consumption intensity and influential factors, *Resource Science*, **30**(9) (2008) 1290–1299.
L. B. Liu, Y. Jiang and L. Fu, Test of a demonstration project using on-off valve regulation and household heat allocation technique based on building heat metering, *Heating Ventilating & Air Conditioning*, **9** (2009) 18–25.

Z. H. Long, Changes to Thermal Power Structure: Five Big Electricity Companies Shut Down Small Thermal Power (2010), http://www.chinanews.com.cn/ny/news/2010/02-10/2118816.shtml (accessed on 22 October 2010).

S. H. Ma, China's Light Engineering and Its Development (2005), http://www.docin.com/p-13839150.html (accessed on 10 October 2010).

S. H. Ma, Consulting Service on Renewable Energy (2008), http://www.fchot.com.cn/Website/index.php?NewsID=6121 (accessed on 12 September 2010).

F. Meng, The impact of urban transportation modes on energy consumption and the environment, Thesis from Department of Environmental Science and Engineering Tsinghua University, Beijing (2010).

Research on the Implementation of Renewable Energy Law and Industry Development (Energy Foundation, Beijing, 2007).

Department of Science, Technology and Education, Ministry of Agriculture, *Rural Energy Statistical Yearbook 1997* (China Agriculture Press, Beijing, 1997).

Department of Science, Technology and Education, Ministry of Agriculture, *Rural Energy Statistical Yearbook 2008* (China Agriculture Press, Beijing, 2008).

J. H. Pan, Y. Zheng, A. H. Zhang *et al.*, Research on Influence of Low-carbon Development to China's Employment (2009), http://www.china.com.cn/international/txt/2009-12/11/content_19050870.htm (accessed on 10 December 2010).

Z. L. Peng, Suggestions on China's Expenditure Statistics (200), http://www.stats.gov.cn/Tjyj/tiggyj/t20091201402604787.htm (accessed on 28 October 2010).

A. H. Piao, Exploration of inter-city railway development, *Comprehensive Transport*, **4** (2009) 16–20.

H. Qi, How to Ensure the Shutdown of 70 GW Small Thermal Power (2010). http://News.163.com/10/0608/08/68L51EUU00025AED.html (accessed on 23 October 2010).

Building Energy Reduction Center, Tsinghua University, *Annual Research Report on China's Building Energy Reduction Development* (China Building Press, Beijing, 2007).

Building Energy Reduction Center, Tsinghua University, *Annual Research Report on China's Building Energy Reduction Development* (China Building Press, Beijing, 2008).

Building Energy Reduction Center, Tsinghua University, *Annual Research Report on China's Building Energy Reduction Development* (China Building Press, Beijing, 2009).

Building Energy Reduction Center, Tsinghua University, *Annual Research Report on China's Building Energy Reduction Development* (China Building Press, Beijing, 2010).

G. P. Qu, *Environmental Protection Guidebook* (Red Flag Press, Beijing, 1999).

D. M. Ren, Z. Y. Wang, H. Gao *et al.*, *Guidebook of Renewable Energy Laws and Regulations* (Chemical Industry Press, Beijing, 2009).

Ministry of Commerce of PRC, Information Management System of Home Appliances to Countryside (2010), http://jdxx.zhs.mofcom.gov.cn/index.shtml (accessed in 28 September 2010).

Shanghai City Transportation Planning Website, *Shanghai City Transportation Yearbook 2008* (2009), http://www.scctpi.gov.cn/reports.asp (accessed on 3 October 2010).

Shenyin Wanguo Securities, Intelligent building energy reduction, Shenyin Wanguo Research Report, Shanghai (2008).

Shenyin Wanguo Securities, Financial subsidy accelerating industry energy reduction, Shenyin Wanguo Research Report, Shanghai (2010).

Y. Song, Task to Shut Down Small Thermal Power Fulfilled Ahead of Time (2010), http://www.ceh.com.cn/ceh/jryw/2010/7/27/66647.shtml (accessed on 22 October 2010).

A. H. Tan, Influences of Fuel Tax to Agriculture and Related Suggestions (2008), http://www.amic.agri.gov.cn/DestopModules/Infos11/infos/ThisInfo.aspx?ItemID=61076&c=23 (accessed on 22 August 2010).

C. Y. Teng, Exploration of green ocean transportation how qingdao ocean shipping Co. Ltd reduced energy, *China Water Transportation*, 8th edn. (2007).

P. Wang, Y. Huang and W. Zhang, Estimation of methane emission in China's rice fields from 1955–2005, *Advances in Climate Change Research*, **5**(5) (2009) 297–297.

Q. Y. Wang, China energy situations and prospect, *China Coal*, **31**(2) (2005) 22–28.

Q. Y. Wang, References of China's renewable energy projects 2009 energy data, Energy Foundation, Beijing (2009).

Q. Y. Wang, References of China's renewable energy projects 2009 energy data, China Renewable Energy Project Team, Beijing (2009).

Q. Y. Wang, Research report on China's energy reduction development 2010 (2010).

W. X. Wang, Review of Energy Usage in China's Steel Industry (2008), http://www.jishujinbu.com:8888/user-loginfo-wangweixing1123-948.html (accessed on 4 December 2010).

X. K. Wang, Z. W. Feng and Z. Y. Ouyang, Vegetation carbon storage and density of forest ecosystems in China, *Chinese Journal of Applied Ecology*, **12**(1) (2001) 13–16.

Y. Wang and X. J. Bai, *Government Policy Reform of China's Transportation Industry* (Economy & Management Publishing House, Beijing, 2009).

Z. X. Wang, Comparison of Chinese and foreign power development, *China Power Enterprise Management*, (10) (2009) 25–28.

H. Wu, Silk Road of Light Engineering to 30,000 Farmers and Herdsmen in XinJiang Province (2005), http://news.sohu.com/20050617/n225982724.shtml (accessed on 12 September 2010).

Y. Wu, C. B. Liu, Y. Z. Liu *et al.*, *Innovation Research on China Building Energy Reduction* (China Building Press, Beijing, 2007).

X. J. Xiao, S. X. Gao and W. K. Han, Advantages, challenges and recommendations for a leap development in national nuclear power industry, *Energy of China*, (3) (2009) 7–9.

Xinhua News Agency, Households Using Biogas over 30 Million in Rural Areas (2009), http://news.xinhuanet.com/fortune/2009-08/31/content_11972075.htm (accessed on 31 August 2010).

Xinhua News, Motor Vehicles over 200 Million in China by 2020 (2010), http://news.cnstock.com/index/gdbb/201009/838037.htm (accessed on 15 October 2010).

Xinhua News, Direct Subsidy to Agricultural Appliances (2007), http://news.xinhuanet.com/misc/2007-03/05/content_5803664.htm (accessed on 6 December 2010).

Xinhua News, Strategic Opportunities for China's Nuclear Power (2010), http://news.xinhuanet.com/zh_cn/news100/11038989/20100-315/15954117_1.htm.

Xinhua News, Benefits of Energy Reduction Appliances to Farmers, http://news.chinanews.com.cn/it//itjdxw/news2010/06-21/2354382.htm 21. (accessed on 21 June 2010).

Xinhua News, Four to Five Years to Achieve Nuclear Power Development Goal (2009), http://news.xinhuanet.com/2010-09/20/c_12591389.htm (accessed on 20 September 2010).

Xinhua News, Key Stage of Five Year Plan for Energy Reduction (2010a), http://finance.people.com.cn/GB/12877669.html (accessed on 9 December 2010).

Xinhua News, Households Using Biogas over 30 Million in Rural Areas (2009), http://news.xinhuanet.com/fortune/2009-08/31/content_1197 2075.htm (accessed on 31 August 2010).

Xinhua News, 11th Five-Year Plan of Energy Reduction Fulfilled Ahead of Time (2010b), http://news.xinhuanet.com/2010-11/30/c_13628755.htm (accessed 9 December 2010).

SinoTrust International Information & Consulting Co. Ltd., China's Automobiles Sales Volume to 15.12 Million in 2010 (2009), http://

bjyouth.ynet.com/article.jsp?oid=61980922 (accessed on 28 December 2010).

SinoTrust International Information & Consulting Co. Ltd., SUV Sales Volume Increase 88.7%, 3.08 Times of Cars' in the Past Eight Months (2010), http://news.hexun.com/2010-09-15/124903836.html (accessed on 28 October 2010).

Xinmin Auto Website, China's Auto Electronics Market Surpassing U.S. Market by 2012 (2010), http://auto.xinmin.cn/calling/2010/10/09/7125972.html (accessed on 28 October 2010).

News Center, Two Billion Investment in 10 Years for Electric Vehicles (2010), http://www.china.com.cn/news/2010-09/16/content_20943721.htm (accessed on 10 October 2010).

H. Q. Xu, China's Electricity Industry: Current Status and Trend (2007), http://www.docin.com/p-22655048.html (accessed on 20 August 2010).

X. C. Xu and Y. G. Wang, Exploration of China's steel consumption, *Metallurgical Information Review*, (1) (2007) 4–10.

H. R. Yang, G. J. Su and M. F. Zeng, Research on energy consumption of sanitary ceramics and energy reduction possibilities, *Ceramics*, (11) (2005) 9–20.

X. Yang, Study of China's Building Energy Efficiency Based on Energy Data, Doctoral Dissertation, Tsinghua University, Beijing (2010).

Y. P. Yang *et al.*, *Thinking on Energy Reduction of China's Motor Vehicles* (China Machine Press, Beijing, 2010).

Y. S. Yang, Chief Leader Dismissed if not Achieving Energy Reduction Goal (2007), http://www.ccin.com.cn/templet/ccin/ShowArticle.jsp?id=30035 (accessed on 4 December 2010).

Z. M. Yang, Plan on China's railway development. *The Report on China Rail Traffic Construction Development* (China Communications Press, Beijing, 2010).

H. Yu, Alternative fuel of small hydro power beneficial for the country and people, China Energy News, 3rd edn., (2010).

L. Zhang, Solving the problem of wind power and power grid needs multipartite, *Electric Power System Equipment*, (8) (2010) 68–69.

L. Y. Zhang, *Biological Low-carbon Agriculture* (China Economic Publishing House, Beijing, 2010).

Q. Y. Zhang, Y. M. Wei and W. L. Tian, Impact of national vehicle emission standards on vehicle pollution emission factors, *Research of Environmental Science*, **23**(5) (2010) 606–612.

R. D. Zhang, 74 Companies not Achieving Energy Reduction Goal (2008), http://www.caijing.com.cn/2008-09-04/110010332.html (accessed on 4 December 2010).

S. Y. Zhang, Study on urban-life energy use and impact of human behaviors through investigations in seven Chinese cities, Master Thesis, Tsinghua University, Beijing (2010).

S. Y. Zhang, X. Yang and Y. Jiang, Comparison and analysis of building energy reduction of China and developed countries in 2004, *Standards and Inspection of HVAC*, **1** (2009) 7–12.

S. L. Zhang, H. N. Cai and G. F. Piao, Credit deficiencies of energy performance contracting projects and their countermeasures, *Construction Economy*, **1** (2009) 55–58.

W. F. Zhang, Y. X. Ji, W. Q. Ma *et al.*, The situation and managing strategy of fertilizer supply and consumption in China, *J. Natural Res.*, **23**(5) (2008) 754–763.

G. Q. Zhao, J. Lv and B. Zhang, Advantages and problems of LPG vehicles, *Technology & Economy in Areas of Communications*, (2006).

M. Zhao and G. S. Zhou, Carbon storage of forest vegetation and its relationship with climatic factors, *Scientia Geograhpica Sinica*, **24**(1) (2004) 50–54.

X. S. Zhao, Appraisal System for China Wind Power Should Be Changed (2010), http://www.ce.cn/cysc.ny/xny/201008/05/t20100815_20453089.shtml (1010).

Y. Z. Zhao, Suggestions to Modern Agricultural Development,. Jinjiao Newspaper, 8th edn. (2010).

K. Y. Zheng, Geothermal power, China Energy News, 7th edn. (2009).

China Semiconductor Industry Association, China Photovoltaic Power Development (2010), http://info.ec.hc360.com/2010/07/230855-294501-3.shtml (accessed on 8 December 2010).

China Semiconductor Industry Association, Photovoltaic power present situation and development trend, *Semiconductor Photovoltaic Power Industry*, (8) (2010) 16–17.

Research Team on Urban Energy Consumption and Reduction, *Characteristics and Energy Reduction Methods in Urban Consumer Area* (China Engineering Publishing, Beijing, 2010).

China Electricity Council, Summary of Electricity Industry Statistics, China Electricity Council, Beijing (2010a).

China Electricity Council, China Electricity Development Report, Yearbook 2010. Beijing: (China Electricity Press, Beijing, 2010b).

China Iron & Steel Association, *China Iron & Steel Yearbook 2009* (China Iron & Steel Yearbook 2009, Editorial Office, Beijing, 2009).

National Bureau of Statistics of China, *China Statistical Yearbook 2006* (China Statistics Press, Beijing, 2006).

National Bureau of Statistics of China, *China Statistical Yearbook 2009* (China Statistics Press, Beijing, 2009).

China Machine Net, Development Trend of Agricultural Equipment and Mechanization in Foreign Countries (2008), http://www.jx.cn/xwzx/viewnew.asp?id=31718 (accessed on 1 December 2010).

China Building Materials Federation, *Almanac of China Building Materials Industry 2009* (China Building Materials Federation, Beijing, 2009).

China Communications and Transportation Association, *Year Book of China Transportation and Communications 2001-2009* (China Transportation and Communications Press, Beijing, 2001–2009).

China Energy Conservation and Environmental Protection Group, *Development Report on China Energy and Emission Reduction* (China Water Publishing House, Beijing, 2009).

China Energy Conservation Association, EMCA Member List (2010), http://www.emca.cn/bg/pxzx/hymd.html (accessed on 1 October 2010).

Power Saving and Green Power Sub-association, CECA, *Present Situation and Trend of China's Energy and Emission Reduction* (Garden Press, Hong Kong, 2009).

China Economy, National Laboratory of Automobiles Energy Saving and Testing Center of Cherry Motor Technology (2010), http://auto.ce.cn/ctzh/qyzx/201007/05/t20100705_20427634.shtml (accessed on 28 October 2010).

Office of China Renewable Energy Development, Research report on China's photovoltaic industry development 2004 (2004).

China Renewable Energy Society, Wind power installation in 2009, *Wind Energy*, (1) (2010) 29.

CAAC News, Green Civil Aviation in Air Transportation Management (2009), http;//news.carmoc.com/list/150/150128.html (accessed on 9 October 2010).

Civil Aviation Administration, Influences of High-speed Railway to Civil Aviation and Related Policy Suggestions, China Civil Aviation, **104**(1) (2009a)15–19.

Civil Aviation Administration, Influences of High-speed Railway to Civil Aviation and Related Policy Suggestions, China Civil Aviation, **105**(2) (2009a) 41–44.

China Nongmo, World's 11 Main Polyethylene Production Technology (2010), http://info.plas.hc360.com/2010/09/121417162181.shtml (accessed on 29 September 2010).

China Agriculture Website, Fertilizer Development Opportunities (2006), http://pfsf.zgny.com.cn/pfsf/zx_show.asp?n_con_id=96039 (accessed on 28 July 2010).

China Association of Automobile Manufacturer, http://www.caam.org.cn/newslist/a40-1.html (accessed on 17 October 2010).

China Auto Info, Motor Vehicle Volume, http://www.autoinfo.gov.cn/autoinfo_cn/cszh/qcbyl/A012101index_1.htm (accessed on 20 October 2010).

Chinese Academy of Social Sciences, *2008–2009 Blue Book: China Provincial Economic Competitiveness* (Social Sciences Academic Press, Beijing, 2009).

Institute of Industrial Economics of CASS, China industrial development report 2009, Economy & Management Publishing House, Beijing, (2009).

China University of Petroleum, Research on Energy Consumption of Petroleum & Gas in Long–Distance Transportation (2009), http://kysbc.cup.edu.cn/ (accessed on 20 October 2010).

CREIN Website, List of Wind Energy Feed-in Tariff in Different Provinces in 2007 (2008), http://www.crein.org.cn/view/viewnews.asp?=20080410111224038 (accessed on 11 October 2010).

China Nonferrous Metals Industry Association, *Year Book of Nonferrous Metals Industry 2009* (China Nonferrous Metals Industry Press, Beijing, 2009).

Press Conference about Shutdown of Small Thermal Power by Energy Bureau, http://www.gov.cn/xwfb/2009-07/30/content_1379745.htm (accessed on 23 October).

China's Reforestation Area Covering 73% of the World, http://finance.sina.com.cn/roll/20100120/10193190041.shtml (accessed on 20 September 2010).

China's Forestry Area to 304 Million Ha According to the Seventh Forest Assessment, http://news.sohu.com/20100611/n272718383.shtml (accessed on 20 September 2010).

China Quality News, Vehicle Fuel Consumption Assessed by a Third Party, http://www.cqn.com.cn/news/zgzlb/diqi/282676.html (accessed on 28 October 2010).

National Development and Reform Commission, Improvement of Legal System of Automobile Energy Reduction (2007), http://www.ndrc.gov.cn/gyfz/gyfz/t20071203_176643.htm (accessed on 31 October 2010).

Ministry of Transport of PRC, Statistical Communiqué of Highway and Waterway Transportation (2009), http://www.chinahighway.com/news/2010/404760.php (accessed on 28 October 2010).

Ministry of Railways of PRC, Railway Introduction 2005–2008 (2005–2008), http://www.china-mor.gov.cn/ (accessed on 20 November 2010).

Statistical Center, MOR, Railway Statistical Communiqué 2006–2009 (2006–2010), http://www.china-mor.gov.cn/ (accessed on 20 November 2010).

Ministry of Housing and Urban-Rural Development of PRC, Six Most Potential Fields of China's Building Energy Reduction — Address by Deputy Minister (2010), http://www.mohurd.gov.cn/ldjh/jsbfld/201004/t20100408_200306.htm (accessed on 8 April 2010).

Ministry of Housing and Urban-Rural Development of PRC, Notice on the 2nd Green Building Evaluation Label and Design Evaluation Label in 2010 (2010), http://www.mohurd.gov.cn/gsgg/gg/jsbgg/201007/t20100721_201614.htm (accessed on 21 July 2010).

F. Q. Zhou and Q. Y. Wang, *50 Years for China's Energy* (China Electricity Press, Beijing, 2002), pp. 150–151.

J. Zhou, Energy and emission reduction and contact energy management, *China Economists*, **1** (2009) 68–69.

A. Bauen, P. Watson and J. Howes, Methodology and guidance for carbon reporting under the renewable transport fuel obligation, E4Tech Consulting Company, U.K. (2008).

Climate Works Foundation and the International Council on Clean Transportation, Policy Efficiency Project-Transportation (2010).

G. Darido, M. Torres-Montoya and S. Mehndiratta, Urban transportation and CO2 emission: Evidence and trend from Chinese cities, Thesis from World Bank, Washington (2009).

ECOFYS, G8 Climate Scorecards 2009, WWF, Switzerland (2009).

Energy Information Administration, *International Energy Outlook 2009* (EIA Publications, USA, 2009).

Ernest Orlando Lawrence Berkeley National Laboratory, Assessment of China's energy–saving and emission–reduction accomplishments and opportunities during the 11th five year plan, Ernest Orlando Lawrence Berkeley National Laboratory, Berkeley (2010).

Europe's Energy Portal, Feed-in-Tariffs (2010), http://www.energy.en/#feedin (accessed on 20 October 2010).

J. Y. Fang, A. P. Chen, C. H. Peng *et al.*, Changes in forest biomass carbon storage in China between 1949 and 1998, *Science*, **292** (2001) 2320–2322.

J. Fuglestved *et al.*, Climate forcing from the transport sectors, *Proc. Natl. Acad. Sci. U.S.A.* (PNAS), **105**(2) (2008) 454–458.

Halcrow Group Ltd, P. Headicar and D. Banister, Land Use and Transport: Settlement Patterns and the Demand for Travel (2009), http://cfit.independent.gov.uk/pubs/2009/sustainable/technical/pdf/pst-technical.pdf (accessed on 10 October 2010).

http://www.drcnet.com.cn/DRCnet.common.web/DocViewSummary.aspx?version=Integrated&docid=2358175&leafid=18365&chnid=4762&gourl=/DRCnet.common.web/docview.aspx (accessed on 28 October 2010).

IEA, Tracking industrial energy efficiency and CO2 emissions, IEA, Paris (2007).

IEA, Energy balances of OECD countries 2009 Edition, IEA, Paris (2009).

International Council of Clean Transportation, Global Passenger Vehicle Fuel Economy and GHC Emissions Standards: April 2010 Update (2010), http://www.theicct.org/2010/04/ghg-fe-standards-update/ (accessed on 12 October 2010).

International Fertilizer Industry Association (IFA), Fertilizers, climate change and enhancing agricultural productivity sustainably, IFA, London (2009).

IPCC, 2006 IPCC guidelines for national greenhouse gas inventories, IPCC, Genera (2006).

Y. Jiang, Does energy follow form, Master Thesis, Massachusetts Institute of Technology, Cambridge (2010).

Korean Energy Economics Institute, Energy use survey 2005, Ministry of Commerce, Industry and Energy, Seoul (2005).

R. Lal, Soil carbon sequestration impacts on global climate change and food security, *Science*, **304** (2004) 1623–1627.

J. Leather, Urban transport in Asia: Toward climate change mitigation, Asian Development Bank, 4th Regional EST Forum in Asia, Seoul (2009).

Ministry of Commerce, Industry and Energy, Korea Energy Economics Institute, *Yearbook of Energy Statistics 2007* (Ministry of Commerce, Industry and Energy, Seoul 2007).

N. Nakicenovic *et al.*, Special report on emissions scenarios, Cambridge University Press, Cambridge (2000).

National Research Council, Assessment of technologies for improving light duty vehicle fuel economy, (2010).

Natural Resources Canada, 2006 Energy use data handbook, Energy Publications Office of Energy Efficiency, Canada (2007).

OECD/ITF, Green house gas reduction strategies in the transport sector, OECD/ITF, Paris (2008).

K. Paustian, O. Andrea, H. Janzen *et al.*, Agricultural soil as a C sink to offset CO_2 emission, *Soil Use Manage*, **13** (1997) 230–244.

K. Paustian, C. V. Cole, D. Sauerbeck *et al.*, CO_2 mitigation by agriculture: An overview, *Climate Change*, **40** (1998) 135–162.

REN21, Renewable 2010 global status report, REN21 Secretariat, Paris (2010).

P. Smith, D. S. Powlson, J. U. Smith *et al.*, Revised estimates of the carbon mitigation potential of UK agriculture land. *Soil Use Manage*, **16** (2000a) 293–295.

W. N. Smith, R. L. Desjardins and E. Pattey, The net flux of carbon from agriculture soil in Canada 1970–2010, *Global Change Biology*, **6** (2000b) 557–568.

Solarbuzz, World Solar Photovoltaic Market Grew to 7.3 Gigawatt in 2009 (2010), http://www.solarbuzz.com/Marketbuzz/2010-intro.htm (accessed on 6 December 2010).

Solarzoom, Comments on the 2nd Round Solar Power Project Bid Quotation (2010), http://www.spvchina.com/news.sap?news_id=3247 (accessed on 15 September 2010).

The Energy Data and Modeling Center, 2007 Handbook of Energy & Economics Statistics in Japan, The Energy Conservation Center, Japan (2007).

The Energy Data and Modeling Center, 2009 Handbook of Energy & Economics Statistics in Japan, The Energy Conservation Center, Japan (2009).

The United States Department of Energy, 2007 buildings energy data book, D&R International, Ltd, USA (2007).

The United States Department of Energy, 2009 buildings energy data book, D&R International, Ltd, USA (2009).

The White House, President Obama Announces National Fuel Efficiency Policy (2009), http://www.whitehouse.gov/the_press_office/President-Obama-Announces-National-Fuel-Efficiency-Policy/ (accessed on 1 December 2010).

United Nations Environment Programme, Bloomberg New Energy Finance, Global trends in sustainable energy investment 2010: Analysis of trends and issues in the financing of renewable energy and energy efficiency, United Nations Environment Programme, Nairobi (2010).

U.S. Department of Transportation, Transportation's role in reducing U.S. Green house gas emissions, Vol. 1, synthesis report, Department of Transportation, Washington D.C.: U.S. (2010).

U.S. Environmental Protection Agency, Global anthropogenic non–CO_2 greenhouse gas emission: 1990–2020, Department of Transportation, Washington D.C.: U.S. (2006).

L. M. Vleeshouwers and A. Verhagen, Carbon emission and sequestration by agriculture land use: A model study for Europe, *Global Change Biology*, **8** (2002) 519–539.

T. O. West and W. M. Post, Soil organic carbon sequestration rates by tillage and crop rotation: A global data analysis, *Soil Sci. Soc. Am. J.*, **66** (2002) 1930–1946.

World Bank, World development indicators database, World Bank, Washington (2007).

T. Yutaka, F. Oki, M. Yoshimitsu *et al.*, A detailed energy demand estimation and CO_2 emission-inventory of residential house by prefecture and housing type in Japan, *J. Environ. Engg.* **592** (2005) 89–86.

S. Y. Zhang, X. Yang, Y. Jiang *et al.*, Comparative analysis of energy use in China building sector: Current status, existing problems and solutions, *Frontiers of Energy and Power Engineering in China*, **4**(1) (2010) 2–21.

PART 3
Cases

CHAPTER 12

Low Carbon Development in Baoding: Plans and Actions[1]

12.1 Background Introduction

Baoding is located in Hebei Province, to the South of Beijing, covering an area of 22,000 km^2, half of which lies in the Taihang Mountains and the other half in the North China Plain. Baoding has a population of 11 million. And its GDP per capita ranks the second to last, with the fiscal revenue per capita the third from the bottom among all the cities in Hebei Province, thus, Baoding is under great stress for economic development. The rapid

[1]The first three parts of this case (12.1–12.3) are a re-writing of the case written by China Case Center for Public Policy and Management — *A Plan Made for Baoding to Develop a Low Carbon City* (CCC-08-89). This case was under the instruction of Director of Tsinghua University Case Center for Public Policy and Management, Mu Ling, and finished by her assistant Bi Engao based on research and interviews. The rest parts (12.4–12.8) of this case are a re-writing of *A Low Carbon Development Plan for Baoding*. As it was the first low carbon development plan for a prefecture-level city in China, the basic research and the text preparation were co-completed by Baoding Municipal Government and School of Public Policy and Management, Tsinghua University. Mr. Yu Qun, who was Mayor of Baoding City at that time and Professor Qi Ye of School of Public Policy and Management, Tsinghua University co-led the research team and the editorial board; while Kang Qing, Deputy Director of Baoding Municipal Commission of Development and Reform and assistant Professor Meng Yanchun of Tsinghua University took the position of Chief Executives of the research team as well as the editorial board. Other personnel who have participated in the research and editorial work of the plan included Li Qian, the current Mayor of Baoding City, Hao Guochi, Director of Baoding Municipal Commission of Development and Reform, Wu Wenxin, Section Chief of the Commission, Wang Xiaoguang, section leader of the Commission; Dai Yixin and Liu Zhilin, lecturers from Tsinghua University, Zhang Huanbo and Li Huimin, postdoctoral candidates from Tsinghua University, Ma Li, Wang Wenwen, Dong Wengui, Huang Jing, Feng Yue, Wang Yaqi and Lv Jiachen, master candidates from Tsinghua University, and Niu Diehai and Liu Jiyan, master candidates from Renmin University of China.

development of some other cities in Hebei Province mainly depends on their resources. However, for the 25 counties, municipal districts and county-level cities of Baoding,[2] there is little resource to explore for development and Baoding has been agriculture-oriented over the past years. Besides, the anemic private sectors in Baoding encounter various restrains from the existing system. Thus, to explore a new way for development has always been the core task of the Municipal Party Committee and the Municipal Government.

12.1.1 *Domestic and international environment for low carbon development*

In recent years, the Chinese Central Government has been stepping up efforts for saving energy, reducing emission and coping with climate change. Meanwhile, the international community was also increasingly engaged in the reduction of carbon emission and the promotion of low carbon economy. As an important national base for manufacturing new energy equipment such as wind energy equipment and solar energy equipment, Baoding needs to figure out how to face those domestic and international challenges and take the opportunities to explore a new way of achieving economic and social development with the "low carbon Baoding" characteristics. It presents an urgent task in front of the Baoding Municipal Party Committee and Government. Only by grasping this "low carbon" development opportunity, can Baoding make a breakthrough and have a voice in the field of low carbon development.

According to the goal of energy conservation and emission reduction set in the *Outline of China's 11th Five-year Plan*, the energy consumption per unit of GDP shall be reduced by about 20%, and the amount of major pollutant emission shall be reduced by about 10%. On March 5, 2006, when giving the *Report on the Work of the Government*, Premier Wen Jiabao stated that the two constraining indicators — the energy consumption per RMB 10,000 of GDP and the major pollutant emission — "are serious and should not be changed and must be realized unswervingly".

[2]25 municipal districts, counties and county-level cities of Baoding: Xinshi District, Beishi District, Nanshi District, Mancheng County, Qingyuan County, Yi County, Xushui County, Laiyuan County, Dingxing County, Shunping County, Tang County, Wangdu County, Laishui County, Gaoyang County, Anxin County, Xiong County, Rongcheng County, Quyang County, Fuping County, Boye County, Li County, Dingzhou City, Zhouzhou City, Anguo City, and Gaobeidian City.

Over the 11th Five-year Period, the Central Government has carried out a series of measures. The yearly energy consumption per RMB 10,000 of GDP in 2006 decreased for the first time since 2003, and the growth of major pollutant emission also slowed down. By the end of 2006, however, the national goal of energy conservation and emission reduction set at the beginning of the year had not been reached, which made it harder to achieve the overall goal of the 11th Five-year Plan. So, one important task for governments at all levels in 2007 was enforcing energy conservation and emission reduction.

On April 25, 2007, the State Council decided to set up a "Leading Group on Energy Conservation and Emission Reduction" with Premier Wen Jiabao as the leader of the group. Three days later, the State Council held a national video-phone conference on energy conservation and emission reduction, and then issued *The Comprehensive Work Plan for Energy Conservation and Emission Reduction*, which proposed 45 specific measures. It was an unprecedented move to promote energy conservation and emission reduction.

On October 15, 2007, the 17th Congress of the CPC made the basic state policy of adhering to resource conservation and environment protection and proposed to implement the accountability system for conserving energy and reducing emissions accountability system accountability system. By November 23, the State Council had approved three plans, including *The Operation Plan for Statistical Indicator System of Energy Consumption per Unit of GDP*, and proposed *The Method for Calculating the Reduction Sum of Major Pollutants*, etc. Those plans and measures provided yardstick for the responsibility system for conserving energy and reducing emission and "veto power" of energy conservation and emission reduction.

12.1.2 *"Double 30" targets in Hebei Province*

As the Central Government was paying more attention to energy conservation and emission reduction and carrying out a series of important measures, the Hebei Provincial Government also followed its steps. Hebei Province's industrial structure has been highlighted by the heavy chemical industry. Its economic development relies much on massive consumption of resources and energy. On November 6, 2006, the Hebei Provincial Government issued *The Decision on Energy Conservation*, and required governments at all levels give priority to energy conservation in their work. On August 14, 2007, Hebei Provincial Commission of Development and Reform and Hebei

Provincial Environmental Protection Bureau, joined other departments, and made *Hebei Province's Comprehensive Implementing Plan for Energy Conservation and Emission Reduction*. The Plan proposed that during the 11th Five-year Period, the amount of major pollutant emission should be reduced by 15%; by 2010, the emission of sulfur dioxide should be reduced to 1.271 million t from 1.491 million t in 2005, and the chemical oxygen demand should be reduced to 561,000t from 661,000t in 2005; the treatment rate of domestic sewage should reach above 70%, and the comprehensive utilization rate of industrial solid wastes should be increased to 60% or above. As a system guarantee, the Plan makes the achievement of the goal set for energy conservation and emission reduction one measuring indicator in the comprehensive evaluation system for economic and social development for governments at all levels. Meanwhile, the achievement of this goal has also become one important standard for assessing the performance of the government officials and companies' executives. The relevant departments of Hebei Provincial Governments would regularly release the performance of energy conservation and emission reduction done by all districts and cities, and make evaluation so as to enforce the accountability system and the "veto" system.

On October 29, 2007, the 7th Hebei Provincial Committee of the CPC held its Third Session, in which Zhang Yunchuan, Secretary of the Provincial Committee said that goals set for energy conservation and emission reduction must be achieved without any excuses. 30 counties (cities) were selected as key bases in the entire province and were required to set settled energy conservation and emission reduction goals, and their performance would be evaluated in line with the provincial standards. At the Hebei Provincial People's Congress in the beginning of 2008, mayors of counties (cities) were required to pledge to the delegates that the goals set for energy conservation and emission reduction would be achieved within three years, otherwise they would resign from their positions. Besides, 30 medium and large sized energy-intensive state-owned enterprises with large high emission in Hebei Province were also picked up as the key target to implement the cause of energy conservation and emission reduction. If their performance in energy consumption and emission did not meet the national standards, the leaders of those enterprises would be removed from office. There was only one enterprise in Baoding, that is Baoding Swan Co. Ltd, — picked as target for the provincial "Double 30" plan. However, there were four counties (county-level cities) included in the "double 30" in Baoding, which shared the first place with Shijiazhuang, Tangshan and Handan, with all the cities having most, i.e., four counties (county-level cities), picked as targets in the "double 30".

On November 7, 2007, Hebei Provincial Government held a Provincial Energy Conservation Conference in order to put the guideline of the National Energy Conservation Working Conference into practice. It was the second conference on energy conservation held by the Provincial Government after the first one on July 6, 2007, in which the Provincial Government signed *The Commitment of Energy Conservation Goals* with 112 national key energy-saving enterprises. Those 112 enterprises were all members of the "1,000 Enterprises' Energy Conservation Campaign" proposed by the Central Government. Guo Gengmao, Acting Governor of Hebei Province, declared the provincial goals for energy conservation at the conference: By the end of the 11th Five-year Plan (the end of 2010) period, the energy consumption per RMB 10,000 of GDP should decline to 1.56 tce, 20% drop from the end of the 10th Five-year Period, and the annual energy conservation rate would reach 4.4%. Fu Shuangjian, Vice-Governor of Hebei, on the behalf of the Provincial Government, signed *The Commitment of Energy Conservation Goals* with leaders from 11 prefecture-level cities[3] and 109 enterprises (key energy-saving enterprises announced by the Provincial Government) with an annual energy consumption of 100,000 tce.

On December 5, 2007, Hebei Provincial Government held a Working Conference on Energy Conservation and Emission Reduction for key counties (cities or districts) and enterprises which had been picked as targets in the "Double 30" plan. Vice-governor Zhang He gave a prep speech and distributed the work assignments. Thus, the "Double 30" plan of energy conservation and emission reduction in Hebei Province officially started.

On January 25, 2008, the Hebei Provincial Government issued *The Suggestions on Energy Conservation and Emission Reduction*, which clarified the goals for energy conservation and emission reduction. The annual goals of energy consumption were as follows: In 2008, 2009, and 2010, the provincial energy consumption per RMB 10,000 of GDP should register a year-on-year drop of 4.5%, 4.9%, and 5% respectively in 2008, 2009, and 2010 and the energy consumption of industrial added value per RMB 10,000 should achieve a respective year-on-year drop of 5.5%, 6% and 6%. And the

[3]By 2007, there are 11 prefecture-level cities in Hebei: Shijiazhuang, Zhangjiakou, Chengde, Tangshan, Qinhuangdao, Langfang, Baoding, Cangzhou, Hengshui, Xingtai, and Handan.

goals of emission reduction were: Compared with the emission in 2005, the provincial chemical oxygen demand should drop by 4.5%, 6.0%, 4.5% respectively in 2008, 2009, and 2010, and the net reduction of sulfur dioxide should be reduced by 5.5%, 6.0%, 2.5% respectively. By 2010, the net reduction of chemical oxygen demand should reach 100,000 t, while that of sulfur dioxide should amount to 224,000 t.

12.2 Building China's Electric Valley and the Solar Energy City

12.2.1 *The concept of China Electric Valley and its follow-up construction*

The Central Government and the Provincial Government have both attached great importance to energy conservation and emission reduction and make it as one important indicator in the government performance evaluation system. Baoding did make a quick response and tried to explore a new pattern for low carbon development based on the analysis to its original industrial foundation.

There is one company in Baoding named Yingli Group which has been manufacturing solar energy equipment since 1998. By 2003, the demand of solar energy equipment began to rise at an explosive rate. Yingli Group's annual output was only RMB 10 million in 2003, and the number had risen above RMB 10 billion by the end of 2008. Later, Baoding Tianwei Group Co., Ltd.[4] had good expectation for the prospect of new energy sector, so it decided to invested into Yingli Group when the latter was in short of fund and held 51% of Yingli's stake. Tianwei Group was famous as the manufacture of traditional transformers. The transformers' category and

[4]Tianwei Group (formerly Baoding Transformer Factory) was founded in 1958. After reform in 1995, it was built into a large high-tech enterprise, of which the major business is to produce electric transformers. Among the top 500 enterprises of China's manufacturing industry in 2005 announced by China Enterprise Confederation, Baoding Tianwei Group Co., Ltd. ranked the 415th place. On September 26, 2007, State-owned Assets Supervision and Administration Commission of People's Government Baoding Branch transferred 100% of Tianwei Group's stake to China South Industries Group Corporation without payment. The "China South Industries Group Corporation — Baoding China Electric Valley Tianwei Wind Power Industrial Park" started its construction in Baoding, with a total investment of RMB 2.7 billion. Tianwei Group changed its identity from an enterprise affiliated to ministries to a local-government-owned enterprise and back to a state-owned one.

ranks have exceeded those ones produced by the German company Siemens. And by holding the stake shares of Yingli Group, Tianwei managed to enter the new energy field.

Besides, as the core of the solar photovoltaic industry is the upstream polycrystalline silicon material, so those who have access to silicon shall dominate in the industry. However, in China, all industrial silicon and solar grade silicon relied on import. There were about 20 experts in Sichuan E'mei Semiconductor Material Factory who had been working on the purification of polycrystalline silicon. And the production of polycrystalline silicon finally made some progress with their efforts. Thus, Tianwei decided to invest in the E'mei Semiconductor Material Factory, becoming its second largest shareholder and founding a semiconductor manufacture company — Xingguang Silicon, with output of 1,600 t in 2008.

Apart from the famous Tianwei Group, there were enterprises which produce helicopter propellers in Baoding. For a long time, there were no competitive industries in Baoding, and it was also difficult for the traditional industries like Changcheng Automobile, textiles, and small agricultural products processing factories to grow into big enterprises. Among the 54 national high-tech districts, the one in Baoding was in urgent need of a clear positioning to build an industrial cluster advantage. Besides, unlike the cities such as Tangshan and Handan in Hebei Province, which are rich in iron ore resource and coal resource, Baoding was a city lack of sufficient natural resource but with large population. Under this condition, there was only one way to make breakthrough that is to explore an economic development path with one's own characteristics. Based on the analysis of its original industries and the market, Baoding Municipal Government decided to spare no efforts in the development of electric equipment manufacturing industry in its high-tech zone. Thus, inspired by the Silicon Valley in California, US, the concept of "China Electric Valley" was created in Baoding.

Since then, with new energy equipment manufacturing being the core industry, Baoding Municipal Government designed six major industries in line with its own foundation and orientation for future development. On March 28, 2007, Hebei Economy and Trade Delegation went to Hong Kong, for inviting investment, The "Baoding China Electric Valley (Hong Kong) Promotion Section" became the highlighted program during the trip. Guo Gengmao, Governor of Hebei Province, personally presented the "Baoding China Electric Valley" project. He stated that by the end of 2006, more than 60 new energy and new energy equipment manufacturing companies had settled in Baoding, among which 28 projects were under construction, 33 about to start, and 34 projects in planning. The total investment was about RMB 40.4 billion. During this trip to Hong Kong, the Baoding Government demonstrated models of wind power and solar panels, and tried to promote the project of "Baoding China Electric Valley as a New Energy sector Base", and attracted a lot of investors.

After the promotion event, a lot of state-owned enterprises such as China South Industries Group Corporation and China Guodian Corporation came to invest in "China Electric Valley". With two years, in "China Electric Valley", six major industries including optic electricity, wind power, electricity conservation, electricity storage, power transmission and electric power automation were formed, and more than 160 new energy enterprises had settled there. Meanwhile, Baoding City also started strategic cooperation with Chinese Academy of Sciences, North China Electric Power University and China Development Bank. In 2007, the sales revenues of "China Electric Valley" reached RMB 17.5 billion, an increase of 56%.

The concept of "China Electric Valley" and the follow-up construction contributed to the initial development foundation of "Low Carbon Baoding". In order to fulfill the goal, Baoding Municipal Government made some plans and policies, intending to invest RMB 40 billion in the construction of the Electric Valley during the 11th Five-year Plan period. The construction contains some major fields, including solar photovoltaic power generation, wind power, biomass energy, and power transmission, and six major industries including electric power automation, new energy storage and highly-efficient energy conservation. It tries to take 10 years to build an influential internationalized manufacturing base of new energy and energy equipment with an area of $20\,km^2$ and an annual sales revenue above RMB 100 billion and to make Baoding a leading city in the new energy sector.

In order to quicken the construction of the Electric Valley, Baoding Municipal Government made some explorations. First, it signed a series of agreements with China South Industries Group Corporation (South Industries), China Guodian Corporation, Tsinghua University and North China Electric Power University, and China South Industries Group Corporation would invest RMB 35 billion in the construction of "China Electric Valley" in the next 10 years. By 2008, the "China Electric Valley" had seen some achievements: First, the construction of key projects went well; second, the South Industries and Tianwei Industry Park with investment totaling RMB 6.7 billion got started of its construction; third, the Tianwei Wind Power Industrial Park with an investment of RMB 2.7 billion and the third phase of Yingli Solar Photovoltaic Power Project with an investment of RMB 3 billion both got under construction.

As for the driving effect of the enterprises, the sales revenues of Yingli Group, as the leading enterprise in photovoltaic power, achieved a several fold increase in the past few years. In 2003, its sales revenue was only RMB 10.39 million and in 2007, its revenue reached RMB 4.35 billion. The sales revenue of Huiteng, the leading enterprise in wind power industry, rose to RMB 1.28 billion in 2007 from RMB 16 million in 2003. New energy sector and new energy equipment manufacturing sector in Baoding really saw quite rapid development in the past few years.

Moreover, in terms of the innovation of enterprises, there were three state-level enterprise technical centers, and four provincial-level enterprise technical centers in the Electric Valley. And there were more than 300 scientific research achievements and over 300 patents, more than 30 of which became national industrial standards. Also, huge breakthroughs for key technologies had been achieved, for instance, the key technology to develop 1.5 MW wind power generator blades and 1.5 MW wind power generator. In 2006, Baoding was listed as one of "The First National Innovative Export Bases".

Finally, it kept extending the industrial chains with the new energy sector as its core industry. Relying on Tianwei and Yingli's solar energy equipment manufacturing industry, over 40 photovoltaic power generator manufacturing enterprises and 20 enterprises involved in wind power equipment manufacturing and other relative companies had already gathered here. There were more than 130 enterprises in Baoding of which the business was related to wind power generation, solar energy application, energy conservation and biomass energy. With the expansion of the industry chain, the new energy industries kept growing, so did the benefits. In 2007, the entire sales revenue of Baoding's new energy sector reached RMB 20.5 billion, profit and taxes at RMB 3.3 billion and export revenue reaching RMB 3366 million ($510 million). The entire new energy sector accounted for about 14% of Baoding's all industries, making a contribution of 23% to the economy of entire city.

12.2.2 *The construction of the solar energy city*

With the increasing development of the new energy sector, Baoding began to put those research achievements into use in the citizens' daily life, consumption and other areas to help conserve energy and reduce emission from all sides. At the beginning of 2007, Baoding Municipal Government proposed that Baoding should take a lead in the application of solar energy and build itself into a "Solar Energy City". Yu Qun, Mayor of Baoding, expressed his expectation that "solar energy shall promise Baoding a warm and bright future".

On March 15, 2007, Baoding Municipal Government put forward *Baoding's Implementation Suggestions on Building Baoding into the "Solar Energy City"* which proposed that within three years the solar energy should be utilized widely in all fields of the industrial production and people's life. The solar energy can be used for upgrading of lighting facilities in the green space in the city, neighborhood communities, streets, tourist attractions and other public places; and it also can be widely adopted in the countryside. Those were some major work in building the "Solar Energy City".

By the end of 2007, the public facilities upgrading via application of solar energy was finished in 50% of Baoding's public places, 40% of communities and 40% of tourist attractions. Besides, Baoding Municipal Government made a regulation that for those newly-built public places, public facilities and residential area, the solar energy is required to be applied.

The eventual purpose of the "Solar Energy City" is to achieve energy conservation and emission reduction. With the application of solar energy, 430 million kWh electricity and 118,000 t of coal could be saved while the emission of CO^2 and sulfur dioxide could be reduced by 420,000 t and 12,900 t respectively per year.

In order to achieve the goal of the "Solar Energy City", Baoding adopted a series of operation ways, such as market operation, enterprises' independent operation, government fiscal subsidies and other financing methods. Meanwhile, Baoding set forth some reformation tasks, clarified some key points in the reform and proposed a project schedule. With a series of policies and measures, the construction of "Solar Energy City" progressed steadily. During the process, the government spared no efforts into the promotion of solar energy application all over the city.

In the short-term, compared with the traditional street lights, the cost of street lights using solar energy would at least rise by two or three times. However, Mayor Yu Qun did not look at the issue this way, and his concern was not focused on the economic account. Actually, even if we consider the issue in the perspective of cost, it still pays to use solar energy lighting facilities. If LED lights were used, the whole set which include the lights and batteries would roughly cost as much as the traditional street light would in four years. The cost for arrangement of wire, maintenance and electric consumption of four years for the traditional light actually is as much as that for the installation of LED lights. So after the first four years, with the application of solar energy lights, the cost of electric power would be conserved. There is only a one-time investment when adopting solar energy lights.

By the end of 2007, Baoding invested almost RMB 50 million into the construction of the "Solar Energy City". A total of 98 signal lights in the entire urban district had been changed into solar energy lights. Then, photovoltaic and lighting upgrading had been finished in 13 roads, eight key square gardens and 84 key neighborhood communities. The solar energy upgrading of the lighting in Baoding's 13 tourist attractions had also been completed. Besides, Yingli Group finished the construction of the 25-storey Electric Valley Jinjiang International Hotel, the first domestic hotel powered by the solar photovoltaic power generation, waste-water heat and heat pump. The Hotel was officially put into use on October 18, 2008 and could annually generate electricity of 260,000 kWh.

Owing to the constructions of "China Electric Valley" and "Solar Energy City", the National Development and Reform Commission and the Ministry of Science and Technology started focus their attention to Baoding. In order to encourage and support the practice of Baoding, after repeated examinations, the National Development and Reform Commission granted Baoding the "National High-tech New Energy Industrial Base" on February 20, 2007; the Ministry of Science and Technology granted Baoding the "Solar Energy Promotion and Demonstration Base" and the "New Energy Industrial Base". Yu Qun Mayor of Baoding, knew deeply what should be behind these honors. At that time, the city = Guiyang in Guizhou Province was planning on developing recycling economy which had not been tried in other places, and Guiyang presented a national stereotype and an excellent demonstration with its initial trial.

12.3 Introduction to the Low Carbon Development Concept

12.3.1 *Cooperation with World Wildlife Fund*

For its efforts in new energy equipment manufacturing and solar energy application, on January 28, 2008, Baoding, together with Shanghai, was selected to be the first pilot cities for the "China Low Carbon

City Development Program"[5] which was initiated by World Wildlife Fund (WWF).

Lei Hongpeng, in charge of WWF's China Low Carbon City Program, wishes to work together with cities to figure out how to achieve the goal of low energy consumption and emission and to promote the transformation of cities' development patterns. Lei used to work for the Ministry of Housing and Urban–rural Development and began to work for WWF from the very beginning of 2007 when he started to plan for the Low Carbon City Program. Director of WWF China's Project Strategy Department, Li Lin hoped to start a pilot trial in Baoding and then extend research on the energy production and utilization as well as new economy development model for the cities to establish a reference model for other cities to learn.

In the *Cooperation Memorandum* that WWF signed with Baoding, both parties agreed that in the principle of "new energy sector promoting the city's low carbon development", the emphasis of the cooperation was to introduce to Baoding the advanced theories and experiences needed for new energy sector and low carbon development, to promote the spreading of Baoding's successful experience at home and abroad, and to improve Baoding's capacity in developing its new energy sector.

As one of the world's biggest independent nongovernment environmental protection organization, WWF has about five million supporters worldwide and has developed networks in more than 90 nations. WWF hoped that the program would help Baoding to increase the export and application of renewable energy and energy efficient products, to advertise and promote the program at home and abroad and to offer financing support for part of the program. And Baoding would provide the program with fund, human resource and material resources to ensure the program proceed smoothly.

The Chinese Central Government also has attached great importance to the low carbon city development since the total amount of the nation's carbon emission kept increasing. Some officials from the National Development and Reform Commission, form the Ministry of Housing and Urban–rural Development, from the Ministry of Science and Technology, as well as from

[5] Please check http://www.wwfchina.org/aboutwwf/whatwedo/climate/lcci.shtm to get more information on the program and learn about WWF China's low carbon city program.

the Ministry of Environmental Protection, and the Ministry of Commerce have taken the position of the technical consultants in the cooperative program between WWF and Baoding. Those officials hoped to pool all the forces and resources to help realize a 20% drop of energy consumption at the end of the 11th Five-year Period through the multi-level and multi-directional cooperation among different departments, research institutes, companies and international organizations.

On April 18, 2008, WWF and Baoding City co-held the "Baoding International Seminar of Low Carbon City Development". The seminar, held in Baoding High-tech Zone, had 40 delegates participated in from WWF, World Organization for Economic Cooperation and Development (OECD), China Council for International Cooperation on Environment and Development (CCICED), and other international organizations. Yu Qun hoped that the seminar would help Baoding to learn more experiences and new theories on low carbon development of cities.

At the seminar, WWF officially released a research report: *Baoding: the World's Sustainable Energy Producer — Electric Valley*. The report demonstrated that the "Baoding Model" had made contributions to the global development of renewable energy, and emphasized the important role the "Baoding Model" had taken in exploring low carbon development in China or even in the world. The report was an important measure to promote Baoding's successful experience in the cooperation.

It is worth noting that actually not only Baoding and Shanghai were selected in the first pilot cities in the WWF's program, however, other chosen cities had decided to quit for worrying the possibility of slowing down their economic development. Frankly speaking, their concerns were reasonable. Though emission reduction had become an increasingly important issue which caught attention both from the Central Government and local governments, it was not so easy to achieve reduction of energy consumption while at the same time still keeping the economic growth and ensuring the employment.

12.3.2 *Plans for low carbon development needed*

In the early 2008, government officials and civilians of Baoding still knew little about low carbon concept yet. Officials in charge of energy management learned from the Internet that low carbon was actually almost equal to energy conservation and emission reduction. The Baoding Municipal Government had waken up to the opportunity of building a better

Baoding by taking advantage of the growing concern in climate change, energy conservation and emission reduction domestic and overseas to make a plan of low carbon development based on Baoding's specific situations. But it was not easy to make such a plan.

The construction of a low carbon city not only requires technology, but also needs support in terms of policy and management, etc. Globally, there are few models of low carbon economies or low carbon cities, and most of them are still a concept. Then how to put those ideas into practice to make a low carbon city or low carbon industrial clusters is a really a question for Baoding to explore. Since Baoding plans to build a low carbon city, then what supportive policies should it make? And what kind of system should be structured? Since Baoding is going to make a plan for low carbon development, then what should be written in the plan? What industries should be developed and what projects should be carried out, etc are all important questions for Baoding to answer clearly in the plan. Undoubtedly, the government would play an important role in building up a low carbon city. So how the government makes decisions and implements the favorable measures is certainly an important issue. All these questions require much consideration. In Mayor Yu Qun's opinion, the governance mechanism and institutional guarantee of a low carbon city was not just a theoretical issue, but it was more a practical issue, which would directly affect the construction of low carbon city.

Another challenge for the Mayor was that there was no theoretical system yet for guiding Baoding to build a low carbon city. Baoding was actually sailing through uncharted waters and had no precedent to learn. The road had to be explored by them alone. Baoding Municipal Commission of Development and Reform also considered making a plan for low carbon development, which in fact was part of institutional construction and management. Let's look at some foreign experiences, for example, German had completed some institutional construction, and the solar energy used in every household would contribute to the on-grid electricity price. So was it the same with the situation in California, US. There were specific measures in those countries. And some cities in the UK were also making plans for low carbon development. So what goals should Baoding achieve and how does it become the guide for the public and enterprises? The further research or exploration in such areas would be likely to help a lot the future development of Baoding. Perhaps, it would also provide reference for the Central Government and other cities. For this purpose, Baoding Municipal Government cooperated with School of Public Policy and Management,

Tsinghua University to formulate the plan for the first low carbon city development in China.

12.4 Goals and Phases for Baoding's Low Carbon Development Program

12.4.1 *Goals*

The goals for Baoding to build a low carbon city are: To enhance much the quality of economic development and its comprehensive economic strength, to further optimize its industrial structure and energy structure, to emphasize the advantage of its low carbon industry, to promote the development of a low carbon society, to gradually shape a healthy, economical and low carbon life style and consumption mode, to improve residents' living quality, to reduce CO_2 emission, and eventually build a conservation-oriented low carbon city in which the economy develops rapidly, the society is prosperous and eco-friendly and the human beings co-exists harmoniously with the nature:

— By 2015, the energy consumption per RMB 10,000 of GDP will drop by around 39% compared with that of 2005;
— By 2020, the energy consumption will achieve a drop about 18% from that of 2015.

1) Further development of low carbon industry: The supporting system of low carbon industry should take shape, with new energy industry and energy equipment manufacturing, new energy vehicle manufacturing, green agriculture, low carbon service, and low carbon tourism as the leading industries.
2) Notable achievements for energy conservation and emission reduction: Low carbon technologies and products are promoted widely while the recycling economy frame should roughly take shape, which is dominated by energy conservation and emission reduction as well as comprehensive utilization of resources.
3) Low carbon management gradually strengthened: The low carbon plans should be fully put into practice. The urban functional zones are with reasonable layout. The operating efficiency of the city is much improved. Also, the low carbon upgrading in key areas such as energy conservation in buildings, development of low carbon community and low carbon

traffic system should be enforced effectively. The model for building and managing a low carbon city should initially take shape.

4) Concept of low carbon life style root deeply in people's mind: The low carbon concept should be promoted widely. And low carbon production, low carbon life style and consumption model should be widely promoted.

12.4.2 *Developing phases*

The cause of constructing a low carbon Baoding includes two phases:

The first phase (2009–2015): To start low carbon development in key areas and establish an early low carbon development model.

Taking the construction of "Baoding China Electric Valley" (national new energy and new energy equipment manufacturing base) as a breakthrough, Baoding will accelerate its development of low carbon pillar industry and form an industrial structure led by low carbon industries; It also will implement a series of demonstration projects for low carbon production and low carbon lifestyle, in order to reduce CO_2 emission effectively.

The second phase (2016–2020): To thoroughly promote the construction of low carbon city and develop a low carbon Baoding.

Baoding will establish the low carbon city standards which also apply to the city planning. It will also further improve its low carbon featured economic structure, achieve low carbon emission in the in basic urban functional facilities such as buildings and traffic, etc and reduce the intensity of CO_2 emission while keeping improving people's living standard.

12.5 Low Carbon Development Plan in the Production Sector

The production sector is the major sector that creates social wealth for Baoding, while it is also the major part that consumes energy and emits CO_2. In 2005, the production sector in Baoding consumed 9,644,300 tce (including the consumption sum of 244,200 tce during processing and transforming), and emitted 24,014,300 t of CO_2, accounting for 75.9% of the Baoding's total energy consumption and CO_2 emission. Looking at the data of CO_2 emission per unit of GDP produced by separate sectors in Baoding, the emission of production sector reached 2.31 t/RMB 10,000, of which the primary, secondary and tertiary industry respectively reached 1.72 t/RMB 10,000, 3.19 t/RMB 10,000 and 0.95 t/RMB 10,000. Thus, the low carbon development of the production sector plays an important role in Baoding's cause of building a low carbon economy.

12.5.1 *Basic principles and approaches*

(1) *Major principles*

1) Principle of development: The low carbon development in the production sector must be based on the basic municipal condition that Baoding is a city in a developing country. Such understanding would help to accelerate new industrialization and urbanization.
2) Principle of innovation: The low carbon development in the production sector must be grounded in technical innovation and based on upgraded industry level.
3) Principle of multi-win: The low carbon development in the production sector must take economic, social and ecological effects all into consideration and help to realize multi-win situation and the harmony of economic development, social improvement and environment protection.

(2) *Major approaches*

1) Accelerate the adjustment of industrial structure and develop low carbon industry — Baoding should work hard to develop industries with low carbon intensity and increase the proportions of those industries accounting for the total economic output; meanwhile, it should restrain the growth of industries with high carbon intensity and gradually reduce the shares of such industries occupying the total economic output or even get them completely replaced. Baoding should try hard to decrease the intensity of carbon emission by adjusting the industrial structure while sustaining the rapid development.
2) Accelerate innovation of energy-saving technology and the upgrading industries with those technologies so as to efficiently conserve energy and reduce carbon emission — Baoding should try hard to upgrade industries with relatively high carbon emission and increase the energy efficiency by adopting energy-saving technologies, so as to reduce the demands for energy of these industries. Baoding also should try to reduce the carbon emission of traditional high-emission industries to the level of low carbon industries.
3) Develop renewable energy and optimize the energy structure — Baoding should increase the proportions of clean energy and renewable energy in

the production of electricity and change the electric supply structure; it should accelerate the application of clean energy and renewable energy to every social and economic field. And the intensity of CO_2 emission would be reduced from the source through the adjustment of energy structure.

12.5.2 *The goals for reducing carbon emission*

By 2015, the intensity of CO_2 emission in the production sector will be 39% lower than that of 2005, and:

— The contribution rate of the adjustment of industrial structure will be 50.1%;
— The contribution rate of the application of energy-saving technologies will be 24.8%;
— The contribution rate of the adjustment of energy structure will be 25.1%.

By 2020, the intensity of CO_2 emission in the production sector will be 18% lower than that of 2015, and:

— The contribution rate of the adjustment of industrial structure will be 65%;
— The contribution rate of the application of energy-saving technologies will be 18.5%;
— The contribution rate of the adjustment of energy structure will be 16.6%.

12.5.3 *Plans for structural adjustment*

The adjustment of industrial structure is one of the most important ways for Baoding to achieve low carbon development in its production sector. Baoding should endeavor to develop industries with relatively low carbon intensity and control or restrain the development of industries with high carbon intensity so as to keep the industrial development based on the ongoing optimization of the industrial structure and the increasing growth of carbon productivity, providing a firm industrial foundation for a low carbon city. According to the measuring results of the carbon intensity in all industries, it was found out that industries with relatively low carbon emission in Baoding mainly includes modern agriculture, new energy and energy equipment manufacturing, automobile and automobile spare parts, electronic information, finance and insurance, technology service, modern freighting service, cultural industry and tourism and the like. Those industries should be treated as key industries to develop and their proportions in the total economic output should be raised through structural adjustment.

1. *Goals*

1) Modern agriculture represented by eco-agriculture should be developed at a faster pace. To improve the scale and intensification of agriculture production; gradually reduce the proportion of agriculture accounting for the GDP on the premise that the grain safety is guaranteed,

    — By 2015, the proportion of the primary industry in GDP will drop to about 14% and its carbon intensity of per unit added value will also drop greatly.
    — By 2020, the proportion will drop to about 12% and the carbon intensity will get close to or reach the national average.

2) The internal structure of the secondary industry should be further optimized. Those industries with relatively low carbon intensity in Baoding, such as new energy and energy equipment manufacture, automobile and automobile spare parts, and electronic information, should be developed with every effort; while industries with higher carbon intensity, such as chemical industry, papermaking, metallurgy, and construction materials, should reduce their proportions in GDP. Thus, the carbon emission intensity of the industrial sector will achieve an overall reduction.

    — By 2015, industries with good foundations, high technical content and lower carbon intensity including industries of new energy and energy equipment manufacturing, automobile and automobile spare parts production, and electronic information service etc., will account for about 38% of the industries above designated size; industries like chemical industry, papermaking, metallurgy, and construction materials will drop their proportions to around 12%; and the comprehensive energy consumption of added value per RMB 10,000 of the industries above designated size in Baoding will drop about 30% from that of 2010.
    — By 2020, industries with low carbon intensity like new energy and energy equipment manufacturing, automobile and automobile spare

parts production, and electronic information service will compose about 43% share of all the industries above designated size; and the comprehensive energy consumption of added value per RMB 10,000 of the industries above designated size will be about 20% lower than that of 2015.

3) The proportions of the tertiary industry will increasingly grow. Modern service such as finance, technology service, modern freighting service, culture industries and tourism will be developed at a much faster pace and the proportions of the tertiary industry in the economy of the city will be increased.

— By 2015, the product of the tertiary industry will account for 36% of GDP.
— By 2020, the product of the tertiary industry will account for 38% of GDP.

2. *Dominant projects*

There are 10 dominant projects in total.

(1) Agriculture

1) The grain production bases: referring to the national campaign of building grain production bases with capacity of 50 billion kg of grains, Baoding should further develop the grain production bases in County Xushui, County Dingxing, County Qiangyuan, City Dingzhou (county-level) and City Gaobeidian (county-level). By 2015, the area of grain production base will reach about 373,000 hectares, and by 2020, it will reach about 380,000 hectares.
2) The protected agriculture: boost the development of the protected agriculture such as greenhouse agriculture. By 2015, the coverage of protected agriculture will reach 100,000 hectares, and by 2020, it will be 133,000 hectares.
3) The large-scale breeding bases: The focus lies in large-scale livestock cultivation projects which mainly breed pigs, cattle, sheep, and chickens. By 2015, the inventories of large-scale livestock breeding will take up 60% of the total inventories of Baoding, and by 2020, the proportion will rise up to 80%.
4) The standards and detection system of green agriculture: The focus here is to promote the conduction of origin environment standards

and production quality standards for green products, and to build a three-level examination and detection network for product quality and safety in cities, counties and markets (or bases). By 2015, there will be 70 bases which produce green pollution-free agricultural products in Baoding, while the examination and detection network for product quality and safety will cover 80% area of the city; and by 2020, there will be 100 pollution-free green agricultural production bases, while the examination and detection network will cover above 95% area of the city.

(2) Manufacturing

1) The new energy and energy equipment manufacturing base (China Electric Valley): The key task is to boost the development of some key projects including Yingli Group's 800 MW solar photovoltaic power generation equipment manufacturing program, the program of purification of 18,000 t silicon materials of the Fine Silicon Co., Ltd., Lightway Solar's 600 MW solar photovoltaic power generation equipment manufacturing, the productivity extension of Guodian United Power's wind turbine blades and its complete set of wind energy equipment, the industrialization of Tianwei Group's wind power equipment, the smart charging system developed by Yuneng Company, and the new type of energy storage material of Fengfan Co., Ltd. By 2015, the revenue of the new energy and energy equipment manufacturing sector in Baoding will reach RMB 150 billion, and its added value will account for more than 16% of that of all the industries above designated size in Baoding. By 2020, the revenue of the new energy and energy equipment manufacturing sector will reach RMB 300 billion and its added value will account for more than 20% of the total of all the industries above designated size in Baoding.
2) The automobile and spare parts manufacturing base (North China Light Vehicle City): the key task is to boost the development of some key projects including program of expanding Changcheng Automobile's capacity for 500,000 automobiles and a corresponding number of spare parts, expanding Chang'an Automobile's capacity for 200,000 miniature automobiles, 300,000 engines and 10,000 electronic vehicles, and expanding Zhongxing Automobile's capacity for 100,000 pickup trucks. By 2015, the entire city will be capable of producing 1.5 million complete sets of vehicles, among which 210,000 vehicles will adopt new energy, and the revenue of complete vehicles and spare parts production in Baoding will

reach RMB 100 billion, while its added value will account for above 20% of that of all the industries above designated size. By 2020, there will be two million complete vehicles produced, of which around 300,000 vehicles will adopt new energy, and the revenue of the industry will reach RMB 180 billion, while its added value will account for more than a quarter of that of all the industries above designated size in Baoding.

3) The electronic information products manufacturing base: The key task is to promote the construction of China Aerospace Science and Industry Corporation (Zhuozhou) Base and its inertial navigation system, BGP's science park and high performance computer, Lucky Film's TAC film production line and magnetic materials, the industrialization of Tianhe Electronics' ultra-red precision laser radar, and the production base of BHC electronic information products in Dingxing County, etc. By 2015, the revenue of electronic information product manufacturing industry will reach above RMB 10 billion, and its added value will roughly account for 5% of that of all the industries above designated size in Baoding. By 2020, the revenue of the industry will reach RMB 20 billion, and its added value will account for about 10% of that of all the industries above designated size in Baoding.

(3) Service

1) The modern logistic industry base: The key point is to develop projects of logistics in production origin and the third party logistics, such as Baoding Dry Port Storage, Baoding Traffic and Logistics Center, Jiuzhoutong Modern Chinese Herbal Medicine Logistic Base in Anguo County, Xintongyuan Logistics Center of Baigou County, and Longhua Storage and Logistics in Xiong County, etc. By 2015, the entire city's added value in modern logistics will achieve an annual growth of more than 18%, and preliminarily establish a modern energy-efficient, environment-friendly, safe and orderly logistic service system with a reasonable structure, advanced technologies, convenience and high efficiency. By 2020, the annual growth of the added value in the industry will reach above 20%, and a network of modern logistics with multi-level and large-scale coverage will be formed and developed into an international competitive industry with. By then, Baoding will become one important pitch point in the logistic system throughout the Beijing–Tianjin–Hebei metropolitan area and the Bohai bay area.

2) The culture, leisure and tourism base: The key task is to develop the projects such as the fitness, leisure and tourism base in Xinshi District, the Yishui Lake in Yi County and Lake Baiyangdian in Anxin County, the cultural scenic area of Drug King Temple in Anguo County, and the Tianqiao Falls in Fuping County, etc. By 2015, the revenue of tourism in Baoding will reach above RMB 28 million, and by 2020, the revenue will reach above RMB 59 billion.
3) The cultural and creative industry base: the focus is to drive the development of projects in cultural and creative industry. By 2015, the revenue of this field in Baoding will reach RMB 1.5 billion, and by 2020, the revenue will reach RMB 2 billion.

12.5.4 *The plans for energy conservation and carbon reduction*

1. *Goals*

According to the identification results of carbon intensity in all industries, it is found that electric and thermal power sector, textile industry and building material industry and the like were industries with higher carbon emission in Baoding City, thus, those industries should be the highlighted fields for conducting energy conservation and carbon reduction.

— By 2015, the energy consumption per RMB 10,000 of GDP in those industries will drop by 40%.
— By 2020, the energy consumption per RMB 10,000 of GDP in those industries will reach the national average.

2. *Dominant projects*

There are three dominant projects in total.

(1) The technological upgrading of electric and thermal power sector for carbon reduction.

By 2015, 0.5 million kW of capacity of small thermal power generators and, 1100 coal fired boilers under 4 t will be eliminated.

In line with *The National Promotion Catalog of Key Technologies to Conserve Energy* and based on the specific situations of Baoding, the technological upgrading of electric and thermal power sector is recommended to adopt six key carbon-reducing technologies: (1) the renovation technology to modernize the flow passage of steam turbine; (2) the renovation technology of the seal gland of steam turbine; (3) the micro oil ignition technology of coal fired boilers; (4) the plasma oil less ignition technology of coal fired boilers; (5) the technology of the cleaning device on the spiral band of condensers; (6) the renovation of improving the thermal system and drainage system.

(2) The technological upgrading of textile and chemical fiber industry for carbon reduction.

Those printing and dyeing enterprises which do not qualify for the sewage treatment will be shut down and integrated, and all the out-of-date spinning machines and looming machines will not be used again.

In line with *The National Promotion Catalog of Key Technologies to Conserve Energy* and based on the specific situations of Baoding, there are three key carbon-reducing technologies which are recommended for the technical renovation in the textile and chemical fiber industry: (1) low-liquor dyeing technology, waterless dyeing technology, jet printing technology, pigment printing and dyeing technology and plasma aided water-saving technology for printing and dyeing enterprises; (2) the energy-saving technology of smart air conditioning system in cotton textile enterprises; (3) the energy-saving and heat-collecting technologies by adopting light-source heat and ground-source heat in dyeing enterprises.

(3) The renovation of carbon-reducing technologies in building material industry.

By 2015, the cement factories and furnaces with capacity under 4.5 million t will be shut down.

In accordance to *The National Promotion Catalogue of Key Technologies to Conserve Energy* and based on the specific situations of Baoding, there are three key carbon-reducing technologies which are recommended for the technical renovation of the building material industry: (1) wall material companies should promote technologies which use industrial slag, coal stone, fly ash and chimney ash and other industrial waste for the production of wall materials like waste residue and blocks; the energy-saving renovation of the out-of-date kilns to produce wall materials by adopting energy-saving technologies; new wall materials and high-quality environmental-friendly energy-saving materials such as heat-resistant and sound-proof materials, water-proof materials and sealing materials should be developed and produced, providing the construction market with qualified energy-saving building materials; (2) cement producing companies could use six-stage cyclone preheated technology, double support rotary kiln technology, new high-efficiency burner technology with light pollution used for cement producing companies, and float glass furnace energy-saving technology; (3) Glass producing companies could use float one-line large-area coating technology and on-line CVD coating technology.

12.5.5 *The plan for energy structure adjustment*

The primary energy consumption in Baoding is mainly coal consumption, which accounts for more than 50% of the terminal energy consumption. Owing to the shortage of Baoding's primary energy sources, its primary and secondary energy, such as coal, electricity, oil and oil products, and natural gas all mainly depends on the importation from other areas. Thus, in order to implement the national energy strategy (by 2020, the nonfossil energy will account for 15% or so in the consumption of primary energy), based on its resource advantages, Baoding will focus on developing solar photovoltaic power, biomass energy, waste incineration power, water power and other new and renewable energy to replace the consumption of regular energy, and increase the renewable energy's proportions, so as to optimize the overall energy structure.

1. *Goals*

By 2015, the installed capacity of new energy, such as solar photovoltaic power, biomass energy, waste incineration power, water power, will reach 140 MW.

By 2020, such capacity will reach 615 MW.

2. *Dominant projects*

There are five dominant projects in total.

1) The projects of solar photovoltaic power generation: The key task is to build the solar photovoltaic grid-connected power generation project which requires the completion of 10 MW in Dingzhou City, 10+1 MW in Laiyuan County, 10 MW in Boye County and so on. By 2015, the entire city's installed capacity of solar photovoltaic grid-connected power generation will reach more than 30 MW; by 2020, the capacity will rise above 80 MW. Baoding will promptly apply solar photovoltaic power generation to buildings, and adopt distributed solar power generation system like solar energy lighting. By 2015, the entire city's installed

capacity of distributed solar energy system will reach above 5 MW, and by 2020, the capacity will rise up above 10 MW.

2) The projects of wind power generation: The key task here is to build the Dianziliang wind power generation plant in County Laiyuan. By 2015, the installed capacity of wind power generation will reach above 10 MW; by 2020, the number will rise up around 180 MW.
3) The projects of waste incineration power generation: The key task is to promote city-level, county-level and regional waste incineration power generation projects such as Lingfeng waste incineration power generation project in Baoding City, waste incineration power generation project in Dingzhou City, etc. By 2015, the installed capacity of waste incineration power generation in Baoding will reach above 35 MW, and by 2020, the capacity will rise up above 100 MW.
4) The projects of biomass energy power generation: the key task is to develop biomass energy power generation in areas with rich resources such as straws and branches, for example, the development of biomass energy power generation in Yi County. By 2015, the installed capacity of biomass energy power generation in Baoding will reach above 25 MW, and by 2020, the capacity will rise up above 100 MW.
5) The projects of water power generation: The key task is to promote the construction of small water power generation plants in areas with rich water resource such as County Yi and County Fuping. By 2015, the installed capacity of water power generation in Baoding will reach above 35 MW, and by 2020, it will rise up above 45 MW.

12.6 Low Carbon Development in the Consumption Department

The consumption of household is a major sector of energy consumption and CO_2 emission in Baoding. In 2005, the household consumption of energy in Baoding was 3.0624 million tce (the sum of CO_2 emission was 7.6346 million t), which accounted for 24.1% of the total social energy consumption and CO_2 emission in Baoding. For the CO_2 emission per unit of GDP, the emission of the consumption sector was 0.73 t/RMB 10,000. With the growth of economy and improvement of people's living standard, energy consumption and CO_2 emission of the households would also keep increasing. Thus, in order to build a low carbon city, it is one important task to reduce the CO_2 emission and energy consumption of households.

12.6.1 *Major principles and approaches*

(1) *Major principles*

1) The principle of cultivating low-carbon concept. In order to reduce CO_2 emission caused by people's household consumption, it is a priority to shape a concept of living a low carbon life. Therefore, a civilized, healthy and scientific lifestyle would help reduce CO_2 emission.
2) The principle of guaranteeing the quality of people's life. The CO_2 emission in household consumption sector should not be reduced at the expense of sacrificing decline of the quality of people's life. We should reduce CO_2 emission through changing consumption pattern, while at the same time ensuring the improvement of people's living standard.
3) The principle of practicing strict economy. Practicing strict economy is a major means to help reduce CO_2 emission in household consumption sector, which should be carried out from every aspect of people's life: food, clothing shelter [housing] and transportation — all the basic necessities of life.

(2) *Major approaches*

1) Work hard to promote energy conservation in buildings — it mainly refers to reduce energy consumption in the utilization of buildings. We will reduce the energy consumption of heating system, air conditioning, hot water supply and lighting in buildings, and reduce CO_2 emission in the utilization of buildings through the renovation of nonenergy-saving buildings with energy-conservation technologies, the popularization of energy-saving appliances, and the utilization of solar energy and geothermal energy.
2) Build central heating supply system — it means that the highly-efficient central heating supply network will replace the traditional low-efficient separated heating supply facilities. Thus, the energy efficiency will be increased and the CO_2 emission will be reduced.
3) Promote energy conservation in the rural areas — we will establish a energy-saving system which adapts to the producing and living environment in the rural areas and reduce the traditional energy consumption and CO_2 emission through popularizing energy-saving measures in the rural households, optimizing the rural energy structure and using new energy such as marsh gas, solar energy, biomass energy, etc.
4) Carry out an energy-saving campaign in traffic sector — it mainly means establishing a low carbon green urban traffic system and reduce the

energy consumption and CO_2 emission of traffic sector by optimizing traffic structure, changing the way people travel, increasing the energy efficiency and adjusting the energy structure.

12.6.2 *Goals for the reduction of carbon emission*

By 2015, in the total CO_2 emission per unit of GDP, the emission produced in household consumption sector will drop by 36% from that of 2005. And in order to reach the goal, the respective contributions of different fields will be as follows:

— The energy conservation in buildings will contribute 17% to the goal;
— The energy conservation in traffic sector will contribute 13%;
— The energy conservation achieved through central heating supply system will contribute 20%;
— The energy conservation in rural areas will contribute 50%.

By 2020, in the total CO_2 emission per unit of GDP, the emission produced in household consumption sector will drop by 18% from that of 2015. And in order to achieve the goal, the respective contributions of different fields will be like the follows:

— The energy conservation in buildings will contribute 14% to the goal;
— The energy conservation in traffic sector will contribute 16%;
— The energy conservation achieved through central heating supply system will contribute 20%;
— The energy conservation in rural areas will contribute 50%.

12.6.3 *Plans for energy conservation in buildings*

1. *Goals*

All the newly-built buildings will have to comply with the energy-saving standards of buildings, and the current non-energy-saving buildings will be gradually renovated. Also, we should popularize the application of energy-saving appliances, promote the comprehensive utilization of solar thermal energy and geothermal energy, and effectively control the growth of energy consumption of buildings.

— By 2015, the growth of energy consumption of buildings will be controlled within 4%;
— By 2020, it will be controlled within 3.1%.

2. *Dominant projects*

There are six dominant projects in total.

1) The renovation of the current nonenergy-saving buildings: The current buildings which do not measure up to the energy-saving standards will be renovated with energy-conservation technologies. By 2015, 2 million km^2 of nonenergy-saving buildings will be renovated in the urban area; by 2020, 80% of the nonenergy-saving buildings in the urban area will be renovated.
2) The application of renewable energy: We should demonstrate and promote the application of renewable energy like solar energy or geothermal energy in buildings. By 2015, the renewable energy technologies will be applied to above 50% of newly-built residential buildings and public buildings; by 2020, this number will rise up above 80%.
3) Green buildings: We should establish and improve the standards and technology system for green buildings, and promote the construction of demonstration projects of green buildings. By 2015, 20% of the newly-built buildings will be green buildings; by 2020, the number will rise up to 80%.
4) The popularization of heat metering in buildings: We should put more efforts into the installation and popularization of heat metering devices in buildings. By 2015, 100% of new buildings will be installed with heat supply metering device, indoor temperature conditioning device and heat supply system conditioning device. And 100% of users will be charged with heat supply fees based on the metering results.
5) The popularization of solar thermal collectors: We should promote the wide application of large-scale solar thermal collectors or household-used ones depending on the local conditions. By 2015, the proportion of solar water heaters accounting for all the newly-installed water heaters will increase by 15.5%, the water heater's energy consumption per square meter will drop by 11.6%, and the proportion of energy conservation by hot water will reach 11.29%; by 2020, the three respective goals are 30.2%, 11.7%, and 11.69%.
6) Spreading use of energy-saving appliances: We should popularize the energy-saving lights, energy-saving air conditioners and other energy efficient home appliances. By 2015, the energy-saving featured lights and air conditioners will cover 30% of households, and the other energy-saving home appliances will cover 35% of all the households. By 2020, energy-saving featured lights and air conditioners will cover 40% of all

the households and the other energy-saving home appliances will cover 45% of all the households.

12.6.4 *Plans for central heating supply*

We will put more efforts into the construction of heat source and heat supply network, promote the central heating supply system by adopting thermoelectricity central heating in urban area, county-level cities and developed towns, and popularize the regional heat central supply in other towns according to the local conditions, and eventually replace the utilization of boilers with capacity below 10 t. Besides, we will promote other heat supply models by adopting industrial waste heat, geothermal energy and biomass fuels, actively explore how to apply and demonstrate the combined cooling heat and power supplying technology, and then effectively control the too fast growth of energy consumption and CO_2 emission caused by the increase of heat used in industries and households.

1. *Major goals*

— By 2015, 55% of the heat supply in the urban area of Baoding, County Qingyuan, County Mancheng, County Xushui, and County Anxin will be through central heating supply system; 40% of the heat supply in all the county-level towns through the central heating supply system; the heat metering devices and temperature controlling devices will be installed in all the newly-built buildings which adopt central heating supply and in the old buildings which take energy-saving reformation and all the fees will be charged based on the heat metering.

— By 2020, 90% of the heat supply in the urban area of Baoding, County Qingyuan, County Mancheng, County Xushui, and County Anxin will be through central heating supply system; 80% of the heat supply in all the county-level towns through the central heating supply system.

2. *Dominant projects*

There are three dominant projects in total.

1) The construction of heat source projects: We will accelerate the construction of heat source point projects in the city, such as the combined heat and power generation projects in the South suburb, the Northwest suburb and the northeast suburb of Baoding, and the expansion of Datang Baoding Heat Power Factory's eighth-phase project, and

encourage the projects in county-level cities and large county-level towns. By 2015, the heat supply by combined heat and power generation in the urban area will cover more than 60 million m^2 and by 2020, the heat supply will cover more than 100 million m^2.

2) The construction of heat supply network: We will appropriately advance the heat supply network, and the key task is to promote the construction of heat transporting pipelines and heat distribution pipelines.; and we will put more efforts into the technology renovation of the old heat supply network, reduce the loss of the heat supply network, and strengthen the adjusting capacity of the heat supply system. By 2015, the heat supply network will cover over 90% of the urban area, over 80% of the county-level cities, and over 70% of the county-level towns. By 2020, the heat supply network will cover all the urban area, over 90% of the county-level cities, and over 80% of the county-level towns.
3) The demonstration projects of heat supply by new energy: We will carry out pilot projects for utilization of geothermal resource in areas such as County Xiong, County Gaoyang, County Anxin, and County Li; we will promote demonstration pilot projects of the comprehensive utilization of water-source heat and shallow underground ground source heat in areas with appropriate conditions for the applications of water-source heat pumps and ground source heat pumps; and with the development of biomass power generation projects, we will promote the comprehensive utilization of biomass heat energy. By 2015, there will be a number of demonstration and trial projects of heat supply produced through geothermal resource, ground-source heat, water-source heat and biomass fuels; by 2020, with the experience of the demonstration and trial projects, certain programs will be steadily promoted.

12.6.5 *Energy conservation in rural residence*

We will integrate the construction of new rural residence with the comprehensive implementation of the technical regulations on construction of energy-saving residence, so as to enhance the renovation of the original rural residence for energy-saving purpose; Besides, taking advantage of the "home appliances going to the countryside" policy, we will popularize the energy-saving home appliances in rural areas; we will also carry out the comprehensive utilization of new energy, such as marsh gas, solar energy and biomass energy, in the process of agricultural production and people's life in rural areas.

1. *Major goals*

— By 2015, energy-saving technologies will be adopted in over 20% of the newly-built rural residence, and energy-saving home appliances will be used in over 60% of the rural households, and new energy such as marsh gas, solar energy, biomass energy will account more than 18% of the total energy consumption in the rural areas.
— By 2020, energy-saving technologies will be adopted in over 30% of the newly-built rural residence, and energy-saving home appliances will be used in all the rural households, and new energy such as marsh gas, solar energy, biomass energy will account more than 20% of the total energy consumption in the rural areas.

Dominant projects

There will be five dominant projects in total.

1) The construction of energy-saving buildings in rural area: The construction of new residential buildings will comply with the tendency of rural residence concentrated into central villages, small towns and counties, and during the construction process, energy-saving building materials and technologies will be promoted to adopt. By 2015, the newly-built residential buildings which measure up to the energy conservation standards will account for about 20% of all the newly-built rural residence; by 2020, the figure will reach about 30%.
2) The application of marsh gas in rural areas: Promote both household used marsh gas facilities and medium and large scale marsh gas projects in rural areas to increase the wide application of marsh gas. By 2015, there will be 10 newly-built medium and large marsh gas projects, including five projects in the breeding community and five residence-connected projects, and there will be a growth of 170,000 households which are installed household marsh gas facilities. The marsh gas will be used in 27.8% of entire Baoding and even utilized in above 50% of those counties, towns and villages which are suitable for the development of marsh gas projects. By 2020, there will be 20 newly-built medium and large marsh gas projects, including 10 projects in the breeding community and 10 residence-connected projects, and there will be 760,000 households in total which adopt marsh gas. The marsh gas will be used in 34.5% of entire Baoding and even utilized in above 60% of those counties, towns and villages which are suitable for the development of marsh gas projects.

3) The promotion of solar energy application in rural areas: the key task is to popularize the installation of solar thermal collecting devices in rural households. By 2015, there will be an increase of 55,000 solar water heaters installed in the rural areas of Baoding. By 2020, there will be an increase of 100,000 solar water heaters.
4) The development of biomass energy in rural areas: It is important to promote the development and application of new fuels such as straw gas, straw blocks made by raw straws. By 2015, regional demonstrational pilot projects will be conducted; by 2020, based on the experience of the trials, the technology will be widely promoted for application.
5) The popularization of energy-saving home appliances in rural areas: With the implementation of "home appliances going to the countryside" policy, the home appliances which enter the rural market will have to pass the energy-saving certification, which will help to promote wide use of energy-saving home appliances in rural areas. By 2015, all home appliances with policy subsidies will have to pass energy-saving certification. By 2020, all home appliances in the rural market will have to pass energy-saving certification.

12.6.6 *Energy conservation in traffic sector*

Based on the principle of improving the city's traffic structure, it is a priority to develop public traffic system. Also, walking and bicycling are encouraged, and cars should be restrained to use. We need to actively adjust the policy of traffic energy structure, develop the public traffic system with larger volume and faster speed, establish a mature slow traffic system involving bicycling and walking, plan in advance a urban rail traffic system which centers around the urban area of Baoding and covers County Xushui, County Qingyuan, County Mancheng and the region of Lake Baiyangdian, promote the adjustment of traffic energy structure which aims at replacing oil for gas or electricity, and eventually build a traffic system which is featured of fast, convenient, high-efficient, safe and energy-saving.

1. *Major goals*

— By 2015, over 15% of people will use public traffic system and all the public transportation vehicles in the urban area will use gas (natural gas) instead of oil. And the pilot projects of replacing oil for electricity will be steadily carried out.

— By 2020, over 30% of people will use public traffic system and all the public transportation vehicles will manage to replace oil for electricity.

2. *Dominant projects*

There are four dominant projects in total.

1) The integration of bus routes network: More efforts should be made to adjust and improve the current bus route network. And in line with goals for the development of the city, we will steadily increase bus routines as well as the density of bus routines and bus stops. By 2012, there will be 32 new bus routines, 540 new buses, and 21 new stops (covering County Qingyuan, County Mancheng, County Xushui and County Anxin); by 2020, there will be 10 new fast bus lines in urban area, nine major lines, 610 new buses, and 10 new stops (covering County Qingyuan, County Mancheng, County Xushui and County Anxin).
2) The slow traffic system: The key tasks are to build pedestrian street around the urban landscapes, to optimize the urban slow traffic and the roads leading to schools, to improve the network for slow traffic system, and to create an urban traffic space which is suitable for walking and bicycling. By 2012, 20 lanes for nonmotorized vehicles and 15 pavements will be built; by 2020, 33 lanes for nonmotorized vehicles and 34 pavements will be built.
3) The rail traffic system: With the increasing expansion of the central urban area and the requirement for an improved urban traffic system, we will catch the opportunity brought by the extension of Beijing's light railway into its South rural area and plan in advance for an urban railway traffic system. By 2015, the urban railway traffic system, which takes the central urban area as its core and reaches out towards four directions respectively to County Qingyuan, County Mancheng, X County ushui and Lake Baiyangdian, will be finished of the designing, and the preliminary work will be carried out; by 2020, the construction of the city's railway traffic system will be initiated.
4) The adjustment of bus energy structure: At the requirements of technological advance, reasonable cost, safety guaranteed, energy efficiency and environmentally friendly, we will try to replace the oil consumed by the buses, taxies and other vehicles for gas in the urban area, and adjust and optimize the bus energy structure. By 2013, 13 new compressed natural gas stations will be built in the central urban area (including two parent

stations), which will meet the needs of gas for urban buses, and at the same time, the trial project which aims at replacing oil for electricity for buses will be started; by 2020, with the growing demand for gas, nine new compressed natural gas stations will be built in the central urban area (including one parent station), and buses will be gradually upgraded and manage to use electricity instead of oil.

12.7 Plan of Carbon Sink Project

The key task for a low carbon pilot city is to achieve energy conservation, reduction of energy consumption and CO_2 emission in its production and consumption sectors. Meanwhile, afforestation which strengthens the vegetation's capacity of gathering and solidifying CO_2 is also one important way to mitigate the effect of global warming through reducing the density of greenhouse gases such as CO_2 in the atmosphere. By the end of 2009, there were 428,300 hectares of forest land in Baoding and the forest coverage rate was 19.27%. The forest land per capita was only 0.04 hectare, which was 0.027 hectare less than the provincial average and 0.11 hectare less than the national average. The carbon sink capacity of Baoding was rather weak. Thus, it was important for a low carbon pilot city to promote afforestation, increase the total volume of forest resources, and increasingly strengthen the capacity of carbon sink.

1. *Major tasks and goals*

The tasks are as follows:

1) Accelerate afforestation in the urban area and increase the urban green ratio. Following the afforestation guideline of "shade style sightseeing style or leisure style", we will build key projects such as community parks, small gardens, green squares and water-forest networks, and carry out large-scale upgrading reformation of green belts along roads so that Baoding will have an urban forest network system which has abundant vegetation, reasonable layout, beautiful landscapes and the capacity of bring people happiness and fun.
2) Promote afforestation in rural areas and increase the ecological and economic benefit of forestry. We should stick to the principle of emphasizing equally on the ecological benefit and the economic benefit, build ecological forestry with economic benefit, forest of high-quality fruits and

farmland-forest network, and quicken the development of forestry; also in addition, we will widely build green land in towns or villages, and improve the quality of afforestation in rural areas.

3) Initiate afforestation projects in key areas and upgrade the ecological functions of afforestation. It is a key task to conduct afforestation along banks of waters, such as rivers, reservoirs and lakes, and along the roads like highways and railways. Also, it is important to consistently strengthen the ecological functions of afforestation in major regions or important areas.

The goals are as follows:

— By 2020, forest coverage rate in Baoding will reach above 25%.
— The green ratio in the constructed urban area will reach above 35%;
— The forest coverage rates in suburban areas would be as follows: It will reach more than 60% in mountains, more than 40% in hills, and more than 20% on plains;
— The green ratio along waters, such as rivers, reservoirs and lakes, will reach above 80%;
— The green ration along roads, such as roadways and railways, will reach above 80%.

2. *Dominant projects*

There are seven dominant projects in total.

1) The ecological screen in Western mountains: Baoding will insist on the combination of aero-sowing, forest reservation and afforestation, put more efforts into aero-sowing in remote mountainous areas and on high mountains, focus on the growth of forestry with a tree–shrub–grass combination and ecological forestry construction; in shallow mountains, Baoding should mainly pursue ecological benefit, while also taking economic benefit into account, by focusing on the construction of economic-ecological forestry belts; in hills, Baoding should take economic forestry as the major part by developing forests of high-quality fruits to establish a base for high-quality fruit production. Also, Baoding will include shrub in the forest resources as an important part, regard both forest reservation and afforestation as the fundamental way to treat wasted mountains, promote technical afforestation, and enhance the effects of afforestation. By 2020, there will be 134,000 hectare of planted

forest, and 201,000 hectare of forest grown by afforestation in closed mountains; also, the forestry coverage rate in western mountains will reach above 30%.

2) The urban forest system: according to the principle of "planting everywhere and green everywhere", Baoding will mainly build parks, small gardens, green squares and the water-forest network, launch large-scale upgrading reformation of road green belts, manage forestry construction in the rural–urban continuum, and build green belts circling the city, suburban forest parks and sightseeing forests. By 2020, the green ratio in the entire city will reach 40%, the green ratio in already-built areas will reach 35%, and green land area per capita will be above $9\,m^2$.
3) The green land in towns and villages: With the new socialism rural construction and comprehensive renovation, Baoding will carry out campaigns like "Green Home, Rich Village" and "Gardens in Countryside" etc. Through plantation and afforestation along villages, houses, roads and waters and the construction of public green land, the residential environment in the countryside will be improved. Also, industries based on forest will be developed.
4) The farmland-forest system on plains: The farmlands will be divided by administrative units, such as villages, and there will be a comprehensive plan to take all the farmland, water, forest, road and channels into an overall layout. Based on the principle which combines "network, lines and nodes", a comprehensive network system of farmland forest will be built depending on specific local conditions. By 2015, the controlling rate of the farmland-forest network in the plain area will reach 85%; by 2020, the rate will rise up above 90%.
5) Forest for cleaning water and protecting dykes: This project puts its emphasis on the green projects along more than 20 rivers which flow through the city, the Baoding reach of South-to-North water diversion project, and the linking area of Wangkuai Reservoir and Xidayang Reservoir. Also, it focuses on the green projects around 94 reservoirs and the urban water network. Baoding will promptly develop forest which cleans water, protects dykes and conserves water. By 2020, the green ratio along the upstream of Lake Baiyangdian, the reach of South-to-North water diversion project and the linking area of Wangkuai Reservoir and Xidayang Reservoir will reach above 80%.
6) The green corridor: Along the highways through Baoding, go-between roads, railways, state highways, provincial highways, county roads and village roads, various types of trees will be planted. Trees, shrubs,

flowers, grasses, and vines will be integrated in plantation. Keeping both economy and ecology in mind with an emphasis on green plantation, we will build high-quality green belts, economic forest belts and profitable forest belts along the roads. By 2020, the green ratio along roads will reach above 85%.

7) The forest parks, wetland and natural reserves: Adhering to the basic guideline of "strict protection, scientific management, overall planning, reasonable development and sustainable utilization", Baoding will protect and make a good use of the precious forest landscape resources, reasonably adjust the major functions of the forest resources, and plant scenic forest and upgrade scenes in the original forest with a plan. By 2020, there will be more than 18 provincial or national forest parks in total built in Baoding.

12.8 Guarantee Measures

(1) *Strengthen the leadership and effective responsibility system*

A coordination organization need to be founded to carry out trial work, with a team leader served by major leaders in the municipal government and members from corresponding departments. This organization will control the implementation of trial plans. Certain contents in trial plans will be included in the plans for national economy and social development. And all the specific plans for economic and social development should be kept adjusting and improving. Also, all the goals and dominant projects proposed will be respectively distributed to the corresponding departments and included in their annual work plan.

(2) *Make guiding policies and strengthen support*

Baoding will make corresponding policies and strengthen the support for the controlling projects. It will actively apply for national fund support, guide financial organizations and social capital to support the promotion and application of low carbon products and new low carbon technologies. When it comes to government purchase and city construction, low carbon products are the first choices. Specific fund for trial work will be arranged in the financial budget and will be used to support controlling projects and carry out trial projects.

(3) *Enhance cooperation and improve platform*

Grounded in Baoding Institute of Low Carbon City, Baoding should establish the cooperation mechanism for developing low carbon economy and building low carbon city through cooperation with relative international organizations, domestic and international developed regions as well as research institutes in terms of funds, technology and human resources. A technology-supporting system for low carbon city should be established and improved by importing, learning and recreating new technologies including technologies for clean production and all management methods which would help improve the low carbon management level. Baoding should attract talented technicians and management personnel, then train and establish a high-level research team for low carbon development. In addition, Boading should bring in the market mechanisms such as "carbon sink", "carbon emission trading", carry out corresponding trials, learn from successful experiences and strengthen market operation, and then gradually develop itself into a regional trade center.

(4) *Enhance promotion and have all citizens participate*

Taking establishing low carbon concepts and transforming production means and life style as the core guideline, the government should rely on press media, to popularize low carbon knowledge, persuade residents to accept the concepts of economy, environment protection and low carbon, and guide people to voluntarily change their high-carbon lifestyle into a low-carbon lifestyle in terms of clothing, food, housing, traffic and other aspects in their daily life. Besides, activities like promoting the development of low carbon communities and schools, and granting low carbon organization with certification and product with identification should be carried out regularly. The government needs to encourage and organize citizens, companies and social organizations to participate in the construction of the low carbon city, and form a "Low Carbon Baoding" construction structure led by government, demonstrated by key projects and widely participated by companies and citizens.

PART 4

Indicators

CHAPTER 13

Energy Consumption and Sum of Carbon Dioxide Emission

Table 13.1. Sum of energy consumption and its constituents.

Year	Electro-thermal equivalent calculation Total energy consumption (10,000 tce)	Calculation of coal consumption for power generation							
			Proportions in the total energy consumption (%)						
						Water, nuclear and other energy for power generation		
		Total energy consumption (10,000 tce)	Coal	Oil	Natural gas	Sum	Water power	Nuclear power
1990	95,384	98,703	76.2	16.6	2.1	5.1	5.1	—
1995	123,471	131,176	74.6	17.5	1.8	6.1	5.7	0.4
2005	225,781	235,997	70.8	19.8	2.6	6.8	5.9	0.8
2006	247,562	258,676	71.1	19.3	2.9	6.7	5.9	0.7
2007	268,413	280,508	71.1	18.8	3.3	6.8	5.9	0.8
2008	277,515	291,448	70.3	18.3	3.7	7.7	6.7	0.8

Source: China's Energy Statistics Yearbook 2009.

Table 13.2. CO_2 emission released by energy consumption (unit: Mt CO_2).

Year	IEA①	EIA②	CDIAC③	WRI④	This research⑤
2000	3,038	2,872	3,107	3,038	3,195
2005	5,068	5,558	5,081	5,060	5,270
2006	5,608	5,862	5,496	5,607	5,791
2007	6,032	6,247	5,863	—	6,235
2008	6,508	6,534	—	—	6,368

Note:① IEA, CO_2 Emissions from Fuel Combustion 2010-Highlights Paris: IEA;② EIA, *International Energy Statistics, Total Carbon Dioxide Emissions from the Consumption of Energy* (EIA, Washington D.C., 2010); ③ T. A. Boden, G. Marland and R. J. Andres, Global, Regional, and National Fossil-Fuel CO_2 Emissions. U. S. A. doi 10.3334/CDIAC/00001_ V2010. Oak Ridge: Carbon Dioxide Information Analysis Center, Oak Ridge National Laboratory, U. S. Department of Energy;④ World Resources Institute, Climate Analysis Indicators Tool (CAIT) Version 7.0, World Resources Institute, Washington D.C., 2010;⑤ According to the emission coefficients of different energies, accounting and data collection will be carried out in different sectors. The emission coefficients of different sorts of energy adopted in this research are as follows, coal: 2.71 tCO_2/tce; coke: 3.15 tCO_2/tce; coal gas: 1.28 tCO_2/tce; other gases: 1.28 tCO_2/tce; raw oil: 2.13 tCO_2/tce; petrol: 2.02 tCO_2/tce; coal oil: 2.09 tCO_2/tce; diesel oil: 2.16 tCO_2/tce; fuel oil: 2.27 tCO_2/tce; liquefied petroleum gas: 1.83 tCO_2/tce; oil refinery oil gas: 1.69 tCO_2/tce; other oil products: 2.13 tCO_2/tce; natural gas: 1.65 tCO_2/tce. The carbon emission per unit electricity is calculated based on the energy structure of the year.

Table 13.3. Carbon sink of forest.

				Carbon uptake rate of forest vegetation (Mt CO_2/a)				Carbon uptake rate of planted forest (Mt CO_2/a)	
Period	Forest coverage rate① (%)	Forest area① (10,000 hm^2)	Forest volume① (10,000 m^3)	Li Kerang, *et al.*②	Feng *et al.*③	Area of planted forest① (10,000 hm^2)	Volume of planted forest① (10,000 m^3)	Li Kerang, *et al.*②	Feng *et al.*③
1973–1976	12.7	12,186	86.6	—	16.3	1,139	1.6	35.3	44
1977–1981	12.0	11,528	90.3	−112	−44.0	1,273	2.7	78.3	139
1984–1988	13.0	12,465	91.4	33	36.7	1,874	5.3	58.3	66
1988–1993	13.9	13,370	101.4	260	132.0	2,137	7.1	120	81
1994–1998	16.6	15,894	112.7	398	88.0	2,914	10.1	140	—
1999–2003	18.2	17,491	124.6	623	—	3,229	15.0	177	—
2004–2009	20.4	19,545	133.6	—	—	6,169	19.6	—	—

Sources: ① All previous survey data of forest; ② K. R. Li, M. Huang, B. Tao, *et al.*, *A Stimulation Research on how China's Land Ecosystems Respond and Adjust to Global Climate Change* (China Meteorological Press, Beijing, 2009). ③ J. Y. Fang, A. P. Chen, C. H. Peng, *et al.*, Changes in forest biomass carbon storage in China between 1949 and 1998, *Science* **292** (2001) 2320–2322.

Table 13.4. Total energy consumption in different sectors (unit: 10,000 tce).

Project	2000	2005	2006	2007	2008
1. Energy consumption of energy sector and loss in processing, transformation and storage	50,962	78,991	87,106	95,206	95,816
1.1 Energy consumption of energy sector	15,602	22,892	23,572	25,087	25,536
1.2 Loss in processing, transformation and storage	35,361	56,100	63,534	70,119	70,280
Including: loss of coal power generation	29,097	46,823	53,836	60,069	58,541
Loss of heat supply	2,214	2,091	2,799	2,734	3,364
2. Terminal energy consumption	88,482	146,789	160,456	173,207	181,699
2.1. Agriculture	2,136	3,592	3,707	3,590	3,514
2.2. Manufacturing	56,340	96,287	105,683	114,298	120,735
2.3. Traffic and transportation	14,671	23,049	25,272	27,376	28,847
2.4. Architecture	15,335	223,862	25,795	27,943	28,603
Sum of one-time energy consumption	139,445	225,781	247,562	268,413	277,515

Note: In the Table of Energy Balance of China, there are seven sectors which take terminal energy consumption. And these seven sectors are integrated into four, including agriculture, manufacturing, traffic and transportation and architecture. The specific methods of integration are as follows: ① According to the amounts of terminal energy consumption in different industries, five industries, including coal mining and washing industry, petroleum and natural gas exploitation industry, petrochemical, coking and nuclear fuel industry, the production and supply industry of electric power and heat power, and gas production and supply industry, are all taken under the category of energy sector. ② According to the Table of Energy Balance of China, agriculture, forestry, husbandry, fishing, and water resources are all included in the category of agriculture sector; other industries apart from energy sector and architecture belong to the manufacturing sector; traffic and transportation, storage and mail and communication service belong to the traffic department, and other industries belong to the architecture sector. ③ Among the sectors, except the household consumption in the manufacturing sector and architecture sector, 95% of their gasoline consumption and 35% of their diesel oil are distributed to the traffic sector; all gasoline consumption and 85% of diesel oil consumed by citizens in the architecture sector are distributed to the traffic sector; all gasoline and 25% of diesel oil consumed in the agriculture sector are distributed to the traffic sector; besides, 15% of the electricity consumption in the traffic department is distributed to the architecture sector. Energy consumption in the category of "traffic and transportation" does not include the consumption by foreign ships and airplanes and that by Chinese ships and airplanes in foreign countries. ④ Energy consumption under the category of "architecture" does not include the consumption by central heating supply system. ⑤ The energy consumption amount is calculated by the means of electro-thermal equivalent calculation.

Table 13.5. Terminal CO_2 emission of different sectors (unit: Mt CO_2).

Categories	2000	2005	2006	2007	2008
Energy sector	485	683	715	768	766
Agriculture	79	123	128	126	122
Manufacturing	1,814	3,208	3,563	3,829	3,943
Traffic and transportation	327	509	557	604	631
Architecture	592	1,007	,116	1,228	1,260

Note: According to the terminal consumption of different sorts of energy in different sectors, the data was obtained through the calculation combined with the carbon emission factors of different sorts of energy. Since there is crossover among sectors, the sum of carbon emission of different sectors does not equal to the total amount of carbon emission that year. In Table 13.5, the emission amount of architecture sector includes the carbon emission released by central heating supply.

Table 13.6. Terminal energy consumption of sub-industries in the energy sector (unit: 10,000 tce).

Project	2000	2005	2006	2007	2008
Coal mining and washing industry	2,499	5,009	5,074	5,591	5,975
Petroleum and natural gas exploitation industry	2,506	2,495	2,505	2,588	3,141
Petrochemical, coking and nuclear fuel industry	6,261	9,496	9,574	10,462	10,273
Production and supply of electric power and heat power	3,985	5,530	6,008	6,009	5,738
Gas production and supply industry	351	398	412	437	410
Sum	15,602	22,892	23,572	25,087	25,536

Table 13.7. CO_2 emission of sub-industries in the energy sector (unit: Mt CO_2).

Project	2000	2005	2006	2007	2008
Coal mining and washing industry	87.02	159.62	161.25	174.87	184.01
Petroleum and natural gas exploitation industry	70.10	54.65	67.75	67.71	76.99
Petrochemical, coking and nuclear fuel industry	155.54	224.95	213.29	248.10	240.85
Production and supply of electric power and heat power	162.46	234.93	262.70	266.94	254.01
Gas production and supply industry	9.99	9.24	9.60	10.65	10.36
Sum	485.11	683.39	714.59	768.26	766.23

Note: Based on the terminal consumption of different sorts of energy in sub-industries, the results are obtained through the calculation combined with carbon emission factors of different sorts of energy.

Table 13.8. Energy consumption of sub-industries in the manufacturing sector (unit: 10,000 tce).

Project	2000	2005	2006	2007	2008
Steel	15,429	33,076	37,184	41,443	43,406
Nonferrous metals	2,125	3,886	4,512	5,360	5,736
Chemical industry	10,671	17,580	19,496	21,313	22,026
Textile	2,074	3,510	3,903	4,185	4,112
Papermaking	1,547	2,494	2,621	2,533	2,824
Food, beverages, and tobacco	1,513	1,896	2,012	2,029	2,112
Others	22,80	33,845	35,953	37,435	40,517
Sum	56,340	96,287	105,683	114,298	120,735

Note: According to in *China's Energy Statistics Yearbook 2009*, the results were calculated through the power equivalent calculation of terminal energy consumption (standardized amount) of different industries. Here, steel stands for "ferrous metallurgical smelting and rolling processing industry"; nonferrous metals stands for "nonferrous metallurgical smelting and rolling processing industry"; chemical industry stands for "chemical raw materials and chemical product manufacturing industry"; textile stands for "textile industry" and "clothes, shoes and hats manufacturing industry"; papermaking stands for "papermaking and paper product manufacturing industry"; food, beverages and tobacco stand for "food manufacturing industry", "beverage manufacturing industry" and "tobacco manufacturing industry"; others include building material industry. The data of energy consumption of "building materials" was calculated by the consumption of per unit building material and its output. Its energy consumption in 2005, 2006, 2007, and 2008 respectively reached 32,257/10,000 tce, 34,005/10,000 tce, 35,932/10,000 tce, and 35,834/10,000 tce.

Table 13.9. CO_2 emission of sub-industries in the manufacture sector (unit: Mt CO_2).

Project	2000	2005	2006	2007	2008
Steel	472.71	1,031.73	1,170.66	1,276.2	1,312.10
Nonferrous metals	90.84	166.75	199.68	241.35	245.24
Chemical industry	324.55	538.11	597.81	650.15	657.11
Textile	76.15	135.88	157.44	167.7	160.99
Papermaking	53.68	86.23	91.92	88.16	95.05
Food, beverages, and tobacco	50.27	61.33	66.23	67.48	69.43
Others	745.66	1,187.71	1,279.69	1,338.10	1,402.62
Sum	1,813.86	3,207.74	3,563.43	3,829.14	3,942.54

Note: Based on the terminal consumption of different sorts of energy in sub-industries, the results are obtained through the calculation combined with carbon emission factors of different sorts of energy.

Table 13.10. Energy consumption of different types of buildings (unit: 10,000 tce).

Project	2000	2001	2002	2003	2004	2005	2006	2007	2008
Central heating in towns of Northern China	3,769	5,638	6,680	7,640	8,503	9,712	10,834	12,038	13,303
Residence in towns and cities (except central heating)	7,983	8,253	8,740	9,763	10,990	12,006	13,364	5,137	15,170
Rural residence	6,183	6,429	6,718	7,459	8,536	9,302	10,101	11,137	11,452
Public buildings (except central heating)	6,780	7,029	7,392	8,699	10,041	11,206	12,527	13,576	14,122
Sum	14,715	27,349	29,531	33,561	38,070	42,227	46,827	51,888	54,047

Note: The methods applied to calculate energy consumption and carbon emission in different buildings are as follows: ① The energy consumption of residence in towns and cities (except central heating system): According to the Table of Energy Balance of China in *China's Energy Statistics Yearbook*, it's the sum of coal, coke oven gas, other gases, kerosene, liquefied petroleum gas, natural gas, electricity in household consumption in towns and cities and 5% consumption of diesel oil. ② Consumption in rural areas: According to the Table of Energy Balance of China in *China's Energy Statistics Yearbook*, it's the sum of coal, coke oven gas, other gases, kerosene, liquefied petroleum gas, natural gas, electricity in household consumption in towns and cities and 5% consumption of diesel oil. ③ Energy consumption in public buildings (except central heating): According to the Table of Energy Balance of China in *China's Energy Statistics Yearbook*, it includes 15% electricity of the "traffic and transportation, storage, and mail and communication industry", coal, coke oven gas, other gases, kerosene, liquefied petroleum gas, natural gas, electricity and 65% of diesel oil consumption and 5% of gasoline consumption in the "wholesale and retail industry and lodging and catering service" and "others". ④ Central heating in towns and cities of northern China: It adopts the results from *2009 China's Annual Development Research Report of Energy Conservation in Buildings*. And the electricity power is conversed to standard coal through the calculation of coal power generation.

Table 13.11. CO_2 emission in different types of buildings (unit: Mt CO_2).

Project	2000	2001	2002	2003	2004	2005	2006	2007	2008
Central heating in towns of Northern China	102	152	180	205	227	259	289	320	353
Residence in towns and cities (except central heating)	175	179	189	213	235	259	288	323	316
Rural residence	157	161	168	187	211	230	249	271	273
Public buildings (except central heating)	158	161	171	203	229	258	290	314	318
Sum	592	653	707	808	902	1,007	1,116	1,228	1,260

Note: Based on the terminal consumption of different sorts of energy in sub-industries, the results are obtained by the calculation combined with carbon emission factors of different sorts of energy.

CHAPTER 14

Energy and Efficiency of Carbon Dioxide Emission

Table 14.1. China's energy consumption per unit of GDP and intensity of CO_2 emission.

Year	Constant GDP of 2005① (RMB 100 million yuan)	Total amount of energy consumption ②(10,000 tce)	Total amount of CO_2 emission released by energy③ (Mt CO_2)	Constant energy consumption per RMB 10,000 yuan of GDP of 2005④ (tce/RMB 10,000)	Constant carbon emission per RMB 10,000 of GDP of 2005⑤ (t CO_2/RMB 10,000)
1990	43,043	98,703	2,289	2.293	5.318
1991	46,994	103,783	2,389	2.208	5.084
1992	53,686	109,170	2,470	2.033	4.600
1993	61,183	115,993	2,649	1.896	4.329
1994	69,187	122,737	2,855	1.774	4.127
1995	76,745	131,176	2,885	1.709	3.760
1996	84,426	135,192	2,917	1.601	3.455
1997	92,275	135,909	3,107	1.473	3.367
1998	99,504	136,184	2,991	1.369	3.006
1999	107,086	140,569	2,909	1.313	2.716
2000	116,114	145,531	2,872	1.253	2.473
2001	125,752	150,406	2,992	1.196	2.380
2002	137,173	159,431	3,492	1.162	2.546
2003	150,925	183,792	4,102	1.218	2.718
2004	166,146	213,456	5,132	1.285	3.089
2005	184,937	235,997	5,558	1.276	3.006
2006	208,381	258,676	5,862	1.241	2.813

(*Continued*)

Table 14.1. (*Continued*)

Year	Constant GDP of 2005① (RMB 100 million yuan)	Total amount of energy consumption ②(10,000 tce)	Total amount of CO_2 emission released by energy③ (Mt CO_2)	Constant energy consumption per RMB 10,000 yuan of GDP of 2005④ (tce/RMB 10,000)	Constant carbon emission per RMB 10,000 of GDP of 2005⑤ (t CO_2/RMB 10,000)
2007	237,893	280,508	6,247	1.179	2.626
2008	260,813	291,448	6,534	1.117	2.505

Note: ① All the results are obtained through the calculations of data from *China's Statistics Yearbook 2010.* ② The results are obtained through the calculation based on coal power generation in *China's Statistics Yearbook 2010.* ③ EIA *Total Carbon Dioxide Emission from the Consumption of Energy* (EIA, Washington D.C., 2010). ④ The total amount of energy consumption/constant GDP of 2005. ⑤ Total amount of CO_2 emission released by energy/constant GDP of 2005.

Table 14.2. Major indicators of energy consumption in the power sector.

Project	Unit	2000	2005	2006	2007	2008
Net coal consumption rate	gce/(kWh)	392	370	367	356	345
Gross coal consumption rate	gce/(kWh)	363	343	342	332	322
Auxiliary power consumption rate	%	6.28	5.87	5.93	5.83	5.90
Thermal power consumption rate	%	7.31	6.80	6.77	6.62	6.79
Loss rate of transmission	%	7.70	7.21	7.04	6.97	6.79

Sources: Synopsis of the statistical materials about the power sector issued by China Electricity Council in 2010.

Table 14.3. Carbon emission in thermal power generation (1 kWh).

Project	Unit	2005	2006	2007	2008
Total amount of carbon emission by thermal power generation①	Mt CO_2	1,915.94	2,211.20	2,490.78	2,477.05

(*Continued*)

Table 14.3. (*Continued*)

Project	Unit	2005	2006	2007	2008
Thermal power generation capacity②	T·Who	2,047.3	2,369.6	2,722.9	2,790
Emission/1kwh	g CO_2/(kWh)	935.8	933.2	914.7	887.8

Note: ① It is calculated based on the energy consumption of thermal power in *China's Energy Statistics Yearbook 2009* being multiplied by emission coefficient. ② Data from *China's Energy Statistics Yearbook 2009*. ③ The total amount of carbon emission of thermal power/Total thermal power generation capacity.

Table 14.4. Power generation capacity from 2005–2008 (unit: TWh).

Project	2005	2006	2007	2008
Thermal power	2047.3	2369.6	2722.9	2790
Nuclear power	53.1	54.8	62.1	68.4
Water power	397	435.8	485.3	585.2
Wind power and others	1.3	2.7	5.6	13.1
Total power generation capacity	2500.3	2865.7	3281.6	3466.9

Sources: *China's Energy Statistics Yearbook 2009*.

Table 14.5. Structure of energy sources for power generation from 2005–2008 (unit: %).

Project	2005	2006	2007	2008
Thermal power	81.9	82.7	83.0	80.5
Nuclear power	2.12	1.91	1.89	1.97
Water power	15.88	15.21	14.79	16.88
Wind power and others	0.05	0.09	0.17	0.38

Sources: *China's Energy Statistics Yearbook 2009*.

Table 14.6. Energy consumption of per unit of industrial added value of sub-industries in the manufacturing sector (unit: tce/ 10,000 yuan).

Project	2005	2006	2007	2008
Steel	6.34	6.62	6.46	6.34
Nonferrous metals	3.58	2.88	2.81	2.66
Chemical industry	4.93	4.98	4.48	4.26

(*Continued*)

Table 14.6. (*Continued*)

Project	2005	2006	2007	2008
Textile	1.18	1.22	1.18	1.07
Papermaking	2.91	2.86	2.44	2.45
Food, beverages, and tobacco	0.54	0.54	0.49	0.46
Average	2.10	2.05	1.93	1.82

Note: The industrial added value is the constant value of 2005, energy consumption is counted by the calculation of coal power generation, and the gasoline and diesel oil used for traffic and communication are not excluded. The classification of industries is the same with that in Table 13.8.

Table 14.7. Carbon emission of per unit of industrial added value of sub-industries in the manufacturing sector (unit: t CO_2/ RMB 10,000).

Project	2005	2006	2007	2008
Steel	16.72	17.48	16.58	16.18
Nonferrous metals	8.02	6.46	6.24	5.79
Chemical industry	10.59	10.62	9.45	8.97
Textile	2.75	2.86	2.76	2.45
Papermaking	7.07	6.96	5.95	5.87
Food, beverages, and tobacco	1.32	1.33	1.20	1.11
Others	3.25	3.07	2.79	2.65
Average	5.08	4.96	4.58	4.30

Note: The industrial added value is the constant value of 2005, and the gasoline and diesel oil used for traffic and communication are not excluded. The classification of industries is the same with that in Table 13.8.

Table 14.8. Added value of sub-industries in the manufacturing sector (unit: 100 million yuan).

Project	2005	2006	2007	2008
Steel	6,180.5	6,704.5	7,708.3	8,121.2
Nonferrous metals	2,064.5	3,061.0	3,831.9	4,190.2
Chemical industry	4,698.8	5,167.6	6,281.9	6,728.5
Textile	4,985.7	5,548.4	6,143.8	6,648.2
Papermaking	1,226.5	1,237.1	1,491.7	1,632.6
Food	4,700.0	5,059.7	5,703.1	6,366.8
Others	47,118.3	54,281.3	63,121.5	70,522.0
Sum	70,974.3	81,149.5	94,283.3	104,209.6

Table 14.9. Structure of added value of sub-industries in the manufacturing sector (unit: %).

Project	2005	2006	2007	2008
Steel	8.71	8.27	8.18	7.79
Nonferrous metals	2.91	3.77	4.06	4.02
Chemical industry	6.62	6.37	6.66	6.46
Textile	7.02	6.84	6.52	6.38
Papermaking	1.73	1.64	1.58	1.57
Food	6.62	6.24	6.05	6.11
Others	66.39	66.89	66.95	67.67

Table 14.10. Comprehensive energy consumption of major industrial products.

Project	2005	2006①	2007	2008	2009	Advanced level in the world②
Steel						
Comprehensive energy consumption of the entire industry (kgce/t)	1,019	998	976	975	973	—
Comprehensive energy consumption of large and medium enterprises③ (kgce/t)	760	745	733	729	717	—
Power consumption of large and medium enterprises (kWh/t)	452	—	433	445	434	—
Comparable energy consumption of large and medium enterprises (kgce/t)	732	729	718	709	697	610
Electrolytic aluminum						
Consumption of alternating currents (kWh/t)④	14,680	14,671	14,488	14,323	14,171	14,100
Copper						
Comprehensive energy consumption (kgce/t)	780	729	610	549	509	500
Consumption of direct currents (kWh/t)	—	—	310	304	309	—
Cement						

(*Continued*)

Table 14.10. (*Continued*)

Project	2005	2006①	2007	2008	2009	Advanced level in the world②
Comprehensive energy consumption⑤ (kgce/t)	167	161	158	151	139	118
Power consumption (kWh/t)	92	—	91	91	90	—
Plate glass						
Comprehensive energy consumption (kgce/weight case)	22	19	17	16.6	16.5	15
Power consumption (kWh/t)	—	—	7.6	7.4	6.9	—
Refined oil						
Comprehensive energy consumption⑥ (kgce/t)	114	112	110	108	106	73
Power consumption (kWh/t)	—	—	60.0	60.3	60.2	—
Ethylene						
Comprehensive energy consumption⑦ (kgce/t)	1,073	1,013	1,026	1,010	976	629
Power consumption (kWh/t)	—	—	124.5	126.4	126.1	—
Synthetic ammonia						
Comprehensive energy consumption⑧ (kgce/t)	1,700	1,700	1,702	1,661	1,591	—
Power consumption (kWh/t)	1,280	—	1,202	1,184	1,172	—
Sodium hydroxide						
Comprehensive energy consumption (kgce/t)						
Diaphragm method	1,448	—	1,272	1,351	1,293	1,250
Ionic membrane	1,067	—	937	922	916	910
Power consumption (kWh/t)						
Diaphragm method	—	—	2,364	2,340	2,309	—

(*Continued*)

Table 14.10. (*Continued*)

Project	2005	2006①	2007	2008	2009	Advanced level in the world②
Ionic membrane	—	—	2,213	2,201	2,186	—
Sodium carbonate						
Comprehensive energy consumption (kgce/t)	396	402	398	355	323	310
Calcium carbide						
Power consumption (kWh/t)	3,450	3,666	3,465	3,440	3,395	3,000
Paper and paper matrix						
Comprehensive energy consumption of the entire industry⑨ (kgce/t)	528	502	476	440	395	—
Power consumption of the entire industry (kWh/t)	745	—	605	578	567	—
Comprehensive energy consumption of self-pulping enterprises (kgce/t)	1,380	1,290	1,225	1,153	1,090	610
Chemical fiber						
Power consumption (kWh/t)	1,396	1,207	1,086	1,094	1,091	980

Note: ① In 2006, when the power consumption in the comprehensive energy consumption of China's products is converted to standard coal, the calculation of gross coal consumption is changed into electro-thermal equivalent calculation. In order to make comparison with the previous unit consumption and the advanced international level (the calculation of gross coal consumption adopted by Japan and EU), the power consumption is calculated based on the gross coal consumption adopted in the Table of Energy Balance of China [about 350 gce/(kWh)], and unit consumption of certain products in 2008 and 2009 is calculated based on the drop from last year. ② Advanced level in the world is an average of those of countries and regions which are at leading positions in the world. ③ The comparable energy consumption of steel in China stands for the situation of large and medium enterprises. There were 70 large and medium steel enterprises in China in 2008 whose output accounted for 83% of the national sum. The power consumption of China's large and medium steel enterprises in 2008 calculated by gross coal consumption and electro-thermal equivalent calculation were respectively 730 kgce/t and 630 kgce/t, comprehensive energy consumption of steel production was 445 kWh/t. Advanced level in the world stood for the level of Japan, which was calculated by crude steel output in 2007 and the energy consumption of steel industry. ④ The power consumption of alternating currents for the production of electrolytic aluminum in Canadian Alma Company was 129,000 kWh/t. ⑤ The comprehensive energy consumption of cement was an average of the entire industry, including those production besides the building material system. In 2007, the energy consumption of China's cement clinker was 126 kgce/t, the comprehensive power consumption of cement was 91 kWh/t. The consumption at the advanced level in the world were 103 kgce/t and 82 kWh/t. ⑥ In 2007, the power consumption of crude oil processing in China was 60 kWh/t. When it comes to the global comparison of energy efficiency of crude oil processing, it is more realistic to adopt "the consumption of energy factors". "Energy factors" are indicators which reflect the phase of crude oil processing, and they are calculated by "unit energy consumption" and the compositions of processing devices in oil refinery factory. In 2005, China's "consumption of energy factors" was 13 kgce/t, and the advanced level in the world was 10 kgce/t. ⑦ The production of ethylene in China takes naphtha as major raw material, and the advanced level in the world stood for the level in Mid East, taking ethane as major raw material. ⑧ The energy consumption of synthetic ammonia stood for the average level of the devices in companies with all sizes which used coal, petroleum and gas as raw materials. ⑨ The energy consumption of sodium hydroxide was the weighted average of diaphragm method and ionic membrane. In 2007, the gross coal consumption respectively by diaphragm method and ionic membrane were 1,272 kgce/t and 937 kgce/t, the consumption calculated by electro-thermal equivalent were respectively 736 kgce/t and 434 kgce/t, and the power consumption were 2,364 kWh/t and 2,213 kWh)t.⑩ In 2008, in the papermaking industry of China, waste-paper pulp took 60% of all the pulp. The amount of waste-paper pulp was 55.49 Mt, the importing proportion of which was 24.21 Mt, and its importing pulp was 9.52 Mt, which took 59% of all the consumption of pulp. 1 t pulp could be used to produce 1.2 t paper. Besides, to renew waste paper would consume 80% less of energy than directly adopting wood to make paper.

Source: National Bureau of Statistics of China. *China's Statistics Abstract 2010*, *China's Statistics Yearbook 2010*; Ministry of Industry and Information Technology of the People's Republic and China; China National Coal Association; China Electricity Council; China Nonferrous Metals Industry Association; China Building Material Council; China Petroleum and Chemical Industry Energy Conservation Association; China Paper Association; China Chemical Fibers Association; the Institute of Energy Economics, Japan, *2010 Handbook of Energy and Economics Statistics in Japan*; Japan Iron & Steel Institute; Korean Iron and Steel Institute; Japan Cement Association; *Data Book of Japan Energy Association*; IEA, Energy Statistics of OECD Countries; (USA) National Mining Association.

Table 14.11. Energy consumption of per unit of building area in different types of architecture (unit: kgce/m^2).

Project	2000	2005	2006	2007	2008
Central heating in towns of Northern China	19.76	18.04	17.73	17.47	17.2-
Residence in towns and cities (except central heating)	18.10	11.15	11.84	12.82	12.30
Rural residence	3.08	4.20	4.46	4.83	4.86
Public buildings (except central heating)	20.87	19.72	20.33	20.43	19.82
Average level in China	8.92	11.05	11.84	13.68	12.84

Table 14.12. Carbon emission of per unit of building area in different types of architecture (unit: kg CO_2/m^2).

Project	2000	2005	2006	2007	2008
Central heating in towns of northern China	53.46	48.17	47.26	46.48	45.70
Residence in towns and cities (except central heating)	39.73	24.06	25.55	27.38	25.65
Rural residence	7.83	10.38	10.98	11.73	11.56
Public buildings (except central heating)	48.70	45.49	47.12	47.72	44.58
Average level in China	21.38	26.09	27.84	29.58	29.27

Table 14.13. Building area of different types of architecture (unit: 100 million m^2).

Project	2000	2005	2006	2007	2008
Residence in towns and cities	44.10	107.69	112.89	118.09	123.29
Public buildings	32.49	56.82	61.63	66.45	71.26
Buildings in rural areas	200.48	221.40	226.39	230.62	235.88

Table 14.14. Percentages of various types of architecture occupying in the total amount of building area (unit: 100 million m^2).

Project	2000	2005	2006	2007	2008
Residence in towns and cities	15.92	27.91	28.16	28.44	28.64
Public buildings	11.73	14.72	15.37	16.01	16.56
Buildings in rural areas	72.36	57.37	56.47	55.65	54.80

Table 14.15. Energy efficiency of different transportation means.

Project	Unit	2000	2005	2006	2007	2008
Commercial vehicles						
Passenger vehicles powered by gasoline	L(10^2 tkm)	11	13.2	12.8	13.1	—
Passenger vehicles powered by diesel oil	L(10^2 tkm)	9	11.6	11.2	11.2	—
Freight vehicles powered by gasoline	L(10^2 tkm)	7	8	7.9	8.3	
Freight vehicles powered by diesel oil	L(10^2 tkm)	5	6.3	6.5	6.3	—
Railway						
Diesel locomotive	kg/(10^4 tkm)	25.8	24.6	24.3	24.6	24.9
Electric locomotive	kwh·h/(10^4 tkm)	113.2	111.8	110	109.5	—
Comprehensive energy consumption of railway transportation	tce/(10^6 tkm)	—	6.48	6.12	5.78	5.6
Aviation	kgoe/(tkm)	0.396	0.336	0.327	0.309	0.312
Water transportation	kgoe/(10^3 tkm)	9	7	5	12	—

Source: *China's Traffic Yearbook* from 2001 to 2009 issued by the Ministry of Railways of the People's Republic of China.

Table 14.16. Volumes of passenger and goods transportation by different transportation means.

Project		2000	2005	2006	2007	2008
Volume of passenger transportation (10,000 people)	Railway	105,073	115,583	125,656	135,670	146,193
	highway	1,347,392	1,697,381	1,860,487	2,050,680	2,682,114
	Water	19,386	20,227	22,047	22,835	20,334
	Aviation	6,031	12,602	14,553	16,884	17,732
Volume of goods transportation (10,000,t)	Railway	178,581	269,296	288,224	314,237	330,354
	highway	1,038,813	1,341,778	1,446,347	1,639,432	1,916,759
	Water	99,442	171,099	194,290	222,296	252,158
	Aviation	147	230	257	285	288
	Pipeline	18,700	31,037	33,436	40,552	43,906

Note: The data of water transportation has excluded the overseas transportation; the data of aviation refers to that of domestic airlines, including Hong Kong and Macau.
Source: *China's Traffic Yearbook* from 2001 to 2009.

Table 14.17. Rotation volume of passenger and goods transportation by different transportation means.

Project		2000	2005	2006	2007	2008
Rotation volume of passenger transportation (100 million people·km)	Railway	4,533	6,062	6,622	7,216	7,779
	highway	6,657	9,292	10,131	11,507	12,476
	Water	101	68	74	78	59
	Aviation	738	1,593	1,847	2,173	2,306
Rotation volume of goods transportation (100 million tkm)	Railway	13,771	20,726	21,954	23,797	25,106
	highway	6,129	8,693	9,754	11,355	32,868
	Water	6,661	11,120	12,909	15,599	17,412
	Aviation	21	34	38	42	42
	Pipeline	636	1,088	1,551	1,866	1,944

Note: The data of water transportation has excluded the overseas transportation; the data of aviation refers to that of domestic airlines, including Hong Kong and Macau.
Source: *China's Traffic Yearbook* from 2001 to 2009.

Table 14.18. Structure of traffic and transportation (unit: %)

Project		2000	2005	2006	2007	2008
Structure of passengers	Railway	7.11	6.26	6.21	6.09	5.10
	Road	91.17	91.96	91.98	92.12	93.57
	Water	1.31	1.10	1.09	1.03	0.71
	Aviation	0.41	0.68	0.72	0.76	0.62

(*Continued*)

Table 14.18. (*Continued*)

Project		2000	2005	2006	2007	2008
Rotation structure of passengers	Railway	37.68	35.63	35.46	34.41	34.39
	Road	55.35	54.61	54.25	54.86	55.16
	Water	0.84	0.40	0.39	0.37	0.26
	Aviation	6.13	9.36	9.89	10.36	10.19
Structure of goods transportation	Railway	13.37	14.85	14.54	14.18	12.99
	Road	77.77	73.99	93.96	73.95	75.36
	Water	7.45	9.44	9.80	10.03	9.91
	Aviation	0.01	0.01	0.01	0.01	0.01
	Pipeline	1.40	1.71	1.69	1.83	1.73
Rotation structure of goods transportation	Railway	50.59	49.75	47.51	45.19	32.45
	Road	22.52	20.87	21.11	21.56	42.48
	Water	24.47	26.69	27.94	29.62	22.50
	Aviation	0.08	0.08	0.08	0.08	0.05
	Pipeline	2.34	2.61	3.36	3.54	2.51

Source: *China's Traffic Yearbook* from 2001 to 2009.
Note: The data of water transportation has excluded the overseas transportation; the data of aviation refers to that of domestic airlines, including Hong Kong and Macau.

CHAPTER 15

Structure of Energy Consumption

Table 15.1. The structure of energy consumption in the manufacturing sector (Unit: %).

Project	2000	2005	2006	2007	2008
Coal	66.26	67.04	65.57	64.09	64.89
Petroleum	10.94	8.21	8.23	8.12	7.62
Natural gas	2.52	2.77	3.03	3.29	3.39
Heat	4.81	4.35	4.39	4.39	4.13
Electricity	15.46	17.63	18.78	20.11	19.96

Table 15.2. The structure of energy consumption in the architecture sector (unit: %).

Project	2000	2005	2006	2007	2008
Coal	47	45	43	41	40
Petroleum	4	3	2	2	2
Gas	12	12	13	14	13
Electricity	37	40	42	43	44

Table 15.3. The structure of energy consumption in the traffic and transportation sector (Unit: %).

Project	2000	2005	2006	2007	2008
Coal	4.24	2.72	2.31	1.97	1.76
Gasoline	34 70	30.75	30.29	29.44	31.11
Kerosene	5.37	6.08	5.88	6.07	5.99
Diesel oil	44.50	48.40	48.63	48.80	49.95
Fuel oil	8.28	7.82	8.37	9.18	5.66
Liquefied petroleum gas	0.17	0.35	0.36	0.34	0.32
Natural gas	0.36	1.78	2.00	1.96	2.91
Electricity	2.00	1.95	1.93	2.03	2.07
Others	0.36	0.16	0.22	0.20	0.21

Table 15.4. Structure of energy consumption.

	Electro-thermal equivalent calculation							Calculation of gross coal consumption						
		Proportions in the total amount of energy consumption (%)							Proportions in the total amount of energy consumption (%)					
					Power generated by water, nuclear, and other energies							Power generated by water, nuclear, and other energies		
Year	Total amount of energy consumption (10,000 tce)	Coal	Petroleum	Natural gas	Sum	Water power	Nuclear power	Total amount of energy consumption (10,000 tce)	Coal	Petroleum	Natural gas	Sum	Water power	Nuclear power
1990	95,384	79.0	17.2	2.1	1.7	1.7	—	98,703	76.2	16.6	2.1	5.1	5.1	—
1995	123,471	77.0	18.6	1.9	2.5	2.4	0.1	131,176	74.6	17.5	1.8	6.1	5.7	0.4
2000	139,445	72.4	23.1	2.3	2.1	2.0	0.1	145,531	69.2	22.2	2.2	6.4	5.9	0.4
2001	142,972	71.9	23.0	2.6	2.6	2.4	0.2	150,406	68.3	21.8	2.4	7.5	7.1	0.4
2002	151,789	71.5	23.4	2.6	2.6	2.4	0.2	159,431	68.0	22.3	2.4	7.5	7.1	0.4
2003	176,074	73.1	22.1	2.6	2.3	2.0	0.3	183,792	69.8	21.1	2.5	6.5	5.7	0.8
2004	204,219	72.8	22.2	2.6	2.5	2.2	0.3	213,456	69.5	21.3	2.5	6.7	5.9	0.8
2005	225,781	74.1	20.7	2.8	2.5	2.2	0.3	235,997	70.8	19.8	2.6	6.8	5.9	0.8
2006	247,562	74.3	20.2	3.0	2.5	2.2	0.3	258,676	71.1	19.3	2.9	6.7	5.9	0.7
2007	268,413	74.3	19.7	3.5	2.6	2.2	0.3	280,508	71.1	18.8	3.3	6.8	5.9	0.8
2008	277,515	74.9	19.2	2.9	3.0	2.6	0.3	291,448	70.3	18.3	3.7	7.7	6.7	0.8

Source: *China's Energy Statistics Yearbook 2009*.

Table 15.5. Carbon emission of per unit of energy.

Project	Unit	2000	2005	2006	2007	2008
The total amount of energy consumption	10,000 tce	145,531	235,997	258,676	280,508	291,148
Carbon emission of energy consumption	Mt CO_2	3,195	5,270	5,791	6,235	6,368
Carbon emission of per unit of energy consumption	t CO_2/tce	2.20	2.23	2.24	2.22	2.18

Table 15.6. Generator volumes in the power sector from 2005 to 2008 (10,000 kW).

Project	2005	2006	2007	2008
Volume of thermal power generators	39,138	48,405	55,442	60,132
Volume of water power generators	11,739	12,857	14,526	17,152
Volume of wind power generators	106	187	403	894
Volume of nuclear power generators	685	685	885	908
The total volume of power generators	51,674	62,142	71,267	79,101

Source: *China's Power Sector Statistics Express* from 2005 to 2008 issued by China Electricity Council.

CHAPTER 16

Comparison among Countries

Table 16.1. Global comparison of gross coal consumption of thermal power plants [unit: gce/(kWh)].

Project	1990	1995	2000	2005	2006	2007	2008
China①	392	379	363	343	342	332	322
Japan②	317	315	303	301	299	300	292
Germany	309	322	309	301	306		
Korea	332	322	311	302	300		

Note: ① Unit capacity above 6 MW. ② Average number of nine major power companies.
Source: *China's Energy Statistics Yearbook 2009.*

Table 16.2. Global comparison of net coal consumption of thermal power plants [unit: gce/(kWh)].

Project	1990	1995	2000	2005	2006	2007	2008
China	427	412	392	370	367	356	345
Japan	332	330.6	316	313.7	312.2	311.8	309.9

Source: *China's Energy Statistics Yearbook 2009.*

Table 16.3. Global comparison of unit consumption of major high-energy-consuming products.

Project	Unit	2000	2005	2006	2007	2008	Advanced level in the world
Comparable energy consumption of steel	kgce/t	784	732	729	718	709	616
Comprehensive energy consumption of cement	kgce/t	181	167	161	158	151	127
Comprehensive energy consumption of ethylene	kgce/t	1,125	1,073	1,013	1,026	1,003	629
Comprehensive energy consumption of electrolytic aluminum	kWh/t	15,480	14,680	14,671	14,488	14,323	14,100
Comprehensive energy consumption of sodium carbonate	kgce/t	406	396	—	398	378	310
Comprehensive energy consumption of calcium carbide	kgce/t	1,946	2,088	1,141	1,107	1,095	1,061
Comprehensive energy consumption of paper and paper matrix	kgce/t	1,540	1,380	1,290	1,225	1,153	640

Table 16.4. Global comparison of energy consumption and carbon emission of the architecture sector in 2006.

Indicator	Total amount of energy consumption	Energy consumption of per unit of building area	Total amount of carbon emission	Carbon emission of per unit of building area	Per capita carbon emission	Structure of energy consumption				
						Coal	Petroleum	Gas	Renewable energy	Electricity
Unit	**Mtce**	**kgce/(m^2·a)**	**MtCO$_2$**	**kgCO$_2$/(m^2·a)**	**kgCO$_2$/(person·a)**	%	%	%	%	%
USA	1,388	54.4	2,225	87.2	7,521	4.9	19.2	0.3	1.3	74.3
OECD (Europe)	1,032		1,317		2,019	12.6	25.1	2.4	0	59.8
Canada	195	85.8	176	77.4	5,454	7.4	20.3	0	0	72.3
Japan	321	49	418	64	3,271	13.9	12.4	0	0	70.8
Korea	97	48.3	138	68.6	2,940	14.8	22.2	0	0	63
Australia	66		136		6,589	0	5.5	0	0	94.5
Brazil	76		33		177	14.3	0	0	0	85.7
Russia	300		487		3,394	6	45.7	4.8	0	70.2
India	121		263		245	20.9	0	8.9	0	70.2
China (urban areas)	367	21	868	49.7	1,504	—	—	—	—	—
China (rural areas)	101	4.5	248	11	336	—	—	—	—	—
China's average	468	11.7	1,116	27.9	849	10.7	5.2	42.4	0	41.7
Africa	140		269		297	23.2	5.2	5.2	0	66.5
World	5,155		8,588		1,326	10.7	19.3	3.6	0.3	66.1
OECD (member countries)	3,147		4,683		3,982	9.3	20.1	0.9	0.6	69.2
BRIC	895		1,795		660	13.7	17.7	15.3	0	53.3

Source: ① The population of the countries and the GDP data: World Bank, World Development Indicators database, (World Bank, Washington D.C., 2007). ② China: National Bureau of Statistics of China, *China's Statistics Yearbook 2009* (China Statistics Press, Beijing, 2009). ③ USA: The United State Development of Energy, *2009 Buildings Energy Data Book*, (D&R International, Ltd., USA, 2009). ④ Canada: Natural Resources Canada, *2006 Energy Use Data Handbook* (Energy Publications Office of Energy Efficiency, Canada, 2007). ⑤ Japan The Energy Data and Modeling Center, *2009 Handbook of Energy & Economic Statistics in Japan* (The Energy Conservation Centre, Japan, 2009). ⑥ Korea: Ministry of Commerce, Industry and Energy & Korea Energy Economics Institute, *Yearbook of Energy Statistics 2009* (Korean Energy Economics Institute, Seoul, 2009). ⑦ Other countries: Energy Information Administration, *International Energy Outlook 2009* (EIA Publications, USA, 2009).

Table 16.5. Global comparison of fertilizer application and mechanization.

Country	Fertilizer application in per unit of area (kg/h m^2)	Nitrogenous fertilizer application in per unit of area (kg/h m^2)	No. of tractors in per unit of area (per 10^3 h m^2)	No. of harvesting-threshing machines in per unit of area (per 10^3 h m^2)
China	297.80	207.10	123.09	71.31
Bangladesh	181.21	134.48	0.36	—
India	133.06	85.00	18.56	2.81
Indonesia	101.99	67.24	0.14	2.95
Iran	79.29	48.87	17.01	0.61
Israel	400.00	147.50	62.50	0.00
Japan	317.91	115.57	400.21	204.05
Kazakhstan	5.42	2.62	1.78	0.84
Korea	398.35	174.18	134.07	46.70
Malaysia	192.36	47.04	0.00	0.00
Mongolia	5.17	5.17	3.45	0.86
Myanmar	10.95	3.10	0.64	1.73
Pakistan	161.89	122.93	21.30	0.09
Philippine	67.29	50.00	5.89	0.09
Thailand	117.30	69.10	46.63	11.80
Vietnam	301.68	183.24	18.21	24.92
Egypt	451.99	376.70	29.26	0.85
South Africa	40.61	28.01	4.01	0.70
Canada	73.00	51.71	14.06	1.55
Mexican	58.15	33.30	8.66	0.83
America	164.65	94.49	24.77	1.96
Argentina	60.79	33.24	8.61	1.69
Brazil	169.84	44.32	11.66	0.81
Venezuela	148.41	97.10	14.20	1.74
Czech	150.15	107.90	25.53	3.04
France	195.01	122.30	57.79	3.92
Germany	190.74	132.23	63.39	7.11
Italy	128.70	75.97	186.09	5.35
Holland	1,381.91	1,291.81	153.19	6.38
Poland	212.38	119.33	124.04	12.78
Russia	13.87	8.36	3.29	0.87
Spain	106.49	54.59	54.54	2.79
Turkey	82.41	60.28	39.05	0.49
Ukraine	31.84	21.92	10.10	1.74
Britain	267.99	174.39	76.64	6.92
Australia	36.67	15.97	6.34	1.15
New Zealand	291.50	81.52	22.58	0.88

Source: *International Statistics Yearbook 2009*. The table contains the data of 2007.

APPENDIX 1

Summary of Low Carbon Technologies

Electricity

1. Supercritical and ultra supercritical pressure unit.
2. Plasma ignition technique.
3. Circulating fluidized bed boiler.
4. Green coal power generation.
5. Third generation nuclear reactor.
6. Distributed energy.
7. Thin film photovoltaic cell.
8. Off-shore wind power generation.
9. Hot dry rock power generation.
10. Ultra-high voltage transmission line.
11. Smart grid.

Industry

12. Advanced coal selecting technique.
13. Coal bed methane mining.
14. Shale gas mining.
15. Gas hydrate mining.
16. Coke dry quenching.
17. Continuous casting.
18. Top gas pressure recovery turbine.
19. Waste plastic injection in blast furnace.
20. Direct reduction iron making process.

21. Third generation iron making technology.
22. Micro-wave, electric are and exothermic heating steelmaking technology.
23. Energy management system for still enterprises.
24. Large capacity preroaster for electrolytic aluminum.
25. New dry technique for cement production.
26. Net waste heat generation.
27. Cement unpackaged.
28. New type wall materials.
29. Ecological cement.
30. Ultra high performance concrete.
31. Low-E membrane plating glass.
32. Caustic soda production technique by ion exchange membrane.
33. Refining-chemical integration.
34. Forest-pulp-paper integration.
35. Net forming manufacturing.
36. Friction stir welding technology.
37. Remanufacturing technique.
38. Regenerated copper.
39. Regenerated aluminum.

Architecture

40. Green building.
41. Ecological city.
42. Ecological wall.
43. Factory made house technology.
44. Compact fluorescent lamps.
45. Light emitting diode.
46. High efficiency room air-conditioners.
47. Ground source heat pumps.
48. Rural biogas.
49. Intellectual lights.
50. Building automation system.

Traffic and transportation

51. High efficiency diesel vehicles.
52. Small-displacement vehicles.

53. Cellulose ethanol.
54. Pure electric vehicle.
55. Hybrid electric vehicle.
56. Electric bicycle.

Agriculture

57. Controlled release fertilizer by contained membrane.
58. Spread manure by measure the soil and make up a prescription.
59. Methods of exempt from tillage.
60. Straw returning cultivated land.
61. Penning raise livestock by fold feed.
62. Ecological agriculture.

Cross functional sectors

63. Permanent magnetic coreless motor.
64. Power-electronic technology.
65. Tri-network integration.
66. Internet of things.
67. Clouding computing.
68. Membrane technology.
69. Super-conduction technology.
70. Carbon capture and sequestration.

Electric Power

1. *Supercritical and ultra supercritical pressure unit*

A thermal power unit with steam pressure of boiler higher than 22.12 MPa is called supercritical unit. When pressure is at the critical value, there is no coexistence of steam and water between saturated water and saturated steam. A thermal power unit with steam pressure of boiler higher than 27.0 MPa is called ultra supercritical unit. Popularization and application of these units will greatly increase the power generation efficiency and reduce the pollutant emission. Thermal efficiency of ultra supercritical coal fired units in Denmark is up to 47%. Maximum capacity of ultra supercritical coal fired units in America is 1,300 MW. Steam parameter of ultra supercritical units newly built in 2000 in Japan is 31 MPa at 610°. In March 2007, the first domestic 1,000 MW ultra supercritical unit was put into commercial operation. Its main steam pressure is 25.25 MPa at 600°, thermal efficiency is 45% and net coal consumption rate is 288.5 gce/(kWh), 77.5 gce/(kWh) less than that in 2006.

By the end of 2008, 120 supercritical units had been built, 100 sets of 600 MW units were under construction and 90 units were planned to construct in China. By the end of 2009, there had been 21 sets of 100 MW ultra supercritical units in operation and 24 sets under construction.

The key point of research and development of ultra supercritical units is high temperature alloy steel. Japan is developing ferritic heat-resisting steel for 650°-level units. America is developing heat-resisting high strength steel alloy at 760° for ultra supercritical coal fired units in Vision 21 Program.

2. *Plasma ignition technique*

It is a kind of oil-free ignition and low load stable combustion technique for pulverized coal boiler with the advantages of saving oil and operation cost. Its operation principle is that plasma generator generates high-power plasma (arc, 50–150 kW for 220 t/h boiler) to directly ignite the pulverized coal with the central temperature up to 6,000°. Primary air is blown into plasma burner. After dense-thin separation, the dense pulverized coal enters arc core, is rapidly ignited within about 1/10s and stably burned; the thin pulverized coal is burned by burned flame after the secondary air is blown in. The burner sprays pulverized coal torch into boiler. When temperature of the furnace rises to a certain value, use main burner and start up the boiler. Until 2009, China had more than 440 large-scale coal-fired units (total capacity over 190 GW). This technique has saved 3 million tons of oil in total.

3. *Circulating fluidized bed boiler (CFBB)*

Its operation principle is that the adsorbent (limestone) is added into bed of combustion chamber and air is blown from bottom of furnace to suspend the bed for fluidized combustion. Fluidization forms the turbulent mixing condition to improve combustion efficiency; limestone desulfurization reduces emissions of SO_2; low combustion temperature (830–900°) greatly reduces NO_x. The high speed air with solid particles enters the boiler and returns to the furnace for auxiliary combustion to totally burn the pulverized coal.

Compared with common boiler, CFBB can save 10% of coal. Compared with the power station utilizing pulverized coal furnace and stack gas desulfurization device, CFBB can reduce 50% of SO_2 and NO_x with no need of stack gas desulfurization device.

China is capable of designing and manufacturing CFBB with capacity of 50 MW, 100 MW, 150 MW, and 300 MW. In 2008, there had been more than 130 sets of 50 plus MW CFBB in use; and the capacity of CFBB over 100 MW was up to 20 GW; there were 13 sets of 300 MW CFBB put into operation and 50 CFBB under planning or construction.

4. *Green coal power generation*

It refers to the coal-based power generation plant with nearly zero emission of pollutants and CO_2. China Huaneng Green Coal Power Generation Project adopts advanced technology with self-owned intellectual property right, including Integrated Gasification Combined Cycle (IGCC), pollutant recovery and carbon separation, utilization or sequestration. The technology is leading the clean coal technologies worldwide. In July 2009, the project was officially started in Tianjin Lingang Industrial District. In the same month, Future Fuels LLC (Pennsylvania, America) introduced two-stage dry feed entrained flow gasification technology independently developed through this project. This is the first time that Chinese coal gasification technology was imported by the developed countries. Its installed capacity is 250,000 kW, total investment is RMB 2.1 billion ($318 million) and it is planned to be put into operation in June 2011. Its efficiency will be 30% higher than that of the most advanced coal-fired power plant. Compared with the regular coal-fired power plants with the same capacity, it can save coal by 70,000 tons and reduce CO_2 emission by 200,000 tons. Its desulfurization efficiency will be up to 99.8%. Finally, near zero emission of pollutant and CO_2 will be realized.

Now, 12 IGCC co-production projects are proposed to be built in North, East, and South China. Three 400,000 kW commercial demonstration plants are expected to be built up in 2015.

5. *The third generation nuclear reactor*

The first generation nuclear reactors are the experimental reactor and the prototype reactor built by Soviet Union and America in 1950s. The second generation nuclear reactors are pressurized water reactor, boiling water reactor and heavy water reactor with capacity over 300,000 kW designed in 1960s based on the first generation nuclear reactor. At that time, the technical feasibility of nuclear power was proved and the economical efficiency of nuclear power can be competitive with thermal power. In 1980s and 1990s, in order to eliminate the adverse impact caused by major accidents of Three-Miles Island and Chernobyl nuclear power stations, the nuclear power field in the world was dedicated to research and making strategy on prevention of the major accident of nuclear power station and release of results. *Advanced Light Water Reactor Utility Requirement Documents* prepared by America in 1989 regulates that "as for the reactor with passive safety system, probability of core melt, radioactive substance leakage and other major accidents shall reduce 1–2 order of magnitude than previous reactors, the service life shall be no longer than 60 years, availability over 87%, building cycle 42–54 months; after 10 years' operation, its economical efficiency shall be 10% more than that of selective power plant and after 30 years, more than 20%." Now, the relatively matured third generation nuclear reactors are AP-1000 (America) and ERP (Europe).

6. *Distributed energy*

It is an energy system in which the power, heat energy (hot water and steam) and cold energy (cold water or air) are provided for users in certain area by small power generation units (several kW to 50,000 kW) installed in place near users.

Compared with regular centralized power supply, the distributed power supply has the following advantages: No or less loss in transmission and distribution; unnecessary of building transformer substation and distribution substation to reduce transmission and distribution cost; adjustable based on demand for heat or electricity to increase annual utilization hours; low cost for installation; independent, controlled by users, no possibility of large scale power supply accident, and high reliability; remote controlling and monitoring quality and performance of power; suitable for village, pastoral area, mountainous area, developing area, commercial area and residential area; high co-production efficiency, and reduction of emission of CO_2 and atmospheric pollutants (no or little CO_2 and atmospheric pollutants from power generation of renewable energy resources). Now, internal combustion engine, micro gas turbine and fuel cell are used for distributed power generation.

The distributed power generation sprang up from America in 1970s. At present, America mainly adopts it in hospital, factory, Internet server, and other places which may subject to great loss in case of the power failure. Europe, America and Japan are studying the distributed power generation and energy accumulation technology and micro grad technology. EU, Japan, and Canada have built the demonstration projects of micro grid.

7. *Thin film photovoltaic cell*

It is photovoltaic cell made of amorphous silicon, cadmium sulphide, gallium arsenide and CIS and other thin films as the basic material. Manufacture processes are glow discharge, chemical vapor deposition, sputtering and vacuum deposition etc. Plastic and glass are used for basic plate of thin film photovoltaic cell. If plastic cement is used as the basic plate, the cell can be flexible and collapsible. Thin film photovoltaic cell can save material and reduce the cost with wide application. If it is used in building, it is easy to blend in with roof or external wall. Thickness of new generation thin film photovoltaic cell is 2–3 μm, only 1/100 of crystalline silicon cell. Its conversion efficiency is 7%–8%, even up to 13%.

Power consumption for producing thin film photovoltaic cell is less than that for polysilicon cell. Its energy payback period (the period of time over which the energy savings of a product equal the amount of energy expended in production) is about one year, while that of the polysilicon cell is about five years.

From 2001 to 2009, global sales volume of thin film cells grew at the rate of 58% from 14 MW to 2141 MW. Market share of photovoltaic cells increased from 2.8% to 25%.

8. *Off-shore wind power generation*

It refers to wind power generation unit established off shore. The off-shore wind turbine has to endure corrosion of salt spray and collision of wave and tide and conquer the technical problems, such as load bearing of ocean floor, anti-draw and horizontal displacement prevention. Its cost is much higher than that of wind power generation on the land. The maximum installed capacity of offshore areas (50 m higher than the sea level and less than 25 m deep below the sea level) is 200 million kW. Shanghai Donghai Bridge off-shore wind power farm had been completed, where thirty-four 3 MW wind turbines were installed, and the total capacity reached 102,000 kW. Its cost per kW is RMB 20,000 ($3,030) while if installed on the land the cost per kW is only RMB 8,000 ($1,212); its on-grid electricity price is RMB 1 yuan ($0.16)/(kWh) while the price of wind power generation on the land is RMB 0.5 yuan ($0.076)/(kWh). Jiangsu Province expects that installed capacity of off-shore wind power in 2020 will be 7 million kW. Until 2009, Europe had built 38 off-shore wind power farms with total installed capacity over 2 million kW. The 5 MW off-shore wind turbine technology is relatively matured. At the beginning of 2010, Germany built the first abyssal sea wind power farm with the installed capacity of 60,000 kW, about 50 km from the coast line. Germany expects that installed capacity of off-shore wind power in 2030 will be 25 million kW. In October 2010, the UK built the largest off-shore wind power farm in the world with 100 wind turbines and total installed capacity of 300,000 kW.

9. *Hot dry rock power generation*

It means utilizing the underground hot dry rock buried underground less than 5,000 m deep where the temperature is higher than 200° for power generation. The hot dry rock is a kind of geothermal resources which contain more energy than hot water, steam and geo-pressured geothermal resource. Hot dry rock power generation is an important technology for reducing emission of CO_2 in energy field. Drilling technology of Potter Drilling Company in California, America can drill through quartzite and other hard rock and drill into the depth where the rock temperature reaching as high as 200°. In August 2010, the drilling depth reached 300 m. The drilling depth of middle-term objective is 1,000 m and the final objective is 10,000 m.

10. *Ultra-high voltage (UHV) transmission line*

In accordance with the GB grid voltage standards, AC standard voltage 1,000 kV (maximum voltage 1,100 kV of the equipment) and rated DC voltage ±800 kV are called ultra-high voltage. If the distance is long and capacity is large, the ultra-high voltage transmission line can reduce the line loss. Transmission power of 1,000 kV AC can be up to 4–5 GW, 4–5 times of transmission power of 500 kV. Theoretical line loss is only 1/4 of 500 kV.

In August 2006, the construction of Southeastern Shanxi–Nanyang–Jingzhou 1,000 kV ultra-high voltage AC transmission test demonstration project was started and on January 6, 2009, it was formally put into operation. In December 2007, Sichuan–Shanghai ±800 kV ultra-high voltage DC transmission demonstration project, the highest DC voltage level in the world was started construction with the rated transmission power of 6.4 GW and total length of 2,000 km and it was completed on November 13, 2009. On December 28, 2009, the first Yunnan–Guangdong ±800 kV ultra-high voltage DC transmission demonstration project in the world was put into operation with rated transmission power of 5 GW and length of 1,373 km.

11. *Smart grid*

After the great power failure accident of America and Canada in 2003, American scientists put forward a new concept "smart grid". The smart grid is combination of information technology and power sector and is a symbol of new energy-technology revolution in the 21st century. As for smart grid, power exchange and application are through the Internet. All power generation and transformation devices, and terminal electrical equipment and devices are equipped with electronic chips. Advanced communication, information and control technologies are adopted to realize informatization, digitization, automation and interaction of grid so as to greatly improve the capacity of grid resource optimal configuration, guarantee safe and reliable operation of large grid and solve the problem of grid-connection of renewable energy. When the grid has severe disturbance or failure, the gird can automatically heal to effectively prevent the large power failure accident. Electricity consumed by users at different hours can be automatically priced. The grid will carry out precision management to save power and improve terminal power consumption efficiency.

American government regards the smart grid as the core project of national energy strategy. The "Economic Recovery Plan" issued in February, 2009 indicated that American government would invest RMB 26.4 billion ($4 billion) to install 40,000,000 smart meters for households within three years to promote the grid modernization. In May 2009, 16 smart grid industrial standards (first batch) were issued and the smart grid project was formally started. America built the first smart gird demonstration city in Colorado. State Grid Corporation of China puts forward the concept of "firm smart grid" and now the pilot work has been started. It is predicted by experts that until the completion of smart grid in 2020, energy saved per year can be up to 190,000,000 tce.

Industry

12. *Advanced coal preparation technology*

It mainly refers to the dense medium coal preparation technology. The technology utilizes dense medium suspension mixed by magnetite powder (its density between coal and waste rock) to separate coal and waste rocks or other impurities. This coal preparation technology is characterized by high separation efficiency, strong adaptability to coal quality, convenient operation and easy self-control. Wash ability of raw coal in China is somewhat lagging behind (uneasy separation between coal and waste rock), so it is significant to develop dense medium coal preparation technology. China has made large three-product dense medium cyclone with treatment capacity of 450–550 t/h and annual treatment capacity per product can be up to 2.1 million tons. At present, productivity of dense medium coal preparation nationwide is up to 150 Mt/a. Proportion of dense medium coal preparation to raw coal steam preparation is 44% in China, 56% in America and 90% in Australia and in South Africa.

13. *Coal bed methane mining*

Coal bed methane is a kind of self-reserved unconventional natural gas existing in coal bed in absorbed or free-state and its methane content is over 90%. It is clean energy and green house gas. Methane from mine is dangerous because of its explosive feature. Therefore, it is a great hidden danger for safe production in coal mine. Total resource quantity of coal bed methane in the world is up to 260 trillion m^3. China has 36.8 trillion m^3 equal to total resources of conventional gas on the land. Coal bed methane can be extracted through underground drilling or ground drilling. Development and utilization of coal bed methane is significant for preventing gas accident, increasing clean energy and reducing the greenhouse gas. In 2009, coal bed methane product in America was up to 70 billion m^3 and in China 7.46 billion m^3 (underground 6.45 billion m^3 and on the ground 1.01 billion m^3).

Important points of research and development are geological theory and resource evaluation of coal bed methane and advanced exploitation and mining technology.

14. *Shale gas mining*

It is a technological process of mining the shale gas from stratum to ground. Generally, drill on the proved gas field, and lead the gas flow; the shale gas will spray to well head under the pressure of stratum.

The shale gas is a kind of unconventional natural gas in argillutite in absorbed or free-state. Predicted reserve of shale gas in the world is up to 456 trillion m^3. In America, it is more than 28 trillion m^3, among which the proved recoverable reserve is about 14 trillion m^3. Horizontal drilling and hydraulic fracturing technology break through the technical barrier in shale gas mining. Output of shale gas was 22.7 billion m^3 in year 2004, and 90 billion m^3 in 2009, accounting for 15% of total output of natural gas (593.4 billion m^3).

The recoverable reserve of shale gas in China is about 26 trillion m^3. Ministry of Land and Resources predicts that recoverable reserves in 2020 will be 1 trillion m^3 and the productivity will be 15 billion m^3/a-30 billion m^3/a.

15. *Gas hydrate mining*

Gas hydrate is a kind of ice type solid crystal composed of natural gas and water molecules, and also called methane hydrate or flammable ice. It exists at low temperature (0–10°C) and high pressure (1–10 MPa). It is mainly in marine bottom sediment at the depth of more than 300–500 m or in cold high latitude area (especially in ever-frozen layer). 1 m^3 saturated gas hydrate can release 164 m^3 of natural gas. The global reserve of gas hydrate is double of total reserve of coal, petroleum and natural gas. At present, many countries including America, Japan, China, India, Germany, Russia, Netherlands, and Brazil are studying the gas hydrate. In 2007, samples were collected in the continental slope of the South China Sea.

The content of methane in global gas hydrate is 3,000 times of methane in atmosphere. Exploitation of gas hydrate confronts with huge environmental risk. In exploitation, once the methane leaks to atmosphere, it will not only cause disaster to global environment, but also evoke sea slide and other geologic hazards. This point is the most difficult technical problem to be solved before exploitation of gas hydrate.

In 2010, Japan carried out the production test and planed to put it into commercial production in 2018. China plans to carry out the production test in 2020.

16. *Coke dry quenching, CDQ*

The principle of coke dry quenching technology is that in closed device, inert gas-nitrogen gas is used as the heat carrier to quench the red coke, and heat energy of high temperature nitrogen gas is utilized to generate steam for power generation. The coke dry quenching device is generally composed of coke quenching groove, exhaust heat boiler, power generation equipment, lifting equipment, belt conveyor, nitrogen gas circulating system, and dust removal system. The whole process can be divided into material flow, nitrogen gas circulation and steam thermodynamic cycle. About 1,500 Nm^3 of nitrogen gas is needed for quenching 1 ton of red coke. The coke is cooled below 250°C in general. Compared with coke wet quenching, it can recover the physical sensible heat of red coke. Recoverable steam per ton of coke is 500–600 kg; it can save energy 40 kgce per ton of red coke, greatly reduce the discharge of coke quenching water and other pollutants, and improve the coke quality. In 2009, popularity rate of CDQ (proportion of CDQ treatment quantity to coke output) in China steel industry was up to 70%.

17. *Continuous casting*

It is a production process of direct casting of billet by continuous caster. It has the following advantages: First, simplify the production process: eliminate the die casting, blooming and cogging procedure; not only reduce the energy consumption, but also shorten the billet forming period. Second, improve the metal making efficiency by 8%–10%, from 84%–88% (die casting) to 95%–96% (continuous casting). Third, save energy; energy for production of 1 ton continuous casting billet is 30 kgce less than billet produced by die casting; adding the energy saved by improving billet production rate, the comprehensive energy consumption per ton can reduce 100 kgce–110 kgce. Fourth, improve working condition and easy for realizing automation. Fifth, quality of billet is good because the cooling speed of continuous casting is fast, and continuous throwing and casting condition are controllable and stable, the internal tissue of billet is even and compact with little liquation. In 2009, continuous casting ratio of large and medium steel enterprises in China was up to 99%.

18. *Top gas pressure recovery turbine, TRT*

It is a device utilizing the top gas pressure for power generation. Heat value of blast furnace gas is 3,350 kJ/Nm^3. It is generally used as fuel without full utilization of top pressure. Blast furnace gas is transmitted to pipe network after decompressed with pressure reducing valve. As to top gas pressure recovery power generation, the blast furnace gas with certain pressure (over 9.8×10^4 Pa) passes through expansion turbine generator to generate power, and then utilize its heat energy. In order to keep the stable production of blast furnace, the expansion turbine with good adjustable performance can automatically switch to operation of generator when blast furnace is lack of gas. Before entering turbine, gas shall be treated through multistage dust removal process to control the dust content below $5\,mg/Nm^3$ to reduce abrasion and avoid jam of turbine. In 2008, 98.6% of blast furnaces larger than $1{,}000\,m^2$ in steel industry were installed with TRT.

19. *Waste plastic injection in blast furnace*

It is a blast furnace iron-making process using waste plastic instead of coke or pulverized coal. Heat value of waste plastic is equal to that of coal. Waste plastic contains no sulfur and less ash. Effective availability of waste plastic injected in blast furnace is up to 80% and the discharge amount is 0.1%–1.0% of incineration amount. According to energy conservation plan of Japan in 2010, waste plastic injected in blast furnace will be over 1 million tons. It is regarded as a clean production technology to be promoted for application in China. Hard waste plastic is directly crushed as pretreatment and soft waste plastic is melt for pelleting. 1 ton of waste plastic is equal to 1.2 ton of pulverized coal. An iron and steel plant with annual capacity of 10 million tons can use 200,000 tons of waste plastic per year.

20. *Direct reduction iron making process*

It is an iron-making process by direct reduction of pellet, lump ore or fine ore in solid state and production of reduced iron (also called sponge iron) in smelter under the action of reducing agent (natural gas, gas or non-coke pulverized coal). At present, COREX manufactured by Siemens VAI Manufacturing Co. Ltd., adopts the process. Baosteel Group introduced COREX. In 2007, pig iron produced from direct reduction and smelting reduction in the world was over 65 million tons. Direction reduction iron making in China is still at the beginning stage. In 2008, the output was only 600,000 tons.

21. *Third generation iron making technology*

It is a kind of iron-making technology beyond blast furnace through fast reduction, carburetion, smelting, and slag separation at relative low temperature. Make pulverized iron ore and pulverized coal as balls, put the balls into furnace and raise the temperature to 1,450°C to separate iron and impurities so as to make slab. The whole process only takes 10 minutes. Iron content of the product is up to 97% higher than that of iron making in blast furnace (96%). The technology can treat the low class iron ore and coal which cannot be treated by blast furnace. Its cost is only 15% of that of blast furnace. The KOBE Steel Group (Japan) and a steel plant in America prepare to build the equipment with annual output of 500,000 tons.

22. *Micro-wave, electric arc and exothermic hearing steel-making technology*

It is a kind of steel-making technology utilizing the heat generated by micro-wave and electric arc plus the heat released in reaction of coal for smelting after the iron ores, the coal and limestone are added into furnace. This new technology can replace the sintering, coking, blast furnace and converter in traditional steel-making technology. It is characterized by short smelting period, less energy loss, low cost, low requirement for coal, great reduction of emission of green house gas and saving 25% of energy. This technology is successfully developed in America.

23. *Energy management system for steel enterprises*

The system can guarantee the stable energy supply for production, monitor the state of energy consuming equipment and realize the centralized management and automatic operation of the energy consuming equipment through energy optimal dispatch and on-line real time monitor of energy consumption through computer network in production of enterprises. Steel industry in China plans to build energy management system in steel enterprises with the annual capacity of 3 million tons within three years (started from 2009) to save 6 million tce per year.

24. *Large capacity preroaster for electrolytic aluminum*

It is a high efficient electrolytic aluminum process. In production of aluminum, from mining, beneficiation, alumina smelting, aluminum electrolysis to aluminum product machining, aluminum electrolysis is a procedure with greatest energy consumption (DC). The principle of aluminum electrolysis is that direct current passes through dielectric composed of alumina (raw material) and cryolite (fluxing agent) to decompose the alumina into aluminum and O_2 at the temperature between 950°C and 970°C; Aluminum liquid separated out in cathode is collected at the bottom of electrolytic cell and CO_2 and CO are separated out in anode; after purification, the aluminum liquid is casted into aluminum pig. Preroaster is an anode cell and the cathode is in electrolytic cell. The large capacity preroaster generally refers to the preroaster with the current strength over 140 kA. Compared with 60 kA self-roaster, 300 kA large capacity preroaster can reduce 2,000 kWh power consumed by one tons of aluminum. In 2007, China thoroughly rejected self-roaster. In 2009, output of 160 kA and above large capacity preroaster was up to 90%.

25. *New dry technique for cement production*

It is also called external decomposition kiln for cement kiln. The suspension preheated kiln equipped with decomposition kiln is a new process for cement production developed in 1970s. The principle of the new process is that drying and preheating procedure (in rotary kiln in traditional technology) are carried out in suspension preheated and decomposition reaction of material is carried out in decomposition kiln; cement kiln only carries out the reaction with little heat consumption, so thermal load of the cement kiln is greatly reduced. The decomposition kiln is installed in kiln end and equipped with fluidized bed burner which changes the situation of low efficiency, i.e., flame contact with surface of material, to realize the grading utilization of energy. Compared with wet kiln, the external decomposition kiln for cement kiln with the same diameter can reduce half of the heat consumption and greatly increase the output. China has many external decomposition kilns for cement kiln with daily output of 2,000 t, 4,000 t, 5,000 t, 8,000 t, and 10,000 t. In 2009, output from new dry technique was accounting for 72.5% of clinker output.

26. *New waste heat generation*

It is a technology utilizing waste heat of the new dry method cement kiln for power generation. Waste heat boilers are equipped in kiln head and kiln end to recover the waste heat. During recovery, requirement of production process shall be met in priority; except for heat required by grinding coal and drying ground material, residual waste heat is recovered by waste heat boiler to generate steam. Generally, the waste heat boiler at the kiln end directly generates the superheated steam to supply to steam turbine for power generation. If the waste heat boiler at the kiln head is equipped with back heating system, it can directly generate the superheated steam; if not, it can generate saturated steam and superheated water to deliver to boiler at the kiln end. As to new dry cement kiln waste heat power generation system with daily output of 2,000 t, installable capacity is 3,000 kW and annual power generation quantity is about 16.20 million kWh. In 2008, its installed capacity in China was 1.662 million kW and annual power generation capacity was 11.2 billion kWh. In 2009, the newly added installed capacity was 1.677 million kW. Waste heat power generation technology researched and developed by Conch Group, the largest cement production enterprise in China (output of 81.36 million tons in 2008), is in the leading position in the world. Its power generated by waste heat can supply 60% of the power this process needs to consume.

27. *Cement unpackaged*

It means that the cement is directly transported to construction site by special purpose truck after production without package. If 100 million tons of cement is transported in such way, 2 billion of packaging paper bags can be saved so as to save 330 m^3 of excellent quality wood, 120 million m^3 of water and 800,000 tons of coal used for producing paper bags; besides, cement loss caused by paper bag damage or residual in paper bag is 5 million tons. Therefore, cement unpackaged method can save 2.37 million tce. In 2009, cement unpackaged rate in China was 46.3%. Generally, the unpackaged cement is mixed with mortar before delivery and is transported to construction site by special purpose truck. Ready mixed mortar can reduce the cement consumption. Computer is used to control composition proportion of mortar to guarantee the quality of building, avoid noise pollution on construction site due to mixing and prevent dust pollution caused by storage in the open air.

28. *New type of wall materials*

They are new wall materials to replace the traditional clay solid brick. The materials include three categories (20 types), i.e., sintered hollow products (hollow brick, aerated concrete, concrete block etc.), sintered products with industrial waste residues (coal waste rock, grey powder etc.) and sludge (sand) in river and lake as main material, and light wall boards (PS plastic foam board, rock-wool board, glass wool board and gypsum board etc.). Compared with clay solid brick, the new type of wall material has the following advantages: Light, good performance, low energy consumption, fast construction and avoid of damaging farm for borrowing soil. Energy for production of new type wall materials is 40% lower than that of clay solid brick. In 2009, output of new type wall materials in China was accounting for 52% of total output of wall materials.

29. *Ecological cement*

The ecological cement is hydraulic binding material made by municipal solid waste incineration ashes, sewage sludge, and limestone as main raw material after burning and grinding and cement with high content of industrial solid waste. The ecological cement is good for protecting ecological environment, reducing energy and resource consumption and reducing emission of pollutants. Output of cement utilizing municipal solid waste incineration ashes in Japan is accounting for 30% of total output of cement. In 2008, building material industry of China utilized coal ash and coal waste stones and other solid wastes in total of 570 million tons and began to produce ecological cement by sludge and solid waste incineration ashes.

30. *Ultra high performance concrete*

Its performance is far beyond the performance of common concrete. It is strengthened by steel fiber instead of reinforcing steel. Compared with common concrete, its compressive strength is 6–8 times higher, rupture strength is 10 times higher and fire resistance is 100 times higher; besides, it has good insulation performance. Under the condition of guaranteeing certain strength, it can be very thin in various colors and shapes like sculpture. If C40–60 concrete for high buildings in China is replaced by C110–137 ultra high performance concrete, 30%–70% of the cement and 15%–25% of the steel can be saved. The concrete is the patented product of Lafarge Group (France). Now, it has been commercialized.

31. *Low-E membrane plating glass*

It is a kind of glass covered with one layer or many layers of membranes composed of silver, copper, tin or their compounds. The glass has high transmissivity to visible light and can reflect 80% of infrared rays radiated by indoor objects to keep infrared rays in the room. Therefore, it has good heat radiation prevention function to avoid pollution of reflected light. In 2009, output of the new type energy saving glass in China was 28 million m^2. It is predicted that the output will be up to 50 million m^2 in 2010. Residential buildings in Dongguan of Guangdong use the glass to save 60% of power consumption per year.

32. *Caustic soda production technique by ion exchange membrane*

It is a process of producing high purity caustic soda, Cl_2 and H_2 by utilizing iron exchange membrane and electrolyte solution. Crude salt is added into anode chamber of electrolytic cell after hydration and refinement; ion membrane between anode chamber and cathode chamber can select specified ion to pass so as to produce high purity soda, Cl_2 and H_2. The ion exchange membrane has the characteristic of repelling anion and attracting cations. In electrolysis, sodium ions with positive charge in anode chamber enter cathode chamber through the ion membrane and combine with OH^- generated by dissociation of pure water in cathode chamber to produce NaOH (caustic soda); meanwhile, H_2 is released from cathode and Cl_2 is released from anode. Compared with diaphragm method, the caustic soda production technique by ion exchange membrane can save 28% of comprehensive power consumption; its advantages is high efficiency, less land occupation, low investment per unit (reducing 25%), stable production and no pollution. In 2009, output of caustic soda made by ion exchange membrane method in China was accounting for 54.7% of total output of caustic soda.

33. *Refining-chemical integration*

It means one enterprise carrying out both the oil refining and the chemical production to fully show the idea of cyclic economy. The Characteristics of the mode is intensive, short in process, safe and environmentally friendly. The production devices are connected by pipelines. Storage tank and vehicle are not used. The raw material can be supplied by each other and the comprehensive utilization rate is high; all fuels are gaseous fuels after desulfurization and purification treatment; waste heat is fully used. Therefore, utilization ratio of raw material and energy is high and pollutant is little. The largest refining-chemical integration project in China was put into operation in November 2009 in City Quanzhou. Its productivity increases to 12 million tons from 4 million tons. It produces 0.8 million tons of ethane, 0.65 million tons of polyphenylene ethylene, 0.4 million tons of polypropylene, and 1 million tons of arene per year.

34. *Forest-pulp-paper integration*

It means to make an integration of forestation, forest management, forest cutting, pulping and paper making. This mode can promote the development of artificial timber forest (fast-growing and high-yield plantation), alleviate shortage of wood pulp in China and improve ecological environment. Compared with grass pulp paper-making process, wood pulp paper-making process is helpful for saving water, saving energy and centralized pollution control. Water consumed with traditional paper-making technology in China is up to 100–200 m^3/t. COD discharge of paper-making industry is accounting for 1/3 of all industries. Gold East Paper in Jiangsu built the forest-pulp-paper integration cyclic economy system with annual productivity of art paper exceeding 1 million tons; the water consumed per ton is only 10.7 m^3, equal to 18% required by GB standard, and the COD discharge is less than 1/6 required by GB standard. Ningxia Meili Paper Industry Co., Ltd. carried out forest-pulp-paper integration project in 2000. It cultivates 500,000 mu of fast-growing forest and produces 300,000 tons of coated white cards per year.

35. *Net forming manufacturing*

It is an advanced material-saved manufacturing technology. If the technology is adopted, the parts can be mechanical component with little machining or no machining treatment after forming. At present, steel consumed in mechanical industry per year in China is about 50 million tons, but utilization ratio of the steel is only 60%–70%. Utilization ratio of advanced technology in the world can be up to 90%–95%. The potential for steel saving in China is great.

36. *Friction stir welding technology*

It is a dead joint method under the mechanical force and friction heat. Friction heat between stirring head and material to be welded makes the material thermoplastic and the surface layer of the work piece forms dead joint under the friction heat and forge. It has been used for jointing and welding of aluminum alloy, copper, zinc, magnesium, and other light metals to replace the traditional fusion welding technology. This welding technology is applied in manufacturing of light structures of space air vehicle, high speed ship, high speed train, and vehicle. The technology eliminates air holes, crack sand other defects caused by fusion weld and improves the joint quality. Its limitation is that the welded equipment shall be rigid, it is lack of welding flexible feature and weld seam may have keyhole. At present, friction stir welding technology is tried to be applied in titanium alloy, stainless steel, common carbon steel and other material with high melting point.

37. *Re-manufacturing technique*

It means utilizing the parts and components of waste mechanical equipment for high-tech repair and performance upgrading in batch so as to reassemble them into products. As to technical performance and quality, the re-manufactured equipment can be same with or surpass the new products. Cost of re-manufactured equipment is only 50% that of the new product with energy conservation of 60% and material conservation of 70%. The re-manufacturing technique is applied in vehicle, engineering, household appliances, office equipment, and other fields. The re-manufacturing technique develops rapidly in China. China develops key techniques with self-owned intellectual property right, such as plasma and high speed arc spraying, nm electro chemical machining and nm self-repairing addictive. In 2010, China can re-manufacture 230,000 sets of vehicle engine, transmission case, steering engine and generator.

38. *Regenerated copper*

It refers to copper produced by recovered waste copper. Compared with production of virgin copper, the regenerated copper can save energy and reduce emission of pollutants. Pure waste copper can be smelted in induction furnace. Regeneration of copper with impurities adopts electro refining process of reflection smelting furnace. In 2009, output of regenerated copper was accounting for 34% of total output of copper in China, 60% in America and 80% in Germany. Energy consumed per unit of regenerated copper is 55% that of virgin copper (production of virgin copper includes mining, preparation and smelting). Compared with virgin copper, production of

10,000 tons of regenerated copper can save water 730 m^3 and reduce CO_2 1,400 t and solid waste 420 tons.

39. *Regenerated aluminum*

It refers to aluminum produced by recovered waste aluminum. At present, in America and Germany, output of regenerated aluminum is accounting for over 50% of total output of aluminum. Production of regenerated aluminum can save much energy.

The technology of regeneration and utilization of waste aluminum in China is mainly the technology of single chamber reflection smelting furnace. In 2009, output of regenerated aluminum in was accounting for 18% of total output of aluminum in China and 1/3 in developed countries. Energy consumed per unit of regenerated aluminum is 3.7% that of virgin aluminum (production of virgin aluminum includes mining, preparation and smelting). Regenerated aluminum is mainly applied in manufacture of vehicle, motorcycle, agricultural machine, and machining of aluminum profile.

Building

40. *Green building*

It refers to the building saving resources (energy, land, water, and material) at the utmost, protecting environment, reducing pollution, providing healthy, proper and highly efficient space and staying harmonious with the nature. It is also called sustainable building, ecological building or energy saving building.

41. *Ecological city*

It refers to the city which achieves harmonious coexistence and development between human and nature as well as human and society, and the sound operation of artificial-ecological combined system shaped by the technical means and cultural mode with ecological features. Its key indicators are environmental quality and natural wetland; proportion of green building and public green space per capita; daily water consumption per capita, domestic wastes per capita and proportion of green travelling; recovery and utilization ratio of wastes, harmless treatment ratio of hazardous wastes and domestic wastes, coverage ratio of municipal pipe network, proportion of communities with free sports facilities within range of 500 m and ratio of the disables friendly facility.

42. *Ecological wall*

It is the external wall of building featured of energy saving and environmentally friendly. It has three layers. The external layer is photovoltaic battery plate and glass, middle layer is sealed cabin and the internal layer is water curtain glass. In winter, the sealed cabin is closed, sunshine irradiated on external layer is converted to electricity, heat generated by radiation of sun and photovoltaic battery heats up the air in closed cabin for warming the room and the water curtain plays the role of cooling down and moisture preservation.

43. *Factory made house technology*

It is a house building technology in the form of prefabricating in factory and assembling on site. Generally, light steel structure and light building material are used with the advantages of anti-seismic, saving material, energy and land, environmental protection, and few construction waste. Factory made house technology of China has three systems for living buildings, hospitals, schools and culture facilities lower than six floors, for middle and high grade living buildings lower than three floors and for living building in rural area with anti-seismic and energy saving function lower than two floors at the cost between RMB 700 ($106)/$m^2$ and RMB 1,500 ($227)/$m^2$. Only one month is demanded for constructing a two-floor living building covering 200 m^2. Its construction period is 1/3 of the construction period of masonry building. Compared with traditional house building method, this technology can save material 30%, energy 70%, and land 20%. In 2010, Beijing started the green low-carbon policy-based house project. In 2010, 500,000 m^2 was constructed and in 2011, 1,000,000 m^2 will be constructed.

44. *Compact fluorescent lamps, CFL*

It is also called energy-saving lamp. It is a new type of high efficient light source with luminous efficiency of 801 m/W and service life of 5,000–10,000 h. Compared with common incandescent lamp, its luminous efficiency is 5–7 times higher, saved electricity is 70%–75% higher and service life is 8–10 times longer. Because of its good light effect, strong color rendering property, small volume, compact structure and convenient application, it is an ideal light source to replace incandescent lamp.

CFL is a kind of low-voltage mercurial vapor discharge lamp. Luminous tube is made of special-functional glass. Both ends are equipped with ligaments which are covered by electronic powder to send electron. The tube is aerated with little mercury and inert gases and the tube wall is covered with rare earth tricolor phosphor powder. The tube is integrated with ballast without starter. The products are various in shape, such as U, D, screw, ball and ring shape and equipped with electronic or inductance ballast. Its illumination principle is basically the same with that of fluorescent lamp. After power on, electrode sends out the electron to impact mercury atom, generate ultraviolet radiation and impact phosphor powder to generate the visible light.

CFL is applicable to illumination in households, hotels, shopping malls, schools, office buildings and public buildings. China is the largest country in the world producing and exporting CFL. The output of CFL increased from 0.2 billion in 1997 to 3.65 billion in 2009 among which 2/3 are exported. The output of CFL in China is accounting for 80% of total global output.

45. *Light emitting diode, LED*

It is also called semiconductor lighting lamp. It is a kind of semiconductor solid light emitting device. Its light emitting principle is that the semiconductor chip is used as the light emitting material; when both sides of chip is powered with forward voltage, holes in P–N junction of semiconductor are combined with electrons to cause photon emission, which directly converts electrical energy to luminous energy. Its advantages are as follows: High luminous efficiency (100 lmw), i.e., electricity consumed is 80% less than incandescent lamp and 50% less than CFL; long service life which is over 50,000 hours, 1,000 times of incandescent lamp; free of ultraviolet, infrared ray, mercury, xenon and other harmful substance; 90% of electricity energy is converted into visible light with little heat generation, while 20% of electricity energy of incandescent lamp is converted to luminous energy, and the other 80% is converted to heat energy; low voltage and current; safe; no lamp line, no glass bulb, maintenance-free and no afraid of lost. Its disadvantages are stroboflash in AC driving, too bright, high brightness in only small angle and high price.

At early stage, light colors of LED were red, yellow, green and blue, and LED was used as indicating light source of instruments; later, it was applied as traffic signal lamp, vehicle signal lamp and large display screen. In 1998, white light LED was launched. At present, greatest power of the white light LED is 60 W. It is an important breakthrough of illumination technology. Half of incandescent lamps sold in America are 60 W. Sales volume is up to 425 million. LED industry in China saw an explosive growth. In 2008, its output value was RMB 70 billion ($10.6 billion) and in 2010, it is predicted to exceed RMB 150 billion ($22.7 billion). In 2008, LED applied in landscape architectures in domestic market was accounting for 28%, display screens of household appliances 27% and back light of computers and mobile phones 22% and the rest LED applied in traffic signal lamps, vehicle lamps and special illumination etc.

Now, there are 3,000 LED enterprises in domestic market, of which 70% are in the downstream of industry chain. 80% of power chips and devices are imported.

46. *High efficiency room air-conditioners*

It is a kind of energy-saving air-conditioner with higher efficiency than common standard air-conditioner. High efficiency air-conditioner adopts compressor, frequency conversion compressor, condenser, evaporator and high efficient fan motor with proper configuration and low power consumption. It adopts overheating control technology (electronic expansion valve) to regulate flow of refrigeration agent for fuzzy control. The variable frequency air-conditioner can save about 30% of energy than constant speed air-conditioner. In 2009, market share of variable frequency air-conditioners was up to 17%.

47. *Ground source heat pumps*

The heat pump is a kind of device to transmit heat from low temperature heat source to high temperature heat source through consuming high quality energy (mechanical energy, electricity energy, and heat energy etc.) as compensation. Because the heat pump can transmit low temperature heat energy to high temperature heat energy and improve the effective utilization ratio of energy, it is an important method to recover low temperature waste heat and utilize the energy existing in environmental medium (underground water, surface water, soil, and air). At present, it is applied in heating, air conditioning, distillation, evaporation, and drying. Heat pump can be divided into compressed heat pump and absorption heat pump based on its form of energy consumption. The compressed heat pump utilizes certain secondary refrigerant (freon, ammonia, or water) to absorb the heat to vaporize at low pressure and release heat by compressor after pressure increase and condensation; the absorption heat pump replace the compressor with absorber, generator, and solution pump.

Economical efficiency of heat pump is measured by heating load by consuming unit quantity of work, and is called heat supply coefficient, i.e., heating load/quantity of work consumed by heat pump. If the heat supply coefficient is four, it means that one heating load can generate four times of heat energy.

Ground source heat pump is a kind of heat pump system of underground heat exchanger which treats the superficial terrestrial heat. The superficial terrestrial heat is from soil, sand and underground water. The pipe of heat exchanger is vertically buried and the drilling depth is less than 50–100 m. Heat supply coefficient of ground source heat pump is 3–4 in general. Heating by ground source heat pump can save 20%–50% energy than coal fired boiler. Cooling by ground source heat pump can save 10%–20% energy than water chiller. In recent years, ground source heat pumps in China have developed rapidly. In 2009, 80 million m^2 of building area were heated by ground source heat pump.

48. *Rural biogas*

Biogas is a kind of combustible gas formed by biomass (feces of human and livestock and agricultural and industrial organic waste) decomposed by microorganisms at the anaerobic condition. It includes 60%–70% of CH_4 and heat value is about 5,500 kcal/m^3(1cal = 4.1868 J).

Popularization and application of biogas in rural area in China make great process. Biogas generating pits increase to 35.07 million in 2009 from 7.64 million in 2000. The output of biogas increases to 13.4 billion m^3 from 2.6 billion m^3. In 2009, there were 39,500 waste biogas generation devices with output capacity of about 2 billion m^3. In 2009, the output of biogas in rural area was 15.4 billion m^3, equal to 12.10 million tce.

The development and utilization mode of rural biogas is from just producing energy to comprehensive utilization by combining domestic sewage purification treatment and aquatic breeding and cropping so as to form many modes suitable for various rural areas. It becomes an important approach for increasing farmers' income, improving environmental sanitation and developing ecological agriculture. In 2005, there were

1.486 million hm^2 of planting area with seed soaked in biogas liquid and the grain output was increased by 996,000 tons 18.12 million pigs were fed with the biogas liquid and 416,000 tons of fodders were saved; 102,000 hm^2 of biogas liquid was used for fish farming and the output was increased by 81,000 tons; 10.89 million m^2 of biogas residual was used for planting mushroom and 20,200 tons of mushrooms were produced. Biogas is also used for fruit preserving and grain storage effectively reducing rot of fruit and grain.

Chinese Government powerfully supports development of rural biogas. From 2002 to 2009, Chinese Government had invested RMB 19 billion ($2.87 billion) in total to subsidize 14.06 million households using biogas, 13,000 breeding communities and joint household biogas projects, 1.77 million large and medium breeding farms and 63,600 service outlets of rural biogas. In 2009, the farmer who builds the biogas generating pit can obtain the subsidiaries of RMB 3,000 ($455) and RMB 4,500 ($682) in Tibet. Biogas generating pits in rural area not only can generate biogas, but also can generate 385 million tons of biogas liquid and residual waste. If the biogas replaces core wood, then it is equal to annual accumulation quantity of 110 million mu of forest. Farmers' income can be increased by RMB 15 billion ($2.27 billion).

49. *Intellectual lights*

It refers to the intellectual control of lighting devices by means of operation of distributed wireless remote control, remote control and communication light control system composed of technologies of computers, wireless communication data transmission, spread spectrum carrier communication by power line, computer intellectual information processing, energy-saving electric control and other technologies. It has the following functions: Automatic regulation of indoor brightness, automatic switch of lighting devices to balance the operation time of devices, timing control, automatic time delay, as well as memory of power failure and locking of ON/OFF state. All the functions guarantee the safeness, energy saving effect, high efficient performance and comfortableness in use.

50. *Building automation system, BAS*

It is an important component of building's intellectual information system. With computer technology and network system, it conducts centralized management and automatic control for the ventilation, air-conditioning, refrigerated water, heating supply, water supply and drainage, illumination and power systems. It can save 25% energy and 50% man power and improve the work efficiency.

Traffic

51. *High efficiency diesel vehicles*

It adopts high efficient combustion engine which is mainly used for carrying cargo. At present, the efficiency of its engine is up to 40%–50% and can be increased to 55% in the future. The high efficiency diesel vehicles adopt technologies including advanced insulation shell, high pressure jet fuel, turbocharged engine, high strength light material, and friction and weight reduction technology.

52. *Small-displacement vehicles*

It is the passenger car with small displacement engine. There is no unified standard at home and abroad. Standards put forward by experts in China are: The displacement is not more than 1.4 L, the length is not longer than 4 m, the engine power is not higher than 45 kW/L, and the gasoline consumption meets the national standard for environmental protection and safety requirement. Now, the criterion for small-displacement vehicles is 1.6 L. Vehicle with displacement less than 1.0 L is called mini-car. In Japan, criterion for mini-car is less than 0.66 L while in European countries, the criterion is less than 1.3 L. In 2008, sales volume of mini-car with displacement less than 1 L was 4.5 million in overseas market. In Japan, sales volume of mini-car with displacement less than 0.66 L was accounting for 60% of total sales of vehicles and 34% of private cars. In July 2009, India Tata Group launched a 0.62 L mini-car-NANO which was priced at RMB 13,860 ($2,100).

The gasoline consumption of 1.3 L vehicles is 40% less than that of 2 L vehicles and 60% less than that of 1.3 L vehicles.

Since January 1, 2009, purchase of vehicles with displacement of 1.6 L or less can save half of the purchase tax. From March 1, 2009 to December 31, 2009, there were financial subsidies for vehicles sold in rural area; people who purchase light truck *via* exchanging discarded agriculture-used tricar or low-speed truck or purchase mini-car with displacement of 1.3 L or less, would be granted subsidies (10% of sales price). In 2009, sales volume of small-displacement vehicles with displacement of 1.6 L or less was up to 7.196 million, increased by 71.3% year on year and was accounting for 69.7% of total sales of passenger cars.

53. *Cellulose ethanol*

It is also called the second generation biological-ethanol produced with agricultural and forest waste and nongrain crops as main raw material. There are many cellulose source including straw of crops, rice hull, bagasse, wood flour, manioca, switchgrass, ricinus communis, pine nut, bamboo, and seaweed etc. Main material of the first generation biological-ethanol is grain, such as sugarcane, corn, wheat, potato, and sugar beet. How it have to consume grain and occupy land and the product is energy intensive, which are the major disadvantages. Proper enzyme which can convert cellulose into sugar is a key for producing cellulose ethanol. Biological-ethanol plus gasoline increases octane number of combustion. It replaces the lead-contained additive and thus can reduce CO and hydrocarbon in the end gas. Furthermore, oxygen content of ethanol is high, so the gasoline can totally burn so as to reduce gasoline consumption. In 2009, civil aviation companies in America, Japan, and New Zealand tried to use aviation fuel with cellulose ethanol added.

54. *Pure electric vehicle*

It refers to the vehicle with automotive rechargeable battery (lead-acid, nickel cadmium, nickel-metal hydride, and lithium ion battery) as the single power source. Its key technology is battery, driving motor and control technology. Pure electric vehicles manufactured in China adopt lithium ion battery and rare-earth permanent magnet brushless motor. There are three charging methods, common charge (AC plug inserted in automotive charger for 2–6 hours), fast charge (20–30 minutes, 50%–80% of battery capacity) and replacement of battery (tenantable battery).

In 2009, BYD Co., Ltd. launched the first lithium ion battery equipped pure electric vehicle in the world. The Science and Technology Department plans to manufacture 1 million pure electric vehicles in 2020.

Pure electric vehicle is the trend for developing new energy vehicles. However, its technology is not matured yet. In China, as 83% of power is generated by coal, compared with gasoline vehicle, there is no obvious advantage of energy efficiency and carbon emission. In accordance with the evaluation of relevant institutions and experts on the life cycle, there is nearly no difference between pure electric vehicle and gasoline vehicle in emission of CO_2. For popularization of pure electric vehicles in China, quantity of charging stations shall be same with the gasoline stations of CPCC. One charging station will cost more than RMB 3 million ($0.45 million) and total charging stations in China will cost RMB 7 billion–9 billion ($1.06 billion–$1.36 billion). Besides, pure electric vehicle is expensive. In China, the subsidiaries for one pure electric vehicle are RMB 60,000 ($9,090), and for hybrid electric vehicle RMB 50,000 ($7,576). The subsidiaries for one pure electric bus are RMB 500,000 ($75,758) and for hybrid electric bus are RMB 420,000 ($63,636).

55. *Hybrid electric vehicle*

It is a kind of vehicle driven by combustion engine and motor with gasoline or diesel as basic fuel. Its power sources are generally the gasoline engine and rechargeable motor. These two kinds of sources operate individually or jointly in different driving state. This combination can reduce the fuel consumption and emission of tail gas. In start up phase or at low speed, the vehicle is completely driven by electricity; when speeding up, gasoline engine and motor run simultaneously; when at high speed, battery supplies power for air-conditioner, audio, headlight and tail light; when speeding down and at brake, motor plays the role of generator to charge the battery. Compared with fuel vehicle, it can save 15%–25% of fuel. Compared with pure electric vehicle, it has following advantages, high dynamic performance, long mileages of continuation and convenient usage. Technology of Prius hybrid electric vehicles manufactured by Toyota is relatively matured; its sales volume worldwide is up to 2 million.

China will sell 60,000 energy-saving and new energy vehicles in 13 cities from 2009 to 2012, of which 95% vehicles are hybrid electric vehicles. The maximum subsidies for one hybrid electric vehicle are RMB 50,000 ($7,576); for one urban hybrid electric bus with length longer than 10 m is RMB 40,000 ($6,061). Since June 1, 2010, if one purchase plug in hybrid electric vehicle in Shanghai, Changchun, Shenzhen, Hangzhou and Hefei, he/she can get subsidies of RMB 50,000 ($7,576) at most.

56. *Electric bicycle*

It is a kind of bicycle driven by electricity generated by battery. Electric bicycle has no noise pollution and no harm to environment. It is convenient in driving and parking with low maintenance cost. Therefore, it can replace the light motorcycle. Electric bicycle develops at the fastest speed in China. 10 years ago, there were only 50,000, while in 2010, there were up to 120 million. At present, 95% of electric bicycles adopt lead-acid battery (4–5 kg). Its price is RMB 1,400–2,500 ($212–$379).

In 2008, there were 125 million motorcycles in China. If all motorcycles are replaced by electric bicycles, 1.8 million tons of gasoline can be saved per year.

The major problem of electric bicycles is traffic accident. In 2007, there were 2,469 deaths caused by electric bicycle accidents.

At present, GB standard for electric bicycle 10 years ago is still effective, i.e., entire weight less than 40 kg and speed lower than 20 km/h.

Agriculture

57. *Controlled release fertilizer by contained membrane*

It can improve the fertilizer utilization ratio by presetting the fertilizer release mode in growth period of crops with contained membrane to make the fertilizer release time synchronous with the nutrient absorption of crops. One contained membrane can be used for one quarter of crop. It can obviously reduce volatilization and leaching loss of nitrogenous fertilizer, greatly increase the utilization ratio of nutrient in fertilizer, save fertilizer, and reduce the pollution to air and water. Proportion of nitrogen, phosphorus, potassium, and micro-elements in organic polymer coating controlled-release fertilizer released by contained membrane is based on the demand of crops and abundant degree in different soils. Compared with common fertilizer, it has the following advantages: saving 1/3–1/2 of fertilizer, increasing utilization ratio by 10%–15% and increasing the output by 10% or more.

58. *Spread fertilizer by measuring the soil and making up a prescription*

By measuring the soil and learning the fertility state of soil, establish scientific manuring system and make up a prescription at demand of crops for fertilizer, The first is to make a prescription of fertilizer that the crop need, and the fertilizer manufactures manufacture the fertilizer in accordance with the prescription and guide farmers to fertilize. It can improve the physicochemical property of soil, increase its moisture and fertility preservation capacity, save fertilizer, increase output of crops, reduce disease of crops and improve quality of products. Grain output in the pilot county in China increases 5% per mu annually. In 2007, it was applied in 30% of farmland in China.

59. *Method of exempting from tillage*

It is also called conservation tillage technology. It is an advanced agriculture technology by covering soil with straw stubble or straw to reduce run-off and evaporation and increase soil fertility and drought-resist ability. Bottom plow is not used in the method. What need to do is guarantee germination of seeds, tillage-free seeding and manuring and spraying herbicide. It is widely applied in China. It can reduce run-off by 50%–60% and soil loss by 80%, increase water utilization ratio by 12%–16%, increase output by 13%–16%, reduce cost by about 20% and increase farmers' income by 20%–30%.

60. *Straw Returning Cultivated Land*

Straw Returning Cultivated Land refers to returning straw into the field. Usually, the straws of corn, broomcorn, and wheat are smashed by smasher on site, scattered on the surface of soil equally and then ploughed into the soil immediately. In the process of decomposing in land, the straws are mineralized, releasing nutrients, and the compounds of organic matters form humus at the same time to improve soil structure, promote the performance of water retention, water absorption, cementation, and ventilation and heat preservation to increase soil fertility and its adjustment to water, fertilizer, temperature and climate to make soil nutrient structure more optimized. Straw returning can also increase the product of wheat by more than 15 kg per mu (a unit of area = 0.067 hectares), and corn by 25 kg. In 2008, 233 million tons of straws were returned into the cultivated land, accounting for 36% of straw production.

61. *Penning Raise Livestock by Fold Feed*

Penning Raise Livestock by Fold Feed refers to breeding livestock like cattle and sheep etc. in the pen. Penning raising livestock by fold feed needs technologies like forage grass processing, feeding facility design, feeding arrangement, the prevention and cure of thinness and sickness and so on. Compared with traditional grazing, the advantages of penning raising are as follows: To protect the ecological environment of grassland, to make full use of straw resources, and to make nutritious and comprehensive forages depending on the physical growth rules of the cattle and sheep, to help spread breeding stock, to provide organic fertilizer, and also to benefit the intensive and specialized production of livestock husbandry to realize sustainable development.

Due to over-grazing, the grassland of China deteriorates seriously and the carbon sink is also greatly reduced. The carbon sequestration competence of grassland of 1 mu is about 0.1 ton of carbon.

62. *Ecological agriculture*

Ecological agriculture refers the modern agriculture operated intensively by using the achievement and management approach of modern science and technology, and the experience of traditional agriculture and following the rules of ecology and eco-economics, and the premises for developing ecological agriculture is protecting and improving agricultural and ecological environment. It is a composite system of agriculture and ecological economy, which follows *Food Chain* principle, organism evolution rules, material recycling, overall coordination principle and the principle of adjusting measures to local conditions. Ecological agriculture will coordinate development of agriculture, forestry, sideline, livestock husbandry and fishery, and agricultural production, processing and sales, to make full use of the overall function of agriculture, protect ecological environment, reduce production cost, promote the safety level of agricultural product and provide more job opportunities for rural areas.

Inter-Department

63. *Permanent Magnetic Coreless Motor*

The rare-earth permanent magnetic coreless motor developed by China Shenzhen Antuoshan Special Motor Co. Ltd, has changed the structure of the silicon-steel sheet and wound stator of traditional motors featured by axial magnetic field structure, unique winding distribution and modeled technology, no coreless in the stator and the rotor, and self-developed electronic intelligent frequency conversion technology. The work efficiency of this motor is up to 95%. Compared with traditional induced motor driven by gears, this motor can save 30% electricity. With a smaller size than the traditional one, it can save 80% steel and 50% bronze, and it is without silicon, excitation, brush and iron core; It is also improved in terms of insulation and is of water-proof, moisture-proof and dust-proof.

As a new world-leading technology, three mass production lines have been built in 2010. If one third of the new added motors in China adopt this new technology, it will save 50 billion kwh of electricity, 500,000 tons of silicon-steel sheet and 20,000 tons of bronze each year.

64. *Power-electronic Technology*

Power-electronic Technology refers the comprehensive technology of electricity, electron and microelectronic technologies, involving semiconductor, circuit, motor, microprocessor and microcomputer, and the control theory, mainly using the exchange function of power semiconductor elements. Power-electronic technology is widely used in the fields of industry, transportation, communication, and domestic appliances, which can save labor and energy, and is smart, high accurate and quick response.

Power-electronic technology performs well in saving energy. For example, it will save 30%–40% electricity once the valve regulator of fan and water pump are regulated to the alternating current control; and it will save more than one third of electricity when the direct current transmission is regulated to the GTO frequency drive; and the high efficient fluorescent light using electrical ballast and new fluorescent can save 75% electricity compared with incandescent lamp. Energy-saving productions produced by power-electronic technology can save 40%–90% materials and it can also improve the production quality, the labor productivity, and the reliability of facility and system.

65. *Tri-network Integration*

Tri-network Integration refers to technically integrating telecommunication network, broadcast-TV network and the Internet together to realize their interconnection and seamless coverage, the integration and interaction in every level of business. Through technology transformation, tri-network can provide comprehensive multimedia communication services with sound, statistics and pictures etc. Just through one network, users can deal with almost all the daily information processing such as surfing the Internet, watching TV, and making a call etc. At present, the telecommunication network and the Internet have been integrated in China, and we are trying to integrate the broadcast-TV network into the tri-network.

66. *Internet of Things*

The concept of ***Internet of Things*** is to make inanimate objects able to communicate and exchange with human beings and also with inanimate objects at anytime and anywhere in the way like human beings communicate with each other. Internet of Things is a kind of social information system including perception technology, information transferring technology and massive information analysis processing, and decision-making technology. It is the third milestone next to computer processing and the Internet in the development of information technology and industries.

Internet of things involves five technologies at least: Item identification technology represented by RFID; sensor technology and transmission technology; network communication technology; data storage and processing technology; and smart objects technology represented by 3C-integration.

The arrangement and practice of Internet of things is already under way in our country. Internet of things technology like RFID has been applied in all kinds of intelligent cards (ID card, telecommunication, social security, transportation, public utility, etc.), ETC, railway locomotive dynamic monitoring, ocean container transport, civil aviation luggage, electron pot, postal parcel, food and drug traceability technique system, coal mine safety production, major pollutant sources automatic measuring system, digital rooms, networking refrigerator and so on. The market size of Internet of things industry IN China is anticipated to be RMB 200 billion in 2010 and RMB 750 billion in 2015. It is estimated by relevant consultancies of the United States that the profit of Internet of things industry will be 30 times of the Internet in 2020.

67. *Cloud Computing*

Cloud Computing is a new computing technology based on the Internet. Its basic concept is to first divide a large and complicated computing and processing program into several smaller subprograms automatically through the Internet and then provide the smaller subprograms to a large and complex system of several servers, and finally to send the processed result back to users after searching and analyzing. In this way, web service provider can process tens of millions of information just in several seconds. It means that the services provided by all the powerful computers in the world including supercomputers can be appreciated as long as the Internet is available. *Cloud* is a computing cluster, each containing hundreds of thousands of even millions of computers. Google has several *clouds* of this kind and is thus powerful in storage and capable to compute global statistics. Its competitive force depends on these *clouds*. Among the web services, the simplest cloud computing like searching engines, network mailbox is easily found. Users are able to have access to large amounts of information as long as they input simple instructions. In the future, users will enjoy the services they need only with a laptop or a mobile phone and will never be worried about losing their documents. *Tianhe-I*, the world fastest supercomputer (which is able to do 47 million calculations per second at its peak speed and 25,660,000 per second at its endurance speed) developed in China, has already connected with the Internet.

68. *Membranes Technology*

Membranes Technology refers to the process of separating the inorganic or high polymer materials with selective separation capability. According to its application characteristics and functions, it falls into several types: Reverse osmosis, nanofiltration, ultrafiltration, separation and pervaporation and so on. Membrane technology is featured by simple facility, high efficiency, low-energy consumption, and less pollution. It has been used in the fields of pharmacy, chemical industry, petroleum, textile, printing and dyeing, light industry, electronics, metallurgy, food, drinks, environment protection, bioengineering, resource reuse etc., and is considered as one of the most potential green and energy-saving technologies in the 21st century. In many industries of energy intensity technology, Membrane technology can replace traditional separation technology to dispose wastes; replace sodium chloride and silver to extract products and chemical materials; improve product quality to reduce energy consumption and pollutant emission. In chemical industry, membrane bioreactor can simplify the work flow, shorten reaction time, and promote convert ratio and product purity. For example, the convert ratio of hydrogen production by palladium film method is nearly 100%, and the purity of the hydrogen is over 99.9999999%.

69. *Super-conduction technology*

The phenomenon of super-conduction appears when there is almost no resistance at extremely low temperature. The temperature when the material begins to be superconducting is represented by absolute temperature K, the absolute temperature of zero equaling $-273°C$ (0 K). The research and application of super-conduction technology will bring a magnificent change into the fields of power transmission, motor, transportation, computer and basic science experiments. With 5–10 times bigger in volume than the conventional one, the superconducting generator is half smaller than it in size, and 50% higher in efficiency. Low-temperature superconducting material can greatly reduce line loss and improve the reliability of electric power system. At present, the first generation of bismuth hydride high-temperature superconducting materials has been commercially demonstrated and the research of second generation superconducting materials of yttrium series is under researching. In 2006, the high-temperature superconducting cable of bismuth series developed in China was successfully grid-connected. And the high-temperature superconducting cable of 500 m developed in Japan was tested in field at the same year. Japan tested its load trial of 365 kw marine high-temperature superconducting motor in September, 2007 when China developed high-temperature superconducting maglev testing train. The *Super Power Transmission Highway* with high-temperature superconducting transmission line, transformer, and current control will occupy a predominant place in the middle of the 21st century.

70. *Carbon Capture and Sequestration, CCS*

Carbon Capture mainly refers to isolate CO_2 out of the gas produced in the burn (or gasification) of fossil fuels, or to isolate CO_2 out of atmosphere. ***Carbon Sequestration*** refers to the technology finally used to dispel CO_2 disposed in the burn of fossil fuels. There are six methods under researching and testing: ① separating and capturing from energy system; ② sequestrating them into the stratum; ③ sequestrating into the ecological system on the land (forests, plants, lands, crops, pasture, tundra, swamps); ④ sequestrating into ocean; ⑤ advanced biological technology; ⑥ advanced chemical method. Finally, it is probable to combine the technologies of capturing, separating and sequestrating together.

The capture technologies of CO_2 in a power plant cover solvent segregation, cryogenic processing, gas separation membrane, adsorption, and so on. Carbon capture consumes more energy and costs more. China Huaneng Beijing Gaobeidian Thermal power plant captured 3,000 tons of CO_2, accounting for 0.75% of the production, to conduct CO_2 capture trial. The trial indicates that if a power plant of 300,000 kw captures 1 million tons of CO_2, the cost of the power plant will be increased to RMB 8,000/kw from RMB 4,000/kw, with a 30% increase of energy consumption and 20%–30%

increase of generation cost. It was estimated from US Department of Energy in 2009 that it took \$60 to capture every ton of CO_2 at that time.

In 2008, the item *Future Power Plant* (power-generating, oxygen-making, CCS) of the US Department of Energy was cancelled due to its over-investment (\$1,8 billion).

It is suitable and economical to store CO_2 in exhausted oil-gas reservoirs and deep saline petroleum reservoir since it does not cost much to prospect for the stored CO_2. What's more, some of the oil and gas extraction facilities can be used to inject CO_2. To inject CO_2 into the oil layer is also a kind EOR (enhanced oil recovery) technology, improving oil recovery. At present, the US has 70 CO_2 EOR miscible displacement items increasing its oil recovery factor by 4%–12%. Petro-Canada Corporation is now conducting the world largest-scaled CO_2 storing and testing item that it delivers the stored CO_2 by a coal gasification factory in the State of North Dakota in US to Weyburn Oil Field in the Province of Saskatchewan in Canada through pipeline to inject them into the oil layer. It is estimated that at the end of each year the coal gasification company can inject 20 million tons of CO_2, with its daily production increasing by 10,000 barrels, profiting \$30 million. China is now doing CO_2 EOR field testing.

APPENDIX 2

Explanations of Terms

Table A.1. Interpretation of the terms.

Terms	Interpretation
Air-cooling unit	Air-cooling generation unit utilizes forced flow air as carrier of heat source for heat emission.
Net coal consumption rate	Also called standard coal consumption rate of power supply, average coal equivalent for thermal power station supplying 1 kWh power.
Gross coal consumption rate	Also called standard coal consumption rate of power generation, average coal equivalent for thermal power station generating 1 kWh power.
Auxiliary power ratio	In starting, operating, stopping and overhaul, mechanical equipment driven by motor operate to guarantee normal operation of primary equipment and auxiliary equipment, such as coal conveyer, coal crusher, ash removal and dust removal machine and water treatment equipment. Motor and electrical equipment for running, operation, testing, overhaul and illumination are auxiliary load, and therefore they are collectively called auxiliary power. Percentage of consumed auxiliary power to total power generated by power station is called auxiliary power ratio.
Average available hour using for power generation	Ratio of power output to average capacity of power generator in report period, an index to reflect utilization ratio of power generator.
Rate of transmission line loss	Percentage of power lost in electric network to power supplied to electric network, for considering the economical performance of power system.
Low voltage	1,000 V or less voltage in AC power system.

(*Continued*)

Table A.1. (*Continued*)

Terms	Interpretation
Power supply structure	Proportion of installed capacity to power output by various modes (thermal power, hydroelectric power and nuclear power)
Small thermal power	Often refer to 50,000 kW or less oil-fired boiler and generating unit with thermal power as main method.
CO_2-capture, utilization and storage	Carbon capture and storage (CCS) is a technology capturing and storing CO_2. CCS technology means separating CO_2 generated by industry and relevant energy sector via carbon capture technology and convey and store CO_2 in seabed, underground or other places isolated with atmosphere via carbon storage technology. Now, CCS technology is still at research and development stage.
Combined heat and power generation	A production method, not only producing power, but also supplying heat to users via steam under applied work by steam turbine; a process producing power and heat at the same time; compared with common production mode, save fuel.
Combined cooling, heating and power generation	Steam of boiler generates power in back pressure steam turbine or extraction steam turbine; it exhaust or extracted steam is not only sufficient for all kinds of loads, but also as the working steam of absorption refrigerator to producing 6–8°C cooled water for air-conditioner and cooling process.
Flexible transmission	AC transmission method to increase the control performance and power transmission ability based on power electronic equipment and other static equipment.
Intermittent power	Discontinuous wind and solar energy power generation mode limited by resource condition.
Membrane method	Membrane electrolysis is one of the most important methods for caustic soda production. A membrane is set between anode and cathode to divide electrolysis products. The membrane is a multi-hole permeable interlayer without preventing the electron transfer and current conveyance and guiding electron and current in certain speed flowing to cathode. However, it can prevent OH^- diffusing to anode so as to avoid mixing products of anode and cathode. At present, vertical membrane electrolytic cell is often used in industry. Anode is graphite or metal and cathode is steel wire mesh or punched steel plate. When electrolysis is

(*Continued*)

Table A.1. (*Continued*)

Terms	Interpretation
	carried out in saline solution after input of AC current, in anode, chlorine ions lose electrons and become chlorine to escape. Sodium ions in anode migrate to cathode. Hydrogen ions in cathode obtain electrons and become hydrogen gas to escape. Because hydrogen ions continuously discharge and hydrogen gas are continuously generated, ionization of water will be continuous. The OH^- and Na^+ in solution combine as soda which is drained from electrolytic cell together with residual NaCl solution. New NaCl solution is continuously supplied for uninterrupted production in anode chamber.
Float glass	In floating production, shaping is completed in tin trough supplied with protective gas. Float glass is widely applied. It includes colored glass, float silver mirror, float white glass for auto windshields, deep-processing, scanner, blooming and mirror making. Ultra-clear float glass is widely applied with bright market prospect. It is mainly used in following industries, high grade building, glass machining, solar energy photoelectric curtain wall, high grade furniture with glass, glass for decoration, artificial crystal product, lamp, precise electronic industry and special building.
Itmk3 smelting technology	New generation smelting technology; mix iron ore powder and pulverized coal in furnace and heat up to 1,450°; after separating iron and impurities, form iron slab. Whole process only spends 10 minutes, faster than blast furnace and Fastmalt method. Iron content is up to 97%, higher than 96% of blast furnace and Fastmalt method. Ore powder and pulverized coal are low grade which cannot be smelted by blast furnace. Iron slab manufactured via Itmk3 smelting technology replaces scrap steel to apply in electric steel-making. The technology can also produce common steel production with high added value. At present, high grade common carbon products for auto are produced by converter. Electric furnace mainly produces steel for building. One Itmk3 smelting device with capacity of 500,000 t/a only costs about 15 billion Japanese Yen with efficiency equal to blast furnace at the cost of 100 billion Japanese Yen. Its construction period is short and investment risk is low.

(*Continued*)

Table A.1. (*Continued*)

Terms	Interpretation
Converter low-pressure saturated steam generation technology	Low-pressure saturated steam generation unit with special design; a technology for power generation by low-pressure saturated steam generated by heating furnace, converter steam water cooling device and other waste heat recovery devices of steel plants.
Technology of hydrogen production from coke oven gas	Hydrogen is important protective gas in producing cold rolled plate in steel plant. As to traditional technology of hydrogen production from water electrolysis, investment is large and operation cost is high. At present, technology of hydrogen production from coke oven gas is popularized. Purpose of the technology is carrying out purification treatment for coke oven gas before pressure swing adsorption stage to remove impurities in gas, reduce the burden of adsorption tower, prolong service life of adsorbent and guarantee quality of hydrogen.
Technology of coal moisture control	Dehydrate and dry the wet coal to guarantee water content lower than 6% before entering furnace. It develops based on coal drying technology. Its core technology is control of water content in coal between 5% and 6% no matter how much water content in coal before drying. Japan calls the technology as technology of coal moisture control.
Technology of waste heat generating system of sinter	Key point of waste heat recovery in sintering procedure is recovery of waste heat of sintering waste (smoke) gas and sensible heat during sintering. Its working principle is that heat up waste heat boiler via recovery of flue gas in ore break down bellows at the sinter tail and closed section of sintering cooler to recover low grade waste heat energy, and push steam turbine generation unit with low parameter by overheat steam of waste heat boiler to apply work for generating power via combining low temperature waste heat power generation technology.
High efficient milling technology	There are vertical milling technology and roll squeezer milling technology. The former adopts powder bed milling principle to effectively improve milling efficiency, reduce over-milling and energy consumption. The latter adopts high pressure squeezing principle to obviously reduce energy consumption with proper grading plant.

(*Continued*)

Table A.1. (*Continued*)

Terms	Interpretation
Steam condensation water close-recycling technology	Via systematic analysis on pressure grade and order of steam pipe network, adopt ladder pressurization and sub-segment recovery to thoroughly eliminate the steam-water mix and pipeline block phenomenon caused by different pressure of pipelines and smooth the recovery of condensed water.
New soda manufacturing technology via alternating gas	In accordance with low temperature soda manufacturing principle, change three-tower to single-tower and change the internal heat exchange to external heat exchange to improve crystalline quality of sodium bicarbonate, reduce scouring water equivalent, prolong service life of soda manufacturing tower, realize zero-discharge of waste liquid of associated-soda system, reduce resistance and save energy.
Enclosure	Enclosure of building and room, such as doors, windows and walls, effectively preventing the impact of adverse environment.
Fuel economy standards	A index to measure the fuel consumption for driving a certain distance.
Converted volume	Volume of passengers and cargos volume calculated by the same measurement unit to convert into tkm.
Passenger and cargo volume	Complex index for calculating passengers, cargos and distance in transportation.
Share rate of transportation modes	Statistics based on traffic time or volume (person tkm or ton tkm), i.e., percentage of volume of certain transportation mode to volume of all transportation modes.
Footprint area of vehicle	Vertical projection area among four wheels of vehicle.
Carbon source	Parent population of carbon released to atmosphere by nature.
Carbon sink	Opposite to carbon source. Processes, activities and mechanisms removing CO_2 from atmosphere.
Forest carbon sink	A process of forest plants absorbing CO_2 in atmosphere and fixing in plants and soil via photosynthesis to reduce the density of CO_2 in atmosphere.
Forestry carbon sink	Processes, activities and mechanisms combining with carbon sink trade in accordance with relevant rules to absorb and fix CO_2 in atmosphere via forestation, strengthening management of forest, reducing forest damage and protecting and restoring the forest plants by utilizing the carbon storage function of forest.

(*Continued*)

Table A.1. (*Continued*)

Terms	Interpretation
Conservation tillage	Conservation tillage takes mechanized operation as main means and adopts less or free tillage method; what need to do is guarantee germination of seeds; cover soil with straw stubble or straw and mainly use pesticide to control weed, plant diseases and insect pests.
No-till	A tillage mode without stirring surface soil but leaving stubble to protect soil. It includes no-tillage, strip tillage, stubble covering and other tillage measures without stirring surface soil. In 1984, Soil Protection Bureau amended definition of no-till, and divided tillage measures into three types, i.e., no-till, less till and common tillage.
Straw returning	Return the straws (corn straws and sorghum straws) which are not suitable as fodder to soil directly or after thoroughly decomposing so as to improve soil features, speed up period from immature soil into cultivated land, and increase fertility.
Spread fertilizer by measure the soil and make up a prescription	Based on soil test and fertilizer field test and proper application of organic fertilizer, in accordance with fertilizer demand rule of crops, soil fertilizer supply performance and fertilizer effect, determine the proportion of nitrogen, phosphorus, potassium and secondary and micro-elements, period and method of applying fertilizer.
Embodied carbon	Any production process will directly or indirectly cause carbon emission. CO_2 emitted in whole production chain is called "embodied carbon".

APPENDIX 3

Unit Comparison Table and Acronyms

Table A.2. Table of Units.

Unit symbol	Meaning
gce	Gram of coal equivalent
kgce	Kilo gram of coal equivalent
tce	Ton of coal equivalent
Mtce	10^6 tons of coal equivalent
$MtCO_2$	10^6 tons of CO_2
GW	10^9 W
kW·h	Kilowatt hour
kg	Kilo gram
MW	10^6 W(megawatt)
TWh	10^9 kWh
€	Euro
tce/10^4 yuan	Ton of coal equivalent per 10^4 yuan (energy consumption per unit of industrial added value)
$tCO_2/10^4$ yuan	Ton of CO_2 per 10^4 yuan (CO_2 emission per unit of industrial added value)
kgce/t	Kilo gram of coal equivalent per ton (comprehensive energy consumption per unit product)
kW·h/t	Kilowatt hour per ton (electricity consumption per unit product)
Mt	10^6 tons
kgce/m^2	Kilo gram of coal equivalent per square meter
kgce/person	Kilo gram of coal equivalent per person
L/100 km	Liter per hundred kilometer
MPG	Mile per gallon
Hm^2	hectare
CO_2-eq	CO_2 equivalent

Table A.3. Acronym.

English acronym	Meaning
BRT	Bus rapid transit
CAFE	Corporate average fuel economy
CDM	Clean development mechanism
CNG	Compressed natural gas
CSIA	China Semiconductor Industry Association
ECOFYS	Ecofys Netherlands B.V.
GHG	Green house gas
IEA	International Energy Agency
LPG	Liquefied petroleum gas
NEDC	New European Driving Cycle
OECD	Organization for Economic Co-operation and Development
REN21	Renewable Energy Policy Network for the 21st Century
Solarbuzz	A world known solar energy research institution, one international solar energy research and consulting company
Solarzoom	Photovoltaic solar energy web site
SUV	Sport utility vehicle
UNEP	United Nations Environment Program
USEPA	United States Environmental Protection Agency
VMT	Vehicle-miles traveled